AF477129

News Trends in Powder Metallurgy: Microstructures, Properties, Durability

News Trends in Powder Metallurgy: Microstructures, Properties, Durability

Editors

Eric Hug
Guy Dirras

MDPI • Basel • Beijing • Wuhan • Barcelona • Belgrade • Manchester • Tokyo • Cluj • Tianjin

Editors
Eric Hug
Normandy University
France

Guy Dirras
Université Paris
France

Editorial Office
MDPI
St. Alban-Anlage 66
4052 Basel, Switzerland

This is a reprint of articles from the Special Issue published online in the open access journal *Metals* (ISSN 2075-4701) (available at: https://www.mdpi.com/journal/metals/special_issues/powder_metallurgy).

For citation purposes, cite each article independently as indicated on the article page online and as indicated below:

LastName, A.A.; LastName, B.B.; LastName, C.C. Article Title. *Journal Name* **Year**, *Volume Number*, Page Range.

ISBN 978-3-0365-1782-7 (Hbk)
ISBN 978-3-0365-1781-0 (PDF)

Contents

Eric Hug and Guy Dirras
News Trends in Powder Metallurgy: Microstructures, Properties, Durability
Reprinted from: *Metals* **2021**, *11*, 1216, doi:10.3390/met11081216 . . . . **1**

Bhupendra Sharma, Guy Dirras and Kei Ameyama
Harmonic Structure Design: A Strategy for Outstanding Mechanical Properties in
Structural Materials
Reprinted from: *Metals* **2020**, *10*, 1615, doi:10.3390/met10121615 . . . . **5**

**Benjamin Guennec, Takayuki Ishiguri, Mie Ota Kawabata, Shoichi Kikuchi, Akira Ueno
and Kei Ameyama**
Investigation on the Durability of Ti-6Al-4V Alloy Designed in a Harmonic Structure via
Powder Metallurgy: Fatigue Behavior and Specimen Size Parameter Issue
Reprinted from: *Metals* **2020**, *10*, 636, doi:10.3390/met10050636 . . . . **21**

**Benoît Fer, David Tingaud, Azziz Hocini, Yulin Hao, Eric Leroy, Frédéric Prima
and Guy Dirras**
Powder Metallurgy Processing and Mechanical Properties of Controlled Ti-24Nb-4Zr-8Sn
Heterogeneous Microstructures
Reprinted from: *Metals* **2020**, *10*, 1626, doi:10.3390/met10121626 . . . . **43**

Yang Song, Zhe Zhang, Hantuo Ma, Masashi Nakatani, Mie Ota Kawabata and Kei Ameyama
Ratcheting-Fatigue Behavior of Harmonic-Structure-Designed SUS316L Stainless Steel
Reprinted from: *Metals* **2021**, *11*, 477, doi:10.3390/met11030477 . . . . **63**

Lucía García de la Cruz, Mayerling Martinez Celis, Clément Keller and Eric Hug
Exploring the Strain Hardening Mechanisms of Ultrafine Grained Nickel Processed by Spark
Plasma Sintering
Reprinted from: *Metals* **2021**, *11*, 65, doi:10.3390/met11010065 . . . . **75**

**Jean-Philippe Monchoux, Alain Couret, Lise Durand, Thomas Voisin, Zofia Trzaska
and Marc Thomas**
Elaboration of Metallic Materials by SPS: Processing, Microstructures, Properties, and Shaping
Reprinted from: *Metals* **2021**, *11*, 322, doi:10.3390/met11020322 . . . . **93**

**Sarah Dine, Elodie Bernard, Nathalie Herlin, Christian Grisolia, David Tingaud
and Dominique Vrel**
SHS Synthesis, SPS Densification and Mechanical Properties of Nanometric Tungsten
Reprinted from: *Metals* **2021**, *11*, 252, doi:10.3390/met11020252 . . . . **113**

**Mathias Moser, Sylvain Lorand, Florian Bussiere, Frédéric Demoisson, Hervé Couque
and Frédéric Bernard**
Influence of Carbon Diffusion and the Presence of Oxygen on the Microstructure of
Molybdenum Powders Densified by SPS
Reprinted from: *Metals* **2020**, *10*, 948, doi:10.3390/met10070948 . . . . **129**

**Awais Ikram, Muhammad Awais, Richard Sheridan, Allan Walton, Spomenka Kobe,
Franci Pušavec and Kristina Žužek Rožman**
Spark Plasma Sintering as an Effective Texturing Tool for Reprocessing Recycled HDDR
Nd-Fe-B Magnets with Lossless Coercivity
Reprinted from: *Metals* **2020**, *10*, 418, doi:10.3390/met10030418 . . . . **147**

Ntebogeng F. Mogale and Wallace R. Matizamhuka
Spark Plasma Sintering of Titanium Aluminides: A Progress Review on Processing, Structure-Property Relations, Alloy Development and Challenges
Reprinted from: *Metals* **2020**, *10*, 1080, doi:10.3390/met10081080 . . . . **165**

Foad Naimi, Jean-Claude Niepce, Mostapha Ariane, Cyril Cayron, José Calapez, Jean-Marie Gentzbittel and Frédéric Bernard
Joining of Oxide Dispersion-Strengthened Steel Using Spark Plasma Sintering
Reprinted from: *Metals* **2020**, *10*, 1040, doi:10.3390/met10081040 . . . . **189**

Jorge Sergio Téllez-Martínez, Luis Olmos, Víctor Manuel Solorio-García, Héctor Javier Vergara-Hernández, Jorge Chávez and Dante Arteaga
Processing and Characterization of Bilayer Materials by Solid State Sintering for Orthopedic Applications
Reprinted from: *Metals* **2021**, *11*, 207, doi:10.3390/met11020207 . . . . **199**

Jesús E. González, Gabriela de Armas, Jeidy Negrin, Ana M. Beltrán, Paloma Trueba, Francisco J. Gotor, Eduardo Peón and Yadir Torres
Influence of Successive Chemical and Thermochemical Treatments on Surface Features of Ti6Al4V Samples Manufactured by SLM
Reprinted from: *Metals* **2021**, *11*, 313, doi:10.3390/met11020313 . . . . **215**

Editorial

News Trends in Powder Metallurgy: Microstructures, Properties, Durability

Eric Hug [1,*] **and Guy Dirras** [2,*]

[1] CRISMAT Laboratory, UMR 6508, Normandy University, 6 Boulevard Marechal Juin, CEDEX 4, 14050 Caen, France
[2] LSPM Laboratory, CNRS-UPR 3407, Sorbonne Paris Nord University, 99 Avenue Jean Baptiste Clément, 93430 Villetaneuse, France
* Correspondence: eric.hug@ensicaen.fr (E.H.); dirras@lspm.cnrs.fr (G.D.)

Citation: Hug, E.; Dirras, G. News Trends in Powder Metallurgy: Microstructures, Properties, Durability. *Metals* **2021**, *11*, 1216. https://doi.org/10.3390/met11081216

Received: 28 July 2021
Accepted: 28 July 2021
Published: 30 July 2021

Publisher's Note: MDPI stays neutral with regard to jurisdictional claims in published maps and institutional affiliations.

1. Context and Scope of the Special Issue

Compared with traditional casting or plastic deformation processes, powder metallurgy-based methods are versatile routes for producing in-demand microstructures of various types. Centimeter-sized, fully dense, coarse-grained, ultrafine-grained, nanostructured, and heterogeneous-controlled structure (architected) materials can indeed be processed in respect to the bottom-up approach through techniques of fast sintering (SPS, microwave sintering...) and the emerging additive manufacturing such as by powder bed fusion (PBF), leading to an original microstructure design. Consequently, the mechanical and physical properties of the alloys, and their durability, are profoundly modified, thus broadening the application range.

In this Special Issue, we have initially proposed to interested authors to contribute to the following topics:

- Toward new microstructures: analysis, properties, and stability;
- Powder properties, nanostructuration, mechanical alloying, and aging;
- Unconventional sintering processes: SPS, microwave, etc.;
- Additive manufacturing by powder bed melting processes;
- Mechanical properties: fatigue, creep, plasticity mechanisms;
- Physical properties: magnetism, electrical conduction;
- Damage, fracture, and effect of the environment: oxidation, electrochemical corrosion.

Finally, a review of the scientific advances in this field have been carried out through a selection of 13 original research papers (progress reviews and articles) on the impact of the microstructure on the mechanical and functional properties of metallic alloys obtained by sintering (SPS) and additive manufacturing (PBF) routes.

2. Presentation of the Contributions

From the point of view of traditional metallurgy, mechanical resistance and ductility are antagonist properties. Therefore, the quest for methodologies and concepts that are susceptible to make them coexist are of great interest. To this end, and among others, the idea of a harmonic structure, first introduced by Prof. Ameyama, has been proven to be of prime interest for addressing the dual problem of improving both the strength and the ductility and other fatigue properties of metallic materials. Four main contributions to this Special Issue by Bhupendra et al. [1], Guennec et al. [2], Fer et al. [3], and Song et al. [4] illustrate the incredible potential of the concept which can be indeed applied to various metallic materials and alloys.

Nevertheless, the microstructural mechanisms at the origin of the coexistence between ductility and mechanical resistance of heterogeneous structures of harmonic types are not yet well understood, particularly in terms of synergy between the different structural entities (shell and core). This understanding begins with identifying the unique plasticity mechanisms of the coarse grains of the core and the ultra-fine grains of the shell. In this Special Issue, a first answer is provided by Garcia de la Cruz et al. [5]. In their study, the authors discuss the work hardening mechanisms of UFG nickel, interestingly considering samples with grain sizes ranging from 0.82 to 25 µm. While a lower strain hardening capability is observed with decreasing grain size, samples in the submicrometric range display the three distinct stages of strain hardening representative of face-centered cubic metals, with a short second stage and the third stage beginning soon after yielding.

Again, in terms of properties (mechanical or physical), and considering durability issues, metallic structures could not be conceived without optimizing the microstructure, which we know has a colossal influence on the subsequent properties, and therefore on the performances. The review by Monchoux et al. [6] provides an overview of the link between process, microstructures, and properties of metallic alloys elaborated by SPS. In this Special Issue, these aspects are also approached and presented systematically by Moser et al. through two studies that focus on the importance of densification on the mechanical properties of tungsten which are made ductile via nanostructuring and densification after SPS, by Dine et al. [7], as well as on the influence of impurities such as oxygen and carbon on SPS-processed Molybdenum. [8]. Finally, considering sustainability, metal recycling is a hot topic. Ikram et al. [9] demonstrate how SPS can be effective with recycled metal powders for obtaining Nd-Fe-B permanent magnets with magnetic properties equivalent to those elaborated using pristine powder.

If it was still needed to convince of the versatility of the processes for the development of products and parts with great added value, based on powder metallurgy routes (in particular spark plasma sintering and additive manufacturing), the reader would find some essential data in this Special Issue. First, a detailed review on titanium aluminides having the potential of substituting nickel-based superalloys (NBSAs) in the aerospace industries is discussed by Mogale et al. [10]. Then, this article further reviews published works on phase constituents, microstructures, alloy developments, and mechanical properties of TiAl alloys produced by SPS. Finally, an overview of challenges as far as the implementation of TiAl in industries of interest is highlighted.

In the same vein, assembly processes to produce structural or functionalized structural materials (or parts to be functionalized later) can become real challenges. Powder metallurgy and SPS make it possible to meet these challenges. Three studies presented in this Special Issue deal with these aspects. First, Naimi et al. [11] propose to use SPS sintering of ODS steels to overcome the difficulties with joining oxide dispersion-strengthened (ODS) steels using traditional welding routes. Then, in the work of Tellez-Martinez et al. [12], ingeniously, a new processing route is proposed to produce graded porous materials by placing particles of Ti6Al4V with different sizes in different configurations to obtain bilayer samples that can be used as bone implants. It is concluded that the proposed elaboration way can produce materials with specific and graded characteristics, with the radial configuration being the most promising for biomedical applications. Finally, mastering the texturing of surfaces is an important issue, in particular for the chemical or biological functionalization of medical implants, which constitutes a societal issue of great importance for which new material solutions are sought. In this quest, additive manufacturing and, in particular, powder bed processes are of particular interest. In this perspective, Gonzales et al. [13] present an original approach for texturing Ti6Al4V samples obtained by SLM subjected to successive treatments: acid etching, chemical oxidation in hydrogen peroxide solution, and thermochemical processing. The effect of temperature and time of acid etching on the surface roughness, morphology, topography, and chemical and phase composition after the thermochemical treatment was studied. These surfaces are expected to generate greater levels of bioactivity and high biomechanics fixation of implants, as well as better resistance to fatigue.

Acknowledgments: As guest editors of this Special Issue, we hope that the present selected papers will be useful to researchers working in the area of powder metallurgy. We would like to thank the authors for their contributions to this Special Issue, and all of the reviewers for their efforts in ensuring high-quality publications. We would also thanks to the Metals editorial assistants for their support during the preparation of this volume. Finally, we would especially thanks Natalie Sun, managing Editor of Metals, for offering us the opportunity to work on this project, and for guiding us through the editorial set-up.

Conflicts of Interest: The authors declare no conflict of interest.

References

1. Sharma, B.; Dirras, G.; Ameyama, K. Harmonic Structure Design: A Strategy for Outstanding Mechanical Properties in Structural Materials. *Metals* **2020**, *10*, 1615. [CrossRef]
2. Guennec, B.; Ishiguri, T.; Kawabata, M.O.; Kikuchi, S.; Ueno, A.; Ameyama, K. Investigation on the Durability of Ti-6Al-4V Alloy Designed in a Harmonic Structure via Powder Metallurgy: Fatigue Behavior and Specimen Size Parameter Issue. *Metals* **2020**, *10*, 636. [CrossRef]
3. Fer, B.; Tingaud, D.; Hocini, A.; Hao, Y.; Leroy, E.; Prima, F.; Dirras, G. Powder Metallurgy Processing and Mechanical Properties of Controlled Ti-24Nb-4Zr-8Sn Heterogeneous Microstructures. *Metals* **2020**, *10*, 1626. [CrossRef]
4. Song, Y.; Zhang, Z.; Ma, H.; Nakatani, M.; Kawabata, M.O.; Ameyama, K. Ratcheting-Fatigue Behavior of Harmonic-Structure-Designed SUS316L Stainless Steel. *Metals* **2021**, *11*, 477. [CrossRef]
5. De La Cruz, L.G.; Celis, M.M.; Keller, C.; Hug, E. Exploring the Strain Hardening Mechanisms of Ultrafine Grained Nickel Processed by Spark Plasma Sintering. *Metals* **2020**, *11*, 65. [CrossRef]
6. Monchoux, J.P.; Couret, A.; Durand, L.; Voisin, T.; Trzaska, Z.; Thomas, M. Elaboration of Metallic Materials by SPS: Processing, Microstructures, Properties, and Shaping. *Metals* **2021**, *11*, 322. [CrossRef]
7. Dine, S.; Bernard, E.; Herlin, N.; Grisolia, C.; Tingaud, D.; Vrel, D. SHS Synthesis, SPS Densification and Mechanical Properties of Nanometric Tungsten. *Metals* **2021**, *11*, 252. [CrossRef]
8. Moser, M.; Lorand, S.; Bussiere, F.; Demoisson, F.; Couque, H.; Bernard, F. Influence of Carbon Diffusion and the Presence of Oxygen on the Microstructure of Molybdenum Powders Densified by SPS. *Metals* **2020**, *10*, 948. [CrossRef]
9. Ikram, A.; Awais, M.; Sheridan, R.; Walton, A.; Kobe, S.; Pušavec, F.; Rožman, K. Spark Plasma Sintering as an Effective Texturing Tool for Reprocessing Recycled HDDR Nd-Fe-B Magnets with Lossless Coercivity. *Metals* **2020**, *10*, 418. [CrossRef]
10. Mogale, N.F.; Matizamhuka, W.R. Spark Plasma Sintering of Titanium Aluminides: A Progress Review on Processing, Structure-Property Relations, Alloy Development and Challenges. *Metals* **2020**, *10*, 1080. [CrossRef]
11. Naimi, F.; Niepce, J.C.; Ariane, M.; Cayron, C.; Calapez, J.; Gentzbittel, J.M.; Bernard, F. Joining of Oxide Dispersion-Strengthened Steel Using Spark Plasma Sintering. *Metals* **2020**, *10*, 1040. [CrossRef]
12. Téllez-Martínez, J.S.; Olmos, L.; Solorio-García, V.M.; Vergara-Hernández, H.J.; Chávez, J.; Arteaga, D. Processing and Characterization of Bilayer Materials by Solid State Sintering for Orthopedic Applications. *Metals* **2021**, *11*, 207. [CrossRef]
13. González, J.; Armas, G.; Negrin, J.; Beltrán, A.; Trueba, P.; Gotor, F.; Peón, E.; Torres, Y. Influence of Successive Chemical and Thermochemical Treatments on Surface Features of Ti6Al4V Samples Manufactured by SLM. *Metals* **2021**, *11*, 313. [CrossRef]

Review

Harmonic Structure Design: A Strategy for Outstanding Mechanical Properties in Structural Materials

Bhupendra Sharma [1],*, Guy Dirras [2] and Kei Ameyama [3]

[1]   Department of Mechanical Engineering, Research Organization of Science and Technology, Ritsumeikan University, 1-1-1 Noji-Higashi, Kusatsu 525-8577, Shiga, Japan

[2]   Universite Sorbonne Paris Nord, LSPM-CNRS, 99 Avenue Jean-Baptiste Clément, 93430 Villetaneuse, France; dirras@univ-paris13.fr

[3]   Faculty of Science and Engineering, Ritsumeikan University, 1-1-1 Noji-Higashi, Kusatsu 525-8577, Shiga, Japan; ameyama@se.ritsumei.ac.jp

*   Correspondence: bhupen@fc.ritsumei.ac.jp; Tel.: +81-80-7793-0128

Received: 6 October 2020; Accepted: 28 November 2020; Published: 1 December 2020

Abstract: Structured heterogeneous materials are ubiquitous in a biological system and are now adopted in structural engineering to achieve tailor-made properties in metallic materials. The present paper is an overview of the unique network type heterogeneous structure called Harmonic Structure (HS) consisting of a continuous three-dimensional network of strong ultrafine-grained (shell) skeleton filled with islands of soft coarse-grained (core) zones. The HS microstructure is realized by the strategic processing method involving severe plastic deformation (SPD) of micron-sized metallic powder particles and their subsequent sintering. The microstructure and properties of HS-designed materials can be controlled by altering a fraction of core and shell zones by controlling mechanical milling and sintering conditions depending on the inherent characteristics of a material. The HS-designed metallic materials exhibit an exceptional combination of high strength and ductility, resulting from optimized hierarchical features in the microstructure matrix. The experimental and numerical results demonstrate that the continuous network of gradient structure in addition to the large degree of microstructural heterogeneity leads to obvious mechanical incompatibility and strain partitioning, during plastic deformation. Therefore, in contrast to the conventional homogeneous (homo) structured materials, synergy effects, such as synergy strengthening, can be obtained in HS-designed materials. This review highlights recent developments in HS-structured materials as well as identifies further challenges and opportunities.

Keywords: powder metallurgy; harmonic structure; severe plastic deformation; strength-ductility; structural characteristics

---

1. Introduction

Metallic materials are significantly crucial for manufacturing industry and structural applications. Metals with an optimized balance of higher strength and ductility are in demand for many industrial applications. Unfortunately, strength and ductility are antinomies in conventional homogeneous materials. Over many years, ultrafine-grained (UFG) metallic materials proved to be attractive structural materials due to higher strength [1–3], especially when compared to their coarse-grained (CG) counterparts. However, the drawback of homo UFG materials is typically a poor elongation due to the lack of sustained strain hardening rate in the early stage of deformation [4,5]. A major challenge, therefore, is to engineer novel microstructures to obtain materials with high synergetic strength and ductility. Therefore, recently, new fabrication and subsequent processing approaches were introduced to achieve overall as well as nano-scale microstructural control via grain engineering. One can now create ultrafine-grained homo and heterostructures in metallic materials. The strength and ductility trade-off became a primary challenge in developing high-performance metallic materials.

There have been several successful research studies to achieve high strength while retaining reasonable ductility through solid solution strengthening, precipitation strengthening, transformation hardening, dispersion hardening, etc., by incorporating multiple phases in alloys with several elements, as well as developing composites [6–10]. In addition, several techniques were used to refine the microstructures and improve the mechanical properties through grain refinement by several means, for example; severe plastic deformation, thermo-mechanical treatments, friction stir processing, etc. These techniques use the interruption in the dislocations' motion, but still make a dislocation activity possible to ensure some ductility. It was also reported that using a combination of the above-mentioned approaches, i.e., compositional change and thermomechanical treatment can be used to achieve balanced mechanical properties [11,12]. The focus of this review is, however, on single-phase metallic materials, such as; pure metals (Ti and Ni) or solid solutions (SUS316L) based on principal elements.

2. The Strength–Ductility Behavior of Homo- and Hetero-Structured Materials

For the conventional UFG homo metallic materials, the slope of the engineering stress–strain curve after the yield point, i.e., during work hardening, is remarkably lower than that of the counterpart materials with comparatively coarse grains. Figure 1 depicts a schematic drawing of a true stress–true strain diagram of conventional homo-structured metallic materials. It is well known that, during the tensile deformation, the localized reduction in the cross-sectional area of the tensile specimen, i.e., necking, occurs when the rate of strain hardening is lower than the flow stress of the material. It can be expressed as $d\sigma/d\varepsilon \leq \sigma$ (σ is the true stress and ε is the true strain).

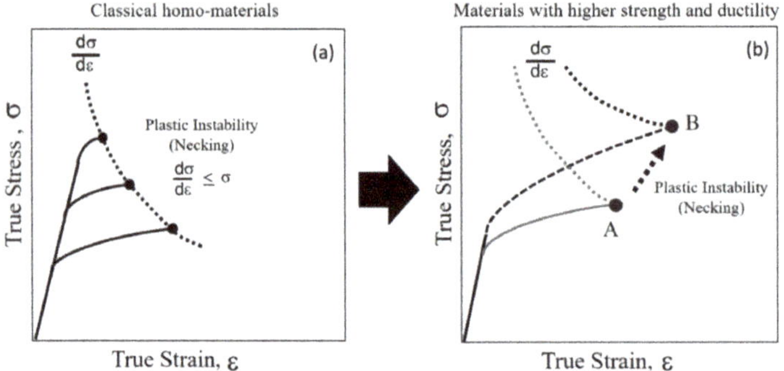

Figure 1. Schematic illustrations of true stress–true strain curves in; (**a**) materials with conventional homogeneous microstructure, and (**b**) advanced materials with a combination of higher strength and ductility.

In general, high strength but poor ductility of the metallic materials with microstructural homogeneity and fine grains is due to the poor strain hardening capabilities (Figure 1a). However, by introducing heterogeneities in the microstructure matrix, the strain hardening rate $d\sigma/d\varepsilon$ of the metallic materials can be improved while preventing the inequality conditions (Figure 1b). In other words, a good combination of superior strength and ductility can be achieved when a larger strain hardening rate is achieved. In addition, at the same time, deformation localization should be suppressed to obtain further elongation.

To address this strength–ductility trade-off issue, heterogeneous structures with large microstructural heterogeneities were introduced in metallic materials [3,13–16], such as bimodal structure [17], gradient structure [18–20], gradient nano-twined structure [21], nano-laminae structure [22], and heterogeneous laminae structure [13]. Such heterogeneous-designed structures were prepared via developing modern processing techniques without changing the composition, which includes surface grinding or rolling, severe plastic deformation, and electrodeposition. The resulting structured heterogeneous materials exhibit excellent mechanical properties.

Many pieces of research are being conducted on hetero-structured materials owing to their controllable and superior mechanical properties that are unattainable in the conventional homogeneous-structured materials. Therefore, structured heterogeneous materials are quickly emerging as a major materials research area, which also presents renewed concepts of materials science that challenged our traditional perceptions and impressions. In this review paper, we primarily focus on the microstructural heterogeneities in the metallic materials. Compared to conventional homo materials, hetero-structured metallic materials contain heterogeneous regions that contrast in their constitutive properties. It was reported that structured heterogeneous materials are advantageous properties owing to the synergetic effects occurring from the interactivity and pairing between structural contrasting. For example, the trade-off in strength and ductility can be alleviated or controlled [13–15]. This intended structural contrasting induces heterogeneous plastic deformation, and the fine-scale constituents stipulate exorbitant strain gradients. As a result, higher strain hardening and consequently uniform tensile ductility at high flow stresses can be achieved. However, the realization of heterogeneous design together with their microstructure and properties control is limited by multiple complex processing steps. Nevertheless, their properties are limited to the uni- or bidirectional (anisotropic characteristics).

Recently, Ameyama et al. proposed a novel concept of "harmonic structure" (HS) design, consisting of a specific spatial distribution of ultra-fine grains (UFG) and coarse grains (CG), that is, the CG areas (core) surrounded by three-dimensionally (3D) continuously connected network of UFG areas (shell) [23–31]. Owing to its unique topological 3D gradient structure, the harmonic-structured materials were reported to exhibit high work hardening that extends to higher strain regions, leading to delay in the initiation of plastic instability. Consequently, a good combination of high strength and high ductility can be achieved [29–31].

In the present review, an overview of the basics of harmonic structure design and highlighting recent developments in harmonic structured materials is presented, as well as identifying perspectives and future challenges and opportunities.

3. Concept of Harmonic Structure Design and Processing Considerations

From the above points, the harmonic structure (HS) design can be a candidate material design, which combines high strength with high ductility at the same time. Figure 2 demonstrates a concept of the HS design. The HS material is composed of a shell/core bimodal particle unit. A harmonic structure designed material is produced by the sintering of the particle units. In contrast to a homo-UFG material, HS material has a heterogeneous microstructure consisting of controlled and uniquely arranged fine and coarse grains with a particular topological distribution in a network. That is, the HS can be visualized as homogeneous on large scale but heterogeneous on a small scale, as shown in Figure 2.

During fabrication processing, it is crucial to control and optimize the characteristics of the shell/core regions, such as; volume fraction, grain size, morphology, and topology. These characteristics may strongly influence the properties of the HS materials. The physical properties of the HS materials can be affected by the grain size distribution and grain gradients and morphology of the shell/core regions. To control these factors in the as-fabricated HS compacts, it is necessary but challenging to invent efficient processing methods. In this regard, the unique three-dimensional network-type microstructure of the HS materials adds considerable complexity to processing. In addition to the wide range of multiple variables, the type of material, i.e., the behavior of a particular class of materials, also affects the microstructure of the as-fabricated HS compacts. However, in general, by varying these parameters, HS microstructure was developed in various metals and alloys. To achieve a tailor-made HS microstructure design, the following efficient and feasible processing method, based on powder metallurgy processing, was developed.

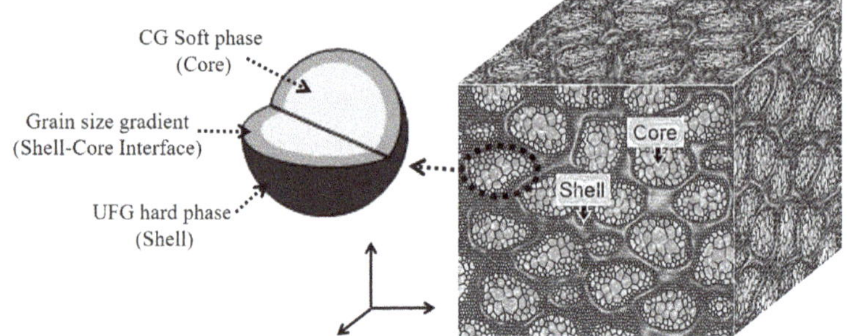

Figure 2. Concept of the harmonic structure (HS) design.

3.1. Mechanical Milling Process (MM Process)

The MM process involves mechanical milling (i.e., ball milling, high-pressure gas milling, etc.) of Plasma Rotating Electrode Processed (PREP) powder of metallic powder and their subsequent sintering. As shown in Figure 3a, powder with a bimodal grain size distribution, i.e., SPD powder surface with fine grains (shell), and comparatively less deformed (or undeformed) powder interior (core), can be fabricated by the MM process. The MM processed powder surface becomes the three-dimensional shell network structure after sintering. As depicted in the schematic diagram of Figure 3a, the depth of plastic deformation, at the powder surface, can be changed by changing the mechanical milling time. Consequently, it is feasible to achieve a tailor-made microstructure with predetermined shell/core fractions. However, it is to be noted that the deformation at the surface and respective microstructure evolution depends on the type of initial material.

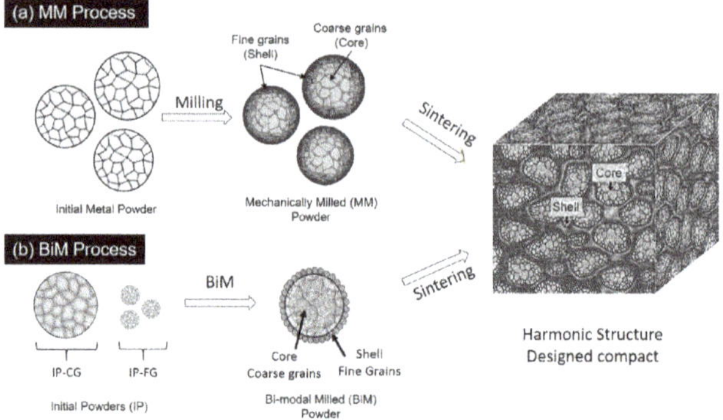

Figure 3. Fabrication processes of the harmonic structure materials; (**a**) Mechanical Milling (MM) and (**b**) Bimodal Milling (BiM) process.

3.2. Bimodal Milling Process (BiM Process)

The BiM process involves milling initial powders of coarse-grained large-sized (IP-CG) and comparatively fine-grained small-sized (IP-FG) powders, wherein the powder fractions and milling time are chosen depending on the desired fractions and grain size of shell/core, respectively [32]. The hierarchical distribution of fine/coarse-grains can be achieved due to repeated coating of plastically deformed IP-FG to the IP-CG powder, during milling. A schematic diagram illustrating the BiM powder metallurgy-based processing approach is demonstrated in Figure 3b. As depicted in the schematic diagram of Figure 3b, the fraction of a tailor-made microstructure with desired shell/core fractions can be achieved either by varying the fractions of IP-FG and IP-CG powders, and/or varying milling time.

The selection of a specific fabrication process was based on many factors such as the type of material, powder particles size, the susceptibility of the material with milling media, the permitted level of contamination from the milling vial and balls, and malleability and work hardening characteristics of the materials being processed [27]. For example, the soft materials can be flattened, i.e., might form flakes, during the severe milling process. Therefore, the BiM processing approach can be utilized to achieve a high shell fraction in a short time milling. On the other hand, less ductile materials can be processed through the MM process.

4. Microstructure of BiM and MM Processed Powders

The effect of controlled milling through BiM and MM processes on the morphology and microstructure of SUS316L alloys powders is shown in Figure 4. In the BiM process, IP-FG (powder particle size ~7 µm) and IP-CG (powder particle size ~142 µm) powders were mixed in a 3:2 ratio, respectively, and BiM processed for 36 and 360 ks (Figure 4a,b). On the other hand, in the MM process, the PREP alloy powder of SUS316L IP-CG was milled for 108 and 540 ks.

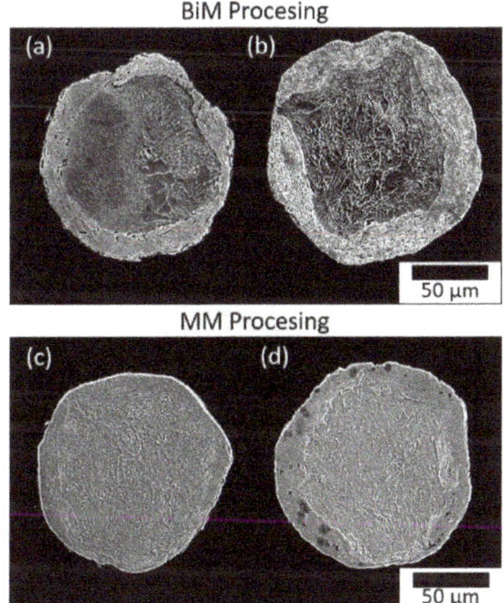

Figure 4. SEM images of SUS316L powders cross-section after BiM and MM processing. (**a**) BiM 36 ks, (**b**) BiM 360 ks, (**c**) MM 108 ks, and (**d**) MM 540 ks.

It is evident that the overall morphology of the spherical shape of the powder does not change as a result of both MM and BiM processing. However, it is to be noted that a highly deformed zone (shell) is formed at the surface, whereas the inner regions (Core) of the powder remain approximately undeformed or partially deformed, depending on the milling conditions. It is interesting to note that the high fraction of shell could be achieved in a shorter time of BiM processing compared to the MM processing. In contrast to the MM powder, the outer surface of BiM powder is composed of several layers of fine powder particles.

Therefore, it can be predicted that the shell evolution mechanism in both BiM and MM process is different. It was reported that the powder particle size of BiM powders increases with BiM processing time, whereas in the case of the MM process, the MM powder particle size remains approximately similar to the starting powder. Therefore, it can be assumed that the surface irregularity of the BiM powder is primarily due to the repeated coating of SPD fine powder to the surface of coarse powder, during milling. On the other hand, the surface irregularity in MM powder is owing to the surface deformation of the starting powder particles. However, it can be ensured that both MM and BiM powders show a bimodal microstructure wherein the center of powder is coarse-grained, whereas the surface composed of SPD sub-micron-sized grains. Similar morphological changes were also reported wherein it was observed that the increase in milling time leads to the formation of the severely deformed layer, consisting of nano-crystallites, near the surface of the powders [33–37]. Moreover, it must be realized that a gradient of the degree of deformation exists in both BiM and MM processed powders, and the severity of the accumulated plastic strain decreases from the particle surface towards the center of the particles.

In particular, the successful application of the new BiM process to achieve controlled deformation in the variety of materials is of important technological interest, which is a feasible, energy-efficient, and cost-effective process, particularly, for industrial-scale production. The effectiveness of the BiM process to create a controlled microstructure at shell and core is owing to the fine and coarse powders, respectively. It must be pointed out that the finer powder particles make the milling process easier, and grain refinement occurs in a shorter time when compared to coarse powder particles. Therefore, in contrast to the prolonged milling in the MM process, significantly shorter milling might result in the development of a large shell fraction with nano grain formation. It naturally follows that one can build a hierarchical microstructure with controlled fractions of shell/core zones by varying amounts of IP-FG and IP-CG powders, and/or milling time. Optimizing such hierarchical grained zones and their fractions for a better synergy of superior strength and ductility, in an as-sintered bulk compact, calls for modern fabrication approaches for precise microstructure control. Hence, it would be appealing to analogize the designs of BiM- and MM-processed SUS316L compacts for a better interpretation and rationalization of the synergistic effects of various combinations of structural contrasts.

Figure 5 shows TEM micrographs near the shell region of the mechanically milled pure titanium powder with enlarged areas of selected rectangular areas indicated as A, B, and C. It could be observed that near to the surface, equiaxed grains (<20 nm) were observed whereas a layer of elongated grains was seen in the inner zones of the milled powder. The grain subdivision and rotation of those elongated grains led to the formation of equiaxed nano-grain structure in milled powder [30,33,37–41].

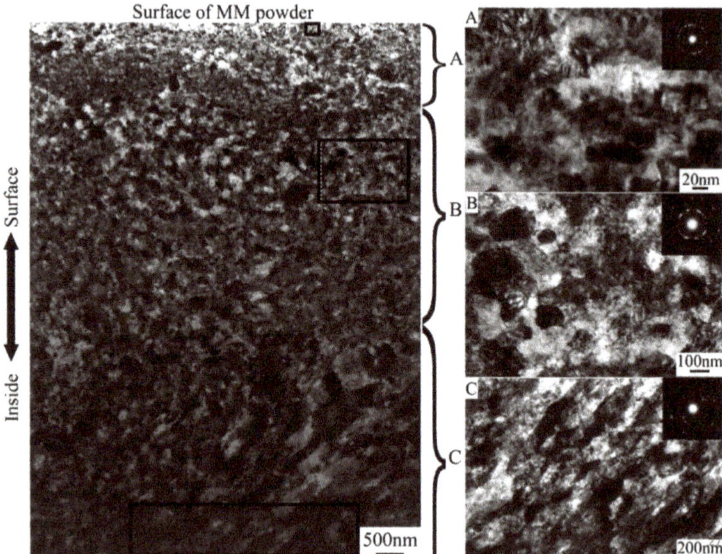

Figure 5. TEM micrographs of the shell area of milled pure Ti powder. Enlargement of rectangular areas of (**A**), (**B**) and (**C**) are separately presented.

5. Microstructure of Harmonic Structure Designed Compacts

The milled powders via the BiM and MM process were sintered at 1173 and 1223 K temperatures, respectively, to prepare bulk compacts. Recently, spark plasma sintering (SPS) was utilized extensively for the consolidation of powders due to fast sintering together with avoidance of significant grain growth even at high-temperature operations. The microstructures of the H-structured compacts (MM and BiM) SUS316L compacts are shown in Figure 6.

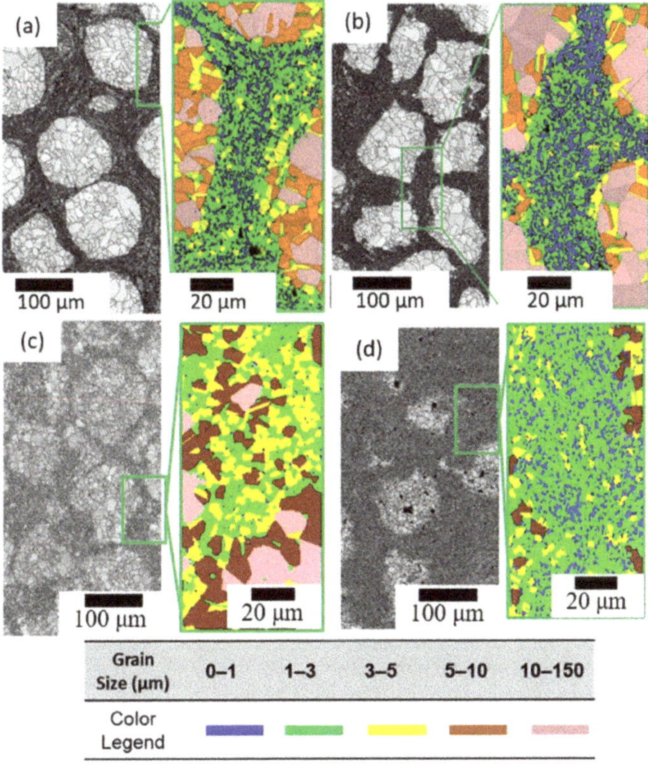

Figure 6. EBSD image quality maps of BiM (**a,b**), and MM compacts (**c,d**) of SUS316L. (**a**) BiM 36 ks, (**b**) BiM 180 ks, (**c**) MM 216 ks, and (**d**) MM 540 ks. The corresponding enlarged grain size images of the shell–core interface are separately shown (the grain size legend is presented at the bottom).

As shown in Figure 6a,b, the BiM compacts exhibited a network-type arrangement of bimodal grains wherein the coarse grains (d > 5.0 μm) were surrounded by comparatively finer grains (d ≤ 5.0 μm). Particularly, it is to be noted that the well-defined and dense shell network could be achieved even after a short time (36 ks) of BiM processing (Figure 6c). The grain size contrast of the shell and core could be observed in the enlarged Elecrtron Backscatter Diffraction (EBSD) grain size images. It should be noted that the core grains are not remarkably deformed even after a long milling time (180 ks). Therefore, it can be envisaged that the shell area of BiM compacts mainly developed due to the coating and frequent plastic deformation of fine particles to the surface of coarse particles, during milling. However, the spherical shape of the core gradually deformed with milling time, suggesting that the prolonged milling leads to the partial deformation of the outer core area. Therefore, it can be envisaged that increasing milling time leads to effective plastic deformation of fine powder, which in turn increases the fraction of fine shell grains (<3 μm), which was obvious in the shell grain size distribution through the color-coded grain size distribution map of BiM compacts. Also, in the case of MM compacts (Figure 6c,d), a harmonic structure was achieved. However, substantial damage apparent to the core area was noticed after a long period of milling (540 ks). In addition, most notable was that even after prolonged milling up to 540 ks, the large fractions of shell grains were in the range of 3–5 μm, which is significantly larger than that of the BiM-processed specimens. Finally, it would be worth emphasizing that, in comparison to the MM process, the BiM approach is highly efficient in achieving ultra-fine shell grains in a shorter period.

6. Mechanical Properties of Harmonic Structure-Designed Compacts

Figure 7 shows the representative engineering stress–strain curves of harmonic structured compacts of SUS316L developed by BiM and MM approaches. All of the HS materials that were examined indicated high strength, high ductility, and high tensile toughness at the same time. That is to say, regardless of the fabrication process, the HS compacts of SUS316L had a good combination of strength and elongation compared to the conventional counterparts developed and/or processed by different methods, such as; casting, powder metallurgy, selective laser melting, and thermomechanically treated compacts of SUS316L alloy (Figure 8) [42–61]. In general, it was reported that the conventional homo-structured compacts of SUS316L exhibited significant strength-ductility trade-off owing to the grain size strengthening effect [42–61]. However, in contrast, as shown in Figure 7, the improvement in the strength of harmonic structured SUS316L compacts were accompanied by a nominal ductility loss. The improvement in strength and nominal ductility loss can be attributed to grain boundary strengthening caused by grain refinement in the shell and higher shell fraction. Particularly, the reduction in ductility is associated with a higher shell fraction, as shown in Figure 6. Plastic deformation in most metallic materials was commonly mediated by the dislocation slip [26,27]. The increased fractions of fine-grained shell from a certain level suppressed the dislocation initiation and slip phenomena, and increased the prospect of dislocation annihilation at boundaries by thermally activated cross slip and climb, which in turn leads to ductility loss [28,29]. The most noticeable was the synergy of high strength and ductility of harmonic-structured BiM compacts, which was achieved in an efficient time compared to that achieved by MM processing. As can be seen, the improvement in mechanical properties by efficient BiM processing was encouraging, which may lead to the synergetic strength–ductility combination by introducing an optimized fraction of the shell network. In the present case, the optimized grain size and shell fractions were approximately 2 μm and 30%, respectively [32].

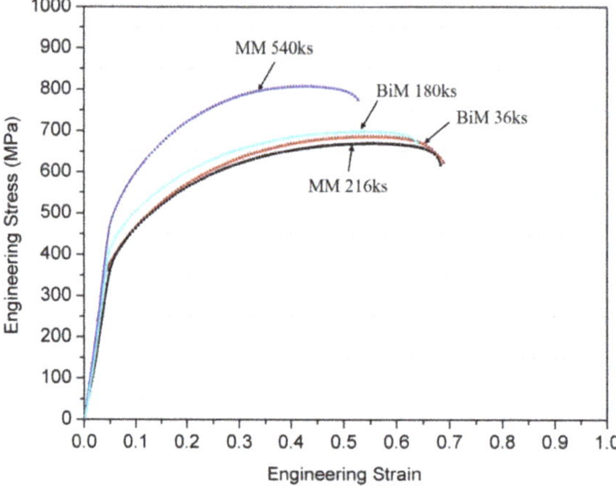

Figure 7. Tensile engineering stress-strain curves of harmonic structured SUS316L compacts developed through BiM and MM processes.

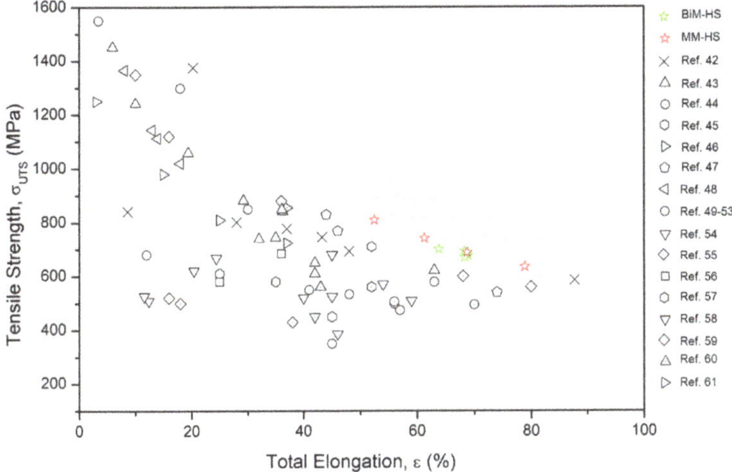

Figure 8. Ultimate tensile strength versus total elongation chart for HS SUS316L compacts, and their comparison with SUS316L counterparts produced by various methods [42–61].

7. Mechanism of the Outstanding Mechanical Properties

7.1. Stress Concentration Effects

Figure 9 shows the microstructure of conventional homogeneous-structured (homo) and harmonic-structured pure Ti compacts after 10% cold rolling. EBSD Kernel Average Misorientation (KAM) images show that, in contrast to the homo Ti with significant KAM values in the matrix, slight KAM value increased especially in the shell region. Since the increase in the KAM value strongly related to the dislocation storage, it suggests that deformation took place in the shell and shell/core interface regions.

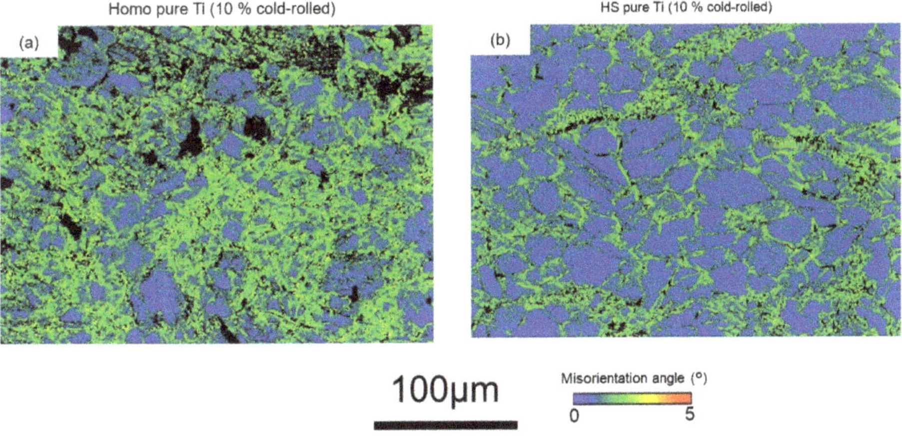

Figure 9. KAM image of; (a) homo and (b) HS structured pure Ti compacts.

As reported by Ameyama et al. [35] using ECCI (Electron Channeling Contrast Image) of Figure 10a: shell area and Figure 10b core area, it can be seen that dislocations were mainly in the shell grains, whereas the core dislocations were cell-like, and comparatively not highly dense

(Figure 10). These results were quite consistent with the KAM distribution observed for pure Ti (Figure 9). The higher dislocation density in the shell region was evidence of stress concentration in the shell. Therefore, it was confirmed that during the early stage of deformation, the shell and its network can be an effective dislocation source. It is to be noted that such a "dislocation-source network" structure is a unique characteristic of the HS materials. Consequently, a large strain hardening rate and higher uniform elongation can be achieved in the HS materials. Interestingly, despite the soft core regions, the hard shell deforms before the core in the early stage of deformation. Such a stress concentration effect was also predicted by a multi-scale-FEM simulation (Figure 11) [62]. From the numerical simulation, the UFG shell regions are the high-stress concentration areas, while the CG core region did not show evidence of accumulation of any significant stress concentration.

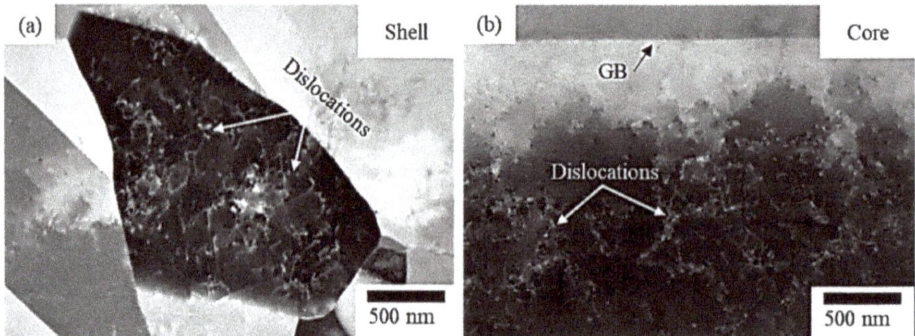

Figure 10. ECCI of pure Ni HS after 5% tensile deformation. (**a**): shell and (**b**): core.

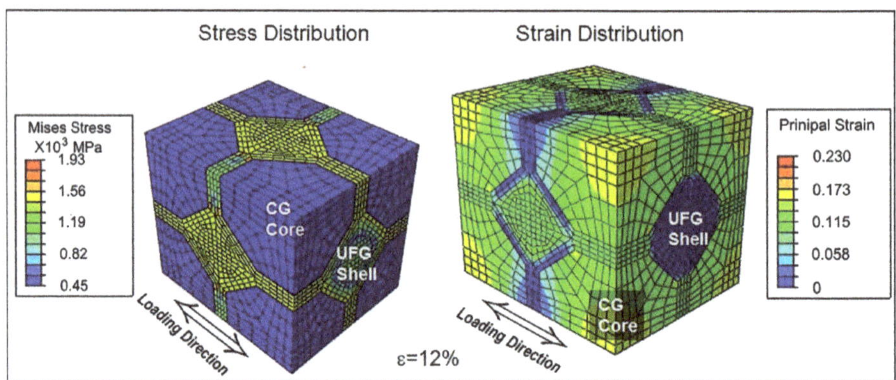

Figure 11. Numerical simulation (FEM) results of HS pure Ti illustrating the stress and strain distribution after ~12% tensile deformation [62].

7.2. Strain Dispersion Effects

It was also reported that the mutual constraints of the unstable UFG area and the comparatively stable CG area made the deformation processing captive and lead to a different stress level [63]. The geometrically necessary dislocations (GNDs) emerged due to the geometrical constraints of soft (core) and hard (shell) areas in the HS specimens' need to accommodate the strain gradient areas; if not, the voids and overlaps may occur between the grain boundaries [64]. As a consequence of the interaction and amassing of dislocations, extraordinary work hardening occurs.

As shown in the numerical simulation results of strain distribution (Figure 11), higher strain concentration was seen in the shell zones aligned parallel to the tensile axis. In contrast, no evidence of accumulated strain concentration was seen in the shell zones aligned perpendicular to the tensile deformation axis [62]. However, interestingly, accumulated strain concentration successively increased from the shell to core. Consecutively, the highest accumulated strain was observed at the center of the core.

In general, the early necking of fine-grained materials is due to high-stress concentration. The crack propagation is hard to block with fine grains owing to suppressed dislocation initiation and slip phenomenon [1,65]. However, when the well-bonded, tri-directional network of hierarchical microstructure was stressed, the fast strain localization tendency of the fine-grained shell can be effectively constrained by the stable coarse-grained core through the interfaces [66]. In addition, the strain partitioning between incompatible shell and core areas may also play a crucial role in alleviating the strain localization [62,67,68]. Such behavior may suggest that the complete shell network structure plays a crucial role in the uniform strain dispersion, and thus, deformation localization is prevented.

8. Fracture Behavior of HS Materials

The superior strength–ductility combination of pure HS materials can be attributed to the three-dimensional topological distribution of connected-type hard shell regions. In particular, the higher toughness can be associated with topology-driven diversion of crack propagation paths as well as an obstacle in their growth and propagation due to the soft core regions [69]. To analyze the fatigue crack propagation and growth mechanism, Kikuchi et al. [70] and Osaki et al. [71] investigated the effect of grain size on fatigue crack propagation in HS pure Ti. It was reported that the fatigue crack propagation and growth resistance of HS Ti were significantly reduced for a high-stress ratio as compared to the counterpart with homogeneous coarse-grained microstructure. Nukui et al. [72] found that the fatigue strength is increased because the resistance of the material to the initiation of fatigue cracks was improved by the grain refinement at the shell. The beginning of fatigue cracks was also observed in the coarse grains core in harmonic-structured designed materials, including CP titanium [72], Ti–6Al–4V alloys [73,74], and austenitic stainless steel [75,76]. However, in alloys, it can be attributed to both the network topology and strain-induced phase transformations, for example; α-fcc to ε-hcp in CCM [28]. However, further exhaustive studies should be carried out to confirm the fracture behavior of different metallic materials.

9. Perspectives of HS Materials and Future Challenges

Harmonic structure-designed materials are emerging structural materials that are becoming an attractive research field. The area of hetero-structured materials was extensively studied. However, in particular, harmonic structure-designed materials are in their infancy, requiring intense studies both in terms of characterization methods, scientific modeling tools, and fabrication approaches.

A complete clarification of the underlying mechanism and availability of any consolidated model regarding the excellent strengthening of HS materials is still a matter of pursuit. Some efforts were made to annotate this direction through experimental documentation as well as theoretical techniques [77–80]. In particular, owing to the unique 3D connected type microstructural design of harmonic-structured materials, rigorous efforts are required to decipher the mechanism of strengthening completely.

In addition, it should be mentioned that development and microstructure controlling of all the customized microstructures, including harmonic structure, are relatively laborious to produce and are not comprehensive. The microstructure control at micro- as well as macro-scale and repeatability in performances are another great challenge. To this aspect, further efforts are required to develop a consolidated design concept and mechanistic approach for harmonic structure-designed materials.

The technological and engineering affairs need to be addressed by both theoretical models and experiments to quantify the correlation between structural heterogeneity and mechanical performance, which is crucial for the development of harmonic structure materials with efficient design and optimized mechanical performance.

10. Conclusions

In this review, a relatively new concept of harmonic structure materials has been introduced, and its basic concept, mechanical performance, and advantages along with present development status and perspectives are discussed. The harmonic structure (HS)-designed metallic materials can be successfully produced by efficient powder metallurgy approaches, such as mechanical milling and bimodal milling (BiM). In contrast to the significant strength–ductility trade-off issues of conventional homogeneous-structured materials, harmonic-structured materials exhibit a superior combination of high synergetic strength and ductility. The superior tensile strength of the HS materials is owing to its unique network-type arrangement of the UFG grains (shell), which provides higher strain hardening due to the stress concentration effect, whereas the superior ductility of the HS materials is attributed to the soft core regions. The results suggested that the higher ductility of the HS materials was due to strain dispersion effects. In other words, microstructural singularity in harmonic structure design plays a crucial role in achieving an outstanding combination of strength and ductility.

Author Contributions: Conceptualization, B.S., G.D., K.A.; methodology, B.S.; formal analysis, B.S.; investigation, B.S.; writing—original draft preparation, B.S.; writing—review and editing, B.S., G.D., K.A.; supervision, K.A.; project administration, B.S., G.D., K.A.; funding acquisition, B.S., G.D., K.A. All authors have read and agreed to the published version of the manuscript.

Funding: This work was supported by the Japan Society for The Promotion of Science (JSPS) Grants-in-Aid for Scientific Research (KAKENHI) Grant Numbers 20K15064 and JP18H05256. This support is gratefully appreciated. The support by the French National Research Agency in the framework of the ANR 14-CE07-0003 "HighS-Ti" program is highly acknowledged.

Conflicts of Interest: The authors declare no conflict of interest.

References

1. Meyers, M.A.; Mishra, A.; Benson, D.J. Mechanical properties of nanocrystalline materials. *Prog. Mater. Sci.* **2006**, *51*, 427–556. [CrossRef]

2. Khan, H.; Farrokh, B.; Takacs, L. Compressive properties of Cu with different grain sizes: Sub-micron to nanometer realm. *J. Mater. Sci.* **2008**, *43*, 3305–3313. [CrossRef]

3. Ma, E.; Zhu, T. Towards strength—Ductility synergy through the design of heterogeneous nanostructures in metals. *Mater. Today* **2017**, *20*, 323–331. [CrossRef]

4. Saito, Y.; Utsunomiya, H.; Tsuji, N.; Sakai, T. Novel ultra-high straining process for bulk materials—Development of the accumulative roll-bonding (ARB) process. *Acta Mater.* **1999**, *47*, 579–583. [CrossRef]

5. Estrin, Y.; Vinogradov, A. Extreme grain refinement by severe plastic deformation: A wealth of challenging science. *Acta Mater.* **2013**, *61*, 782–817. [CrossRef]

6. Manzoni, A.M.; Glatzel, U. New multiphase compositionally complex alloys driven by the high entropy alloy approach. *Mater. Charact.* **2019**, *147*, 512–532. [CrossRef]

7. Kumar, K.; Van Swygenhoven, H.; Suresh, S. Mechanical behavior of nanocrystalline metals and alloys11The Golden Jubilee Issue—Selected topics in Materials Science and Engineering: Past, Present and Future, edited by S. Suresh. *Acta Mater.* **2003**, *51*, 5743–5774. [CrossRef]

8. Zhao, Y.H.; Lavernia, E.J.; Whang, S.H. *Nanostructured Metals and Alloys*; Woodhead Publishing: Cambridge, UK, 2011; pp. 375–376.

9. Hofinger, M.; Turk, C.; Ognianov, M.; Leitner, H.; Schnitzer, R. Precipitation reactions in a Cu-Ni-Al medium carbon alloyed dual hardening steel. *Mater. Charact.* **2020**, *160*, 110126. [CrossRef]

10. Wegst, U.G.K.; Bai, H.; Saiz, E.; Tomsia, A.P.; Ritchie, R.O. Bioinspired structural materials. *Nat. Mater.* **2015**, *14*, 23–36. [CrossRef]

11. Gwalani, B.; Soni, V.; Lee, M.; Mantri, S.; Ren, Y.; Banerjee, R. Optimizing the coupled effects of Hall-Petch and precipitation strengthening in a Al 0.3 CoCrFeNi high entropy alloy. *Mater. Des.* **2017**, *121*, 254–260. [CrossRef]

12. Cheng, H.; Wang, H.Y.; Xie, Y.C.; Tang, Q.H.; Dai, P.Q. Controllable fabrication of a carbide-containing FeCoCrNiMn high-entropy alloy: Microstructure and mechanical properties. *Mater. Sci. Technol.* **2017**, *33*, 2032–2039. [CrossRef]

13. Wu, X.; Yang, M.; Yuan, F.; Wu, G.; Wei, Y.; Huang, X.; Zhu, Y. Heterogeneous lamella structure unites ultrafine-grain strength with coarse-grain ductility. *Proc. Natl. Acad. Sci. USA* **2015**, *112*, 14501–14505. [CrossRef] [PubMed]

14. Embury, D.; Bouaziz, O. Steel-based composites: Diving forces and Classifications. *Annu. Rev. Mater. Res.* **2010**, *40*, 213–241. [CrossRef]

15. Wu, X.; Zhu, Y.T. Heterogeneous materials: A new class of materials with unprecedented mechanical properties. *Mater. Res. Lett.* **2017**, *5*, 527–532. [CrossRef]

16. Lu, K. Stabilizing nanostructures in metals using grain and twin boundary architectures. *Nat. Rev. Mater.* **2016**, *1*, 16019. [CrossRef]

17. Wang, Y.; Chen, M.; Zhou, F.; Ma, E. High tensile ductility in a nanostructured metal. *Nat. Cell Biol.* **2002**, *419*, 912–915. [CrossRef]

18. Fang, T.H.; Li, W.L.; Tao, N.R.; Lu, K. Revealing Extraordinary Intrinsic Tensile Plasticity in Gradient Nano-Grained Copper. *Science* **2011**, *331*, 1587–1590. [CrossRef]

19. Wu, X.; Jiang, P.; Chen, L.; Yuan, F.; Zhu, Y.T. Extraordinary strain hardening by gradient structure. *Proc. Natl. Acad. Sci. USA* **2014**, *111*, 7197–7201. [CrossRef]

20. Thevamaran, R.; Lawal, O.; Yazdi, S.; Jeon, S.-J.; Lee, J.-H.; Thomas, E.L. Dynamic creation and evolution of gradient nanostructure in single-crystal metallic microcubes. *Science* **2016**, *354*, 312–316. [CrossRef]

21. Cheng, Z.; Zhou, H.; Lu, Q.; Gao, H.; Lu, L. Extra strengthening and work hardening in gradient nanotwinned metals. *Science* **2018**, *362*, eaau1925. [CrossRef]

22. Liu, X.C.; Zhang, H.; Lu, K. Strain-Induced Ultrahard and Ultrastable Nanolaminated Structure in Nickel. *Science* **2013**, *342*, 337–340. [CrossRef] [PubMed]

23. Zhang, Z.; Vajpai, S.K.; Orlov, D.; Ameyama, K. Improvement of mechanical properties in SUS304L steel through the control of bimodal microstructure characteristics. *Mater. Sci. Eng. A* **2014**, *598*, 106–113. [CrossRef]

24. Vajpai, S.K.; Ota, M.; Watanabe, T.; Maeda, R.; Sekiguchi, T.; Kusaka, T.; Ameyama, K. The Development of High Performance Ti-6Al-4V Alloy via a Unique Microstructural Design with Bimodal Grain Size Distribution. *Met. Mater. Trans. A* **2015**, *46*, 903–914. [CrossRef]

25. Sharma, B.; Nagano, K.; Kawabata, M.; Ameyama, K. Microstructure and mechanical properties of hetero-designed Ti-25Nb-25Zr alloy fabricated by powder metallurgy route. *Lett. Mater.* **2019**, *9*, 511–516. [CrossRef]

26. Ota, M.; Vajpai, S.K.; Imao, R.; Kurokawa, K.; Ameyama, K. Application of High Pressure Gas Jet Mill Process to Fabricate High Performance Harmonic Structure Designed Pure Titanium. *Mater. Trans.* **2015**, *56*, 154–159. [CrossRef]

27. Vajpai, S.K.; Ota, M.; Zhang, Z.; Ameyama, K. Three dimensionally gradient harmonic structure design: An integrated approach for high performance structural materials. *Mater. Res. Lett.* **2016**, *4*, 191–197. [CrossRef]

28. Vajpai, S.K.; Sawangrat, C.; Yamaguchi, O.; Ciuca, O.P.; Ameyama, K. Effect of bimodal harmonic structure design on the deformation behavior and mechanical properties of Co-Cr-Mo alloy. *Mater. Sci. Eng. C* **2016**, *58*, 1008–1015. [CrossRef]

29. Fujiwara, H.; Oda, E.; Ameyama, K. Mechanical Milling Process as Severe Plastic Deformation Method. *Tetsu-to-Hagane* **2008**, *94*, 608–615. [CrossRef]

30. Ameyama, K.; Horikawa, N.; Kawabata, M. Unique Mechanical Properties of Harmonic Structure Designed Materials. *Tetsu-to-Hagane* **2019**, *105*, 124–126. [CrossRef]

31. Fujiwara, H.; Inomoto, H.; Sanada, R.; Ameyama, K. Nano-ferrite formation and strain-induced-ferrite transformation in an SUS316L austenitic stainless steel. *Scr. Mater.* **2001**, *44*, 2039–2042. [CrossRef]

32. Yagi, K.; Sharma, B.; Kawabata, M.; Ameyama, K. Fabrication of SUS316L Harmonic Structure Compacts by Bi-Modal Milling Process. *J. Jpn. Soc. Powder Powder Met.* **2020**, *67*, 239–244. [CrossRef]

33. Nagata, M.; Horikawa, N.; Kawabata, M.; Ameyama, K. Effects of Microstructure on Mechanical Properties of Harmonic Structure Designed Pure Ni. *Mater. Trans.* **2019**, *60*, 1914–1920. [CrossRef]

34. Sawangrat, C.; Yamaguchi, O.; Vajpai, S.K.; Ameyama, K. Application of Harmonic Structure Design to Biomedical Co-Cr-Mo Alloy for Improved Mechanical Properties. *Mater. Trans.* **2014**, *55*, 99–105. [CrossRef]

35. Ameyama, K.; Kawabata, M.; Sharma, B. ISOPE-I-19-101. In Proceedings of the 29th International Ocean and Polar Engineering Conference, Honolulu, HI, USA, 16–21 June 2019; ISBN 978-1-880653-85-2.

36. Sawangrat, C.; Kato, S.; Orlov, D.; Ameyama, K. Harmonic-structured copper: Performance and proof of fabrication concept based on severe plastic deformation of powders. *J. Mater. Sci.* **2014**, *49*, 6579–6585. [CrossRef]

37. Fujiwara, H.; Akada, R.; Noro, A.; Yoshita, Y.; Ameyama, K. Enhanced Mechanical Properties of Nano/Meso Hybrid Structure Materials Produced by Hot Roll Sintering Process. *Mater. Trans.* **2008**, *49*, 90–96. [CrossRef]

38. Fujiwara, H.; Inomoto, H.; Ameyama, K. Formation of (α + γ) Nano-duplex Structure by Severe Plastic Deformation in an SUS316L Stainless Steel. *Tetsu-to-Hagane* **2005**, *91*, 839–845. [CrossRef]

39. Mohamed, F.A. A dislocation model for the minimum grain size obtainable by milling. *Acta Mater.* **2003**, *51*, 4107–4119. [CrossRef]

40. Chaubey, A.K.; Scudino, S.; Khoshkhoo, M.S.; Prashanth, K.G.; Mukhopadhyay, N.K.; Mishra, B.; Eckert, J. Synthesis and Characterization of NanocrystallineMg-7.4%Al Powders Produced by Mechanical Alloying. *Metals* **2013**, *3*, 58–68. [CrossRef]

41. Oda, E.; Fujiwara, H.; Ameyama, K. Nano Grain Formation in Tungsten by Severe Plastic Deformation-Mechanical Milling Process. *Mater. Trans.* **2008**, *49*, 54–57. [CrossRef]

42. Zheng, R.; Zhang, Z.; Nakatani, M.; Ota, M.; Chen, X.; Ma, C.; Ameyama, K. Enhanced ductility in harmonic structure designed SUS316L produced by high energy ball milling and hot isostatic sintering. *Mater. Sci. Eng. A* **2016**, *674*, 212–220. [CrossRef]

43. Li, J.; Cao, Y.; Gao, B.; Li, Y.; Zhu, Y.T. Superior strength and ductility of 316L stainless steel with heterogeneous lamella structure. *J. Mater. Sci.* **2018**, *53*, 10442–10456. [CrossRef]

44. Chen, X.; Li, J.; Cheng, X.; Wang, H.; Huang, Z. Effect of heat treatment on microstructure, mechanical and corrosion properties of austenitic stainless steel 316L using arc additive manufacturing. *Mater. Sci. Eng. A* **2018**, *715*, 307–314. [CrossRef]

45. Yin, Y.; Sun, J.; Guo, J.; Kan, X.; Yang, D. Mechanism of high yield strength and yield ratio of 316 L stainless steel by additive manufacturing. *Mater. Sci. Eng. A* **2019**, *744*, 773–777. [CrossRef]

46. Long, Q.; Lu, J.; Fang, T. Microstructure and mechanical properties of AISI 316L steel with an inverse gradient nanostructure fabricated by electro-magnetic induction heating. *Mater. Sci. Eng. A* **2019**, *751*, 42–50. [CrossRef]

47. Lu, K.; Yan, F.; Wang, H.; Tao, N. Strengthening austenitic steels by using nanotwinned austenitic grains. *Scr. Mater.* **2012**, *66*, 878–883. [CrossRef]

48. Xiong, L.; You, Z.; Lu, L. Enhancing fracture toughness of nanotwinned austenitic steel by thermal annealing. *Scr. Mater.* **2016**, *119*, 55–59. [CrossRef]

49. Chen, X.; Lu, J.; Lu, L.; Lu, K. Tensile properties of a nanocrystalline 316L austenitic stainless steel. *Scr. Mater.* **2005**, *52*, 1039–1044. [CrossRef]

50. Wang, Y.; Huang, C.; Wang, M.; Li, Y.; Zhu, Y. Quantifying the synergetic strengthening in gradient material. *Scr. Mater.* **2018**, *150*, 22–25. [CrossRef]

51. Ueno, H.; Kakihata, K.; Kaneko, Y.; Hashimoto, S.; Vinogradov, A. Enhanced fatigue properties of nanostructured austenitic SUS 316L stainless steel. *Acta Mater.* **2011**, *59*, 7060–7069. [CrossRef]

52. Poling, W.A. Grain Size Effects in Micro-Tensile Testing of Austenitic Stainless Steel. Master's Thesis, Colorado School of Mines, Golden, CO, USA, 2012.

53. Vinogradov, A.; Yasnikov, I.; Matsuyama, H.; Uchida, M.; Kaneko, Y.; Estrin, Y. Controlling strength and ductility: Dislocation-based model of necking instability and its verification for ultrafine grain 316L steel. *Acta Mater.* **2016**, *106*, 295–303. [CrossRef]

54. Zhong, Y.; Rannar, L.E.; Liu, L.; Koptyug, A.; Wikman, S.; Olsen, J.; Cui, D.; Shen, Z. Additive manufacturing of 316L stainless steel by electron beam melting for nuclear fusion applications. *J. Nucl. Mater.* **2017**, *486*, 234–245. [CrossRef]

55. Röttger, A.; Geenen, K.; Windmann, M.; Binner, F.; Theisen, W. Comparison of microstructure and mechanical properties of 316L austenitic steel processed by selective laser melting with hot-isostatic pressed and cast material. *Mater. Sci. Eng. A* **2016**, *678*, 365–376. [CrossRef]

56. Casati, R.; Lemke, J.; Vedani, M. Microstructure and Fracture Behavior of 316L Austenitic Stainless Steel Produced by Selective Laser Melting. *J. Mater. Sci. Technol.* **2016**, *32*, 738–744. [CrossRef]

57. Bartolomeu, F.; Buciumeanu, M.; Pinto, E.; Alves, N.; Carvalho, O.; Silva, F.; Miranda, G. 316L stainless steel mechanical and tribological behavior—A comparison between selective laser melting, hot pressing and conventional casting. *Addit. Manuf.* **2017**, *16*, 81–89. [CrossRef]

58. Suryawanshi, J.; Prashanth, K.; Ramamurty, U. Mechanical behavior of selective laser melted 316L stainless steel. *Mater. Sci. Eng. A* **2017**, *696*, 113–121. [CrossRef]

59. Wang, S.; Li, J.; Cao, Y.; Gao, B.; Mao, Q.; Li, Y. Thermal stability and tensile property of 316L stainless steel with heterogeneous lamella structure. *Vacuum* **2018**, *152*, 261–264. [CrossRef]

60. Ma, M.; Wang, Z.; Zeng, X. A comparison on metallurgical behaviors of 316L stainless steel by selective laser melting and laser cladding deposition. *Mater. Sci. Eng. A* **2017**, *685*, 265–273. [CrossRef]

61. Peng, P.; Wang, K.-S.; Wang, W.; Han, P.; Zhang, T.; Liu, Q.; Zhang, S.; Wang, H.; Qiao, K.; Liu, J. Relationship between microstructure and mechanical properties of friction stir processed AISI 316L steel produced by selective laser melting. *Mater. Charact.* **2020**, *163*, 110283. [CrossRef]

62. Vajpai, S.K.; Yu, H.; Ota, M.; Watanabe, I.; Dirras, G.; Ameyama, K. Three-Dimensionally Gradient and Periodic Harmonic Structure for High Performance Advanced Structural Materials. *Mater. Trans.* **2016**, *57*, 1424–1432. [CrossRef]

63. Yang, X.; Zhang, J.; Gong, Y.; Nakatani, M.; Sharma, B.; Ameyama, K.; Zhu, X. A superior strength-ductility combination in gradient structured Cu-Al-Zn alloys with proper stacking fault energy and processing time. *Mater. Sci. Eng. A* **2020**, *789*, 139619. [CrossRef]

64. Kundu, A.; Field, D.P. Influence of plastic deformation heterogeneity on development of geometrically necessary dislocation density in dual phase steel. *Mater. Sci. Eng. A* **2016**, *667*, 435–443. [CrossRef]

65. Yang, M.; Pan, Y.; Yuan, F.; Zhu, Y.; Wu, X. Back stress strengthening and strain hardening in gradient structure. *Mater. Res. Lett.* **2016**, *4*, 145–151. [CrossRef]

66. Lu, K. Making strong nanomaterials ductile with gradients. *Science* **2014**, *345*, 1455–1456. [CrossRef] [PubMed]

67. Zheng, R.; Liu, M.; Zhang, Z.; Ameyama, K.; Ma, C. Towards strength-ductility synergy through hierarchical microstructure design in an austenitic stainless steel. *Scr. Mater.* **2019**, *169*, 76–81. [CrossRef]

68. Fan, G.; Geng, L.; Wu, H.; Miao, K.; Cui, X.; Kang, H.; Wang, T.; Xie, H.; Xiao, T. Improving the tensile ductility of metal matrix composites by laminated structure: A coupled X-ray tomography and digital image correlation study. *Scr. Mater.* **2017**, *135*, 63–67. [CrossRef]

69. Ueno, A.; Fujiwara, H.; Rifai, M.; Zhang, Z.; Ameyama, K. Fractographical Analysis on Fracture Mechanism of Stainless Steel Having Harmonic Microstructure. *J. Soc. Mater. Sci. Jpn.* **2012**, *61*, 686–691. [CrossRef]

70. Kikuchi, S.; Mori, T.; Kubozono, H.; Nakai, Y.; Kawabata, M.O.; Ameyama, K. Evaluation of near-threshold fatigue crack propagation in harmonic-structured CP titanium with a bimodal grain size distribution. *Eng. Fract. Mech.* **2017**, *181*, 77–86. [CrossRef]

71. Osaki, K.; Kikuchi, S.; Nakai, Y.; Kawabata, M.O.; Ameyama, K. The effects of thermo-mechanical processing on fatigue crack propagation in commercially pure titanium with a harmonic structure. *Mater. Sci. Eng. A* **2020**, *773*, 138892. [CrossRef]

72. Nukui, Y.; Kubozono, H.; Kikuchi, S.; Nakai, Y.; Ueno, A.; Kawabata, M.O.; Ameyama, K. Fractographic analysis of fatigue crack initiation and propagation in CP titanium with a bimodal harmonic structure. *Mater. Sci. Eng. A* **2018**, *716*, 228–234. [CrossRef]

73. Kikuchi, S.; Takemura, K.; Ueno, A.; Ameyama, K. Evaluation of the 4-points bending fatigue properties of Ti-6Al-4V alloy with harmonic structure created by mechanical milling and spark plasma sintering. *J. Soc. Mater. Sci. Jpn.* **2015**, *64*, 880–886. [CrossRef]

74. Kikuchi, S.; Kubozono, H.; Nukui, Y.; Nakai, Y.; Ueno, A.; Kawabata, M.O.; Ameyama, K. Statistical fatigue properties and small fatigue crack propagation in bimodal harmonic structured Ti-6Al-4V alloy under four-point bending. *Mater. Sci. Eng. A* **2018**, *711*, 29–36. [CrossRef]

75. Kikuchi, S.; Nakatsuka, Y.; Nakai, Y.; Nakatani, M.; Kawabata, M.O.; Ameyama, K. Evaluation of Fatigue Properties under Four-point Bending and Fatigue Crack Propagation in Austenitic Stainless Steel with a Bimodal Harmonic Structure. *Frattura ed Integrità Strutturale* **2019**, *13*, 545–553. [CrossRef]

76. Kikuchi, S.; Nukui, Y.; Nakatsuka, Y.; Nakai, Y.; Nakatani, M.; Kawabata, M.O.; Ameyama, K. Effect of bimodal harmonic structure on fatigue properties of austenitic stainless steel under axial loading. *Int. J. Fatigue* **2019**, *127*, 222–228. [CrossRef]

77. Wang, X.; Cazes, F.; Li, J.; Hocini, A.; Ameyama, K.; Dirras, G. A 3D crystal plasticity model of monotonic and cyclic simple shear deformation for commercial-purity polycrystalline Ti with a harmonic structure. *Mech. Mater.* **2019**, *128*, 117–128. [CrossRef]
78. Liu, J.; Dirras, G.; Ameyama, K.; Cazes, F.; Ota, M. A three-dimensional multi-scale polycrystalline plasticity model coupled with damage for pure Ti with harmonic structure design. *Int. J. Plast.* **2018**, *100*, 192–207. [CrossRef]
79. Dirras, G.; Tingaud, D.; Ueda, D.; Hocini, A.; Ameyama, K. Dynamic Hall-Petch versus grain-size gradient effects on the mechanical behavior under simple shear loading of β-titanium Ti-25Nb-25Zr alloys. *Mater. Lett.* **2017**, *206*, 214–216. [CrossRef]
80. Mompiou, F.; Tingaud, D.; Chang, Y.; Gault, B.; Dirras, G. Conventional vs. harmonic-structured β-Ti-25Nb-25Zr alloys: A comparative study of deformation mechanisms. *Acta Mater.* **2018**, *161*, 420–430. [CrossRef]

Publisher's Note: MDPI stays neutral with regard to jurisdictional claims in published maps and institutional affiliations.

Article

Investigation on the Durability of Ti-6Al-4V Alloy Designed in a Harmonic Structure via Powder Metallurgy: Fatigue Behavior and Specimen Size Parameter Issue

Benjamin Guennec [1],*, Takayuki Ishiguri [2], Mie Ota Kawabata [3], Shoichi Kikuchi [4], Akira Ueno [3] and Kei Ameyama [3]

[1]   Department of Mechanical System Engineering, Faculty of Engineering, Toyama Prefectural University, Kurokawa 5180, Imizu, Toyama 959-0398, Japan
[2]   Lead Frame Business Unit, Mitsui High-tec Inc., 965-1 Oaza Nakaizumi, Nogata, Fukuoka 822-0011, Japan; takayuki-ishiguri@mitsui-high-tec.com
[3]   Department of Mechanical Engineering, Faculty of Science and Engineering, Ritsumeikan University, 1-1-1 Noji-higashi, Kusatsu, Shiga 525-8577, Japan; mie-ota@fc.ritsumei.ac.jp (M.O.K.); ueno01@fc.ritsumei.ac.jp (A.U.); ameyama@se.ritsumei.ac.jp (K.A.)
[4]   Department of Mechanical Engineering, Faculty of Engineering, Shizuoka University, 3-5-1 Johoku, Naka-ku, Hamamatsu, Shizuoka 432-8561, Japan; kikuchi.shoichi@shizuoka.ac.jp
*   Correspondence: bguennec@pu-toyama.ac.jp; Tel.: +81-766-56-7500

Received: 28 April 2020; Accepted: 10 May 2020; Published: 14 May 2020

Abstract: In the present work, the four-point bending loading fatigue properties of a heterogeneously distributed grain size microstructure consolidated from Ti-6Al-4V alloy powder are studied. The microstructure involved here, a so-called "harmonic structure", possesses quasi-spherical large grain regions ("cores") embedded in a continuous fine grain region ("shell"). Unlike the previous reports dealing with this issue, the effect of the specimen size on the fatigue characteristics is also probed, since two distinct specimen configurations are considered. Furthermore, the obtained experimental data are compared with the corresponding fatigue results derived from homogeneous coarse grain counterparts. Contrary to homogeneous structure material, discrepancies on both the fatigue strength and the fatigue crack initiation aspects are found for the harmonic structure material. Consequently, the present work aims to clarify the underlining phenomena involved in the specimen size effect detected for Ti-6Al-4V designed in the harmonic structure. A less active interface surface between the core and the shell combined with a wider critical volume in the large size specimen should be the main reasons of the fatigue strength discrepancy.

Keywords: fatigue; titanium alloy; bimodal structure; size effect; EBSD investigation; fatigue crack growth

---

1. Introduction

As the mechanical properties of metallic materials are highly influenced by their microstructures, advanced material processes have been developed to acquire superior strengths. Among them, the grain refinement of polycrystalline metallic materials is an effective way to increase yield stress [1–3]. Nevertheless, fine-grained structural materials also lead to a critical loss of ductility. To reach both high strength and high ductility simultaneously, numerous microstructural designs have been proposed lately [4–11]. The most common strategy to achieve such superior mechanical properties consists in the design of bimodal grain size structures, possessing coarse and fine grains. Such structures are reported to enhance the back-stress hardening of metallic materials [11]. Fabrication of such heterogeneous microstructures can be controlled by several techniques, such as severe plastic deformation processes [8,11,12] or the recently developed cold spray additive manufacturing [13].

Another process for the creation of heterogeneous structure materials is given by the flexibility of the powder metallurgy route. Taking advantage of this characteristic, a peculiar bimodal structure having approximately spherical coarse grain regions (hereafter referred as the "cores") surrounded by a continuous fine grain network (denoted as "shell") has been developed [14–19]. This so-called "harmonic structure", obtained from spark plasma sintering (SPS), enhances the tensile strength of Ti-based materials [14–21] compared with homogeneous grain size counterparts without significant loss of ductility. Nevertheless, tensile properties are not sufficient to assess the reliability of harmonic structured compounds for structural purposes. Indeed, the investigation of the fatigue behavior is one of the crucial aspects to probe before attempting any application as structural material. Therefore, several works have already been carried out to estimate the properties of materials tailored into harmonic structures under cyclic loading [20–27]. An increase of the fatigue endurance is reported, in line with the static tensile strength, despite the decrease of the fatigue stress intensity factor threshold. However, these reports involved miniature sized specimens (specimen thickness of 1 mm) due to the limited diameter of initial sintered compacts considered. Therefore, it is interesting to undertake investigation on the fatigue behavior of harmonic structured materials based on thicker specimens to acquire a refined image of its behavior from a macroscopic viewpoint, as examination of the specimen size effect on the fatigue properties of material presenting bimodal microstructure is very scarce.

In the present work, the four-point bending fatigue aspects of Ti-6Al-4V designed in a harmonic structure have been investigated using two specimen size configurations. In particular, the present report involves fatigue tests performed with the specimens' size following the ISO 22214 Standard, which is significantly larger than the one accepted in our previous work [20]. The experimental results will be compared with the fatigue behavior derived from miniatured sized specimens, accepting a configuration similar to [20]. Furthermore, the effect of the harmonic structure itself is also examined, since homogeneous coarse grains microstructure material is involved in this work, for both specimen size configurations. The experimental results highlight a significant influence of the specimen size on the fatigue strength of Ti-6Al-4V designed in a harmonic structure, whereas homogeneous material shows a similar fatigue strength for both size configurations. In addition, noticeable differences on the fatigue crack initiation mechanisms are observed. Regardless of the size of the specimen, the fatigue crack propagation path in the harmonic structure material is indifferent of the location of the shell and core regions. In this study, a series of electron backscatter diffraction (EBSD) data acquisitions are undertaken to assess the geometrically necessary dislocations (GNDs) accumulation in the fatigued harmonic material, revealing an effect of the specimen size. Based on these experimental results, coupled to the introduction of a simplified model for comparison of the interface activity between the shell and the cores, the main reasons for the size effect arisen in the present report can be grasped.

2. Experimental Procedure

2.1. Material Processing

Even though the material processing is following the same procedure already described elsewhere [21], this section will sum up the main processing stages. Raw material is 186 μm-particle diameter Ti-6Al-4V powder, prepared by a plasma rotating electrode process (PREP). The material inside the initial powder presents an acicular microstructure, which is a typical characteristic of α' martensitic transformation induced by PREP rapid cooling, with an average hardness of 326 Hv [17]. The chemical composition of this powder is introduced in Table 1. This powder is subjected to mechanical milling (MM) for 90 ks in an argon atmosphere, to generate a several tens micron-thick layer of fine grains from the powder surface, resulting in 397 and 369 Hv average hardness values inside the fine grain layer and powder center region, respectively [17]. The obtained powder is then consolidated by SPS at 1123 K for 1.8 ks in a vacuum to obtain the MM material. For the sake of comparison, sintered samples from the as-received initial powder (IP) are considered, leading to IP material. The sintered compacts of MM and IP materials consist of a cylindric geometry with a 50 mm diameter and a height of 11 mm. Post SPS examination indicates a relative density of 99.9% of the compacts, which suggests a limited number of pores inside the examined materials, as already reported [14]. The compacts are large enough to be able to cut off every needed specimen from the same compact for each material. In other words, only one compact for each material has been sintered to provide all required specimens.

Table 1. Chemical composition of the Ti-6Al-4V plasma rotating electrode process (PREP) powder (mass %).

Al	V	Fe	H	N	O	C	Ti
6.24~6.37	4.10~4.15	0.10~0.13	0.002	0.005~0.006	0.108~0.116	0.0024	Bal.

2.2. Mechanical Testing Procedure

The uniaxial tensile properties of both the MM and IP materials have been investigated through AG-X plus a tensile testing machine (Shimadzu Corp., Kyoto, Japan). To this end, tensile specimens accepting the dimensions depicted in Figure 1a are subjected to crosshead displacement control tests at an initial strain rate of 5.6×10^{-4} s^{-1}. The strain measurement is assured by two strain gauges attached on opposite surfaces to each other to optimize the recording accuracy.

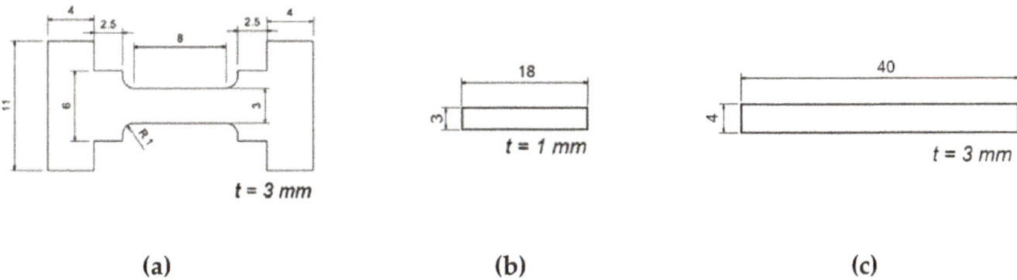

(a)   (b)   (c)

Figure 1. Shapes and dimensions of the specimens. (**a**) Tensile specimen; (**b**) fatigue miniature size (MS) specimen; and (**c**) fatigue standard size (SS) specimen.

The fatigue characteristics of the investigated compounds are conducted as follows. Load-controlled fatigue tests under four-point bending loading have been performed using two distinct specimen sizes:

- The miniature size (hereafter denoted MS) specimen with length, width and thickness dimensions of 18, 3 and 1 mm, respectively (see Figure 1b). To this end, an electro-dynamic fatigue testing machine using a 9514-AN/SD actuator (EMIC Corp., Tokyo, Japan), operating at a loading frequency of 10 Hz, is considered. The frame geometry of the four-point bending loading adopts an inner span of 5 mm and an outer span of 15 mm. Such a configuration is identical to the one accepted in our previous work [20].
- The standard size (SS) specimen, presenting the length, width and thickness values of 40, 4 and 3mm, respectively (see Figure 1c). The dedicated frame possesses inner and outer spans of 10 and 30 mm, respectively, in accordance with the ISO 22214 Standard. To this end, a servo-hydraulic fatigue testing machine (SMH201, SAGINOMIYA, Tokyo, Japan), accepting a loading frequency of 20 Hz, has been used.

Regardless of the specimen size, small chamfers have been introduced by #2000 grinding paper on the edges of every fatigue specimen to relieve the stress concentration there. The tensile surface of the fatigue specimens is finished by 1 μm alumina suspension buffing. All fatigue tests have been carried out in air and at room temperature environment, considering a stress ratio R (= $\sigma_{min}/\sigma_{max}$) of 0.1. Fatigue tests were terminated at 10^7 cycles for unbroken specimens.

2.3. Analysis Tools

Observations of the fracture and the tensile surfaces of the failed specimens were carried out using a SU6600 SEM (Hitachi, Tokyo, Japan), equipped with an HKL NordlyF camera for EBSD acquisition, with AZtec 3.1 software (Oxford Instrument, Abington, UK) for data processing. A critical angle of 5° has been accepted to define grain boundaries in the present work. Some investigations included in this work involve kernel average misorientation (KAM) analysis to determine the local lattice rotation, based on the orientation of the fifth nearest neighbor pixels, with a distance between two neighbors (i.e. step size) of 0.25 μm. The same microscope is also equipped with an energy dispersive X-ray spectrometry (EDS) device for chemical composition analysis. Every fatigue strength regression showed in *S-N* diagrams in the present work is obtained by the JSMS-SD-11-08 standard [28].

3. Experimental Results

3.1. Characterization of the Materials' Microstructure

The microstructures of the obtained compacts were investigated prior to the mechanical testing campaign by the EBSD technique, over a surface of 500 μm × 400 μm regions. The experimental results are displayed through grain size (GS) maps in Figure 2, where the black lines highlight high angle boundaries having a misorientation angle higher than 15°. One example of the microstructure found from the IP material is shown in Figure 2a, where a large predominance (approximately 80%) of acicular large grains, appearing in red color, is found. The remaining consists of some fine grains, presumably induced by a thin layer of such grain at the initial powder surface. According to these experimental results, the average grain size in the IP material is 11.6 μm.

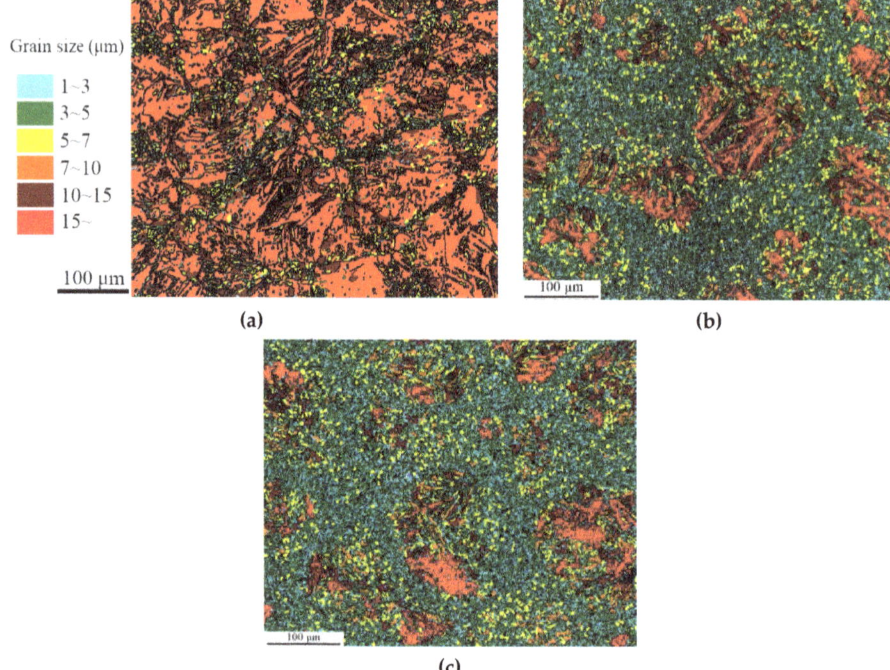

Figure 2. Microstructure characterization of the investigated materials displayed in the form of grain size and high angle boundaries maps. (**a**) IP material; (**b**) MM material in the inner region; and (**c**) MM material in the outer region of the sintered compact. Step size of 0.25 μm.

The corresponding microstructure characterization for the MM material has been undertaken at two different spots of the sintered compact. Indeed, since the present report manipulates a sintered compact having larger dimensions than the previous reports [20,21], a temperature gradient throughout the compact may arise in the SPS process. In such a case, the microstructure of the MM material should reveal significant discrepancies depending on its relative position in the compact. Therefore, Figure 2b,c present examples of the microstructures found in the MM material in the inner and outer regions of the sintered compact. These regions are located at a maximum distance of 3 mm from the center and the edge of the sintered compact, respectively. For both GS maps, MM material consists of coarse grains regions accepting mainly acicular geometry, surrounded by a continuous network of fine equiaxed grains. Consequently, the MM material highlights a harmonic structure.

In addition, the Figure 2b,c outlines the absence of significant microstructure differences. Grain possessing an equivalent diameter larger than 5 µm is denoted as "coarse" (and referred to as "fine" in the opposite case), and a refined analysis of the MM material's microstructures in the inner and outer regions is given in Table 2, based on two distinct EBSD acquisitions. The numerical results highlight very close figures, in both regions, which assure similar microstructures of the MM material throughout the entire compact.

Table 2. Microstructure of mechanical milling (MM) material in the inner and outer regions of the sintered compact.

Position in Compact	Inner Region	Outer Region
Areal fraction of coarse grains (%)	31	32
Average coarse grain size (µm)	7.1	7.1
Overall average grain size (µm)	4.3	4.3

3.2. Tensile Tests Results

Static mechanical properties on average values, based on nominal stress calculation, are listed in Table 3. One can note that MM material highlights both a larger ultimate tensile strength σ_{UTS} and elongation at fracture δ than IP counterpart. This trend, which has been reported for material designed in a harmonic structure [14–19], is usually induced by the ability to reach larger elongation before emergence of necking phenomenon. To reconfirm this feature, an example of the true stress–true strain curve is presented in Figure 3 for each material investigated. The MM material shows both superior yield stress and enhanced capability to deform uniformly. This combination results in a higher tensile strength than the IP counterpart, in addition with a postponed necking phenomenon.

Table 3. Average mechanical properties of the MM and initial powder (IP) materials based on nominal stress–strain curve.

Material	Tensile strength σ_{UTS} (MPa)	Yield Stress at 0.2% Offset σ_y (MPa)	Reduction of Area φ (%)	Elongation at Fracture (%)
IP	1092	984	34.0	22.7
MM	1124	1089	33.3	23.3

It is noteworthy that the tensile strength levels obtained in the present investigation from both the IP and MM materials are significantly larger than in our previous study [20], outlining a gap of approximately 150 MPa. Indeed, the microstructure observed in this previous report revealed significantly larger grain sizes, with an overall average grain size of 26.3 and 16.9 µm for IP and MM materials, respectively. Thus, the higher tensile strength reported in the present work is likely caused by the finer microstructures.

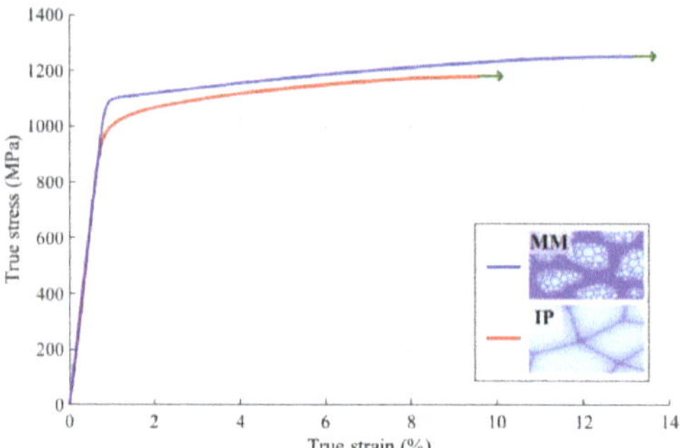

Figure 3. Example of the true stress–true strain curves from the IP and MM materials. Green arrows underline the beginning of necking phenomenon.

3.3. Fatigue Tests Results

3.3.1. Fundamental Fatigue Strength from the MS Configuration

Firstly, let us focus on the fatigue test results related to the MS configuration only. The corresponding fatigue strength results are depicted in the form of an *S-N* curve, in Figure 4a. According to the *S-N* diagram, IP and MM compounds show equivalent strength in the $N_f < 10^6$ cycles region, whereas the MM material highlights a slight increase of fatigue strength compared to its IP counterpart for the $N_f > 10^6$ cycles region. Therefore, the fatigue endurance at 10^7 cycles of the MM material (738 ± 15 MPa) is higher than the IP one (683 ± 14 MPa). A similar trend has been reported in our previous work dedicated to Ti-6Al-4V alloy [20], as emphasized in Figure 4a. The fatigue strength offset between data from [20] and reported in the present study is caused by the tensile strength discrepancy of the investigated materials, as mentioned in Section 3.2.

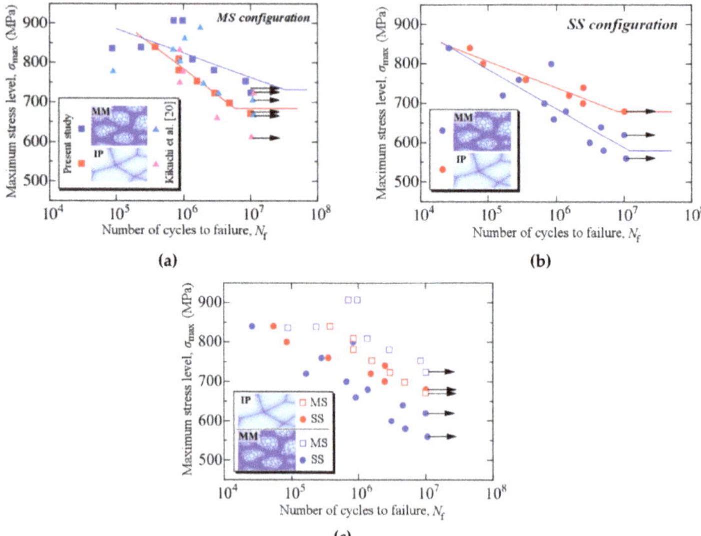

Figure 4. Fatigue data obtained from IP and MM materials in the form of *S-N* diagrams. (**a**) Case of MS configuration only and comparison with experimental results from our previous work, data from [20]; (**b**) case of SS configuration only; and (**c**) entire collected data in the same diagram. Arrows indicate run-out tests.

3.3.2. S-N Characteristics from SS Specimens

The *S-N* diagram comparing the fatigue strength measured in the SS configuration for both the IP and MM materials is presented in Figure 4b. One can note that the fatigue strength is almost equivalent in the $N_f < 2 \times 10^5$ cycles region, in a way rather similar to the results found in Section 3.3.1 for the MS configuration. However, in a clearly opposite trend to the MS configuration, the MM compound shows a distinct decrease of fatigue strength in the $N_f > 2 \times 10^5$ cycles region, compared to its IP counterpart. This decrease is particularly visible in the region of the $10^6 < N_f < 10^7$ cycles, where several specimens made of the MM material highlight fatigue failure at low stress levels ranging from 580 to 640 MPa. It results in a 570 ± 10 MPa fatigue endurance for the MM material, significantly lower than the 680 ± 20 MPa one found for the IP counterpart. Therefore, the experimental data obtained from the SS configuration are in discrepancy with the data collected from the MS configuration.

In order to better grasp the situation here from a strictly fatigue strength viewpoint, Figure 4c gathers the *S-N* data obtained in the present study from both specimen size configurations in the same diagram. On the one hand, even though a slight discrepancy is found from the IP material fatigue strengths from the MS and SS configurations in the limited fatigue region, the fatigue endurances at 10^7 cycles are similar. Thus, no significant specimen size effect on the fatigue properties of Ti-6Al-4V in the IP structure is detected here. On the other hand, the MM material highlights a clear specimen size effect, where the fatigue endurance decreased from 738 ± 15 to 570 ± 10 MPa for the MS to SS configurations, respectively.

3.3.3. Fatigue Crack Initiation Features

In the case of the MS configuration, the typical features of micrographs focused on the fatigue crack initiation region are depicted in Figure 5. In a way similar to the observations carried out in our previous work [20] involving only experiments on the MS configuration, the IP compound showed a crack initiation at the specimen's tensile surface revealing clear facets, as depicted in Figure 5a. Due to the size of facets observed, the crack nucleated from acicular coarse grains of the structure. For the MM material, the fatigue crack initiation has occurred at three different possible positions underlined by white ellipses in Figure 5b–d: (i) at the tensile surface (Figure 5b); (ii) showing facets in contact with the surface (Figure 5c); and (iii) inside the specimen body (Figure 5d). In accordance to the related micrographs, regardless the position of the crack initiation region, one can detect clear facets. According to our previous work, such facets are certainly caused by the basal slip in the α-phase grain [20]. Due to the size of the formed facets at the initiation site, cracks are certainly nucleated inside a core of the harmonic structure, in agreement with our previous work [20]. An analysis of the occurrences of these three nucleation patterns is depicted in the form of an *S-N* diagram, in Figure 5e. Clearly, the distinct patterns are well distributed over the *S-N* diagram. Thus, rather than an effect of the fatigue loading level or fatigue life region, the pattern occurrence is presumably governed by the intrinsic local resistance against fatigue crack nucleation of the specimen. Several literature reports [29–32] have underlined similar fatigue strengths from interior and surface-induced crack mechanisms in Ti-6Al-4V alloy, in line with the experimental results found in the present work.

Let us now introduce the fatigue crack initiation features observed in the SS configuration. Two nucleation patterns were found for the IP material, as depicted in Figure 6a,b. In Figure 6a, the nucleation site takes place at the specimen's tensile surface, whereas the interior induced initiation is shown in Figure 6b. Regardless of the position of the initiation, similar facets to the one depicted in Figure 5 can be observed, thus reasonably induced by the same basal slip mechanism discussed earlier in this section. In the case of the MM material, four distinct fatigue crack initiation modes have been identified and relevant micrographs are presented in Figure 6c–f. In Figure 6c, the crack initiation takes place close to the chamfer of the specimen. However, an orthogonal viewpoint of the fracture surface is not suitable to observe the nucleation site. Thus, a tilted micrograph taken in the orientation highlighted by the black arrow is introduced as an inset of Figure 6c, where facets similar to Figure 5d are found. The size of the facets, probably induced by the basal slip mechanism, suggests that the initiation is nucleated from the core of the harmonic structure, in a way similar to the MM material in the MS configuration. On the other hand, Figure 6d underlines intergranular crack initiation, which occurred from a core in regard to the size of the intergranular facets thus formed. For a unique specimen case, the crack was nucleated far from the tensile surface, as depicted in Figure 6e, where pores can be observed (see inset). These kinds of defects are certainly induced by SPS processing, leading to the fatigue crack nucleation. Furthermore, occurrences of cracks induced by inclusions, located certainly in the shell of the MM material, were found. Indeed, Figure 6f shows a typical example of a fatigue crack initiation site, where an inclusion has been detected by the means of an EDS analysis carried out in the white-squared region highlighted on the micrograph. Local high concentrations of Ca, Mg and O chemical elements are detected, in this particular case. Some other inclusions also present Si atoms. Since titanium alloys are known to be relatively clean compounds [30,33], such foreign atoms in Ti-6Al-4V alloy are certainly induced by a contamination of the powder previous to SPS processing. Furthermore, the IP counterpart is not prone to such a fatigue crack initiation phenomenon, leading to the suggestion that the mechanical milling process likely causes this contamination.

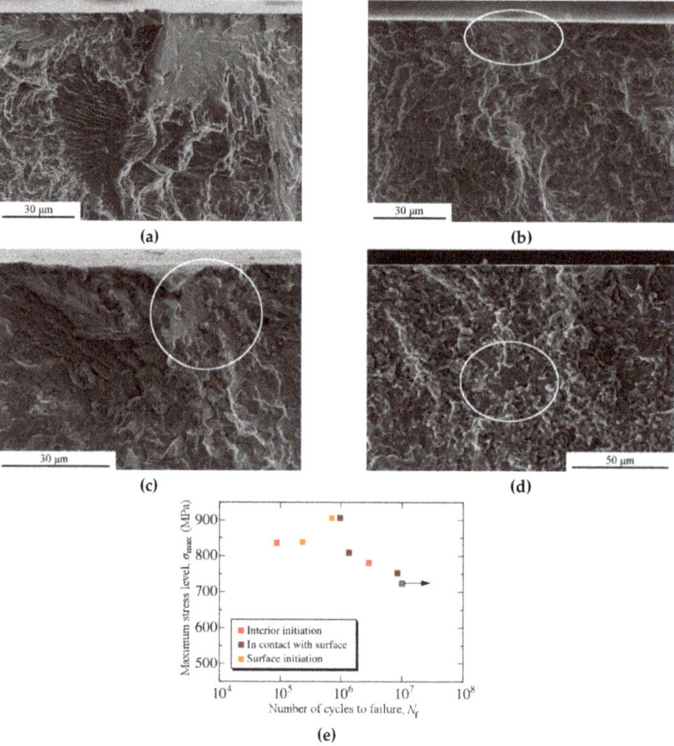

Figure 5. Typical fatigue crack initiation features in the MS configuration. The white ellipse highlights the crack initiation site. (**a**) Intragranular facets at the tensile surface for IP material (σ_{max} = 753 MPa, N_f = 1.59 × 10^6 cycles); (**b**) facets at the tensile surface for MM material (σ_{max} = 907 MPa, N_f = 7.14 × 10^5 cycles); (**c**) facets in contact with the tensile surface for MM material (σ_{max} = 907 MPa, N_f = 9.71 × 10^5 cycles); (**d**) interior-induced initiation feature with formation of facets for MM material (σ_{max} = 836 MPa, N_f = 8.76 × 10^4 cycles); and (**e**) fatigue crack initiation phenomenon occurrence for MM material in the form of an *S-N* diagram.

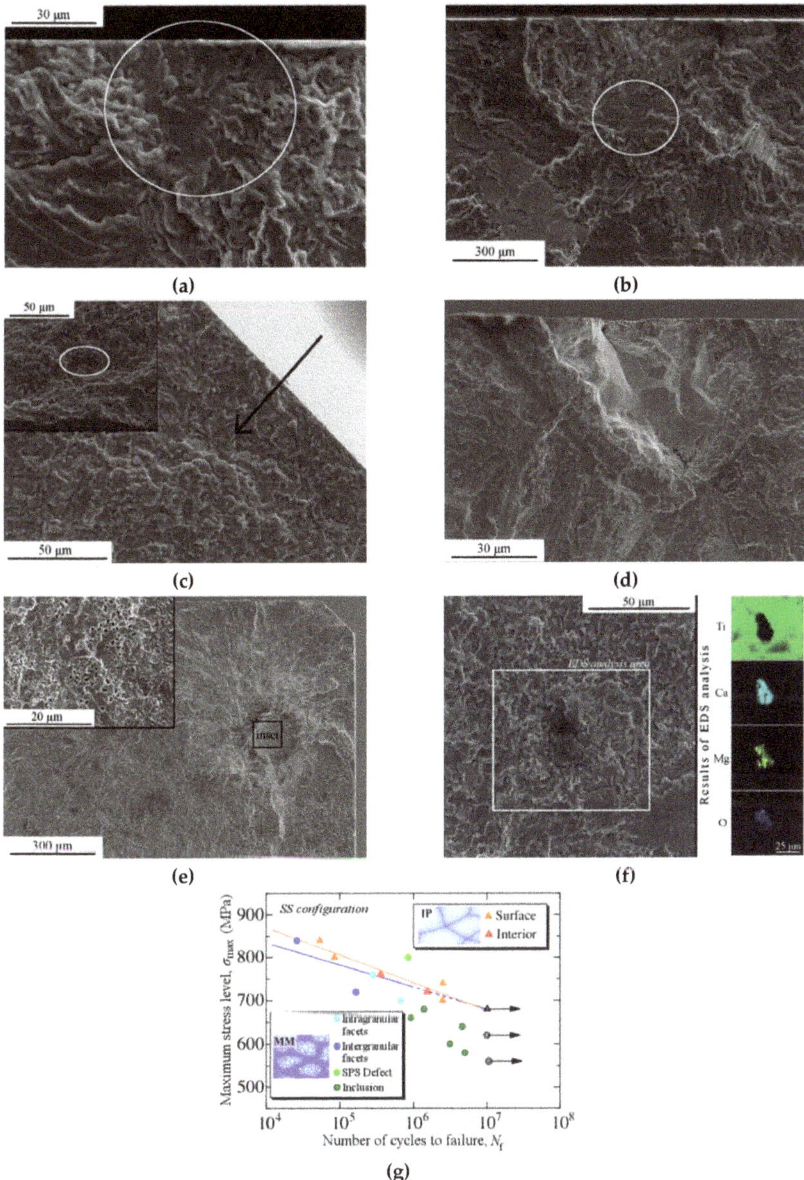

Figure 6. Typical fatigue crack initiation features found in SS configuration, for IP material (**a**,**b**) and MM material (**c**–**f**). The white ellipse highlights the initiation site. (**a**) Initiation at the tensile surface with facets (σ_{max} = 700 MPa, N_f = 2.48 × 10^6 cycles); and (**b**) interior-induced initiation with facets (σ_{max} = 720 MPa, N_f = 1.54 × 10^6 cycles); (**c**) intragranular initiation mechanism from a core forming facets (σ_{max} = 700 MPa, N_f = 6.65 × 10^5 cycles); (**d**) intergranular initiation from a core (σ_{max} = 840 MPa, N_f = 2.61 × 10^4 cycles); (**e**) initiation from pore defects (σ_{max} = 800 MPa, N_f = 8.40 × 10^5 cycles); (**f**) initiation from an inclusion detected by EDS analysis (σ_{max} = 660 MPa, N_f = 9.18 × 10^5 cycles); and (**g**) fatigue crack initiation phenomenon occurrence in the form of an *S-N* diagram, where the regression in blue excludes fatigue data from inclusion-induced nucleation phenomenon.

3.3.4. Fatigue Crack Propagation Aspects in the MM Material

In the case of the MM material, typical features of the fatigue crack propagation zone are depicted in Figure 7a,b, corresponding to the MS and SS specimen configurations, respectively. One can notice formation of apparent dark zones in the fatigue crack propagation region, regardless of the specimen size. The most preeminent zones are highlighted by white-dotted ellipses in Figure 7a,b. In order to clarify the nature of these zones, a refined analysis of the region highlighted by a solid ellipse in Figure 7b will be carried out. A high magnification micrograph of the corresponding zone is shown in Figure 7c.

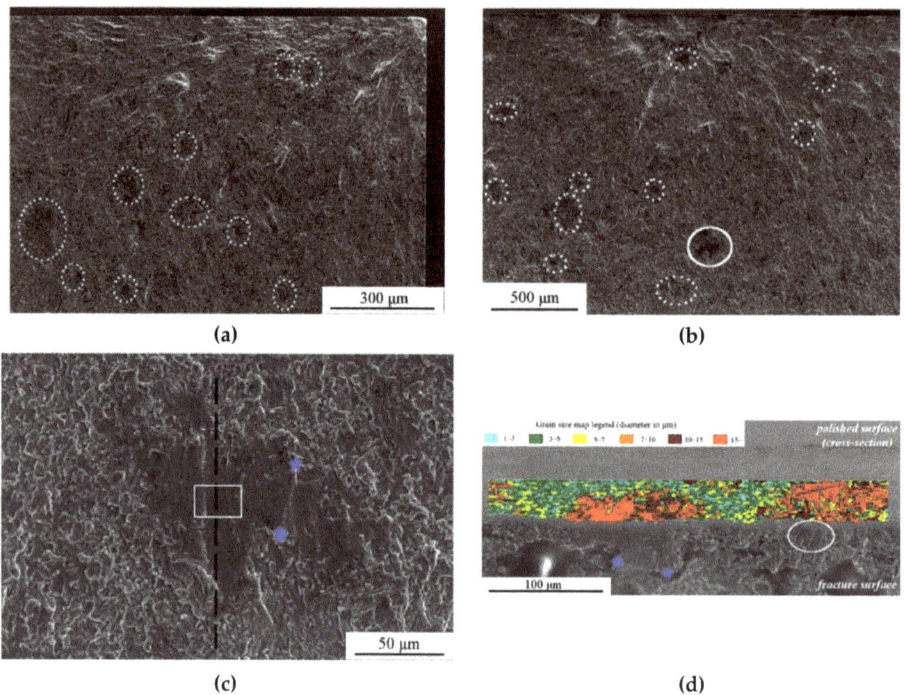

Figure 7. Analysis of the fatigue crack propagation features in MM material. (**a**) Overview of the fatigue crack propagation region, MS configuration (σ_{max} = 781 MPa, N_f = 2.89 × 10^6 cycles); (**b**) overview of the fatigue crack propagation region, SS configuration (σ_{max} = 680 MPa, N_f = 1.38 × 10^6 cycles); (**c**) high magnification micrograph of the dark zone highlighted by a solid white ellipse in (**b**); and (**d**) after 90° clockwise rotation, acquisition results from the electron backscatter diffraction (EBSD) technique in the form of a grain size map showing the microstructure beneath the dark zone analyzed. Blue dots in (**c**,**d**) indicate the same locations before and after rotation.

The specimen has been grinded along the vertical black broken line displayed in Figure 7c, and then the newly created surface was finished by oxide polishing suspension (OPS). The prepared surface is then subjected to an EBSD analysis to monitor the microstructure beneath the dark zone pattern. Acquisition results thus obtained are depicted in Figure 7d, where the specimen orientation has been rotated to 90° clockwise and tilted at an angle of 70°, showing the prepared cross-section surface and the fracture surface on the upper and lower sides of the micrograph, respectively. In order to understand more easily the respective locations on the fracture surfaces depicted in Figure 7c,d, two distinct location spots in both photographs have been highlighted by two reference points appearing in blue color.

The results from the EBSD analysis conducted on the prepared surface are presented in the form of a GS map, in Figure 7d. This map underlines the presence of the coarse grain structure beneath the investigated dark zone. This analysis demonstrates that the dark zone is induced by the fatigue crack propagation through the core of the MM material. Furthermore, on the right-hand side of the same map, one can also point out another coarse grain region beneath the fracture surface. Even though a dark zone as large as the initially investigated one cannot be observed on the fracture surface, a tiny region that depicts similar features is also highlighted by a white ellipse. The reason for the size discrepancy between these two zones observed on the fracture surface results from the relative locations of the crack path and the two cores. Indeed, the crack path almost gets through the center of the core in the initial studied region, whereas the crack path propagates only close to the outer region of the core located at the right-hand side of the EBSD analyzed region. Therefore, these experimental results demonstrate that the fatigue crack path is unaffected by the microstructure features of the harmonic structure, at least in locations far from the fatigue crack initiation site. Similar behaviors for Ti-based materials designed in a harmonic structure have been recently reported in [34,35].

3.3.5. Characterization of the Fatigue Behavior of MM Material by Local Misorientation Analysis

The Section 3.3.2 emphasized distinct fatigue resistance levels of the MM material depending on its specimen size configuration, i.e., MS or SS specimens. This fatigue behavior discrepancy is mandatory to clarify. To this end, this section aims at identifying possible divergences between one individual specimen of each size (MS and SS) prone to the same stress level $\sigma_{max} = 780$ MPa. According to the *S-N* regression curves introduced in Figure 4a,b, the respective fatigue lives N_f for the MS and SS configurations are 5.0×10^6 and 9.0×10^4 cycles, respectively. Since both values are remarkably distinct to each other, the present discussion will consider four arbitrary fatigue stages of 0% (initial state), 5%, 20% and 60% of the respective fatigue lives N_f.

For each specimen considered, prior to testing, the tensile surface was finished by OPS to obtain an observation condition suitable for EBSD data acquisition. Then, EBSD analyses were performed at every fatigue stage described earlier in this section at the same peculiar location prone to the maximum bending moment. To ensure meaningful data processing, the lattice rotation over a region covering roughly ten cores and its surrounding shell network has been probed. EBSD data results are presented in the form of KAM maps in Figure 8. Figure 8a–d depicts the KAM maps for the SS specimen at 0%, 5%, 20% and 60% of N_f, respectively. In a similar trend to the results observed from harmonic materials prone to uniaxial tension tests, the lattice misorientation increases in the shell network [22,36,37]. Furthermore, Figure 8e,f displays the corresponding KAM maps for the MS specimen at the initial stage and at 60% of N_f, respectively. Even though the lattice rotation shows a similar trend to the SS specimen, the shell network in Figure 8f presents a KAM value at 60% of N_f significantly larger in the MS configuration than the SS one, in accordance with the color scale.

To further discuss these collected data, the present analysis considers the KAM average value, as a hint of the concentration of GNDs [38–41]. The average KAM value is calculated on the basis of KAM distribution considering classes width of 0.05° over the 0° to 5° range. The obtained values, for both the MS and SS specimen size configurations, are presented through bar diagrams in Figure 9. First, let us introduce in Figure 9a the KAM average values obtained from the whole analyzed surface (i.e., the entire surface shown in Figure 8). Whereas the SS specimen shows a slight increase of the KAM average value at approximately 0.86° after 5% N_f and then keeps almost constant up to 60% N_f, the analyzed surface from the MS specimen emphasized a steady increase of the KAM average value from approximately 0.85° to 0.99°. Since the core regions are concentrating high KAM values, such data processing is obviously dependent on the core fraction in the considered areas. Furthermore, in accordance with Figure 8, a change of lattice misorientation occurs essentially in the shell of the harmonic structure. Thus, KAM average values were also investigated in one selected shell location for each specimen size configuration, highlighted by a white square in Figure 8a,e, for the SS and MS configuration, respectively. The corresponding results are shown in a bar diagram in Figure 9b. In an analogous trend to Figure 9a, the collected data from the shell location emphasized a clear increase of the KAM average value in the MS specimen, whereas the increasing trend in the SS configuration is rather limited, reaching a saturation level after only 5% N_f stage. This discrepancy suggests that the MM material designed in a harmonic structure does not undergo the same hardening behavior under cyclic loading in the MS and SS configurations. Further discussions will be elaborated in Section 4.4.

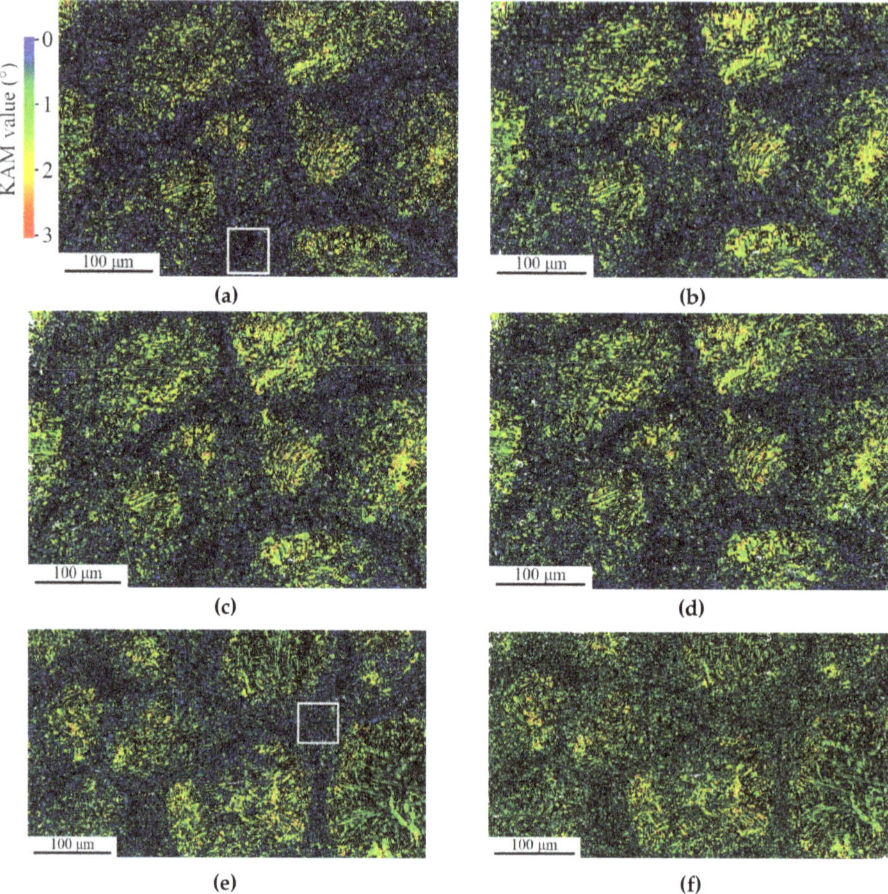

Figure 8. Kernel average misorientation (KAM) maps obtained from the fatigue specimen on the tensile surface, MM material. Results from the SS configuration at initial state (0% N_f) and 5% N_f, 20% N_f and 60% N_f are displayed in (**a–d**), respectively. Results from MS configuration at initial state and 60% N_f are included in (**e,f**), respectively. Step size of 0.25 μm. Color range from blue (0°) to red (3° and higher).

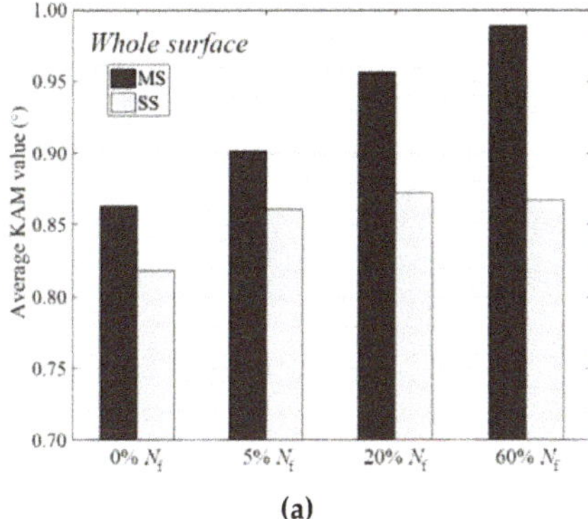

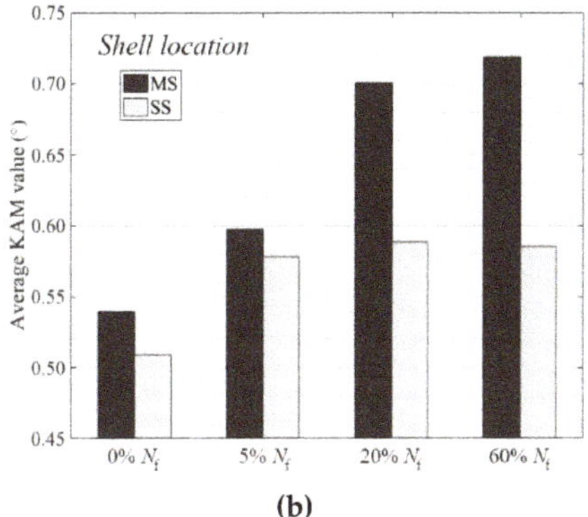

(a)   (b)

Figure 9. Comparison of the KAM average values in both MS and SS specimen configurations on the whole surface and in the shell location, in (**a**,**b**), respectively. "Whole surface" is dealing with the analysis of the entire surface shown in Figure 8; "shell location" is corresponding to the data inside the white-squared zone outlined in Figure 8a,e.

4. Discussion

The present discussion will further analyze some remaining interrogations relative to the fatigue behavior of Ti-6Al-4V designed in a harmonic structure. Therefore, these questions will be addressed in the following Sections 4.1–4.4.

4.1. Why the Inclusion Induced Initiation Mechanism Occurs Exclusively in the 10^6~10^7 Cycles Region?

In order to answer to this question, a refined analysis of the inclusion-induced fatigue fracture needs to be carried out, from fracture surface observations. Consequently, several critical aspects relating to these inclusions, such as its depth from the tensile surface d_{inc} or its projected size on the fracture surface $area^{1/2}$ have been listed in Table 4. In this table, the stress range at the inclusion site $\Delta\sigma_{at}$ takes into consideration the bending stress gradient. According to the numerical values introduced in Table 4, one can undertake the calculation of the stress intensity factor induced by the inclusion ΔK_{inc}, accepting the following Equation (1), proposed by Murakami and Endo [42].

$$\Delta K_{inc} = 0.5 \cdot \Delta\sigma_{at} \sqrt{\pi \cdot area^{1/2}} \tag{1}$$

Table 4. Characteristics of inclusions generating fatigue crack initiation phenomenon (MM material, SS configuration).

Number of Cycles to Failure N_f (Cycles)	Depth of Inclusion d_{inc} (μm)	Stress Range at Inclusion $\Delta\sigma_{at}$ (MPa)	Projected Size of Inclusion $area^{1/2}$ (μm)
9.18×10^5	100	553	22.6
1.38×10^6	141	553	17.4
4.99×10^6	84	492	18.3
3.12×10^6	149	485	18.3
4.58×10^6	72	547	12.4

The obtained results are displayed in Figure 10, where ΔK_{inc} is represented in the y-axis, as a function of the number of cycles to failure, with N_f as the x-axis. One can see that the fatigue life tends to increase when the stress intensity factor range induced by the inclusion decreases. In accordance with the experimental results from [21,24], this diagram also indicates the values of the effective stress intensity factor range threshold $\Delta K_{eff,th}$ = 2.9 MPam$^{1/2}$ for long cracks in the MM material prone to the same stress ratio of R = 0.1. Obviously, the stress intensity factor range induced by inclusions ΔK_{inc} takes place in the region lower than the threshold one for long cracks. Fatigue failure induced by small-size inclusion requires numerous stressing cycles to generate a propagating crack, as usually discussed in literature reports dealing with very high cycle fatigue regime [43,44]. In particular, experimental results similar to Figure 10 have been highlighted for high strength steels investigated in very high cycle regime [45,46]. Thus, due to the size of the involved inclusions, the related fatigue crack nucleation can reasonably occur only in the vicinity of the high cycle fatigue region limit ($N_f \sim 10^7$ cycles).

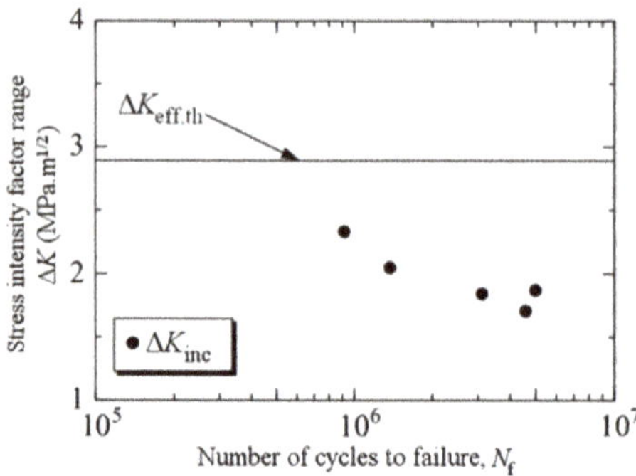

Figure 10. Relationship between stress intensity factor ranges induced by inclusions ΔK_{inc} and fatigue life, in comparison with effective threshold value for long cracks $\Delta K_{eff,th}$ of MM material.

4.2. Why the MS Configuration did not Show Any Inclusion-Induced Fatigue Rupture?

Unlike the SS configuration, none of the MM material specimen fatigued in the MS one was prone to inclusion-induced crack nucleation. This observation is certainly induced by the difference of critical volume between the specimen size configurations investigated here.

Indeed, it is well known that defects in the microstructure of a metallic part have a higher probability to be found in large size specimens than in small size ones. According to Table 4, the inclusion-induced crack nucleation happened at a maximum depth of approximately 150 μm in the SS configuration. Considering the nominal specimen thickness of 3 mm, the local stress level at 150 μm depth corresponds to 90% of the maximum one. Therefore, let us consider a specimen critical volume with a threshold level of 90% of the maximum stress in the rest of the present report. Denoting the critical volume height of the SS configuration h_{SS} = 150 μm, the reciprocal critical volume height for MS configuration becomes h_{MS} = 50 μm. In regard to the inclusion projected size $area^{1/2}$ ranging from 12.4 to 22.6 μm (see Table 4), it seems unlikely that such inclusions were formed in a 50 μm thick layer. Besides, one can note the limited number of MS specimens ruptured in the $N_f = 10^6 {\sim} 10^7$ cycles region, decreasing statistically the probability of the occurrence of inclusion-induced initiation in the present work. From these viewpoints, the absence of inclusion-induced fatigue crack initiations can be reasonably grasped.

4.3. Why the Fatigue Nucleation Site Tends to Move from a Surface-Induced to an Interior-Induced Pattern When Transitioning from the MS to SS Configuration?

According to the literature, the transition of the surface- to interior-induced fatigue crack initiation pattern in Ti-alloys depends on various parameters, such as the material microstructure [30] or the stress ratio [32,33,47]. In addition to these parameters, the type of fatigue loading has undoubtedly an influence on the fatigue crack initiation site, since bending loading induces a stress gradient with a maximum value at the tensile surface, whereas no-stress gradient axial loading should favor the interior-induced mechanism.

The transition from the MS to SS configuration goes hand in hand with a notable change of the stress gradient exerted on the specimen, leading to an increase of the critical volume's height, as already discussed in Section 4.2. Thus, it is reasonable to observe more interior-induced nucleation occurrences in the SS configuration, regardless of the microstructure of the studied materials. Furthermore, the surface to interior transition is more distinct in the case of the MM material, since the critical volume's height h_{SS} becomes larger than the core diameter value. Therefore, the crack is way more likely to initiate from the "second row" of the core regions from the tensile surface, as schematically illustrated in Figure 11. This second row of cores offers a significantly larger number of potential fatigue crack initiation sites inside the specimen, resulting in the heavy interior-induced fatigue crack nucleation trend observed for the MM material in SS configuration.

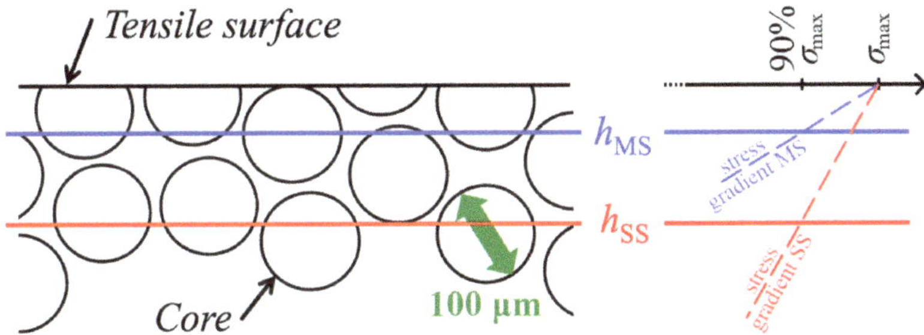

Figure 11. Schematic 2D illustration of the harmonic structure prone to two different critical volume heights h_{MS} and h_{SS}, corresponding to the MS and SS configurations, respectively.

In addition to the stress gradient induced by the fatigue loading type, Adachi et al. [29] have proposed a mechanism taking into consideration the influence of compressive residual stress induced by the localized plastic deformation due to the tension-tension fatigue loading for Ti-alloys. Similar discussions have been evocated in [33,48] to discuss the transition from surface- to interior-induced crack initiation trend. Even though the authors cannot assure that this phenomenon is at stake here, it may contribute to the interior-induced initiation pattern observed in the present report.

4.4. Putting Aside the Inclusion-Induced Ruptures, Why MM Material did not Show Equivalent Fatigue Strength in MS and SS Configurations?

Since contamination of the powder during the mechanical milling process generates the inclusions found in the MM material, this feature is not intrinsically caused by the harmonic structure itself. However, even though fatigue failure data induced by inclusions is taken aside, Figure 12 underlines a fatigue strength in the MS configuration 100 MPa higher than the SS configuration's one. As a consequence, the present work demonstrated an effect of the specimen size on the fatigue strength of the Ti-6Al-4V material designed in a harmonic structure.

Further investigations on the behavior of the MM material have been undertaken through KAM analysis from a series of EBSD data acquisitions, as introduced in Section 3.3.5. Both configurations underlined an increase of the KAM values in the shell network of the harmonic structure. Nevertheless, the increasing trend was notably more distinct in the MS configuration compared with the SS one. This significant increase of the average KAM value in the shell network in the MS configuration is interpreted as an intense accumulation of GNDs there, i.e., a hint of back-stress hardening.

Indeed, heterogenous structured materials in general are usually said to enhance the back-stress hardening in comparison with homogeneous counterparts. This feature of heterogeneous materials was noted a long time ago in a study by Ashby [49], which pointed out the essential role of GNDs to mitigate the voids and overlaps between neighboring grains having different mechanical properties. Such a trend has been experimentally reported lately for harmonic structured materials. Park et al. [50] compared the back-stress levels in homogeneous and harmonic structured SUS 304L stainless steel derived from the strain partitioning method. The heterogeneous harmonic material showed a significantly larger back-stress component than the homogeneous counterpart. From another viewpoint, several analyses based on the finite element method (FEM) have been conducted. Yu et al. [51] reported that most of the stress is distributed on the shell, whereas most of the strain is concentrated in the core of the harmonic structure. Similar conclusions have been drawn from another FEM analysis conducted by Wang et al. [52] simulating monotonic and cyclic shear loading, leading to the conclusion that "the shell regions form a thin but hard skeleton enveloping the multicrytalline soft core region." Mompiou et al. [37] have performed TEM observations on harmonic β-titanium alloy, leading to the following behavior: (i) at low strain, the soft core tends to yield first, resulting in formation of slip bands there; (ii) these slip bands eventually extend up to the hard shell, which activate dislocation sources from GBs in highly constrained grains in the shell; (iii) such sources generate an accumulation of dislocations in the form of piles-up or tangles, inducing an increase of the GND density in the shell. It should be noted that similar dislocation nucleation in the hard phase of an heterogeneous Ti-6242Si alloy has been reported [53]. This mechanism hardens the shell network of the harmonic structure, allowing it to bear the majority of the external load in the process. Analogous behavior is reported more generally in the heterogeneous structure materials [9].

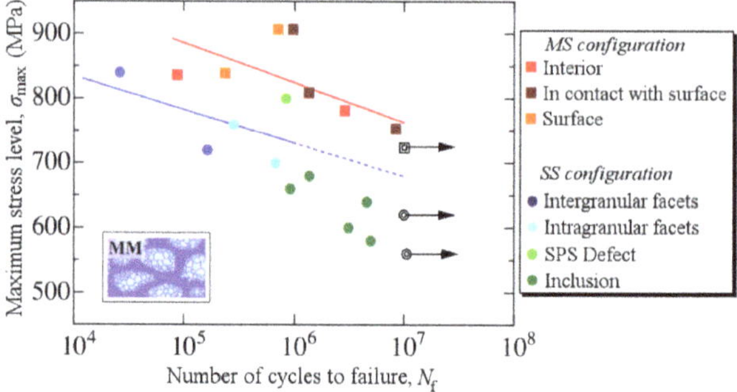

Figure 12. Occurrences of the different fatigue crack initiation features for MM material in MS and SS configurations. The red line denotes the regression for the MS configuration, whereas the corresponding regression for the SS configuration in blue excludes fatigue data from inclusion-induced crack nucleation phenomenon.

As pointed out in Section 3.3.5, experimental results from the average KAM values suggest that the size of the specimen impacts the ability of the MM material to generate GNDs in the shell. Which feature could reasonably suppress the GNDs accumulation in the SS configuration? According to the previous TEM observations [37], the interface between the hard and soft regions fundamentally contributes in the back-stress hardening of harmonic materials. Therefore, the present work proposes a simplified model for assessment of the active interface in both the MS and SS configurations. This model is based on the following assumptions:

- MM material consists of an "assembly" of $u = 150$ μm edge length cubic pattern, where the core takes place in a $r_c = 50$ μm-radius spherical region at its center (i.e., the region outside the sphere corresponds to the shell region). It leads to a harmonic structured material having a core areal fraction of 35%, close to the actual value in the MM material investigated here, as written in Table 2.

- The active interface takes place only inside the critical volume region, i.e., for a depth lower than h_{MS} and h_{SS} for the MS and SS configurations, respectively. Since the bending stress is maximum in this region, core grains inside the critical volume are very likely to yield there, generating the GNDs accumulation at the core interface.

- A core region generates an active interface only if its center is inside the critical volume region. Indeed, it is reasonable to consider that a sufficient volume of the core region is needed to cause dislocation pile-ups and tangles at the interface between the core and shell regions. For the sake of simplicity, the authors propose that this sufficient core volume threshold is half of the total core region, resulting in this third assumption.

Based on these assumptions, a 2D illustration of the model is displayed in Figure 13, where the active interface is highlighted in purple color. Accordingly, the active interface length L_i depends on the relative locations of the core region and the tensile surface. Therefore, let us consider that c is the distance between the tensile surface and the center of the core region. Furthermore, introduce α, β and β' angles defined as $\sin \alpha = c/r_c$, $\sin \beta = (h_{SS}-c)/r_c$ and $\sin \beta' = (h_{MS}-c)/r_c$, as shown in Figure 13a,b. As h_{SS} is larger than the diameter of the core region, the active interface length is $(L_i)_{SS} = r_c(\pi + 2\alpha)$ for $0 < c \leq r_c$, $(L_i)_{SS} = 2\pi r_c$ for $r_c \leq c \leq 2r_c$ and $(L_i)_{SS} = r_c(\pi + 2\beta)$ for $2r_c \leq c < 3r_c$. In the case of the MS configuration, $(L_i)_{MS} = 2r_c(\alpha + \beta')$ for $0 \leq c \leq r_c$. After noticing that critical volume heights h_{SS} and h_{MS} correspond to $3r_c$ and r_c values respectively, the collected interface length function traces are plotted in Figure 13c,d for the SS and MS configurations, respectively. Comparing the average values of the obtained functions $\overline{(L_i)_{SS}}$ and $\overline{(L_i)_{MS}}$, and taking into consideration the critical volume discrepancy between the MS and SS configurations, the active length interface indexes, denoted $(i_L)_{MS}$ and $(i_L)_{SS}$, are defined by Equations (2) and (3) for the MS and SS configurations, respectively.

$$(i_L)_{MS} = \frac{\overline{(L_i)_{MS}}}{h_{MS}} \tag{2}$$

$$(i_L)_{SS} = \frac{\overline{(L_i)_{SS}}}{h_{SS}} \tag{3}$$

Developing the Equations (2) and (3) results in the following numerical values: $(i_L)_{MS} = 2\,(\pi - 2)$ and $(i_L)_{SS} = 2\,(3\pi - 2)/9$. Therefore, the active length interface index in MS configuration is nearly 40% higher than the one derived from SS configuration. One can also carry out a similar approach involving the spherical interface surface in a 3D viewpoint, in substitution to the length in the 2D one. The corresponding activated interface surface S_i functions are represented in Figure 13e,f for the MS and SS configurations, respectively. In a way comparable to Equations (2) and (3), let us define the active surface interface indexes $(i_S)_{MS}$ and $(i_S)_{SS}$ by the ratios of the activated surface average values over the critical surfaces in both configurations involved, i.e., $u \times h_{MS}$ and $u \times h_{SS}$ for MS and SS, respectively. Numerical applications of these two surface indexes gives $(i_S)_{MS} = 2\pi/3$ and $(i_S)_{SS} = 10\pi/27$, which emphasize a value 80% larger for MS configuration. The index gaps reported from activated lengths and surfaces viewpoints are induced by two features: (i) the very narrow critical volume height h_{MS}, which tends to maximize the active interface fraction for MS configuration, and (ii) the low fraction of the core region in the investigated MM material, which increases the distance between each core region and decreases the active interface index for the SS configuration.

Even though this model consists in a simplified overview of the interaction at the interface between the core and shell regions, it draws the main reasons for the lower ability of the shell region at the specimen surface to generate GNDs in the SS configuration, in agreement with the experimental results discussed in Section 3.3.5 outlining a low saturation level value of KAM in shell region. As reported for other heterogeneous structure materials [54,55], and in accordance with theoretical discussion by Ashby [49], insufficient GNDs accumulation leads to strain localization phenomenon. In the case of the harmonic structure, Park et al. [50] outlined that strain localization may take place inside grains of the core region, in line with the fatigue crack initiation site observed in Figure 6. Such strain localization phenomenon is very likely to generate early fatigue crack nucleation.

Lastly, as discussed in Section 4.3, the larger critical height in the SS configuration than the MS one results in a subsequent larger amount of core region locations in the critical volume. Since accumulation of GNDs is very limited in the SS configuration, each new core region inside the critical volume can be considered as a "weak point", where the crack can potentially initiate, due to strain localization phenomenon. Therefore, the decreasing fatigue strength of the MM material in the SS configuration, highlighted by the *S-N* curves in Figure 12, is caused by the combination of two features: (i) the limited

GNDs accumulation resulting in possible strain localization phenomenon inside the core regions, and (ii) the larger amount of core regions inside the critical volume, which highly increases the probability of an early fatigue crack nucleation.

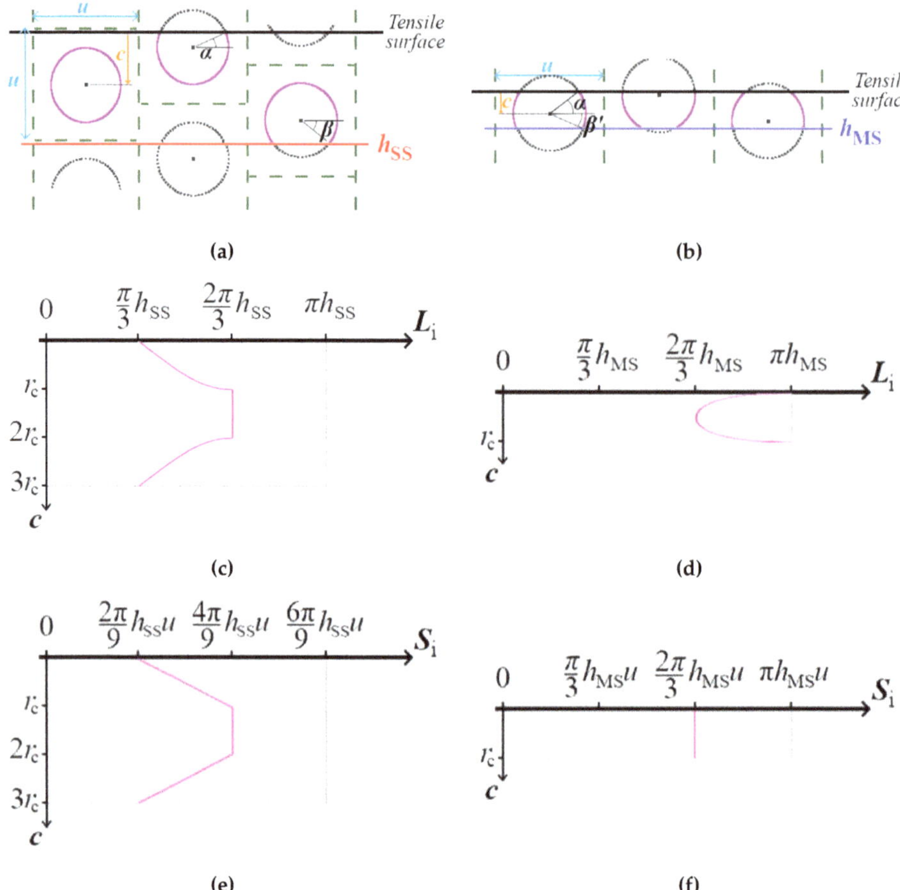

Figure 13. Schematic illustration for active interface calculation based on the proposed model. The active interface between the shell and cores is highlighted in purple color for the SS and MS configurations in (**a**,**b**), respectively. Corresponding active interface length functions L_i are represented in (**c**,**d**), whereas active interface surface functions S_i are depicted in (**e**,**f**), for the SS and MS configurations, respectively.

5. Conclusions

In the present investigation, the four-point bending fatigue properties of Ti-6Al-4V alloy designed in a bimodal harmonic structure (MM material) have been investigated. Unlike the previous reports dealing with this issue, the present work investigates the fatigue properties using the standard specimen (SS) configuration to better grasp the reliability of macroscopic parts. Furthermore, fatigue properties of the miniature specimen (MS) size configuration were also carried out, for the sake of comparison with previous experimental results. Lastly, fatigue data from homogeneous coarse grain IP material were collected to assess the impact of the harmonic microstructure on the mechanical characteristics. The main findings of the present report are summarized below:

1. Analysis of the microstructure of 50 mm diameter sintered compact made from powder prone to mechanical milling revealed a harmonic microstructure, without significant discrepancies throughout the entire compact.

2. Fatigue test results obtained from the MS configuration underlined fatigue endurance at 10^7 cycles of 738 ± 15 MPa, distinctly higher than the IP one of 683 ± 14 MPa. This trend is similar to results found in our previous work. An opposite situation is reported for fatigue tests accepting the SS configuration, where fatigue endurances at 10^7 cycles are 570 ± 10 and 680 ± 20 MPa, for the MM and IP materials, respectively.

3. Excepting the case of the MM material in the SS configuration, fatigue crack nucleation of both IP and MM materials revealed facets, presumably induced by the basal slip of α-Ti grains. More particularly, the crack nucleates at the core region of the MM material in the MS configuration. However, the MM material highlighted other initiation features in the SS configuration, since inclusion-induced nucleation from the shell is found exclusively in the $N_f > 10^6$ region. According to the chemical compositions of the involved inclusions, they are likely caused by powder contamination during the mechanical milling process.

4. A simplified model has been proposed to assess the activity between the core and shell regions, which outlined a significantly larger active interface fraction in the MS configuration than the SS one. Such a trend is in line with the more distinct increase of GNDs accumulation in the shell in the MS configuration, according to KAM analysis. GNDs accumulation mechanism mitigated, heterogeneous structure materials cannot keep their cohesion, leading to the generation of a local zone prone to considerable strain concentration, usually referred to as strain localization phenomenon.

5. The fatigue strength loss observed in the MM material for the SS configuration, in comparison with the MS one, should be induced by two main reasons: mitigated GNDs accumulation resulting in strain localization in core regions; coupled to a larger critical volume increasing notably the number of cores vulnerable to generate early fatigue crack nucleation.

Even though further analyses should be undertaken to refine our comprehension on the effect of the specimen size on the fatigue properties of harmonic structure materials, experimental results of the present work urge us to consider possible effects of the specimen size on various mechanical properties of such heterogeneous structure materials. In accordance with the simplified model, fatigue performance from harmonic structure materials with a large shell areal fraction should be altered, if the smallest dimension of the critical volume is out of the same order as the core/shell unit size. On the contrary, the fatigue strength of harmonic structure materials possessing a low shell areal fraction should hardly be affected, due to the limited influence of the critical volume size on the active interface between the core and shell regions. Lastly, it should be noted that future investigations on harmonic structured materials should pay additional attention on the material fabrication processing in order to keep such compounds as clean as possible. Refined examinations dealing with the fatigue behavior on such advanced tailored materials are at this cost.

Author Contributions: Conceptualization, K.A. and A.U.; methodology, B.G., S.K. and A.U.; software, B.G.; validation, B.G. and A.U.; formal analysis, B.G.; investigation, T.I., B.G., S.K. and A.U.; resources, K.A. and M.O.K.; data curation, B.G.; writing—original draft preparation, B.G.; writing—review and editing, S.K., A.U. and K.A.; visualization, B.G.; supervision, A.U., S.K. and B.G.; project administration, A.U., S.K. and B.G.; funding acquisition, K.A. All authors have read and agreed to the published version of the manuscript.

Funding: This work was supported by the French national research agency in the framework of ANR 14-CE07-0003 "HighS-Ti" program. This study is carried out with the aid of JSPS Grant-in-Aid for Scientific Research JP18H05256 and JP19H02024, along with The Light Metal Educational Foundation, Inc.

Acknowledgments: Authors would like to thank Takuya Nobori and Hirotaka Kuwahara (Graduate school of Mechanical Science and Engineering, Ritsumeikan University) for their kind helps in mechanical testing operations. A warm thank is dedicated to Guy Dirras (Université Paris 13, France) for our communications on various subjects dealing with this manuscript.

Conflicts of Interest: The authors declare no conflict of interest.

References

1. Hall, E.O. The deformation and ageing of mild steel: III Discussion of results. *Proc. Phys. Soc. B* **1951**, *64*, 747–752. [CrossRef]
2. Petch, N.J. The cleavage strength of polycrystals. *J. Iron Steel Inst.* **1953**, *174*, 25–28.
3. Tsuji, N.; Ito, Y.; Saito, Y.; Minamino, Y. Strength and ductility of ultrafine grained aluminum and iron produced by ARB and annealing. *Scr. Mater.* **2002**, *47*, 893. [CrossRef]
4. Ma, E. Instabilities and ductility of nanocrystalline and ultrafine-grained metals. *Scr. Mater.* **2003**, *49*, 663–668. [CrossRef]
5. Morita, T.; Hatsuoka, K.; Iizuka, T.; Kawasaki, K. Straightening of Ti-6Al-4V by short-time duplex heat treatment. *Mater. Trans.* **2005**, *46*, 1681–1686. [CrossRef]
6. Szabo, P.J.; Field, D.P.; Joni, B.; Horky, J.; Ungar, T. Bimodal grain size distribution enhances strength and ductility simultaneously in a low-carbon low alloy steel. *Metall. Mater. Trans. A* **2015**, *46*, 1948–1957. [CrossRef]
7. Terada, D.; Inoue, M.; Kitahara, H.; Tsuji, N. Change in mechanical and microstructure of ARB processed Ti during annealing. *Mater. Trans.* **2008**, *49*, 41–46. [CrossRef]
8. Wang, Y.; Chen, M.; Zhou, F.; Ma, E. High tensile ductility in a nanostructured metal. *Nature* **2002**, *419*, 912–915. [CrossRef]
9. Ma, E.; Zhu, T. Towards strength-ductility synergy through the design of heterogeneous nanostructures in metals. *Mater. Today* **2017**, *20*, 323–331. [CrossRef]
10. Fang, T.H.; Li, W.L.; Tao, N.R.; Lu, K. Revealing extraordinary intrinsic tensile plasticity in gradient nano-grained copper. *Science* **2011**, *331*, 1587–1590. [CrossRef]
11. Wu, X.; Yang, M.; Yuan, F.; Wu, G.; Wei, Y.; Huang, X.; Zhu, Y. Heterogeneous lamella structure unites ultrafine-grain strength with coarse-grain ductility. *Proc. Natl. Acad. Sci. USA* **2015**, *112*, 14501–14505. [CrossRef] [PubMed]
12. Wu, X.; Jiang, P.; Chen, L.; Yuan, F.; Zhu, Y. Extraordinary strain hardening by gradient structure. *Proc. Natl. Acad. Sci. USA* **2014**, *111*, 7197–7201. [CrossRef] [PubMed]
13. Bagherifiard, S.; Astaraee, A.H.; Locati, M.; Nawaz, A.; Monti, S.; Kodas, J.; Singh, R.; Guagliano, M. Design and analysis of additive manufactures bimodal structures obtained by cold spray deposition. *Addit. Manuf.* **2020**, *33*, 101131.
14. Sekiguchi, T.; Ono, K.; Fujiwara, H.; Ameyama, K. New microstructure design for commercially pure titanium with outstanding mechanical properties by mechanical milling and hot roll sintering. *Mater. Trans.* **2010**, *51*, 39–45. [CrossRef]
15. Fujiwara, H.; Sekiguchi, T.; Ameyama, K. Mechanical properties of pure titanium and Ti-6Al-4V alloys with a new tailored nano/meso hybrid microstructure. *Int. J. Mater. Res.* **2009**, *100*, 796. [CrossRef]
16. Kikuchi, S.; Nakamura, Y.; Ueno, A.; Ameyama, K. Low temperature nitring of commercially pure titanium with harmonic structure. *Mater. Trans.* **2015**, *56*, 1807–1813. [CrossRef]
17. Vajpai, S.K.; Ota, M.; Watanabe, T.; Maeda, R.; Sekiguchi, T.; Kusaka, T.; Ameyama, K. The development of high performance Ti-6Al-4V alloy via a unique microstructural design with bimodal grain size distribution. *Mater. Metall. Mater. Trans. A* **2015**, *46*, 903–914. [CrossRef]
18. Ota, M.; Vajpai, S.K.; Imao, R.; Kurokawa, K.; Ameyama, K. Application of high pressure gas jet mill process to fabricate high performance harmonic structure designed pure titanium. *Mater. Trans.* **2015**, *56*, 154–159. [CrossRef]
19. Vajpai, S.K.; Ota, M.; Zhang, Z.; Ameyama, K. Three-dimensionally gradient harmonic structure design: An integrated approach for high performance structural materials. *Mater. Res. Lett.* **2016**, *4*, 191–197. [CrossRef]
20. Kikuchi, S.; Hayami, Y.; Ishiguri, T.; Guennec, B.; Ueno, A.; Ota, M.; Ameyama, K. Effect of bimodal grain size distribution on fatigue properties of Ti-6Al-4V alloy with harmonic structure under four-point bending. *Mater. Sci. Eng. A* **2017**, *687*, 269–275. [CrossRef]
21. Kikuchi, S.; Imai, T.; Kubozono, H.; Nakai, Y.; Ota, M.; Ueno, A.; Ameyama, K. Effect of harmonic structure design with bimodal grain size distribution on near-threshold fatigue crack propagation in Ti-6Al-4V alloy. *Int. J. Fatigue* **2016**, *92*, 616–622. [CrossRef]

22. Zhang, Z.; Ma, H.; Zheng, R.; Hu, Q.; Nakatani, M.; Ota, M.; Chen, G.; Chen, X.; Ma, C.; Ameyama, K. Fatigue behavior of a harmonic structure designed austenitic stainless steel under uniaxial stress loading. *Mater. Sci. Eng. A* **2017**, *707*, 287–294. [CrossRef]
23. Kikuchi, S.; Mori, T.; Kubozono, H.; Nakai, Y.; Kawabata, M.; Ameyama, K. Evaluation of near-threshold fatigue crack propagation in harmonic-structured CP Titanium with a bimodal grain size distribution. *Eng. Fract. Mech.* **2017**, *181*, 77–86. [CrossRef]
24. Kikuchi, S.; Imai, T.; Kubozono, H.; Nakai, Y.; Ueno, A.; Ameyama, K. Evaluation of near-threshold fatigue crack propagation in Ti-6Al-4V alloy with harmonic structure created by mechanical milling and spark plasma sintering. *Frat. Ed Integr. Strutt.* **2015**, *34*, 261–270. [CrossRef]
25. Kikuchi, S.; Nakatsuka, Y.; Nakai, Y.; Nakatani, M.; Kawabata, M.; Ameyama, K. Evaluation of fatigue properties under four-point bending and fatigue crack propagation in austenitic stainless steel with a bimodal harmonic structure. *Frat. Ed Integr. Strutt.* **2019**, *48*, 545–553. [CrossRef]
26. Kikuchi, S.; Nukui, Y.; Nakatsuka, Y.; Nakai, Y.; Nakatani, M.; Kawabata, M.; Ameyama, K. Effect of bimodal harmonic structure on fatigue properties of austenitic stainless steel under axial loading. *Int. J. Fatigue* **2019**, *127*, 222–228. [CrossRef]
27. Osaki, K.; Kikuchi, S.; Nakai, Y.; Kawabata, M.; Ameyama, K. The effects of thermo-mechanical processing on fatigue crack propagation in commercially pure titanium with a harmonic structure. *Mater. Sci. Eng. A* **2020**, *773*, 13892. [CrossRef]
28. JSMS Committee on Fatigue of Materials; JSMS Committee on Reliability Engineering. Standard Evaluation Method of Fatigue Reliability for Metallic Materials—Standard Regression Method of S-N Curves. JSMS-SD-11-08. *J. Soc. Mater. Sci. Jpn.* **2005**. [CrossRef]
29. Adachi, S.; Wagner, L.; Lutjering, G. Influence of microstructure and mean stress on fatigue strength of Ti-6Al4V. In Proceedings of the Fifth International Conference on Titanium, Munich, Germany, 10–14 September 1984; pp. 2139–2146.
30. Zuo, J.H.; Wang, Z.G.; Han, E.H. Effect of microstructure on ultra-high cycle fatigue behavior of Ti-6Al-4V. *Mater. Sci. Eng. A* **2008**, *473*, 147–152. [CrossRef]
31. Golden, P.J.; John, R.; Porter, W.J., III. Variability in room temperature fatigue life of alpha + beta processed Ti-6Al-4V. *Int. J. Fatigue* **2009**, *31*, 1764–1770. [CrossRef]
32. Liu, X.; Sun, C.; Hong, Y. Effects of stress ratio on high-cycle and very-high-cycle fatigue behavior of a Ti-6Al-4V. *Mater. Sci. Eng. A* **2015**, *622*, 228–235. [CrossRef]
33. Tokaji, K.; Kariya, H. Mean stress dependence of fatigue strength and subsurface crack initiation in Ti-15Mo5Zr-3Al. *Mater. Sci. Eng. A* **2000**, *281*, 268–274. [CrossRef]
34. Kikuchi, S.; Kubozono, H.; Nukui, Y.; Nakai, Y.; Ueno, A.; Kawabata, M.; Ameyama, K. Statistical fatigue properties and small fatigue crack propagation in bimodal harmonic structured Ti-6Al-4V alloy under four point bending. *Mater. Sci. Eng. A* **2018**, *711*, 29–36. [CrossRef]
35. Nukui, Y.; Kubozono, H.; Kikuchi, S.; Nakai, Y.; Ueno, A.; Kawabata, M.; Ameyama, K. Fractographic analysis of fatigue crack initiation and propagation in CP titanium with a bimodal harmonic structure. *Mater. Sci. Eng. A* **2018**, *716*, 228–234. [CrossRef]
36. Dirras, G.; Ueda, D.; Hocini, A.; Tingaud, D.; Ameyama, K. Cyclic shear behavior of conventional and harmonic structure-designed Ti-25Nb-25Zr β-titanium alloy: Back-stress hardening and twinning inhibition. *Scr. Mater.* **2017**, *138*, 44–47. [CrossRef]
37. Mompiou, F.; Tingaud, D.; Chang, Y.; Gault, B.; Dirras, G. Conventional vs Harmonic-structured β-Ti-25Nb-25Zr alloys: A comparative study of deformation mechanisms. *Acta Mater.* **2018**, *161*, 420–430. [CrossRef]
38. Calcagnotto, M.; Ponge, D.; Demir, E.; Raabe, D. Orientation gradients and geometrically necessary dislocations in ultrafine grained dual-phase steels studied by 2D and 3D EBSD. *Mater. Sci. Eng. A* **2010**, *527*, 2738–2746. [CrossRef]
39. Allain-Bonasso, N.; Wagner, F.; Berbenni, S.; Field, D.P. A study of the heterogeneity of plastic deformation in IF steel by EBSD. *Mater. Sci. Eng. A* **2010**, *548*, 56–63. [CrossRef]
40. He, W.; Ma, W.; Pantleon, W. Microstructure of individual grains in cold-rolled aluminum from orientation inhomogeneities resolved by electron backscattering diffraction. *Mater. Sci. Eng. A* **2008**, *494*, 21–27. [CrossRef]

41. Britton, T.B.; Birosca, S.; Preuss, M.; Wilkinson, A.J. Electron backscatter diffraction study of dislocation content of a macrozone in hot-rolled Ti-6Al-4V alloy. *Scr. Mater.* **2010**, *62*, 639. [CrossRef]

42. Murakami, Y.; Endo, M. Effects of defects, inclusions and inhomogeneities on fatigue strength. *Int. J. Fatigue* **1994**, *16*, 163–182. [CrossRef]

43. Chapetti, M.D.; Tagawa, T.; Miyata, T. Ultra-long cycle fatigue of high-strength carbon steels part I: Review and analysis of the mechanism of failure. *Mater. Sci. Eng. A* **2003**, *356*, 227–235. [CrossRef]

44. Sakai, T.; Nakagawa, A.; Oguma, N.; Nakamura, Y.; Ueno, A.; Kikuchi, S.; Sakaida, A. A review on fatigue fracture modes of structural metallic materials in very high cycle regime. *Int. J. Fatigue* **2016**, *93*, 339–351. [CrossRef]

45. Sakai, T.; Takeda, M.; Tanaka, N.; Kanemitsu, M.; Oguma, N.; Shiozawa, K. S-N property and fractography of high carbon chromium bearing steel over ultra wide life region under rotating bending. *Trans. Jpn. Soc. Mech. Eng. Ser. A* **2001**, *67*, 1805–1812. (In Japanese) [CrossRef]

46. Shiozawa, K.; Morii, Y.; Nishino, S.; Lu, L. Subsurface crack initiation and propagation mechanism in high-strength steel in a very high cycle fatigue regime. *Int. J. Fatigue* **2006**, *28*, 1521–1532. [CrossRef]

47. Shiozawa, K.; Kuroda, Y.; Nishino, S. Effect of the stress ratio on subsurface fatigue crack initiation behavior of beta-type titanium alloy. *Trans. Jpn. Soc. Mech. Eng. A* **1998**, *64*, 2528–2535. (In Japanese) [CrossRef]

48. Li, S.; Xiong, B.; Hui, S.; Ye, W.; Yu, Y. Comparison of the fatigue and fracture of Ti-6Al-2Zr-1Mo-1V with lamellar and bimodal microstructures. *Mater. Sci. Eng. A* **2007**, *460*, 140–145. [CrossRef]

49. Ashby, M.F. The deformation of plastically non-homogeneous materials. *Philos. Mag. Ser.* **1970**, *21*, 399–424. [CrossRef]

50. Park, H.K.; Ameyama, K.; Yoo, J.; Hwang, H.; Kim, H.S. Additional hardening in harmonic structured materials by strain partitioning and back stress. *Mater. Res. Lett.* **2018**, *6*, 261–267. [CrossRef]

51. Yu, H.; Watanabe, I.; Ameyama, K. Deformation behavior analysis of harmonic structure materials by multi-scale finite element analysis. *Adv. Mater. Res.* **2015**, *108*, 853–857. [CrossRef]

52. Wang, X.; Cazes, F.; Li, J.; Hocini, A.; Ameyama, K.; Dirras, G. A 3D crystal plasticity model of monotonic and cyclic simple shear deformation for commercial-purity Ti with a harmonic structure. *Mech. Mater.* **2019**, *128*, 117–128. [CrossRef]

53. Joseph, S.; Bantounas, I.; Lindley, T.C.; Dye, D. Slip transfer and deformation structures resulting from the low cycle fatigue of near-alpha titanium alloy Ti-6242Si. *Int. J. Plast.* **2018**, *100*, 90–103. [CrossRef]

54. Zhang, Y.; Topping, T.D.; Yang, H.; Lavernia, E.J.; Schoenung, J.M.; Nutt, S.R. Micro-strain evolution and toughening mechanisms in a trimodal Al-based metal matrix composite. *Metall. Mater. Trans. A* **2015**, *46*, 1196–1204. [CrossRef]

55. Zan, Y.N.; Zhou, Y.T.; Liu, Z.Y.; Ma, G.N.; Wang, D.; Wang, Q.Z.; Wang, W.G.; Xiao, B.L.; Ma, Z.Y. Enhancing strength and ductility synergy through heterogeneous structure design in nanoscale Al_2O_3 particulate reinforced Al composites. *Mater. Des.* **2019**, *166*, 10762–10769. [CrossRef]

Article

Powder Metallurgy Processing and Mechanical Properties of Controlled Ti-24Nb-4Zr-8Sn Heterogeneous Microstructures

Benoît Fer [1,2,*], David Tingaud [1], Azziz Hocini [1], Yulin Hao [3], Eric Leroy [4], Frédéric Prima [2] and Guy Dirras [1]

[1] Université Sorbonne Paris Nord, Laboratoire de Sciences des Procédés et des Matériaux, CNRS, UPR 3407, 93430 Villetaneuse, France; david.tingaud@univ-paris13.fr (D.T.); hocini@univ-paris13.fr (A.H.); dirras@univ-paris13.fr (G.D.)

[2] PSL University, Chimie ParisTech, CNRS, Institut de Recherche de Chimie Paris, 75005 Paris, France; frederic.prima@chimieparistech.psl.eu

[3] Shi-Changxu Innovation Center for Advanced Materials, Institute of Metal Research, Chinese Academy of Sciences, Shenyang 110016, China; ylhao@imr.ac.cn

[4] Université Paris Est Créteil, CNRS, ICMPE, UMR 7182, 2 rue Henri Dunant, 94320 Thiais, France; leroy@icmpe.cnrs.fr

* Correspondence: benoit.fer@lspm.cnrs.fr; Tel.: +33-648218555

Received: 7 November 2020; Accepted: 1 December 2020; Published: 4 December 2020

Abstract: This paper gives some insights into the fabrication process of a heterogeneous structured β-metastable type Ti-24Nb-4Zr-8Sn alloy, and the associated mechanical properties optimization of this biocompatible and low elastic modulus material. The powder metallurgy processing route includes both low energy mechanical ball milling (BM) of spherical and pre-alloyed powder particles and their densification by Spark Plasma Sintering (SPS). It results in a heterogeneous microstructure which is composed of a homogeneous 3D network of β coarse grain regions called "core" and α/β dual phase ultra-fine grain regions called "shell." However, it is possible to significantly modify the microstructural features of the alloy—including α phase and shell volume fractions—by playing with the main fabrication parameters. A focus on the role of the ball milling time is first presented and discussed. Then, the mechanical behavior via shear tests performed on selected microstructures is described and discussed in relation to the microstructure and the probable underlying deformation mechanism(s).

Keywords: heterogeneous microstructure; titanium alloy; powder metallurgy; mechanical properties

---

1. Introduction

β-metastable titanium alloys are widely used among titanium alloys in aerospace and biomedical application domains. These alloys contain enough β-stabilizer chemical elements (Mo, V, Nb …) to retain metastable β phase at ambient temperature after water quenching from the β phase stability field. Because of the metastability of the β phase, it is possible to reach various microstructures according to the fabrication process or thermo-mechanical treatments and therefore it gives birth to a wide range of mechanical properties [1].

In this study, we focus on Ti-24Nb-4Zr-8Sn (wt%, thereafter called Ti-2448). The weight fraction of niobium in the alloy is high enough to classify it as a β-metastable type alloy. Zirconium and tin are considered as neutral chemical elements. This alloy is of high interest for biomedical applications because:

- Ti-2448 is only composed of non-cytotoxic chemical elements [2].
- Its Young modulus is lower than the one of the other β or α + β titanium alloys, and significantly lower than the Young's modulus of stainless steel or Cr-Co-based alloys generally used in

biomedical applications. Values of Young's modulus around 40 to 50 GPa are quoted for Ti-2448 in multiple studies [3–5].

As a result, it is possible to significantly reduce "stress-shielding" effects which usually occur when the Young's modulus of the implant material is drastically superior to the Young's modulus of human cortical bone. In that case, all mechanical stresses applied on the implanted bone are undertaken by the implant instead of the surrounding bone, which may finally induce bone resorption—according to Wolff's law [6]. This phenomenon has been, for example, studied using a comparison between Ti-2448 and TA6V by Qu et al. on porcine models, resulting in improved results with Ti-2448 implants [7].

Moreover, depending on its initial microstructure, Ti-2448 may benefit from a super-elastic mechanical behavior. Super-elastic alloys are attractive for biomedical application fields, as they allow a better integration of the implant in its new environment. This elastic behavior is due to the stress-induced β to α'' phase transformation, which is reversible and occurs during the first stage of deformation. It allows high values of reversible elastic strain. Stress-induced β to α'' phase transformation in Ti-2448 has been studied several times [8,9]. Besides, twinning may occur when Ti-2448 is plastically deformed. Nevertheless, the exact nature of twins in the deformed microstructure and the full sequence of deformation in conventional Ti-2448 are not completely understood [10–12].

In order to limit implant replacement surgery, yield strength and fatigue strength are critical cursors to be maximized for orthopedic alloys. Several strategies may be followed to improve yield strength of Ti-2448 [13,14]. Processes based on the precipitation of a second phase of the microstructure (usually α phase in β-metastable titanium alloys which is related to ω phase precipitation at low temperature [15,16]) are efficient to significantly improve strength of the β-metastable alloys [17]. Other hardening strategies are based on the chemical design of the alloy. Using Bo-Md diagrams, it is possible to control the deformation mode of the alloys [18], and to induce transformation-induced plasticity (TRIP) or twinning-induced plasticity (TWIP) effects [19,20]. Grain refinement could also be used as a hardening strategy, by the development of ultrafine-grained (UFG) structures, which can be obtained via severe plastic deformation or powder metallurgy processing routes. However, these structures generally suffer from a significant lack of ductility [21]. To face the problems of ductility losses, heterogeneous grain structures have been elaborated on several studies, and they have proved to be an efficient way to maximize the strength-ductility synergy of different materials [22–24].

One particular heterogeneous microstructure is the so-called harmonic structure, which is composed of an interconnected and homogeneous 3D network of fine grains regions called "shell" and coarse grain regions called "core." The fabrication process of harmonic microstructure is based on powder metallurgy, including low energy mechanical ball milling (BM) and sintering (spark plasma, hot isostatic pressing) of pre-alloyed and spherical shaped powder particles [25–28]. Although a range of titanium powders have been consolidated using SPS/FAST [29–31], a very few attempts have been published in literature to fabricate Ti-2448 via powder metallurgy route. On the one hand, sintering of Ti-2448 has been made from elemental powder particles by Li et al. [32] but the sintered material could not achieve full density. Metal Injection Molding of pre-alloyed Ti-2448 particles was achieved by Kafkas et al. [33]. The results are promising despite the precipitation of Ti_xC carbides along the grain boundaries. Additive manufacturing (including both Electron Beam Melting and Selective Laser Melting) was also successfully applied on Ti-2448 [34–36]. On the other hand, the elaboration process of harmonic microstructures has been successfully performed on several materials, including steel [27,37], CP α titanium and $\alpha + \beta$ TA6V alloy [25] and β-stable Ti-25Nb-25Zr alloy [38]. In many cases, it results in an improvement of both yield strength and ultimate tensile strength almost without a ductility loss, even when deformation modes switch from mechanical twinning to dislocation glide [39].

In this work, the effect of heterogeneous grain structures and precipitation hardened structures are combined by processing a specific dual-phase harmonic microstructure based on the Ti-2448 alloy. Heterogeneous structured Ti-2448 samples with various microstructural features (including α precipitates volume fraction, core and shell fraction, grain sizes) have been fabricated via the control of mechanical BM parameters. The link between the resulting microstructures and the mechanical behavior of the alloy is further discussed.

2. Materials and Methods

Ti-2448 spherical shaped powder particles which have been argon atomized at the Institute of Metal Research (IMR) from the Chinese Academy of Sciences have been used as raw materials for this study. A sieving stage has been performed after gas atomization to sort powder particles according to their size. For this work, powder particles whose diameter is in the range 100–160 µm have been chosen, as they allow a good balance between core and shell in the final heterogeneous microstructures. Mechanical BM has been performed by the use of a Pulverisette 7 premium line planetary ball mill (FRITSCH, Idar-Oberstein, Germany), using both jar and 10 mm diameter balls made of steel. No significant iron or carbon contamination was detected after BM in a steel environment. Rotational speed of the mill was fixed at 200 rpm, using a ball-to-powder ratio (BPR, weight ratio) of 2:1. BM time is the varying parameter of the study (0 h, 12 h, 25 h, 50 h, and 100 h). Names of the as-processed samples are the following: IP (for initial powder) when the starting powder was directly sintered, BM 12 h for a BM time of 12 h, BM 25 h for a BM time of 25 h, etc. For the consolidation step, Spark Plasma Sintering (SPS) was performed using Dr. Sinter 515 S Syntex (Fuji Electronic Industrial co., Tsurugashima, Japan) setup belonging to the "Plateforme de Frittage Île-de-France" (Thiais, France). Cylindrical samples whose diameter is 20 mm and height are 2 mm were fabricated under argon atmosphere, and sintering parameters were fixed and chosen in order to ensure complete densification of the powder. Therefore, a sintering temperature of 800 °C and a compressive pressure of 100 MPa were applied for 30 min. Heating rate is 50 °C/min. For illustration purposes, a typical sintering cycle is displayed on Figure 3.

The microstructural characterization of the samples was performed using a scanning electron microscope SUPRA 40VP (ZEISS, Iena, Germany), equipped with a field-emission gun and an Electron BackScatter Diffraction (EBSD) acquisition system. Chemical analysis was performed using Wavelength Dispersion Spectroscopy (WDS) or Energy Dispersive Spectroscopy (EDS) techniques on an electron probe microanalyser SX100 (CAMECA, Gennevilliers, France). The WDS analyses were recorded at a voltage of 15 kV and an electric current of 40 nA. For EBSD acquisition 400 μm^2 or 800 μm^2 zones have been analyzed most of the time, with a respective step size of 0.25 µm or 0.5 µm. Both β and α phase data have been used to index patterns using OimDC NORDIF software (version VS2005, EDAX, Weiterstadt, Germany). OIM Analysis software (version 5.3, EDAX, Weiterstadt, Germany) was used to analyze EBSD data. X-ray diffraction (XRD) measurements were acquired using a diffractometer Equinox 1000 (INEL, Artenay, France) with a copper (λ = 0.1540 nm) or cobalt (λ = 0.1789 nm) monochromatic radiation. Data from the cobalt radiation diffractometer are converted to data which would have been collected using a copper experimental setup in order to ensure coherency of the results. Phase identification was based on Match software database. Density measurements were performed using Archimedes' principle in o-xylene liquid.

Mechanical properties were investigated via simple shear tests, using a home-made shear setup mounted on an MTS M20 testing machine (MTS Systems Corporation, Eden Prairie, MN, USA). The load capacity of the shearing device is 100 kN. Rectangular samples whose thickness is in the range 1.6–1.8 mm and width in the range 16–17 mm were cut along the radial direction of the cylindrical pads which are obtained after SPS. They are settled between two jaws, including one of them which is fixed and the other one which is moving vertically to shear the sample. The sheared volume is similar for every mechanical test. The deformation is followed directly on the sample, by using a video extensometer and computing the variation of the shear angle between a black line drawn on the surface of the sample and its initial horizontal position. These tests were performed at room temperature, at a shear strain rate of 10^{-3} s^{-1}.

3. Results and Discussion

3.1. Powder Metallurgy Processing Route

The fabrication process is based on powder metallurgy. The powder particles which have been used for this study have been obtained via argon atomization of a cylindrical pre-alloyed Ti-2448 alloy ingot. The initial morphology of the powder particles is shown in Figure 1a. They are mostly of spherical shape and their diameters are in the range 100–160 µm. The largest particles are surrounded by satellite particles which have not been shown to play any role during the process. Moreover, each powder particle is composed of single polycrystalline β phase, as it can be seen in the Inverse Pole Figure (IPF) map of a selected representative powder particle (Figure 1b) or on the X-ray diffraction pattern (XRD) of the initial powder (IP on Figure 2e). The average grain size in a representative initial powder particle is 15 µm (Figure 1b).

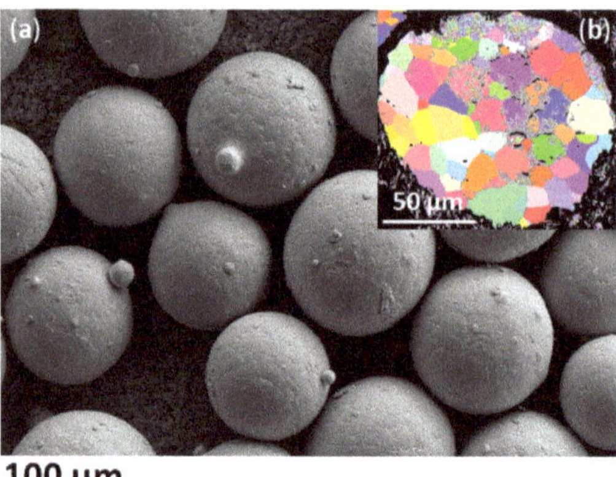

Figure 1. Characterization of initial powder particles (**a**) Scanning Electron Microscope (SEM) secondary electron image of the powder, (**b**) Inverse Pole Figure (IPF) image of a selected powder particle highlighting its microstructure.

The first stage of the fabrication process of this dual phase heterogeneously structured Ti-2448 is the mechanical planetary BM of the powder. Low energy parameters are selected for BM, as the objective of this stage is to induce superficial plastic deformation on the powder particles without affecting the core.

Figure 2 displays the morphology of the powder particles after BM for different BM times. The effect of BM on the morphology of powder particles is clear from a BM time of 12 h after which powder particles have lost their spherical shape. For a longer BM time (25 h and 50 h), some particles are severely affected by BM and are indicated by white arrows on Figure 2b–d. Furthermore, after BM for 100 h, fragmentation of some particles occurs: the fragments are indicated by red arrows on Figure 2d. XRD patterns are also plotted in Figure 2e for powders after BM. The results confirm the high plastic deformation which is induced on powder particles during BM stage. Indeed, a significant broadening of the β phase peaks—especially the main β {110} peak—takes place when increasing the BM time. Inset in Figure 2e plots the evolution of the full width at half maximum (FWHM) of the main {110} peak. A global increase in the FWHM occurs, but it is worth noting that the increase is not linear. A clear broadening of the peak is noted after a BM of 12 h. The evolution is then smoother between 12 h and 50 h, before increasing again for a BM time of 100 h. It means that plastic deformation of the powder is very high at that last BM level. Moreover, we can clearly observe the appearance of an additional peak in the XRD pattern of the 50 h—BM and 100 h—BM powders, which is indicated by a black star on Figure 2e and could be indexed as hexagonal or orthorhombic titanium phase. However, because mechanical BM is not supposed to induce the precipitation of hexagonal α phase in the powder, it is possible that reversible stress-induced martensitic transformation β to α″ occurs during BM. Incomplete reversion of this transformation (because α″ precipitates may be trapped by dislocations or other deformation traces) might explain the presence of this extra peak [12]. Nevertheless, the thermo-elastic behavior of the orthorhombic martensitic phase allows the complete reversion of it when heated at a moderate temperature, as it is the case during the first stage of sintering. As a result, this extra phase might not have a significant impact on the processing route.

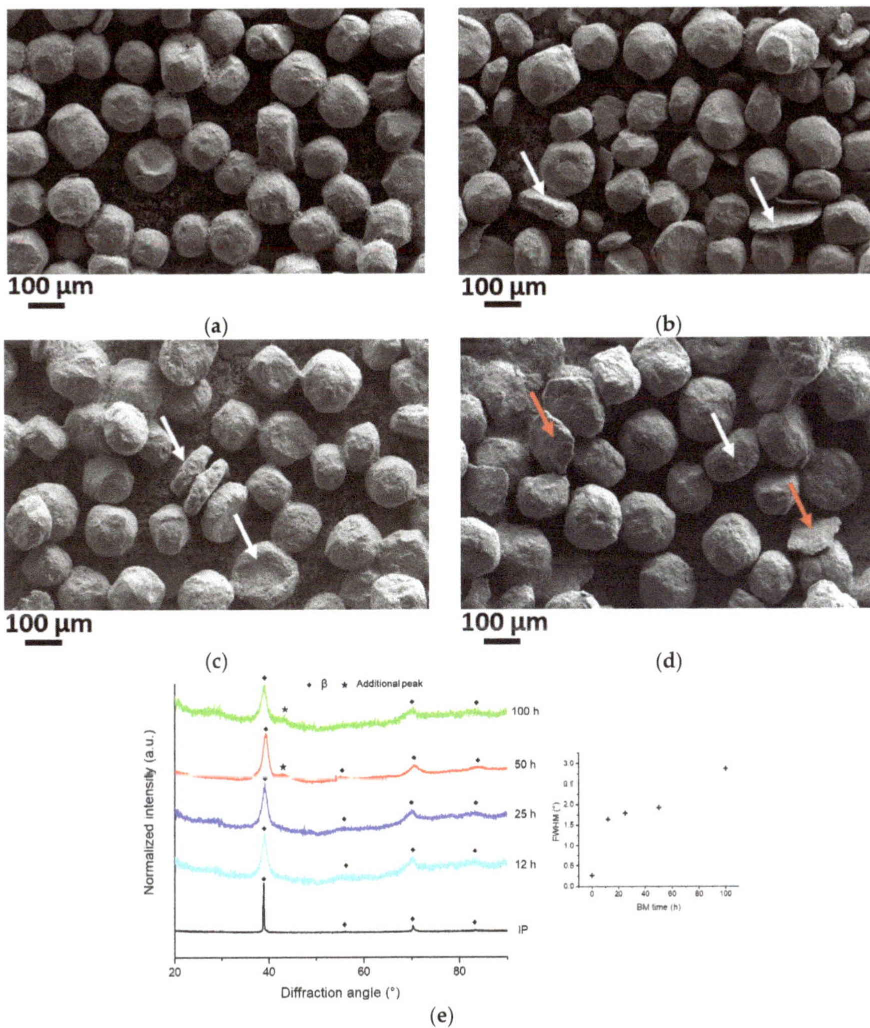

Figure 2. Characterization of ball-milled powders (**a**) SEM secondary electrons image after 12 h ball milling (BM) (**b**) after 25 h BM (**c**) after 50 h BM and (**d**) after 100 h BM. (**e**) X-ray diffraction (XRD) results of Ti-2448 powders after BM, including diffractograms for each BM time and the corresponding evolution of the full width at half maximum (FWHM) of the main {110} β phase peak.

All samples have been sintered using the same SPS cycle, so that only BM time has been modified between samples during the whole fabrication process. Figure 3 is a comparative plot of the SPS responses of IP, BM 12 h, and BM 50 h. It shows the evolution against time of the temperature, pressure, and derivative of the displacement of the piston during sintering of the three samples. This last parameter is useful to follow the densification of the samples during SPS, which takes place in two steps. The first one is associated with cold compaction of the powders, at ambient temperature once the pressure is applied. The second one occurs when the temperature reaches the beginning of sintering temperature. That time is indicated with arrows for each sample on Figure 3 and is associated with the effective sintering of the material. As we can see, BM affects the sintering reactivity of the powders, as the sintering temperatures are not the same for each sample. The longer the BM time is, the lower the sintering temperatures are. It means that the sintering reactivity of the powder particles has been increased by mechanical BM via storing mechanical energy in the material and increasing the density of structural defects in it. Therefore, diffusion phenomena are favored during sintering which consequently occurs at a slightly lower temperature.

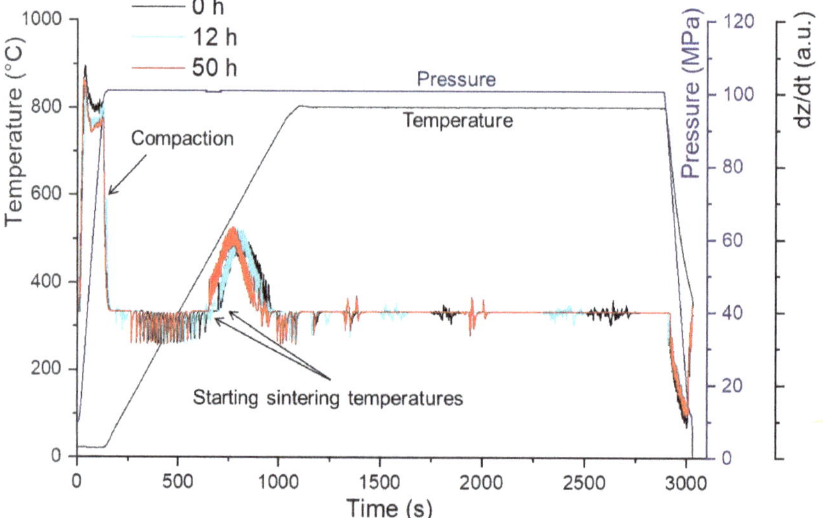

Figure 3. Spark Plasma Sintering (SPS) cycle showing the evolution against time of temperature, pressure and the derivative of the displacement of the piston of IP, BM 12 h, and BM 50 h.

The SPS stage has two main goals: the first one is to obtain dense materials and the second one is to allow recrystallization of the heavily deformed surface of powders, whose volume is directly connected to the BM time (Figure 2e). This recrystallization is therefore the key phenomenon to induce the formation of the fine grain region called "shell."

3.2. Microstructural Features of the As-Sintered Samples

The fabrication process—which includes both BM and SPS—has resulted in dense samples with a relative density of 100%. For each processed sample, the density is measured via Archimede's method and compared with the theoretical density of the material. The theoretical density of single β-phase Ti-2448 has been calculated using XRD data of an as-sintered single-phase IP sample. The cell parameter of the single phase β-BCC structure, which has been computed via Rietveld refinement of the XRD results, is 3.300 Å and the corresponding calculated theoretical density of a full β Ti-2448 structure is 5.44 g/cm^3. This value is confirmed in literature by Li et al., who have found a close value of 5.46 g/cm^3 [32].

Illustrations of a typical heterogeneous microstructure, which is obtained at the end of the processing route, are given in Figure 4. As it is illustrated, the as-sintered microstructures are heterogeneous and bimodal:

- Firstly, they contain two populations of β phase grains homogeneously dispersed in the whole volume of the sample. The first one consists in fine grains—with an average grain size of about 4 to 5 μm depending on the sample—and is located at every zone where plastic deformation occurs during BM. It constitutes the "shell" of the harmonic structure. The other one is made of coarse grains—the average grain size in the core against BM time is plotted for every sample on Figure 7b—and is formed in the core of the powder particles, where no plastic deformation was produced by the BM stage. It constitutes the "core" of the harmonic structure. The presence of fine β grains is clearly observed on the close-up views of Figure 4. The shell region may also be separated into the outer—consisted of micron-wide β grains—and inner shell consisted of grains about hundreds of nanometers wide.

- Secondly, they are bimodal from a chemical point of view, as they contain α phase precipitates near the shell. In most samples, the core is exclusively consisted of β phase and no α phase has been detected (except for BM 100 h). As we can see in Figure 4b, the average grain size of α precipitates is 1 μm and they are located at the β grain boundaries in the shell vicinity. They benefit from the excess of energy due to grain boundaries to precipitate during sintering.

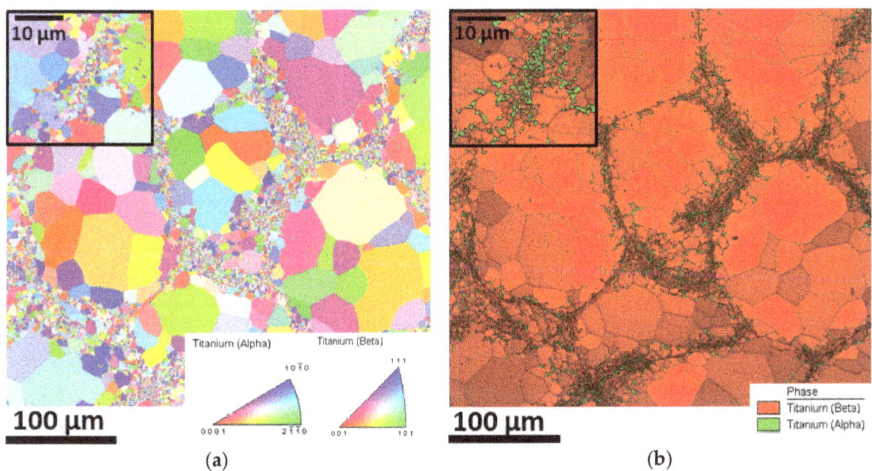

Figure 4. The microstructural characterization of BM 25 h. (**a**) 400 μm^2 IPF image and a close-up view of the shell region framed in black (**b**) corresponding phase image and a close-up view of the shell framed in black.

However, chemical composition of the alloy at this stage of the fabrication process could also be the cause of α phase precipitation. Figure 5 shows WDS maps of chemical elements repartition in the harmonic microstructure of BM 50 h sample. The shell which is composed of both α and β phases is enriched in titanium (Figure 5b) and depleted in niobium (Figure 5c), which is a β-stabilizer chemical element. Besides, Figure 5d clearly shows a local enrichment in oxygen in the shell vicinity, which probably enhances local α phase precipitation. Oxygen is indeed a strong α-stabilizer chemical element [40,41]. The oxygen contamination probably occurs during the mechanical BM stage, as the longer BM is processed, the higher the oxygen content in the alloy is.

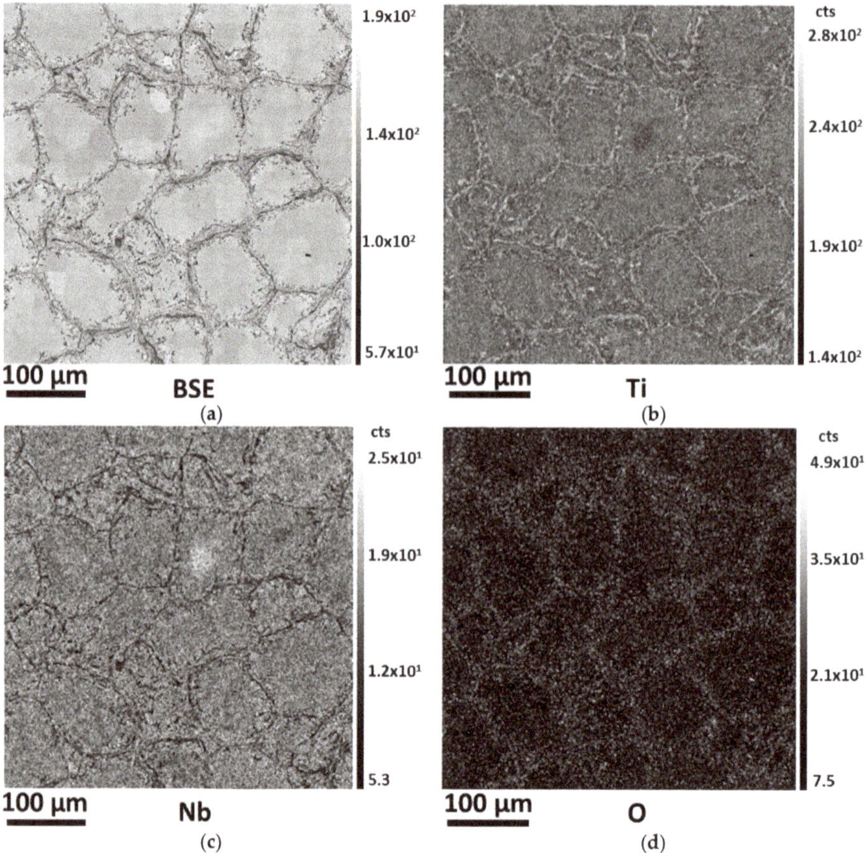

Figure 5. Chemical mapping of a BM 50 h sample. (**a**) Back-scattered electron image and corresponding (**b**) Ti repartition map, (**c**) Nb repartition map, and (**d**) O repartition map.

This investigation shows that it is possible to modify the microstructural features of the harmonic microstructures by playing with the processing parameters. The focus was made on the influence of BM time on the final microstructures. BM is indeed of primary importance in the fabrication process: it must be low energetic to avoid fracture of powder particles and high enough to induce local plastic deformation on the surface of the powder particles. Figure 6 shows IPF and grain size maps of 5 different samples fabricated using various BM times of 0 h, 12 h, 25 h, 50 h, and 100 h. Grains are considered to belong to the shell when their diameter is less than 10 μm. The other grains are considered to belong to the core of the microstructure. All the parameters which have been calculated, including shell surface fraction, α phase surface fraction and grain sizes in the core and in the shell are evaluated thanks to EBSD data. Note that an average grain size in the shell has been calculated for both β and α phases. The main microstructural features of the as-sintered samples are summarized in Table 1.

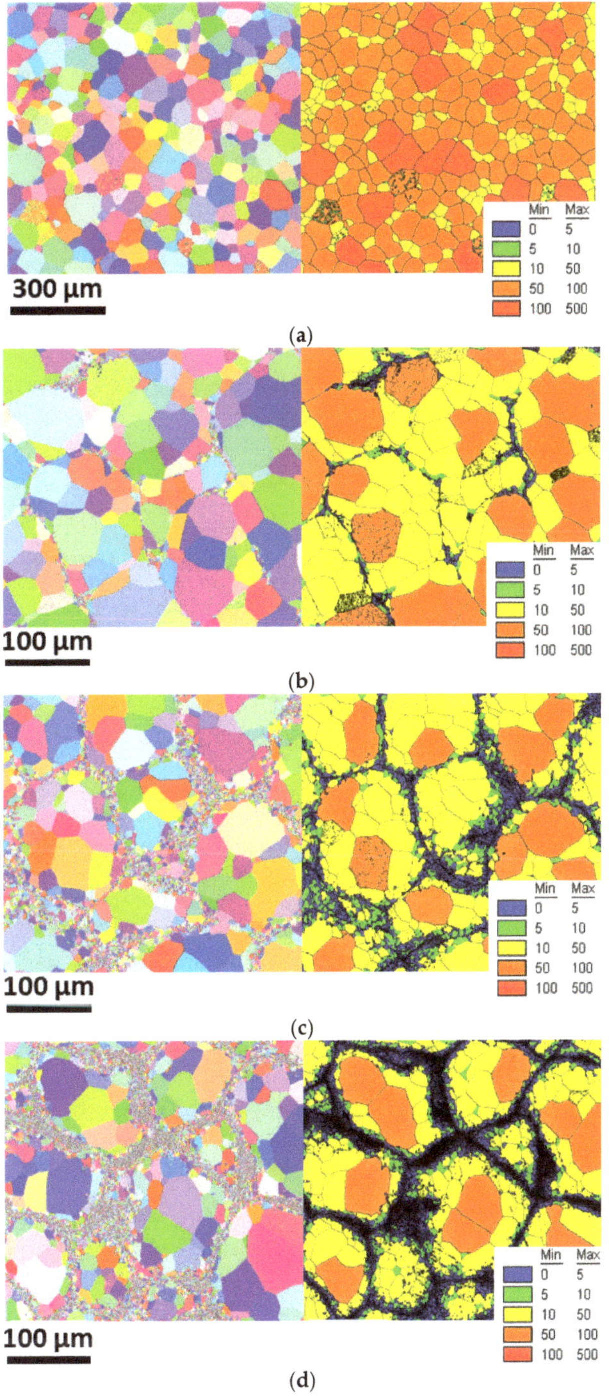

Figure 6. *Cont.*

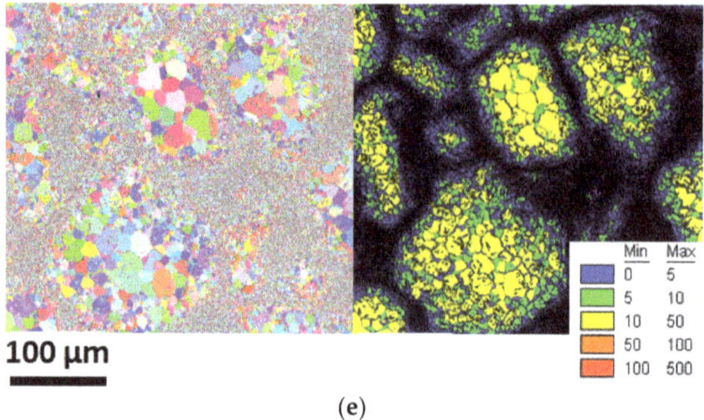

(e)

Figure 6. IPF and grain size maps of as-sintered samples (**a**) IP (no BM), (**b**) BM 12 h, (**c**) BM 25 h, (**d**) BM 50 h, and (**e**) BM 100 h. Black lines are drawn as a representation of grain boundaries on grain size maps. Small grains areas appear thus in black, especially in the shell of (**d,e**).

Table 1. Microstructural features. A cross X is marked when the sample is not concerned by the microstructural feature.

Name of the Sample	Surface Fraction of the Shell (%)	α Phase Fraction (%)	Average β Grain Size of the Core (μm)	Average β Grain Size of the Shell (μm)	Average α Grain Size of the Shell (μm)
IP	X	0	73.7	X	X
BM 12 h	6.6	2.0	50.2	5.2	1.0
BM 25 h	23.7	4.4	40.2	4.1	1.1
BM 50 h	32.1	12.7	39.7	3.4	1.1
BM 100 h	77.2	39.2	19.1	3.8	1.2

From the above data, it is obvious that BM time plays a key role on the shell volume fraction by modifying its width. The longer mechanical BM is processed, the wider the shell region is. The evolution of shell and α phase fractions are plotted versus BM time on Figure 7a and indeed, the shell fraction increases from 6.6% for BM 12 h to 77.2% for a BM 100 h. In return, α phase fraction also increases from 2.0% (for BM 12 h) to 39.2% (for BM 100 h) with BM time. This result could be expected as α phase precipitates on grain boundaries in the shell vicinity. Moreover, as it was said earlier, a long BM stage implies a significant oxygen enrichment of the material, which could favor α phase precipitation. Besides, the evolution of the grain size in the core versus BM time is also plotted on Figure 7b. The average grain size in the core slightly decreases when BM time increases. This decrease is notably obvious for BM 100 h, which has an average grain size in the core of about 20 μm, close to the average value of the initial grain size in the powder particles.

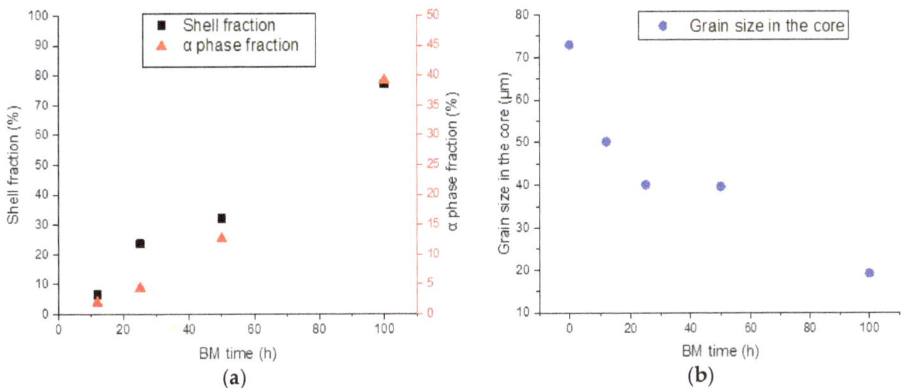

Figure 7. Evolution of (**a**) both shell fraction and α phase fraction against BM time and (**b**) average grain sizes in the core against BM time.

To explain the unusual decrease of grain size in the core with BM time, a particular observation is pointed out for BM 100 h. Indeed, intergranular α phase has precipitated inside the core of this microstructure, which has not been observed for samples processed with a lower BM time. That is probably why β grain growth in the core during SPS is limited for BM 100 h and the average value of grain size in the core is low for this sample.

A comparison of XRD results for all samples is shown on Figure 8. For every BM time, β phase peaks are detected on diffractograms. For BM 25 h, BM 50 h, and BM 100 h, additional α peaks are detected and the intensity of these peaks increases with BM time. It confirms the fact that more and more α phase is contained in the samples when BM time is increased. For BM 12 h, the volume fraction of α phase is too low to be detected on XRD, whereas 2.2% of α phase has been detected thanks to EBSD data. No other unindexed peak is detected on diffractograms.

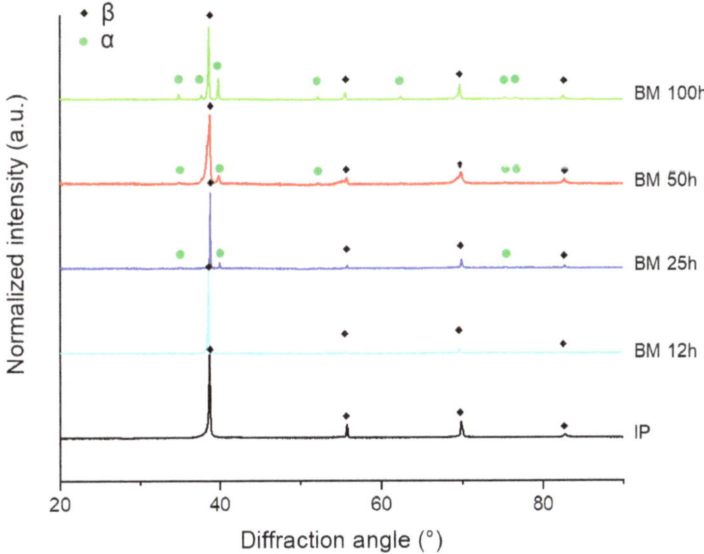

Figure 8. XRD results showing an increase in α phase peak intensities with BM time.

3.3. Mechanical Properties of the As-Sintered Ti-2448 Samples

The samples introduced in the previous part of the study have been mechanically tested using room temperature simple shear tests. Equivalent Von-Mises tensile stress and strain have been computed as: $\sigma = \tau \times \sqrt{3}$ and $\varepsilon = \gamma / \sqrt{3}$, where τ and γ are respectively the shear stress and the shear strain. Figure 9a shows the equivalent tensile true stress-true strain curves of the simple shear tested samples. The corresponding strain hardening rate evolution against true strain is presented in the insert as Figure 9b.

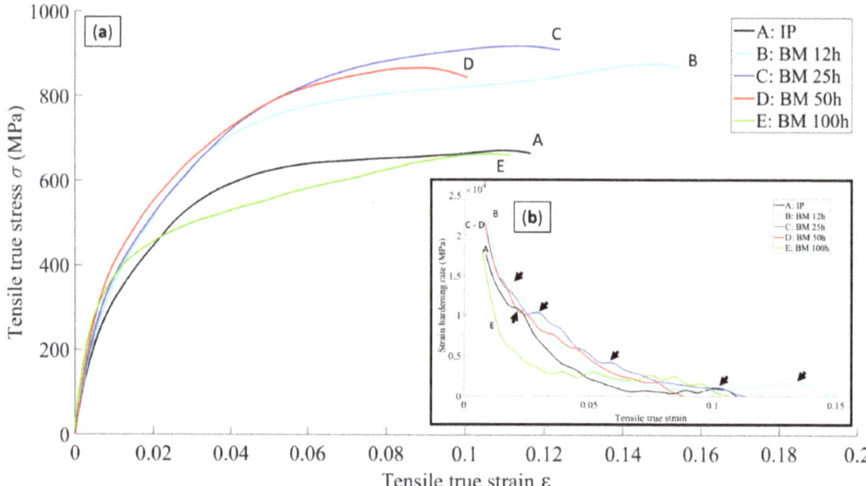

Figure 9. (**a**) Tensile curves and (**b**) corresponding strain-hardening rate versus true tensile strain curves. Curves are indexed with letters for clarity: A for the conventional microstructure IP, B for BM 12 h, C for BM 25 h, D for BM 50 h, and E for BM 100 h.

The mechanical tests have confirmed the high interest of heterogeneous microstructure for a better mechanical performance of materials and processing such heterogeneous materials is promising in order to increase the strength-ductility compromise of Ti-2448. Indeed, when compared with the conventional microstructure IP, both BM 12 h, BM 25 h, and BM 50 h curves reach a higher ultimate tensile stress while keeping comparable values of ductility and the heterogeneity of the microstructure contributes to the global increase of the flow stress. Only BM 100 h whose microstructure is mainly consisted of sub-micrometric α and β grains (about 70%) reaches a lower ultimate tensile stress value than IP.

At low strain values, i.e. until 2%, heterogeneous microstructures reach higher stress values than the conventional one. However, from this first threshold point, BM 100 h differentiates from the other heterogeneous samples by starting a stationary regime where the increase of the flow stress is quite linear. Then, BM 12 h, BM 25 h and BM 50 h display a similar mechanical behavior until 4% strain, which is characterized by several strain hardening rate rebounds—defined as local minima followed by brief increases of the curve. Most of them are marked with black arrows on Figure 9b. The IP sample also benefits from a similar strain hardening rate rebound (also identified with a black arrow on Figure 9b). From this strain value of 4%, BM 12 h tends to soften, whereas both BM 25 h and BM 50 h keep hardening. At the same time, strain hardening rate of BM 100 h tends to stabilize at about 2.5 GPa, between strain values of 4 to 10%; however, stress levels which are reached by this sample are lower than every other sample. Finally, above 6% strain, BM 25 h keeps hardening with a last stain hardening rate rebound, whereas BM 12 h and BM 50 h do not. Yet, it is interesting to note that for BM 12 h, a strain hardening increase starts from a strain of 10% and keeps going until 14%. It could be the cause of the higher true strain, which is finally reached by BM 12 h.

To explain the differences between shear behaviors, a first series of observations may be made on the elastic domains. Figure 10 is a comparative plot of experimentally and theoretically calculated Young's moduli of the samples. Error bars are calculated thanks to the data of the two mechanical tests which have been performed for each sample and the error on stress and strain measurements. The experimentally calculated modulus corresponds to the slope of the experimental curves whereas the theoretical Young's modulus is calculated using a rule of mixture considering both α and β phase fractions and respective Young's moduli. A clear increase of Young's modulus of the samples with BM time is observed, which fits the evolution of the calculated Young's modulus. The evolution of the Young's modulus may be associated with the fact that α phase volume fraction increases when BM time increases. As Young's modulus variations are due to the strength of chemical bonds between atoms, the change in chemistry of the alloy induces a change in the Young's modulus of the alloy. It increases when α phase fraction increases but keeps a rather low value of around 50 GPa for a BM time of 50 h. By interpolating the experimental curve, it may therefore be possible to control the elastic modulus of the as-processed material by controlling BM time. In any case, it remains close to the Young's modulus of human bones (around 30 GPa for cortical human bones) and fulfills one of the requirements for biomedical and orthopedic implants.

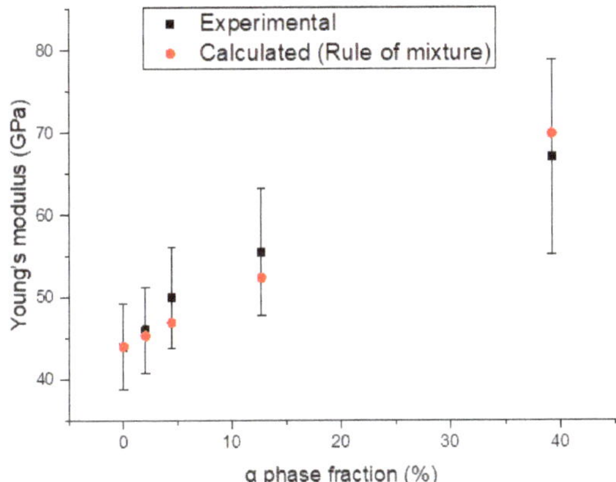

Figure 10. Evolution of Young's modulus (experimentally and theoretically calculated) against the α phase fraction.

Apart from the increase of the Young's modulus with BM time, another global observation could help explaining the differences between the tensile curves of the samples. Indeed, for every sample except BM 100 h, the plots display wide elastic to plastic transition domains. Several hypotheses could be made to explain this unusual behavior. Firstly, it could be linked with the stress-induced martensitic phase transformation which may occur during straining of Ti-2448. For $<110>_\beta$ Ti-2448 single crystals, the stress-induced β to α'' phase transformation takes place at a constant stress and is thus characterized by a plateau on the stress–strain curve [10,11,42]. However, for isotropic polycrystals, the stress-induced transformation might occur together with other deformation mechanisms of β phase and be characterized by a nonlinear elastic domain and a wide elastic to plastic transition zone [8,43,44]. Besides, Ti-2448 is also liable to deform via {332} $<113>_\beta$ mechanical twinning [10–12,45], and this deformation mechanism could be the cause of the nonlinear elastic domain of the alloy. Finally, it could be linked with the heterogeneity of the as-processed microstructures. There is indeed a significant mechanical contrast between core and shell regions in the harmonic structure.

In order to get a better understanding of the mechanical behavior of every sample, three different mechanical characteristics have been calculated using experimental data: the conventional yield stress at 0.2% offset, a chosen 1% plastic strain offset (due to the difficulty of finding a linear part of the elastic domain from Figure 9) and finally the ultimate tensile stress (UTS). If the stress-induced phase transformation actually takes place during straining, the 0.2% plastic strain yield stress could also be interpreted as the stress required to induce the martensitic transformation of the β phase. They are plotted against BM time on Figure 11. It shows a global increase of the characteristic stresses when increasing BM time until BM 50 h is reached. BM 12 h, BM 25 h, and BM 50 h have a close 1% offset stress. For BM 100 h, both yield stress and ultimate tensile stress are lower than the one of the other samples.

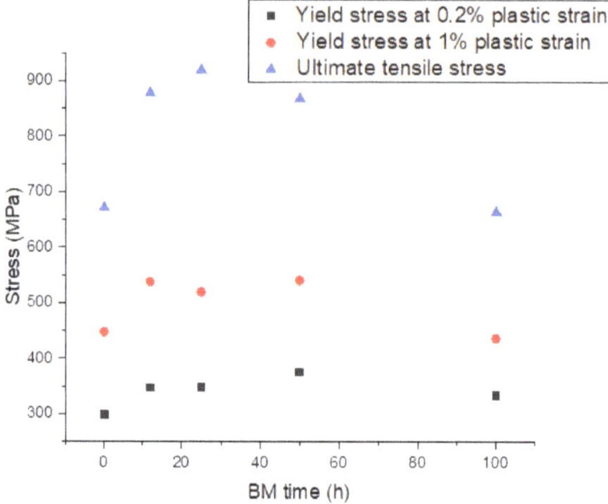

Figure 11. Evolution of yield stresses (at 0.2% and 1% plastic strain) and ultimate tensile stress against BM time.

These characteristic values (Figure 11) could be compared with the one in literature, especially those when Ti-2448 is processed by powder metallurgy. Kafkas et al. found UTS values of 650 to 720 MPa for Ti-2448 processed with Metal Injection Molding with coarse grain sizes and carbides precipitates [33]. When Ti-2448 is sintered using elemental powders by Li et al., UTS reaches 750 to 900 MPa according to the microstructure features [32]. Ti-2448 processed with Selective Laser Melting also display similar values of UTS, from 840 to 950 MPa with a heterogeneous microstructure in Yang et al.'s work to 665 MPa in Zhang et al.'s study. This last value may be improved by thermo-mechanical treatments [35,36]. In our work, UTS value which is reached by IP sample is comparable to UTS values of as-processed samples in literature, whereas heterogeneous samples have UTS values close to the ones of samples whose microstructure seems to be optimized.

The shear results also suggest a surprising behavior of BM 100 h. Indeed, the mechanical curve of BM 100 h is completely opposed to what would be expected based on the Hall-Petch relationship. Yet, the Hall-Petch relationship is mainly used for single phase pure metals and alloys whose grains are micron-wide. It is not the case here and consequently it could be an explanation of the differences which are observed between the experimental data and those expected by respect of the Hall-Petch relationship. Besides, the hypothesis which is made here is that β to α'' stress-induced martensitic transformation occurs during shear deformation of IP, BM 12 h, BM 25 h, and BM 50 h. However, both the drastic decrease of grain sizes and the notable increase of the fraction of α precipitates could have inhibited the TRIP and/or TWIP phenomena for BM 100 h. For IP, the stress-induced phase transformation occurs early, but it is delayed when the shell and α fraction are increased in BM 12 h,

BM 25 h, and BM 50 h. On the contrary, the BM 100 h material behaves like ultra-fine grain materials (UFG). Indeed, strain hardening rate curve of BM 100 h is characteristic of the behavior of UFG materials: strain hardening rate is high at the beginning of plastic deformation, but its decrease is very sharp once the second stage of plastic deformation is reached. This behavior is usually associated with a significant lack of ductility [46,47]. Fracture surface of BM 100 h is by the way characteristic of a brittle material.

The other samples—including IP, BM 12 h, BM 25 h, and BM 50 h—display local minima (followed by either a plateau or an increase) and fluctuations of the strain hardening rate. For IP, the local minimum of the strain hardening rate is followed by a plateau, whereas for heterogeneous ones (BM 12 h, BM 25 h, and BM 50 h), local minima are followed by a short increase of the strain hardening rate. These have been associated in the literature with the activation of a secondary deformation mechanism, which could be twinning or phase transformation [38,48]. However, no twin or martensite traces have been observed for the corresponding deformed post-mortem microstructures, and even for the post-mortem IP conventional microstructure, which is more susceptible to benefit from TRIP and/or TWIP deformation mechanisms. More investigations, including in situ tensile tests and/or cyclic mechanical tests, must be carried out for a better understanding of the global deformation mechanism. Other studies have associated the appearance of fluctuations in strain hardening rate curves with the formation of deformation gradients inside the microstructure [49]. Indeed, in heterogeneous and harmonic structures, the presence of both fine and coarse grains inside the microstructure induces hardness gradients in the microstructure. Coarse grains are easier to deform and start plastic deformation earlier than fine grains regions. Consequently, deformation gradients are formed at the core-shell interfaces which must be accommodated by geometrically necessary dislocations (GNDs). Piling up of GNDs at the core-shell interfaces results in high back stresses acting against the propagation of newly formed dislocations: it is a form of kinematic hardening contributing to the global work hardening of the alloy [50–52]. Fluctuations of the strain hardening rate of BM 12 h, BM 25 h, and BM 50 h might be the signatures of this kind of deformation mechanism.

The differences in mechanical behavior between samples are also illustrated by the characterization of the post-mortem deformed microstructures which has been carried out using XRD, microscopy, and EBSD. Figure 12 presents post-mortem optical images of the deformed surface of BM 12 h, BM 25 h, BM 50 h, and BM 100 h, whereas Figure 12e presents a SEM secondary electrons close-up view of the deformation bands on IP. Deformation marks are visible on the surface of the samples. For IP and BM 12 h, a very high number of marks is clearly identified. The number of these bands decreases with BM time (and consequently with shell and α precipitates fraction). Indeed, for BM 25 h and BM 50 h, only a few bands are visible, and no trace is identified for BM 100 h. EBSD investigation has shown that these bands are not twins, even if some similar deformation features have been identified as mechanical twins in recent studies [11,12]. Therefore, it may be slip bands. These slip bands are straight and parallel in some grains, whereas in others, they intersect and overlap. It means that several gliding systems are successively activated. Moreover, cross-slip of dislocations also occurs, because of the wavy shape of these deformation traces. Their intersection shows shearing effects (clearly visible on Figure 12e), which could give us information about the timing of the appearance of these bands during deformation. The progressive disappearance of deformation marks with BM time confirms the fact that a change in deformation mechanism of Ti-2448 occurs when it is heterogeneously processed.

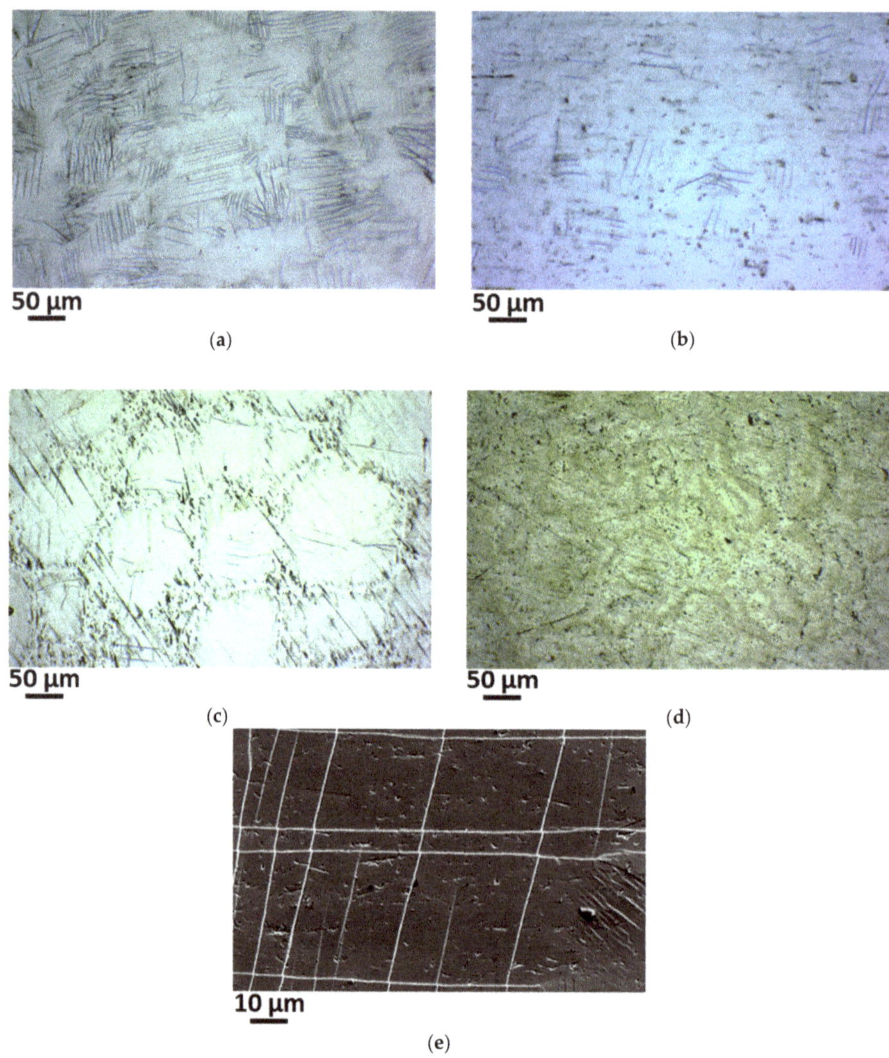

Figure 12. Optical images of the surface of the samples after simple shear (**a**) BM 12 h, (**b**) BM 25 h, (**c**) BM 50 h, (**d**) BM 100 h, and (**e**) SEM secondary electrons image of deformation bands on the surface of the IP sample.

4. Conclusions

This study has highlighted different mechanical responses of Ti-2448 samples according to the powder metallurgy processing route. These differences may reasonably be figured out. Indeed, several microstructural features have been modified by playing with the processing parameters, including grain sizes in both core and shell, chemical concentrations, and α phase precipitation. Most of these features have been shown to play a role in the deformation mechanism of β-metastable titanium alloys. The chemical composition of the alloy, and most particularly the oxygen content, is indeed of primary importance on the stress-induced β to α″ phase transformation [53]. Grain sizes also influence the appearance of phase transformation and mechanical twinning in the deformation sequence of β-metastable alloys [54,55]. Finally, precipitation of α and/or ω phases also significantly influences the mechanical behavior of Ti-2448 [43,56].

Author Contributions: Conceptualization, D.T. and G.D.; investigation, B.F., A.H. and E.L.; methodology, B.F., D.T. and Y.H.; resources, Y.H. and F.P.; supervision, D.T., F.P. and G.D.; writing—original draft, B.F.; writing—review & editing, F.P. and G.D. All authors have read and agreed to the published version of the manuscript.

Funding: This research received no external funding.

Conflicts of Interest: The authors declare no conflict of interest.

References

1. Banerjee, D.; Williams, J.C. Perspectives on Titanium Science and Technology. *Acta Mater.* **2013**, *61*, 844–879. [CrossRef]

2. Steinemann, S.G. Metal implants and surface reactions. *Injury* **1996**, *27*, S/C16–S/C22. [CrossRef]

3. Hao, Y.L.; Li, S.J.; Sun, S.Y.; Zheng, C.Y.; Yang, R. Elastic deformation behaviour of Ti–24Nb–4Zr–7.9Sn for biomedical applications. *Acta Biomater.* **2007**, *3*, 277–286. [CrossRef] [PubMed]

4. Zhang, Y.W.; Li, S.J.; Obbard, E.G.; Wang, H.; Wang, S.C.; Hao, Y.L.; Yang, R. Elastic properties of Ti–24Nb–4Zr–8Sn single crystals with bcc crystal structure. *Acta Mater.* **2011**, *59*, 3081–3090. [CrossRef]

5. Hao, Y.L.; Li, S.J.; Sun, S.Y.; Yang, R. Effect of Zr and Sn on Young's modulus and superelasticity of Ti–Nb-based alloys. *Mater. Sci. Eng. A* **2006**, *441*, 112–118. [CrossRef]

6. Arifin, A.; Sulong, A.B.; Muhamad, N.; Syarif, J.; Ramli, M.I. Material processing of hydroxyapatite and titanium alloy (HA/Ti) composite as implant materials using powder metallurgy: A review. *Mater. Des.* **2014**, *55*, 165–175. [CrossRef]

7. Qu, Y.; Zheng, S.; Dong, R.; Kang, M.; Zhou, H.; Zhao, D.; Zhao, J. Ti-24Nb-4Zr-8Sn Alloy Pedicle Screw Improves Internal Vertebral Fixation by Reducing Stress-Shielding Effects in a Porcine Model. Available online: https://www.hindawi.com/journals/bmri/2018/8639648/ (accessed on 23 March 2020).

8. Yang, Y.; Castany, P.; Cornen, M.; Prima, F.; Li, S.J.; Hao, Y.L.; Gloriant, T. Characterization of the martensitic transformation in the superelastic Ti–24Nb–4Zr–8Sn alloy by in situ synchrotron X-ray diffraction and dynamic mechanical analysis. *Acta Mater.* **2015**, *88*, 25–33. [CrossRef]

9. Yang, Y.; Castany, P.; Cornen, M.; Thibon, I.; Prima, F.; Gloriant, T. Texture investigation of the superelastic Ti-24Nb-4Zr-8Sn alloy. *J. Alloys Compd.* **2014**, *591*, 85–90. [CrossRef]

10. Yao, T.; Du, K.; Wang, H.; Huang, Z.; Li, C.; Li, L.; Hao, Y.; Yang, R.; Ye, H. In situ scanning and transmission electron microscopy investigation on plastic deformation in a metastable β titanium alloy. *Acta Mater.* **2017**, *133*, 21–29. [CrossRef]

11. Yang, Y.; Castany, P.; Bertrand, E.; Cornen, M.; Lin, J.X.; Gloriant, T. Stress release-induced interfacial twin boundary ω phase formation in a β type Ti-based single crystal displaying stress-induced α″ martensitic transformation. *Acta Mater.* **2018**, *149*, 97–107. [CrossRef]

12. Liang, Q.; Kloenne, Z.; Zheng, Y.; Wang, D.; Antonov, S.; Gao, Y.; Hao, Y.; Yang, R.; Wang, Y.; Fraser, H.L. The role of nano-scaled structural non-uniformities on deformation twinning and stress-induced transformation in a cold rolled multifunctional β-titanium alloy. *Scr. Mater.* **2020**, *177*, 181–185. [CrossRef]

13. Li, S.J.; Jia, M.T.; Prima, F.; Hao, Y.L.; Yang, R. Improvements in nonlinear elasticity and strength by grain refinement in a titanium alloy with high oxygen content. *Scr. Mater.* **2011**, *64*, 1015–1018. [CrossRef]

14. Zhang, S.Q.; Li, S.J.; Jia, M.T.; Prima, F.; Chen, L.J.; Hao, Y.L.; Yang, R. Low-cycle fatigue properties of a titanium alloy exhibiting nonlinear elastic deformation behavior. *Acta Mater.* **2011**, *59*, 4690–4699. [CrossRef]

15. Prima, F.; Vermaut, P.; Texier, G.; Ansel, D.; Gloriant, T. Evidence of α-nanophase heterogeneous nucleation from ω particles in a β-metastable Ti-based alloy by high-resolution electron microscopy. *Scr. Mater.* **2006**, *54*, 645–648. [CrossRef]

16. Zheng, Y.; Williams, R.E.A.; Sosa, J.M.; Wang, Y.; Banerjee, R.; Fraser, H.L. The role of the ω phase on the non-classical precipitation of the α phase in metastable β-titanium alloys. *Scr. Mater.* **2016**, *111*, 81–84. [CrossRef]

17. Li, C.; Chen, J.; Li, W.; Ren, Y.J.; He, J.J.; Song, Z.X. Effect of heat treatment variations on the microstructure evolution and mechanical properties in a β metastable Ti alloy. *J. Alloys Compd.* **2016**, *684*, 466–473. [CrossRef]

18. Abdel-Hady, M.; Hinoshita, K.; Morinaga, M. General approach to phase stability and elastic properties of β-type Ti-alloys using electronic parameters. *Scr. Mater.* **2006**, *55*, 477–480. [CrossRef]

19. Lai, M.J.; Tasan, C.C.; Raabe, D. On the mechanism of {332} twinning in metastable β titanium alloys. *Acta Mater.* **2016**, *111*, 173–186. [CrossRef]

20. Marteleur, M.; Sun, F.; Gloriant, T.; Vermaut, P.; Jacques, P.J.; Prima, F. On the design of new β-metastable titanium alloys with improved work hardening rate thanks to simultaneous TRIP and TWIP effects. *Scr. Mater.* **2012**, *66*, 749–752. [CrossRef]

21. Li, Z.; Zheng, B.; Wang, Y.; Topping, T.; Zhou, Y.; Valiev, R.Z.; Shan, A.; Lavernia, E.J. Ultrafine-grained Ti–Nb–Ta–Zr alloy produced by ECAP at room temperature. *J. Mater. Sci.* **2014**, *49*, 6656–6666. [CrossRef]

22. Shin, S.; Zhu, C.; Zhang, C.; Vecchio, K.S. Extraordinary strength-ductility synergy in a heterogeneous-structured β -Ti alloy through microstructural optimization. *Mater. Res. Lett.* **2019**, *7*, 467–473. [CrossRef]

23. Lu, K. Making strong nanomaterials ductile with gradients. *Science* **2014**, *345*, 1455–1456. [CrossRef] [PubMed]

24. Yang, M.; Yan, D.; Yuan, F.; Jiang, P.; Ma, E.; Wu, X. Dynamically reinforced heterogeneous grain structure prolongs ductility in a medium-entropy alloy with gigapascal yield strength. *Proc. Natl. Acad. Sci. USA* **2018**, *115*, 7224–7229. [CrossRef]

25. Vajpai, S.K.; Ota, M.; Zhang, Z.; Ameyama, K. Three-dimensionally gradient harmonic structure design: An integrated approach for high performance structural materials. *Mater. Res. Lett.* **2016**, *4*, 191–197. [CrossRef]

26. Vajpai, S.K.; Sawangrat, C.; Yamaguchi, O.; Ciuca, O.P.; Ameyama, K. Effect of bimodal harmonic structure design on the deformation behaviour and mechanical properties of Co-Cr-Mo alloy. *Mater. Sci. Eng. C* **2016**, *58*, 1008–1015. [CrossRef] [PubMed]

27. Zhang, Z.; Vajpai, S.K.; Orlov, D.; Ameyama, K. Improvement of mechanical properties in SUS304L steel through the control of bimodal microstructure characteristics. *Mater. Sci. Eng. A* **2014**, *598*, 106–113. [CrossRef]

28. Kikuchi, S.; Hayami, Y.; Ishiguri, T.; Guennec, B.; Ueno, A.; Ota, M.; Ameyama, K. Effect of bimodal grain size distribution on fatigue properties of Ti-6Al-4V alloy with harmonic structure under four-point bending. *Mater. Sci. Eng. A* **2017**, *687*, 269–275. [CrossRef]

29. Weston, N.S.; Derguti, F.; Tudball, A.; Jackson, M. Spark plasma sintering of commercial and development titanium alloy powders. *J. Mater. Sci.* **2015**, *50*, 4860–4878. [CrossRef]

30. Weston, N.S.; Jackson, M. FAST-forge of Titanium Alloy Swarf: A Solid-State Closed-Loop Recycling Approach for Aerospace Machining Waste. *Metals* **2020**, *10*, 296. [CrossRef]

31. Bodunrin, M.O.; Chown, L.H.; Omotoyinbo, J.A. Development of low-cost titanium alloys: A chronicle of challenges and opportunities. *Mater. Today Proc.* **2020**, S2214785320317685. [CrossRef]

32. Li, X.; Ye, S.; Yuan, X.; Yu, P. Fabrication of biomedical Ti-24Nb-4Zr-8Sn alloy with high strength and low elastic modulus by powder metallurgy. *J. Alloys Compd.* **2019**, *772*, 968–977. [CrossRef]

33. Kafkas, F.; Ebel, T. Metallurgical and mechanical properties of Ti–24Nb–4Zr–8Sn alloy fabricated by metal injection molding. *J. Alloys Compd.* **2014**, *617*, 359–366. [CrossRef]

34. Liu, Y.; Li, S.; Hou, W.; Wang, S.; Hao, Y.; Yang, R.; Sercombe, T.B.; Zhang, L.-C. Electron Beam Melted Beta-type Ti–24Nb–4Zr–8Sn Porous Structures with High Strength-to-Modulus Ratio. *J. Mater. Sci. Technol.* **2016**, *32*, 505–508. [CrossRef]

35. Yang, C.L.; Zhang, Z.J.; Li, S.J.; Liu, Y.J.; Sercombe, T.B.; Hou, W.T.; Zhang, P.; Zhu, Y.K.; Hao, Y.L.; Zhang, Z.F.; et al. Simultaneous improvement in strength and plasticity of Ti-24Nb-4Zr-8Sn manufactured by selective laser melting. *Mater. Des.* **2018**, *157*, 52–59. [CrossRef]

36. Zhang, L.C.; Klemm, D.; Eckert, J.; Hao, Y.L.; Sercombe, T.B. Manufacture by selective laser melting and mechanical behavior of a biomedical Ti–24Nb–4Zr–8Sn alloy. *Scr. Mater.* **2011**, *65*, 21–24. [CrossRef]

37. Zheng, R.; Zhang, Z.; Nakatani, M.; Ota, M.; Chen, X.; Ma, C.; Ameyama, K. Enhanced ductility in harmonic structure designed SUS316L produced by high energy ball milling and hot isostatic sintering. *Mater. Sci. Eng. A* **2016**, *674*, 212–220. [CrossRef]

38. Dirras, G.; Ueda, D.; Hocini, A.; Tingaud, D.; Ameyama, K. Cyclic shear behavior of conventional and harmonic structure-designed Ti-25Nb-25Zr β-titanium alloy: Back-stress hardening and twinning inhibition. *Scr. Mater.* **2017**, *138*, 44–47. [CrossRef]

Metals **2020**, *10*, 1626

39. Mompiou, F.; Tingaud, D.; Chang, Y.; Gault, B.; Dirras, G. Conventional vs harmonic-structured β-Ti-25Nb-25Zr alloys: A comparative study of deformation mechanisms. *Acta Mater.* **2018**, *161*, 420–430. [CrossRef]

40. Combres, Y. *Propriétés Du Titane Et De Ses Alliages*; Techniques de l'ingénieur: Paris, France, 2010.

41. Chou, K.; Marquis, E.A. Oxygen effects on ω and α phase transformations in a metastable β Ti–Nb alloy. *Acta Mater.* **2019**, *181*, 367–376. [CrossRef]

42. Yao, T.; Du, K.; Wang, H.; Qi, L.; He, S.; Hao, Y.; Yang, R.; Ye, H. Reversible twin boundary migration between α″ martensites in a Ti-Nb-Zr-Sn alloy. *Mater. Sci. Eng. A* **2017**, *688*, 169–173. [CrossRef]

43. Sun, F.; Hao, Y.L.; Zhang, J.Y.; Prima, F. Contribution of nano-sized lamellar microstructure on recoverable strain of Ti–24Nb–4Zr–7.9Sn titanium alloy. *Mater. Sci. Eng. A* **2011**, *528*, 7811–7815. [CrossRef]

44. Hao, Y.L.; Zhang, Z.B.; Li, S.J.; Yang, R. Microstructure and mechanical behavior of a Ti–24Nb–4Zr–8Sn alloy processed by warm swaging and warm rolling. *Acta Mater.* **2012**, *60*, 2169–2177. [CrossRef]

45. Yang, Y.; Castany, P.; Hao, Y.L.; Gloriant, T. Plastic deformation via hierarchical nano-sized martensitic twinning in the metastable β Ti-24Nb-4Zr-8Sn alloy. *Acta Mater.* **2020**, *194*, 27–39. [CrossRef]

46. Shumilin, S.E.; Janecek, M.; Isaev, N.V.; Homola, P.; Kral, R. Low Temperature Plasticity of an Ultrafine-Grained Al-3Mg Alloy Prepared by Accumulative Roll Bonding. *Adv. Eng. Mater.* **2012**, *14*, 35–38. [CrossRef]

47. Krasilnikov, N.; Lojkowski, W.; Pakiela, Z.; Valiev, R. Tensile strength and ductility of ultra-fine-grained nickel processed by severe plastic deformation. *Mater. Sci. Eng. A* **2005**, *397*, 330–337. [CrossRef]

48. Ueda, D.; Dirras, G.; Hocini, A.; Tingaud, D.; Ameyama, K.; Langlois, P.; Vrel, D.; Trzaska, Z. Data on processing of Ti-25Nb-25Zr β-titanium alloys via powder metallurgy route: Methodology, microstructure and mechanical properties. *Data Brief* **2018**, *17*, 703–708. [CrossRef]

49. Jiang, F.; Zhao, C.; Liang, D.; Zhu, W.; Zhang, Y.; Pan, S.; Ren, F. In-situ formed heterogeneous grain structure in spark-plasma-sintered CoCrFeMnNi high-entropy alloy overcomes the strength-ductility trade-off. *Mater. Sci. Eng. A* **2020**, *771*, 138625. [CrossRef]

50. Wu, X.; Zhu, Y. Heterogeneous materials: A new class of materials with unprecedented mechanical properties. *Mater. Res. Lett.* **2017**, *5*, 527–532. [CrossRef]

51. Wang, Y.F.; Wang, M.S.; Fang, X.T.; Guo, F.J.; Liu, H.Q.; Scattergood, R.O.; Huang, C.X.; Zhu, Y.T. Extra strengthening in a coarse/ultrafine grained laminate: Role of gradient interfaces. *Int. J. Plast.* **2019**, *123*, 196–207. [CrossRef]

52. Yang, M.; Pan, Y.; Yuan, F.; Zhu, Y.; Wu, X. Back stress strengthening and strain hardening in gradient structure. *Mater. Res. Lett.* **2016**, *4*, 145–151. [CrossRef]

53. Obbard, E.G.; Hao, Y.L.; Talling, R.J.; Li, S.J.; Zhang, Y.W.; Dye, D.; Yang, R. The effect of oxygen on α″ martensite and superelasticity in Ti–24Nb–4Zr–8Sn. *Acta Mater.* **2011**, *59*, 112–125. [CrossRef]

54. Bhattacharjee, A.; Bhargava, S.; Varma, V.K.; Kamat, S.V.; Gogia, A.K. Effect of β grain size on stress induced martensitic transformation in β solution treated Ti–10V–2Fe–3Al alloy. *Scr. Mater.* **2005**, *53*, 195–200. [CrossRef]

55. Danard, Y.; Sun, F.; Gloriant, T.; Freiherr Von Thüngen, I.; Piellard, M.; Prima, F. The Influence of Twinning on the Strain–Hardenability in TRIP/TWIP Titanium Alloys: Role of Solute–Solution Strengthening. *Front. Mater.* **2020**, *7*. [CrossRef]

56. Coakley, J.; Rahman, K.M.; Vorontsov, V.A.; Ohnuma, M.; Dye, D. Effect of precipitation on mechanical properties in the β-Ti alloy Ti–24Nb–4Zr–8Sn. *Mater. Sci. Eng. A* **2016**, *655*, 399–407. [CrossRef]

Publisher's Note: MDPI stays neutral with regard to jurisdictional claims in published maps and institutional affiliations.

Article

Ratcheting-Fatigue Behavior of Harmonic-Structure-Designed SUS316L Stainless Steel

Yang Song [1], Zhe Zhang [2,3,*], Hantuo Ma [2], Masashi Nakatani [4], Mie Ota Kawabata [4] and Kei Ameyama [4]

1   Tianjin Key Laboratory for Advanced Mechatronic System Design and Intelligent Control, National Demonstration Center for Experimental Mechanical and Electrical Engineering Education, School of Mechanical Engineering, Tianjin University of Technology, Tianjin 300384, China; songyang2661@sina.com
2   School of Chemical Engineering and Technology, Tianjin University, Tianjin 300350, China; hantuoma@hotmail.com
3   Tianjin Key Laboratory of Chemical Process Safety and Equipment Technology, Tianjin 300350, China
4   Department of Mechanical Engineering, Faculty of Science and Engineering, Ritsumeikan University, Shiga 525-8577, Japan; rm005034@ed.ritsumei.ac.jp (M.N.); mie-ota@fc.ritsumei.ac.jp (M.O.K.); ameyama@se.ritsumei.ac.jp (K.A.)
*   Correspondence: zhe.zhang@tju.edu.cn

Abstract: Stainless steels with harmonic-structure design have a great balance of high strength and high ductility. Therefore, it is imperative to investigate their fatigue properties for engineering applications. In the present work, the harmonic-structured SUS316L stainless steels were fabricated by mechanical milling (MM) and subsequent hot isostatic pressing (HIP) process. A series of ratcheting-fatigue tests were performed on the harmonic-structured SUS316L steels under stress-control mode at room temperature. Effects of grain structure and stress-loading conditions on ratcheting behavior and fatigue life were investigated. Results showed that grain size and applied mean stress had a significant influence on ratcheting-strain accumulation and fatigue life. Owing to the ultrafine grained structure, tensile strength of the harmonic-structured SUS316L steels could be enhanced, which restrained the ratcheting-strain accumulation, resulting in a prolonged fatigue life. A higher mean stress caused a faster ratcheting-strain accumulation, which led to the deterioration of fatigue life. Moreover, a modified model based on Smith–Watson–Topper (SWT) criterion predicted the ratcheting-fatigue life of the harmonic-structured SUS316L steels well. Most of the fatigue-life points were located in the 5 times error band.

Keywords: stainless steel; harmonic structure; ratcheting; fatigue; fatigue-life prediction

Citation: Song, Y.; Zhang, Z.; Ma, H.; Nakatani, M.; Kawabata, M.O.; Ameyama, K. Ratcheting-Fatigue Behavior of Harmonic-Structure-Designed SUS316L Stainless Steel. *Metals* **2021**, *11*, 477. https://doi.org/10.3390/met11030477

Academic Editor: Eric Hug

Received: 25 January 2021
Accepted: 5 March 2021
Published: 13 March 2021

Publisher's Note: MDPI stays neutral with regard to jurisdictional claims in published maps and institutional affiliations.

1. Introduction

Austenitic stainless steels have excellent mechanical properties and corrosion resistance, and are often used as structural materials in petrochemical equipment, nuclear equipment, and medical equipment [1–3]. Therefore, it is imperative to realize high strength and high ductility in stainless steels for structural safety. Usually, these types of equipment are often subjected to cyclic loading in service. Thus, cyclic deformation and fatigue properties of stainless steels need to be considered [4–6].

In recent years, structural materials with heterogenous microstructures have been proposed to achieve excellent mechanical properties and good fatigue resistance [7–15]. Mechanical properties and deformation mechanisms of some typical heterogenous microstructures, such as gradient nanograined (GNG) structure [9–11], lamellar structure [12–14], hierarchical and laminated grains and twins structure [15], and harmonic structure [16,17], have been investigated. As indicated in Figure 1a, the conventional bimodal structure has an irregular coarse grain (CG) and ultrafine grain (UFG) distribution. By contrast, harmonic structure has a regular bimodal grain-size distribution, whereas UFG structure (shell) exhibits a three-dimensional, continuously connected network structure (Figure 1b).

63

The regular network structure could relieve strain localization during tensile deformation, which causes an extra strain hardening. Therefore, the harmonic-structure-designed materials exhibit a superior strength–ductility synergy [18–23].

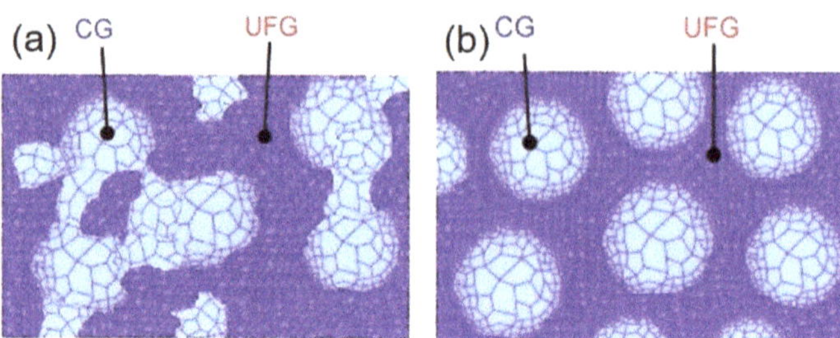

Figure 1. Schematic of (**a**) bimodal structure and (**b**) harmonic structure.

For engineering applications in service, the cyclic response and fatigue properties of heterogenous-structured stainless steels have been studied in the past decades. GNG SUS316L steels demonstrated a superior balance of high strength and large uniform elongation, as well as good fatigue resistance. Due to the increased tensile strength by GNG surface layer, high-cycle fatigue resistance could be improved in GNG steels compared to CG or UFG steels. By contrast, it is remarkable that abnormal grain coarsening could suppress surface roughening and fatigue cracking. Thus, low-cycle fatigue resistance was also improved in GNG materials [9,10]. It was found that lamellar structure impeded fatigue-crack growth, which restrained fatigue-crack growth rates [13,14]. Therefore, the lamellar-structured steels also demonstrated good fatigue resistance. Furthermore, the steels with hierarchical and laminated grains and twins also induced a friction stress that acted on the crack surface and decelerated fatigue-crack opening and growth [15]. The high-cycle fatigue properties and fatigue-crack growth behavior of harmonic-structured SUS316L steels have been investigated [24,25]. Fatigue limits of the harmonic-structure SUS316L steels were higher than that of the CG SUS316L steels, which was related to the enhanced tensile strength [24]. As expected, the UFG structure had worse crack-growth resistance than the CG structure. Therefore, the harmonic-structured 316L steels had higher fatigue-crack growth rates and a lower threshold stress intensity factor range for crack growth (ΔK_{th}) compared to the homogeneous CG SUS316L steels. Fatigue cracks tended to propagate along UFG network structure in harmonic-structured SUS316L steels [25,26].

As mentioned, stainless steels are often used as structural materials in thin-wall piping systems, which are often subjected to cyclic loading and alternating temperature or internal pressure, especially during start/stop operation. Therefore, pipelines are subjected to an asymmetric stress cycling, which can cause an accumulation of plastic strain known as ratcheting strain [27–31]. For example, in the early 1990s, the ratcheting-fatigue behavior of nuclear pipes and its detrimental effect on the nuclear-reactor structure were contained in the American ITER design code and ASME NB 32xx code [27]. As reported, mean stress had a much higher influence on ratcheting-strain accumulation than stress amplitude or stress-loading rate. Therefore, the effects of mean stress on ratcheting strain and low-cycle fatigue life were investigated in stainless steels. In general, a larger mean stress induced a rapid -train accumulation, resulting in deterioration of fatigue life. Moreover, many stress-based criteria have been proposed to correct the effect of mean stress on fatigue-life prediction in the past decades, such as the Goodman criterion, Gerber criterion, Smith–Watson–Topper (SWT) criterion, and so on. It also has been reported that the models based on the SWT criterion possessed a higher prediction accuracy for ratcheting-fatigue life of steels subjected to asymmetrical stress cyclic loading [4,28]. However, comprehensive understanding of ratcheting behavior and fatigue-life prediction for harmonic-structured stainless steels still remains unclear.

In the present work, the ratcheting-fatigue properties of the harmonic-structured SUS316L steels were investigated. The effect of grain structure and applied mean stress on ratcheting strain and fatigue life are revealed. Moreover, a modified model based on the SWT criterion is proposed for the prediction of fatigue life of harmonic-structured SUS316L steels.

2. Materials and Experiments

SUS316L stainless steels with a harmonic-structure topology were fabricated using horizontal ball-milling and hot isostatic pressing (HIP) processes. The chemical composition of the SUS316L gas-atomized powders is shown in Table 1. The powder features, mechanical milling, and sintering process were introduced in our previous study [32]. In the present work, gas-atomized SUS316L steel powders were milled at room temperature for 10 h and 30 h. The compacts were sintered from initial powders and milled powders using a hot isostatic pressing (HIP) process with a pressure of 200 MPa at 900 °C for 4 h. The compact sintered from initial powders was denoted as the MM0h compact. The compacts sintered from milled powers for 10 h and 30 h were denoted as the MM10h compact and the MM30h compact, respectively. The grain structures of sintered compacts are shown in Figure 2. It can be seen in Figure 2a that the MM0h compact had a partial harmonic structure, which was attributed to an irregular initial powder size ranging from 1 μm to 20 μm. As indicated in Figure 2b,c, it was seen that the volume fraction of ultrafine grains increased as the mechanical milling time increased from 10 h to 30 h, resulting in a reduced grain size. Moreover, a continuous UFG network structure was observed, which indicated that a harmonic structure was produced in the compacts sintered from milled powders. Grain size and its volume fraction of sintered compacts are presented in Table 2. As indicated in Table 3, both yield strength (YS) and ultimate tensile strength (UTS) were enhanced with decreasing grain size [32].

Table 1. Chemical composition of SUS316L gas-atomized powders (mass%).

C	Si	Mn	P	S	Ni	Cr	Mo	Fe
0.018	0.9	1.07	0.032	0.017	12.44	17.31	2.11	Bal.

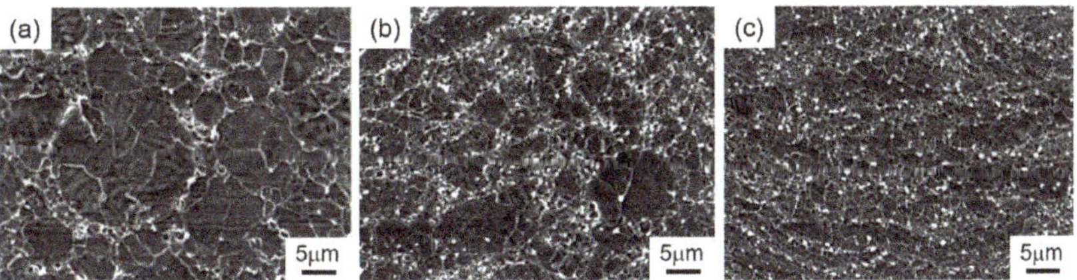

Figure 2. Microstructure of sintered SUS316L compacts: (**a**) MM0h compact; (**b**) MM10h compact; (**c**) MM30h compact.

Table 2. Grain size of harmonic-structured SUS316L compacts.

Materials	Core Grain Size (CG), mm	Core Fraction	Shell Grain Size (UFG), mm	Shell Fraction
MM0h	3.94	86.7%	1.28	13.3%
MM10h	1.84	73.5%	0.64	26.5%
MM30h	1.16	45.5%	0.42	54.5%

Table 3. Loading conditions and fatigue life of ratcheting-fatigue tests for harmonic-structured SUS316L steels.

Materials	Mean Stress σ_m (MPa)	Stress Amplitude σ_a (MPa)	Stress Rate (MPa/s)	Fatigue Life N_f (Cycles)
MM0h	300			29,774
	400			14,440
	450			432
MM10h	400	300	300	17,854
	450			14,666
MM30h	400			41,227
	450			18,599

Figure 3 shows the setup of the ratcheting-fatigue test. As shown in Figure 3a, fatigue tests were conducted with an in-situ fatigue-testing machine (IBTC-5000, CARE, Tianjin, China) under stress-control mode at room temperature. A triangle waveform stress loading with a constant loading rate 300 MPa/s was carried out. In order to prevent buckling of the specimen, the stress ratio R was more than zero. Displacement of gauge area was recorded by a noncontact CCD camera (Figure 3c). The typical stress–strain response is shown in Figure 3b. The dimensions of specimens for the ratcheting fatigue tests are shown in Figure 3b. The specimens were ground using a 5000 grit SiC paper before fatigue tests. The ratcheting strain (ε_r) is defined as:

$$\varepsilon_r = \frac{1}{2}(\varepsilon_{max} + \varepsilon_{min}) \tag{1}$$

where ε_{max} and ε_{min} are demoted as the maximum strain and minimum strain in each cycle, respectively. The loading conditions of ratcheting fatigue tests and observed fatigue life are listed in Table 3. Some high-cycle fatigue data of the harmonic-structured 316 L steels from our previous study were obtained [24]. These data were also used to analyze the fatigue-life-prediction model of harmonic-structured 316 L steels. Grain structure and fatigue fracture surface were examined by field emission-scanning electron microscopy (FE-SEM, SU4800, Tokyo, Japan) at 15 kV.

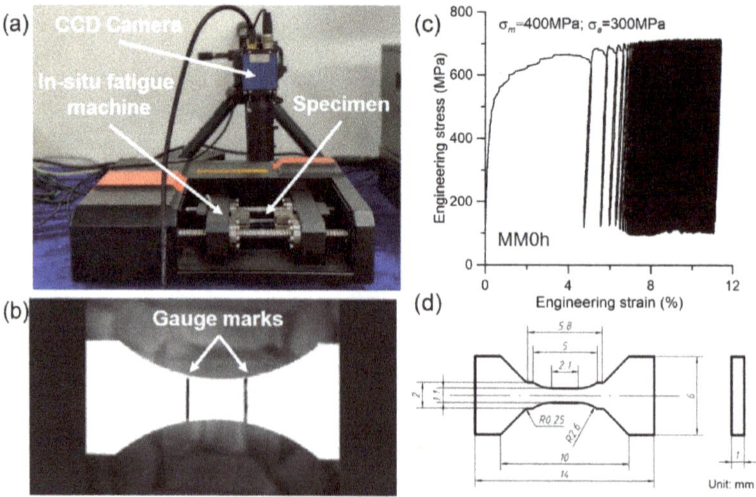

Figure 3. (**a**) Experiment setup; (**b**) typical stress-strain response; (**c**) displacement of gauge area measurement; (**d**) dimensions of specimen for ratcheting-fatigue test.

3. Results and Discussion

3.1. Effects of Grain Structure on Ratcheting-Fatigue Behavior of Harmonic-Structured SUS316L Steels

Figure 4 summarizes the effects of grain structure on ratcheting strain and fatigue life of the harmonic-structured SUS316L steels. As shown in Figure 4a,b, the ratcheting-fatigue tests of the harmonic structured SUS316L steels with different grain sizes were performed under the same loading condition (σ_m = 400 MPa, σ_a = 300 MPa). It is noted that the grain structure had a great influence on both ratcheting strain and fatigue life. Figure 4b shows the ratcheting strain versus N/N_f. It can be seen in Figure 4b that the ratcheting strain accumulated rapidly in the early stage of cyclic deformation. The ratcheting strain of MM0h compact was approximately 10.6% at 0.1N_f. By contrast, the values were approximately 4.2% and 2.5% for the MM10h compact and the MM30h compact, respectively. It is noteworthy that the ratcheting strain decreased with decreasing grain size. Subsequently, the increment of ratcheting strains was stable between 0.1N_f and 0.8N_f. At the final stage, the ratcheting strain increased rapidly to failure as N/N_f was over than 0.8. Moreover, the number of cycles to failure for the MM0h compact was 14,440 cycles, while the values for the MM10h compact and the MM30h compact were 17,854 cycles and 41,227 cycles, respectively. It was indicated that the harmonic-structured SUS316L steel with lower grain size possessed a longer fatigue life. A similar tendency was also observed in the results shown in Figure 4c,d at σ_m = 450 MPa, σ_a = 300 MPa.

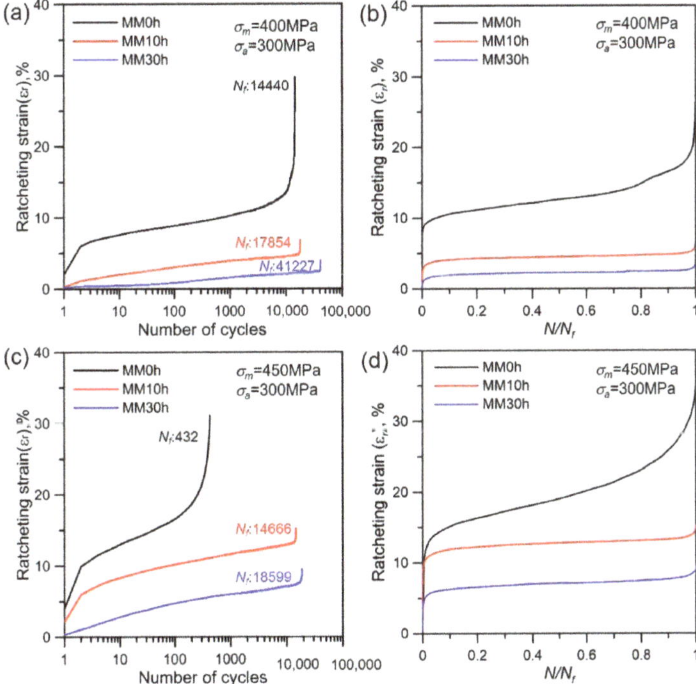

Figure 4. Effects of grain structure on fatigue life (**a,c**) and ratcheting strain (**b,d**) of harmonic-structured SUS316L steels: (**a,b**) σ_m = 400 MPa, σ_a = 300 MPa; (**c,d**) σ_m = 450 MPa, σ_a = 300 MPa.

As indicated in Figure 4, the variation of ratcheting strain could be divided into three stages, which appeared to be creep strain. It is known that creep strain is caused by microvoid formation and grain boundary sliding. However, the mechanism of ratcheting strain is different. Specifically, ratcheting strain is thought to be related to the progressive development of plastic deformation. Active dislocations decrease in the early stage of cyclic deformation, which leads to a rapid ratcheting strain accumulation. The dislocation structure becomes stable in the following stage, resulting in a stable ratcheting-strain rate. At the last stage, due to microcracks or main fatigue-crack growth, the true stress level increases, resulting in an accelerated ratcheting-strain rate [33].

Owing to the shell region in harmonic-structured SUS316L steels, the strength increases with increasing shell volume fraction. Moreover, the ratcheting strain is determined by accumulative plastic deformation. Therefore, it can be seen in Figure 4 that the ratcheting strains were restrained in the MM10h compact and the MM30h compact compared to the MM0h compact. A lower strength produced a higher plastic strain under the same stress level. The ratcheting strain obviously was produced in the first stage of cyclic deformation, and decreased with increasing strength of materials. Moreover, the ratcheting-strain rates of the harmonic-structured SUS316L steels were nearly stable in the secondary stage, which may indicate that the dislocation structure in the harmonic structure was stable under stress cycling [33].

The morphology of ratcheting-fatigue fracture surfaces of the harmonic-structured SUS316L steels are presented in Figure 5. It can be seen in Figure 5a that an obvious necking was observed in the ratcheting-fractured MM0h compact. This indicates that a large plastic deformation occurred before the fatigue fracture. In contrast, the fracture surface area of the MM10h compact and the MM30h compact was larger than that of the MM0h compact. As grain size decreased, the necking became insignificant. Therefore, this also showed that the accumulative plastic deformation became lower as the grain size decreased. Moreover, it can be seen in Figure 5 that the crack initiated from the specimen surface in the harmonic-structured SUS316L steels subjected to ratcheting-fatigue tests.

Figure 5. Morphology of ratcheting-fatigue fracture surface of harmonic-structured SUS316L steels: (**a**) MM0h compact; (**b**) MM10h compact; (**c**) MM30h compact.

As shown in Figures 4 and 5, the fatigue life was prolonged as the grain size decreased. It has been reported that ratcheting strain is produced from accumulative plastic deformation, which has detrimental effects on the fatigue life of materials [31]. In the present work, it is worthy to note that the ratcheting strain could be restrained in the harmonic-structured SUS316L steels due to their enhanced strength, which has great benefits for fatigue life. Therefore, it is thought that the harmonic-structure design could improve fatigue resistance under asymmetric stress cycling.

3.2. Effects of Mean Stress on Ratcheting-Fatigue Behavior of Harmonic Structured SUS316L Steels

As mentioned, the ratcheting strain increases with the increase of mean stress, which also affects fatigue life [29–31]. Therefore, the ratcheting behavior and fatigue life of the harmonic-structured SUS316L steels under the same stress amplitude (σ_a = 300 MPa) but different mean stresses are presented in Figure 6. The results of ratcheting strain versus N/N_f for MM0h compact, MM10h compact, and MM30h compact are shown in Figure 6b,d,f, respectively. As expected, an increased mean stress level brought about a higher ratcheting strain for harmonic-structured SUS316L steels. It can be seen from Figure 6b that the ratcheting-strain rate of the MM0h compact was nearly zero in the secondary stage at σ_m = 300 MPa. In contrast, the stable ratcheting-strain rate increased at high stress levels. Both the ratcheting strain and ratcheting-strain rate increased as the mean stress increased from 300 MPa to 450 MPa. As indicated in Figure 6d,f, the higher mean stress also induced the higher ratcheting strain for the MM10h compact and the MM30h compact. However, the variation in the ratcheting-strain rate in the stable region was not significant.

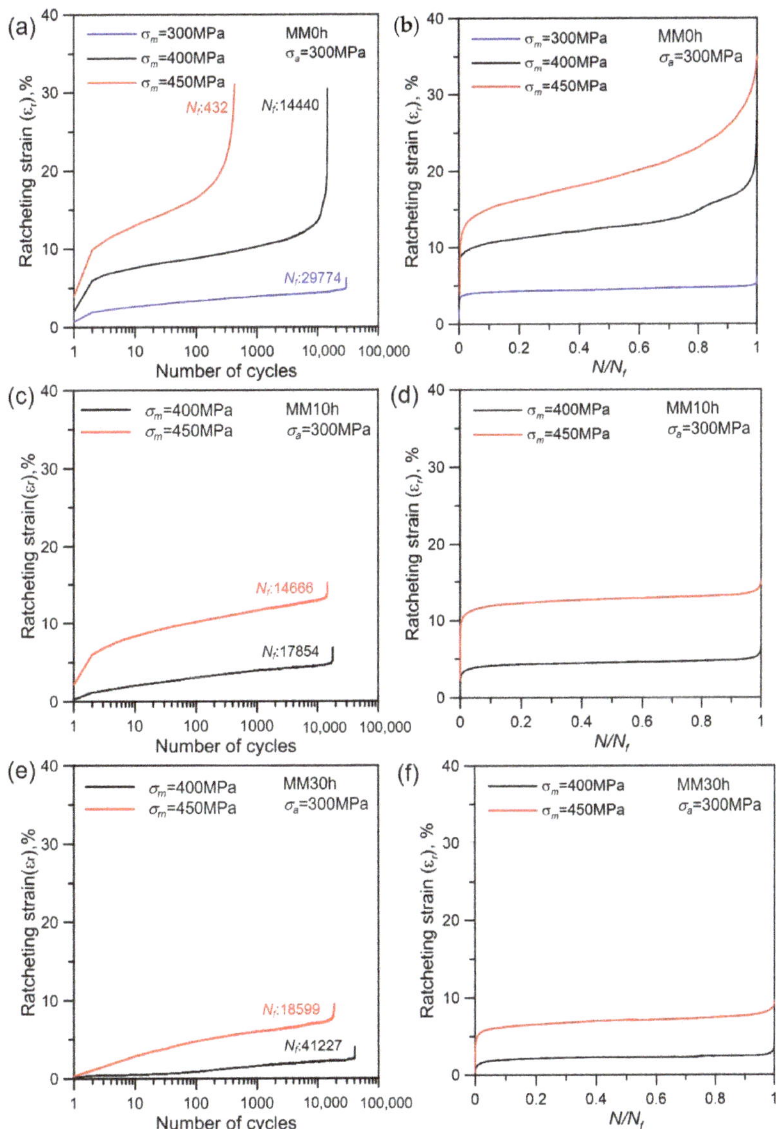

Figure 6. Effects of mean stress on fatigue life (**a**,**c**,**e**) and ratcheting strain (**b**,**d**,**f**) of harmonic-structured SUS316L steels: (**a**,**b**) MM0h compacts; (**c**,**d**) MM10h compacts; (**e**,**f**) MM30h compacts.

As expected, the increased mean stress produced a higher stress level, which promoted ratcheting-strain accumulation. The rapidly accumulative ratcheting strain was harmful for fatigue life. As indicated in Figure 6, the mean stress also played an important role in the ratcheting-fatigue life of harmonic-structured SUS316L steels subjected to asymmetric stress cycling. As the applied mean stress increased, the fatigue life decreased gradually. Therefore, the effects of mean stress on fatigue life need to be considered in fatigue-life prediction.

3.3. Ratcheting-Fatigue-Life Prediction of Harmonic-Structured SUS316L Steels

In the past decades, Basquin's equation has been used as a classical fatigue-life-prediction model for fully reversed cyclic stress loading, as given in Equation (2) [28]:

$$\sigma_a = \sigma_f' \left(2N_f\right)^b \tag{2}$$

where σ_a and N_f are the stress amplitude and fatigue life, respectively. σ_f' is the fatigue strength coefficient, and b is the fatigue exponent. However, if the mean stress is not zero, ratcheting strain could appear, which is harmful for fatigue life. Therefore, equivalent stress amplitude (σ_a^{eq}) has been proposed in many modified models for considering mean stress effects. As mentioned, the modified model based on the SWT criterion showed a higher prediction accuracy for ratcheting-fatigue life of materials performed under asymmetrical cyclic stress loading, as given in Equation (3):

$$\sigma_a^{eq} = \sqrt{\sigma_{max}\sigma_a} = \sigma_a\sqrt{1 + \frac{\sigma_m}{\sigma_a}} = \sigma_f'\left(2N_f\right)^b \tag{3}$$

where σ_{max}, σ_m, and σ_a are ultimate tensile strength, mean stress, and stress amplitude, respectively. In our previous work, a part of fatigue lives of harmonic-structured SUS316L steels were obtained under pulsating cyclic stress loading [24]. Although both mean stress and stress amplitude were changed in these experiments, the fatigue-life data can also be used to fit the parameters of the SWT fatigue-life-prediction model. Therefore, all fatigue life data under asymmetric stress cycling were used to analyze the fatigue life of the harmonic-structured SUS316L steels.

Figure 7 shows the relationship between equivalent stress amplitude based on the SWT criterion and fatigue life. Overall, the fatigue-life deviation of harmonic-structured SUS316L steels was significant. The fatigue life decreased with an increase of equivalent stress amplitude. Compared with the CG bulk SUS316L steels, the harmonic-structured SUS316L steels had a higher fatigue strength. The fatigue strength coefficient σ_f' and fatigue exponent b in the SWT model are shown in Table 4. As indicated in Table 4, no appreciable influence of grain structure on σ_f' and b was observed. Figure 8 shows the comparison between the observed and predicted fatigue life determined from parameters of the SWT model for the MM0h compact. Due to the large deviation in fatigue life, it is seen in Figure 8a that the fatigue lives of MM0h compacts were almost located in the 5 times error band. However, as shown in Figure 8b,c, the fatigue lives of the MM10h and MM30h compacts could not be predicted well using the same parameters of the MM0h compacts. Therefore, this indicated that the deviation of the fatigue life of harmonic-structured SUS316L steels was significant, and fatigue life was influenced significantly by grain structure. The parameters for the MM0h compacts were not suitable for the MM10h and MM30h compacts.

Table 4. Tensile strength and SWT parameters of harmonic-structured SUS316L steels.

Materials	Yield Strength σ_s (MPa)	Ultimate Tensile Strength σ_{UTS} (MPa)	Fatigue Strength Coefficient σ_f'	Fatigue Exponent b
MM0h	512	721	681	−0.046
MM10h	580	776	891	−0.069
MM30h	675	794	735	−0.041
Bulk [24]	315	607	499	−0.073

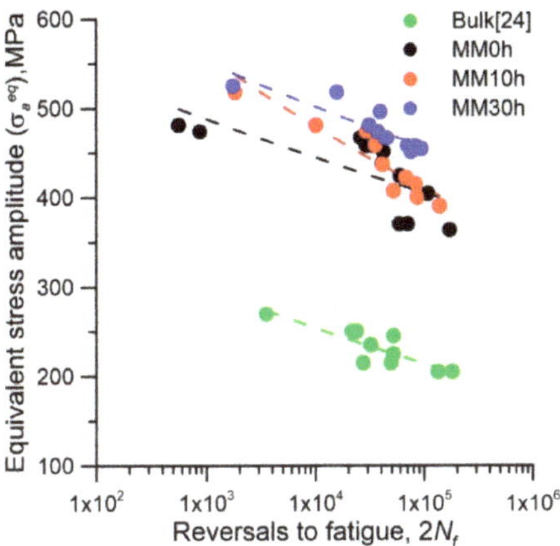

Figure 7. The dependence of the fatigue life ($2N_f$) on the equivalent stress amplitude (σ_a^{eq}) of harmonic-structured 316LN steels.

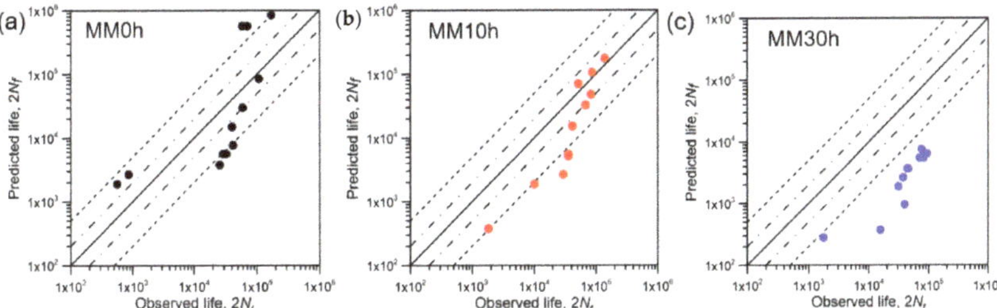

Figure 8. Comparison between observed life and predicted life based on the SWT fatigue-life-prediction model of harmonic-structured SUS316L steels: (**a**) MM0h compacts; (**b**) MM10h compacts; (**c**) MM30h compacts.

As mentioned, the strength of harmonic-structured SUS316L steels increased with increasing shell volume fraction, which also enhanced the fatigue resistance. Therefore, the strength variation caused by different grain sizes needs to be considered in fatigue-life prediction. To improve the response of grain structure on fatigue-life prediction of harmonic structured SUS316L steels, a modified equivalent stress amplitude for the SWT criterion is proposed in Equation (4):

$$\sigma_a^{eq*} = \sigma_a^{eq}\left(1 - \frac{\sigma_u^* - \sigma_{u,\,MM0h}}{\sigma_{u,\,MM0h}}\right) \tag{4}$$

where σ_a^{eq} and σ_a^{eq*} are the as-reserved equivalent stress amplitude determined by the SWT equation and the modified equivalent stress amplitude of the MM10h or MM30h compacts. σ_u^* is the ultimate tensile strength of the MM10h compact or the MM30h compact. $\sigma_{u,\,MM0h}$ is the ultimate tensile strength of the MM0h compact. Therefore, by introducing the effect of the enhanced ultimate tensile strength into the SWT criterion, the fatigue-life-prediction model for the harmonic-structured SUS316L steels under asymmetric stress cycling is proposed in Equation (5). The same parameters σ_f' and b for MM0h compacts are also used in the modified model:

$$\sigma_a^{eq*} = \sigma_a^{eq}\left(1 - \frac{\sigma_u^* - \sigma_{u,\,MM0h}}{\sigma_{u,\,MM0h}}\right) = \sigma_f'\left(2N_f\right)^b \tag{5}$$

The comparison between observed fatigue life and predicted fatigue life determined by the modified model based on the SWT criterion is presented in Figure 9. As indicated in Figure 9b,c, most of the fatigue-life points of the harmonic-structured SUS316L steels were located in the 5 times error band. The predicted results for the MM10h compacts and MM30h compacts by the modified model was better than that predicted by the as-received SWT fatigue-life-prediction model.

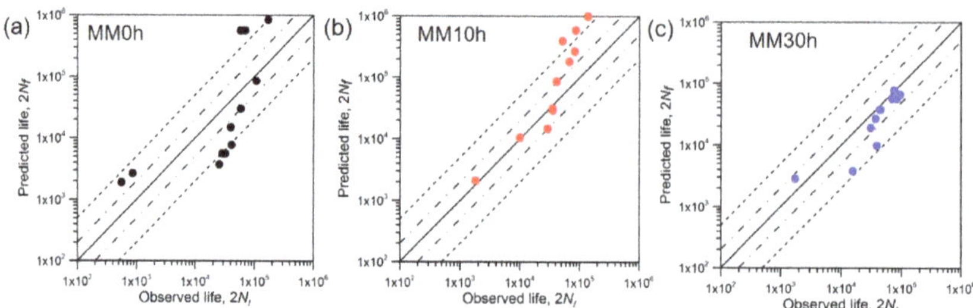

Figure 9. Comparison between observed life and predicted life based on the modified SWT fatigue-life-prediction model of harmonic-structured SUS316L steels: (**a**) MM0h compacts; (**b**) MM10h compacts; (**c**) MM30h compacts.

4. Conclusions

In the present work, the ratcheting-fatigue behavior of harmonic-structured SUS316L stainless steels was investigated under asymmetrical stress cycling at room temperature. The effects of grain structure and mean stress on ratcheting strain and fatigue life were analyzed. The main conclusions are summarized as follows:

(1) The ratcheting behavior of the harmonic-structured SUS316L steels subjected to under asymmetrical stress cycling was significant. The ratcheting strain was produced rapidly in the early stage of cyclic deformation, and became stable in the following stage. Finally, it increased near fatigue failure. The ratcheting-strain accumulation was highly influenced by grain structure and mean stress level. A lower grain size or higher mean stress tended to induce a higher ratcheting strain. The increased shell volume fraction enhanced the strength of materials, which restrained the ratcheting-strain accumulation under the same stress level.

(2) The ratcheting strain had detrimental effects on the fatigue life of the harmonic-structured SUS316L steels. A rapid ratcheting-strain accumulation caused a shorter fatigue life. As the shell volume fraction increased or mean stress decreased, the ratcheting strain was restricted, which prolonged the fatigue life.

(3) A proposed fatigue-life model based on the SWT criterion was employed to predict the fatigue life of the harmonic-structured SUS316L steels under asymmetrical stress cycling. The variation of ultimate tensile strength induced by grain refinement was considered in the modified fatigue-life-prediction model, which showed good predicting accuracy. Most of the fatigue-life points were located in the 5 times error band.

(4) Compared with homogeneous CG materials, the harmonic-structure-designed materials had a good balance of high strength and high ductility. Meanwhile, the method also had great benefits for improving fatigue resistance under asymmetrical stress cycling.

Author Contributions: Conceptualization, Z.Z. and K.A.; methodology, Z.Z.; software, Y.S.; validation, Z.Z.; formal analysis, Z.Z.; investigation, Z.Z.; resources, Y.S., H.M. and M.N.; data curation, Y.S. and H.M.; writing—original draft preparation, Y.S.; writing—review and editing, Z.Z. and M.O.K.; visualization, Z.Z.; supervision, Z.Z. and K.A.; project administration, Z.Z., M.O.K. and K.A.; funding acquisition, Z.Z. All authors have read and agreed to the published version of the manuscript.

Funding: This work was partially supported by the National Key Research and Development Program of China (2018YFC0808600), and the National Natural Science Foundation of China (52075368, 51605325).

Institutional Review Board Statement: Not applicable.

Informed Consent Statement: Not applicable.

Data Availability Statement: Data is contained within the article.

Acknowledgments: The authors would like to thank Xu Chen and Chao Zhang (School of Chemical Engineering and Technology, Tianjin University) for their kind help in mechanical testing and modification of the fatigue-life-prediction model.

Conflicts of Interest: The authors declare no conflict of interest.

References

1. Kikuchi, S.; Nukui, Y.; Nakatsuka, Y.; Nakai, Y.; Nakatani, M.; Kawabata, M.O.; Ameyama, K. Effect of bimodal harmonic structure on fatigue properties of austenitic stainless steel under axial loading. *Int. J. Fatigue* **2019**, *127*, 222–228. [CrossRef]
2. Chadha, K.; Tian, Y.; Spray, J.G.; Aranas, C. Effect of Annealing Heat Treatment on the Microstructural Evolution and Mechanical Properties of Hot Isostatic Pressed 316L Stainless Steel Fabricated by Laser Powder Bed Fusion. *Metals* **2020**, *10*, 753. [CrossRef]
3. Zheng, R.; Liu, M.; Zhang, Z.; Ameyama, K.; Ma, C. Towards strength-ductility synergy through hierarchical micro-structure design in an austenitic stainless steel. *Scr. Mater.* **2019**, *169*, 76–81. [CrossRef]
4. Yuan, X.; Yu, W.; Fu, S.; Yu, D.; Chen, X. Effect of mean stress and ratcheting strain on the low cycle fatigue behavior of a wrought 316LN stainless steel. *Mater. Sci. Eng. A* **2016**, *677*, 193–202. [CrossRef]
5. Li, B.; Zheng, Y.; Shi, S.; Zhang, Z.; Chen, X. Cyclic deformation behavior and failure mechanism of S32205 duplex stainless steel under torsional fatigue loadings. *Mater. Sci. Eng. A* **2020**, *786*, 139443. [CrossRef]
6. Li, B.; Zheng, Y.; Shi, S.; Zhang, Z.; Chen, X. Cyclic deformation and cracking behavior of 316LN stainless steel under thermomechanical and isothermal fatigue loadings. *Mater. Sci. Eng. A* **2020**, *773*, 138866. [CrossRef]
7. Ueno, H.; Kakihata, K.; Kaneko, Y.; Hashimoto, S.; Vinogradov, A. Enhanced fatigue properties of nanostructured austenitic SUS 316L stainless steel. *Acta Mater.* **2011**, *59*, 7060–7069. [CrossRef]
8. Ovid'Ko, I.; Valiev, R.; Zhu, Y. Review on superior strength and enhanced ductility of metallic nanomaterials. *Prog. Mater. Sci.* **2018**, *94*, 462–540. [CrossRef]
9. Lei, Y.; Wang, Z.; Xu, J.; Lu, K. Simultaneous enhancement of stress- and strain-controlled fatigue properties in 316L stainless steel with gradient nanostructure. *Acta Mater.* **2019**, *168*, 133–142. [CrossRef]
10. Pan, Q.; Lu, L. Improved fatigue resistance of gradient nanograined metallic materials: Suppress strain localization and damage accumulation. *Scr. Mater.* **2020**, *187*, 301–306. [CrossRef]
11. Pan, Q.; Long, J.; Jing, L.; Tao, N.; Lu, L. Cyclic strain amplitude-dependent fatigue mechanism of gradient nanograined Cu. *Acta Mater.* **2020**, *196*, 252–260. [CrossRef]
12. Mine, Y.; Katashima, S.; Ding, R.; Bowen, P.; Takashima, K. Fatigue crack growth behaviour in single-colony lamellar structure of Ti–6Al–4V. *Scr. Mater.* **2019**, *165*, 107–111. [CrossRef]
13. Chakraborti, P.; Mitra, M. Room temperature low cycle fatigue behaviour of two high strength lamellar duplex ferrite?martensite (DFM) steels. *Int. J. Fatigue* **2005**, *27*, 511–518. [CrossRef]
14. Nalla, R.K.; Boyce, B.L.; Campbell, J.P.; Peters, J.O.; Ritchie, R.O. Influence of microstructure on high-cycle fatigue of Ti-6Al-4V: Bimodal vs. Lamellar structures. *Metall. Mater. Trans. A* **2002**, *31A*, 899–918. [CrossRef]
15. Koyama, M.; Zhang, Z.; Wang, M.; Ponge, D.; Raabe, D.; Tsuzaki, K.; Noguchi, H.; Tasan, C.C. Bone-like crack re-sistance in hierarchical metastable nanolaminate steels. *Science* **2017**, *355*, 1055–1057. [CrossRef] [PubMed]
16. Orlov, D.; Ameyama, K. Critical Assessment 37: Harmonic structure materials—Idea, status and perspectives. *Mater. Sci. Tech-Lond.* **2020**, *36*, 517–526. [CrossRef]
17. Wu, X.; Zhu, Y. Heterogeneous materials: A new class of materials with unprecedented mechanical properties. *Mater. Res. Lett.* **2017**, *5*, 527–532. [CrossRef]
18. Zhang, Z.; Vajpai, S.K.; Orlov, D.; Ameyama, K. Improvement of mechanical properties in SUS304L steel through the control of bimodal microstructure characteristics. *Mater. Sci. Eng. A* **2014**, *598*, 106–113. [CrossRef]
19. Zhang, Z.; Orlov, D.; Vajpai, S.K.; Tong, B.; Ameyama, K. Importance of Bimodal Structure Topology in the Control of Mechanical Properties of a Stainless Steel. *Adv. Eng. Mater.* **2014**, *17*, 791–795. [CrossRef]
20. Vajpai, S.K.; Ota, M.; Watanabe, T.; Maeda, R.; Sekiguchi, T.; Kusaka, T.; Ameyama, K. The Development of High Performance Ti-6Al-4V Alloy via a Unique Microstructural Design with Bimodal Grain Size Distribution. *Met. Mater. Trans. A* **2015**, *46*, 903–914. [CrossRef]
21. Vajpai, S.K.; Sawangrat, C.; Yamaguchi, O.; Ciuca, O.P.; Ameyama, K. Effect of bimodal harmonic structure design on the deformation behavior and mechanical properties of Co-Cr-Mo alloy. *Mater. Sci. Eng. C* **2016**, *58*, 1008–1015. [CrossRef] [PubMed]

22. Li, G.; Liu, M.; Lyu, S.; Nakatani, M.; Zheng, R.; Ma, C.; Li, Q.; Ameyama, K. Simultaneously enhanced strength and strain hardening capacity in FeMnCoCr high-entropy alloy via harmonic structure design. *Scr. Mater.* **2021**, *191*, 196–201. [CrossRef]
23. Dirras, G.; Ueda, D.; Hocini, A.; Tingaud, D.; Ameyama, K. Cyclic shear behavior of conventional and harmonic structure-designed Ti-25Nb-25Zr β-titanium alloy: Back-stress hardening and twinning inhibition. *Scr. Mater.* **2017**, *138*, 44–47. [CrossRef]
24. Zhang, Z.; Ma, H.; Zheng, R.; Hu, Q.; Nakatani, M.; Ota, M.; Chen, G.; Chen, X.; Ma, C.; Ameyama, K. Fatigue be-havior of a harmonic structure designed austenitic stainless steel under uniaxial stress loading. *Mater. Sci. Eng. A* **2017**, *707*, 287–294. [CrossRef]
25. Zhou, G.; Ma, H.; Zhang, Z.; Sun, J.; Wang, X.; Zeng, P.; Zheng, R.; Chen, X.; Ameyama, K. Fatigue crack growth behavior in a harmonic structure designed austenitic stainless steel. *Mater. Sci. Eng. A* **2019**, *758*, 121–129. [CrossRef]
26. Kikuchi, S.; Mori, T.; Kubozono, H.; Nakai, Y.; Kawabata, M.O.; Ameyama, K. Evaluation of near-threshold fatigue crack propagation in harmonic-structured CP titanium with a bimodal grain size distribution. *Eng. Fract. Mech.* **2017**, *181*, 77–86. [CrossRef]
27. Kang, G. Ratchetting: Recent progresses in phenomenon observation, constitutive modeling and application. *Int. J. Fatigue* **2008**, *30*, 1448–1472. [CrossRef]
28. Lim, C.-B.; Kim, K.; Seong, J. Ratcheting and fatigue behavior of a copper alloy under uniaxial cyclic loading with mean stress. *Int. J. Fatigue* **2009**, *31*, 501–507. [CrossRef]
29. Lin, Y.; Chen, X.-M.; Chen, G. Uniaxial ratcheting and low-cycle fatigue failure behaviors of AZ91D magnesium alloy under cyclic tension deformation. *J. Alloy. Compd.* **2011**, *509*, 6838–6843. [CrossRef]
30. Paul, S.K.; Sivaprasad, S.; Dhar, S.; Tarafder, S. Ratcheting and low cycle fatigue behavior of SA333 steel and their life prediction. *J. Nucl. Mater.* **2010**, *401*, 17–24. [CrossRef]
31. Xia, Z.; Kujawski, D.; Ellyin, F. Effect of mean stress and ratcheting strain on fatigue life of steel. *Int. J. Fatigue* **1996**, *18*, 335–341. [CrossRef]
32. Zheng, R.; Zhang, Z.; Nakatani, M.; Ota, M.; Chen, X.; Ma, C.; Ameyama, K. Enhanced ductility in harmonic structure designed SUS316L produced by high energy ball milling and hot isostatic sintering. *Mater. Sci. Eng. A* **2016**, *674*, 212–220. [CrossRef]
33. Gaudin, C.; Feaugas, X. Cyclic creep process in AISI 316L stainless steel in terms of dislocation patterns and internal stresses. *Acta Mater.* **2004**, *52*, 3097–3110. [CrossRef]

Article

Exploring the Strain Hardening Mechanisms of Ultrafine Grained Nickel Processed by Spark Plasma Sintering

Lucía García de la Cruz [1], Mayerling Martinez Celis [1], Clément Keller [2] and Eric Hug [1,*]

[1] Laboratoire CRISMAT, UNICAEN, Normandie University, CNRS, 6 Bvd du Maréchal Juin, 14050 Caen, France; lucia.garcia9@um.es (L.G.d.l.C.); mayerling.martinez@ensicaen.fr (M.M.C.)

[2] Groupe de Physique des Matériaux, CNRS-UMR6634, Université de Rouen, INSA de Rouen, Avenue de l'Université, 76800 Saint-Etienne du Rouvray, France; clement.keller@insa-rouen.fr

* Correspondence: eric.hug@ensicaen.fr

Abstract: Ultrafine grained (UFG) materials in the bigger grain size range (0.5–1) µm display a good combination of strength and ductility, unlike smaller size UFG and nanostructured metals, which usually exhibit high strength but low ductility. Such difference can be attributed to a change in plasticity mechanisms that modifies their strain hardening capability. The purpose of this work is to investigate the work hardening mechanisms of UFG nickel considering samples with grain sizes ranging from 0.82 to 25 µm. Specimens processed combining ball milling and spark plasma sintering were subjected to monotonous tensile testing up to fracture. Then, microstructural observations of the deformed state of the samples were carried out by electron backscattered diffraction and transmission electron microscopy. A lower strain hardening capability is observed with decreasing grain size. Samples in the submicrometric range display the three characteristic stages of strain hardening with a short second stage and the third stage beginning soon after yielding. Microstructural observations display a low fraction of low angle grain boundaries and dislocation density for the sample with d = 0.82 µm, suggesting changes in plasticity mechanisms early in the UFG range.

Keywords: nickel; spark plasma sintering; ultrafine grained microstructure; plasticity mechanisms; deformed state; dislocation structures

Citation: García de la Cruz, L.; Celis, M.M.; Keller, C.; Hug, E. Exploring the Strain Hardening Mechanisms of Ultrafine Grained Nickel Processed by Spark Plasma Sintering. *Metals* **2021**, *11*, 65. https://doi.org/10.3390/met11010065

Received: 9 December 2020
Accepted: 28 December 2020
Published: 30 December 2020

Publisher's Note: MDPI stays neutral with regard to jurisdictional claims in published maps and institutional affiliations.

1. Introduction

Ultrafine grained (UFG) metals (with an average grain diameter d = 0.1–1 µm), situated between nanostructured metals (NsM) (d < 0.1 µm) and coarse grained (CG) materials display intriguing mechanical properties and plasticity mechanisms [1–6]. In CG polycrystals plastic deformation is controlled by dislocation dynamics, where dislocation sources can be found in grain boundaries and the grain core [7]. Meanwhile, in UFG/NsM metals, grain boundaries have been reported to be exclusively the dislocation sources and sinks, limiting their mean free path and consequently modifying their mechanical properties and hardening behavior [8,9]. The investigation of strain hardening in UFG/NsM metals is not straightforward for different reasons. The low ductility often displayed by such materials [10–12], with fracture or heterogeneous deformation after yielding, renders the study of the plasticity mechanisms and the determination of the classical stages of strain hardening challenging. In addition, the use of nonstandard tensile specimens [6,13,14] can generate confusing results if not analyzed correctly. F. Dalla Torre et al. [15] succeeded in the investigation of the evolution of the strain hardening rate and the hardening stages of two nanostructured Ni samples processed by electrodeposition (d = 21 ± 9 nm) and by high pressure torsion (d = 105 ± 69 nm). Both samples displayed a lower hardening rate than CG Ni, characterized by a steep decrease in the beginning of plastic deformation, and no transition between the first and second stages of work hardening. The reduced strain hardening capability has been explained in terms of a lower ability of the specimens to accumulate dislocations, which act as obstacles to dislocation glide. It has been reported [8,9]

75

that in the UFG/NsM regime, dislocation storage is lessened by dynamic recovery and annihilation of dislocations at grain boundaries. Considering the continuous decrease in hardening rate with increasing strain for NsM, some authors consider that the third stage is directly reached whereas others suggest that no correlation exists between the hardening stages in conventional polycrystals and in NsM/UFG materials [15,16]. To help clarify these different positions, studies addressing the microstructure of NsM/UFG materials after tensile deformation are needed.

Despite the large bibliography concerning the mechanical properties of NsM and the smaller grain size range of UFG metals developed in the last three decades [17–24], information concerning strain hardening of larger grain sized UFG metals (also called submicrometric metals) is still lacking. The study of such samples could bring light to the transition, in terms of plasticity mechanisms, between conventional CG metals and NsM. Investigations on the mechanical properties of UFG copper [25], aluminum [26] and austenitic stainless steel [27] display an intriguing transition of tensile deformation behavior in samples with grain sizes in the range 0.4–4 µm. In this face centered cubic metals, a Lüders-type deformation, preceded in some cases by a yield drop, was observed in the beginning of plastic deformation. Moreover, the yield strength of specimens displaying the Lüders-type deformation was reported higher than the expected value from the Hall–Petch relation for aluminum by Yu et al. [26]. It was shown that such deviation corresponds to inhomogeneous yielding associated with a lack of mobile dislocations in the smaller grains of the UFG microstructures in the beginning of plastic deformation. This feature has also been observed by the authors in a previous article dedicated to the mechanical properties assessment of UFG Ni polycrystals [28]. The persistence of the deviation in UFG Ni, at a lower extent, at higher deformation states, suggests changes in the strain hardening mechanisms.

A large bibliography is available for pure Ni, often used as model material to investigate different properties such as diffusivity, mechanical strength, resistivity, etc. [29–33]. The lack of complex phases, precipitates or other structural features simplifies the study of such properties enabling the investigation of the effect of single parameters (grain size for instance). Thus, Ni was chosen as a starting point to understanding plastic deformation in UFG materials that are more complex.

This paper aims to study the evolution of strain hardening mechanisms for samples with grain sizes in the range 0.82–25 µm and to contribute to the knowledge of plastic deformation in the UFG range. Uniaxial tensile testing up to fracture was performed using standardized specimens and the obtained results analyzed focusing in strain hardening. Finally, electron backscattered diffraction (EBSD) and transition electron microscopy (TEM) were used to examine microstructures deformed at fracture and investigate dislocation accumulation and structures, shedding light on the hardening mechanisms.

2. Materials and Methods

2.1. Sample Synthesis and Characterization

Powder metallurgy was chosen for the synthesis of the specimens, combining high energy ball milling (BM) and consolidation by spark plasma sintering (SPS), to produce specific fine grained (FG) and UFG microstructures. High purity nickel powder (>99.8%) was used as the starting material, with Fe 100 ppm, C 200–600 ppm, O 1000 ppm and S 10 ppm as the main impurities. To produce FG/UFG samples the as-received powder was modified by BM in a Fritsh planetary pulverisette 7, using WC vials (80 mL) and grinding balls (5 mm in diameter). A ball to powder ratio was fixed to 10:1 and anhydrous methanol was used as a process control agent. To avoid oxidation of the powders, vials were filled with high purity argon. Rotation speed and milling time were selected to produce highly deformed UFG powders with different characteristics, to process specimens with distinct microstructures after sintering. The synthesis parameters and grain size for each sample are displayed in Table 1. More information concerning the impact of BM parameters in the microstructure of consolidated samples can be found elsewhere [28].

Table 1. Synthesis parameters of the samples (BM: ball milling).

Sample	Grain Size (μm)	ρ_{rel} (%)	Type of Powder	BM Time (h)	BM Speed (rpm)	Amount of Methanol (%)	Sintering Temperature (K)
1	0.82 ± 0.67	98.5	BM	10	350	66.7	1023
2	1.11 ± 0.84	99.5	BM	3	350	16.7	1023
3	1.39 ± 1.11	99.3	BM	3	300	66.7	1023
4	4.0 ± 2.4	97.6	As-received	–	–	–	1023
5	25 ± 17	98.7	As-received	–	–	–	1273

The powders were sintered by SPS with a FCT System GmbH, HD25 SPS apparatus (Figure 1a left), under medium dynamic vacuum (0.5 mbar), using graphite dies, punches and foils (Figure 1a right). To limit grain growth, the sintering temperature was set to 1023 K, with a heating rate of 100 K/min by continuous current. The heating cycle was accompanied by a uniaxial pressure cycle throughout the sintering process, with a maximum value of 75 MPa. Maximum pressure and temperature were maintained for 10 min to ensure high density. Four disc-like specimens were processed using this program, with a diameter of approximately 50 mm and a thickness of 6 mm, to extract tensile specimens for mechanical testing (Figure 1b). To assess the impact of grain size on strain hardening mechanisms, an extra sample with a grain size in the CG range was produced from the as-received powder using the same sintering conditions, with the exception of temperature, 1273 K (Table 1). To confirm good homogeneity, hardness maps of the surface were performed in all samples. The density of the bulk samples was determined by an immersion densitometry method, with a mass accuracy of 1 mg.

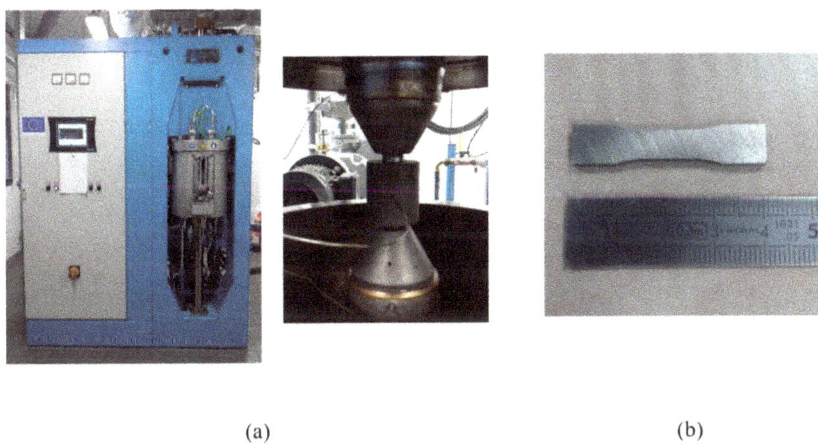

(a) (b)

Figure 1. FCT System GmbH, HD25 SPS apparatus (**left**) and sintering set-up (**right**) (**a**). The metallic wire positioned in a hole in the die is a thermocouple used to control the sintering temperature. Tensile specimen extracted from a 50 mm sintered disc (**b**).

2.2. Microstructure Analysis and Mechanical Properties

The microstructure of the sintered samples was investigated in the initial state and the deformed state at fracture by means of EBSD. Prior to measurements, samples were mechanically polished with SiC paper and electropolished using a solution of perchloric acid, 2-butoxyethanol and ethanol at a voltage of 24 V, for an optimal surface condition. The EBSD acquisition system used was a QUANTAX EBSD (Bruker, Berlin, Germany) mounted on a Zeiss Supra 55. An operating voltage of 15 kV, aperture size of 60 μm and a working distance (WD) of 14.5–16.5 mm were the acquisition parameters. For each sample, at least 1000 grains were covered using a step size ensuring more than 4 points per grain along a given direction. Data was analyzed with the TSL OIM 6 × 64 Analysis™ v 6.2 software (EDAX, Ametek Inc., Berwyn, PA USA) to estimate average grain size, grain boundary character distribution (GBCD), grain orientation, grain orientation spread (GOS) and Kernel average misorientation (KAM) of the sintered specimens. A misorientation angle of 2–15° [34] was chosen to define low angle grain boundaries (LAGBs), and Brandon's criterion [35] was used to determine coincidence site lattice (CSL) fractions, calculated on a length basis. Basic clean-up was performed before each analysis to eliminate artifacts. In addition, TEM observations of dislocation structures were performed on selected samples with a JEOL 2010 at an operating voltage of 200 kV. To this aim, specimens were prepared from 3 mm discs mechanically polished down to approximately 100 µm and then electropolished using a solution of 17% perchloric acid in ethanol by the twin jet method in a Tenupol 5.

The influence of grain size in the UFG range on the strain hardening mechanisms was investigated using uniaxial tensile testing up to fracture. Dog-bone tensile samples with a gauge length of 11 mm and a thickness of 1 mm were extracted from the 50 mm discs (Figure 1b), using electrical discharge machining to avoid sample damage. More information concerning the dimensions of the tensile specimens and their position in the sintered sample can be found elsewhere [36,37]. Prior to testing the surfaces of the specimens were carefully polished with a diamond suspension of 6 µm to remove the roughness induced by the machining and to eliminate any contamination in the near surface area. A Zwick device with a load cell of 50 kN maximum capability was used for uniaxial tensile testing at room temperature. The strain rate was set to 10^{-3} s^{-1}, and the deformation was recorded with the ARAMIS® software (Mentel Co., Ltd., Bangkok, Thailand) by measuring displacement fields using digital image correlation. The stress–strain measurements reported in the following paragraphs correspond to the average value from three different samples.

3. Results and Discussion

3.1. Strain Hardening in FG and UFG Ni

To investigate the influence of grain size in the FG/UFG range on the mechanisms of plasticity, five samples were prepared with grain sizes ranging from 0.82 to 25 µm. The synthesis parameters (Table 1) were selected to produce specimens with a small fraction of low angle grain boundaries (LAGBs) and a high fraction of Σ3 grain boundaries, and random high angle boundaries (HABs) [28]. The high relative density for all samples (Table 1) confirms the good consolidation of powders by SPS. In addition, specimens show overall low internal stress in the initial state, quantitatively estimated from a grain orientation spread (GOS) below 1° [28,38].

The respective true stress (σ)/true strain (ε) curves are plotted in Figure 2. Regardless the grain size, all samples display high strength and ductility in tensile deformation, with the best combination found for sample 1 (d = 0.82 µm).

As discussed in our previous work [28,37], samples 1 (d = 0.82 µm) and 2 (d = 1.11 µm), display an unusual Lüders-type plateau in the beginning of plastic deformation. The origin of this phenomenon seems to be an inhomogeneous yielding associated with a microstructure displaying grains of different sizes, which present specific deformation behaviors. The lack of mobile dislocations in the smallest grains of the UFG microstructures increases the yielding strength, followed by a Lüders-type deformation before homogeneous deformation begins. Despite the differences in grain size, covering three ranges, CG, FG and UFG, all samples display considerable strain hardening after yielding. However, if the ratio of maximum stress over yield stress is considered, bigger grain size samples display higher ratios with a value of seven observed for sample 5 (d = 25 µm), against a ratio of two displayed by sample 1 (d = 0.82 µm).

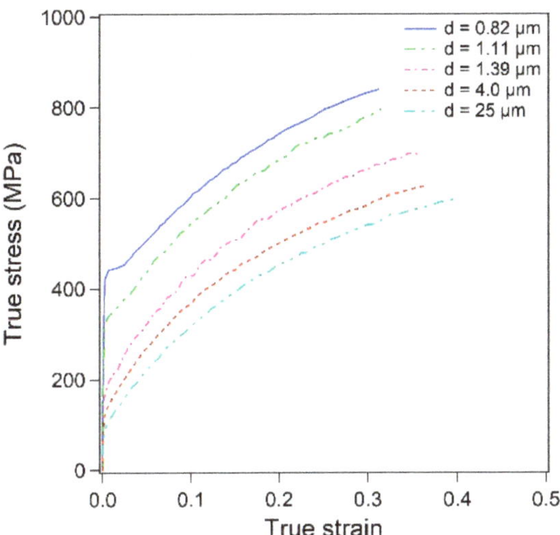

Figure 2. True stress–true strain tensile curves for Ni samples 1–5 processed by spark plasma sintering (SPS).

The evolution of the strain hardening rate $\theta = \left(\frac{d\sigma}{d\varepsilon}\right)$, with true stress normalized by yield strength (σ_y) is plotted in Figure 3. At low flow stress, a strong decrease in the hardening rate can be observed, which evolves towards a more stabilized hardening rate, irrespective of the sample. However, a clear difference can be observed between UFG and FG/CG samples. Samples 3–5 display the same hardening behavior at the beginning of plastic deformation, a steep decrease in θ followed by a smoother decrease. Then, for $\sigma/\sigma_y = 2$, hardening rates diverge as sample 5 (d = 25 μm) displays a higher hardening capability than samples 3 (d = 1.39 μm) and 4 (d = 4.0 μm). Samples 1 (d = 0.82 μm) and 2 (d = 1.11 μm) display, on the other hand, a much steeper decrease of θ in the beginning of plastic deformation followed by a much lower strain hardening rate, compared to the FG/CG counterparts. The lower hardening capability assessed for the UFG samples in this study, is in agreement with the tendency observed by F. Dalla Torre et al. [15]. Nevertheless, the transition to a more stabilized decrease in θ for UFG, contrasting with their observations on nanocrystalline Ni, suggests different hardening mechanisms in our samples.

Work hardening in polycrystals can be divided into three stages that are related to different plasticity mechanisms that can be investigated by physical models [39]. Using the Mecking–Kocks model, the relation between flow stress (σ) and total dislocation density (ρ) can be expressed as [40–42]:

$$\sigma = \alpha\mu b \mathrm{M}\sqrt{\rho}, \tag{1}$$

where α is a coefficient that depends on the type of interaction between dislocations and their configuration, μ is the shear modulus, b is the norm of the Burgers vector and M is the Taylor factor [43,44]. Furthermore, for monophasic and recrystallized face centered cubic metals, the evolution of ρ with deformation can be described by the following equation [40,45,46]:

$$\frac{d\rho}{d\varepsilon} = \frac{\mathrm{M}\sqrt{\rho}}{\beta b} + \frac{k_g \mathrm{M}}{bd} - \frac{2MPy_a\rho}{b}. \tag{2}$$

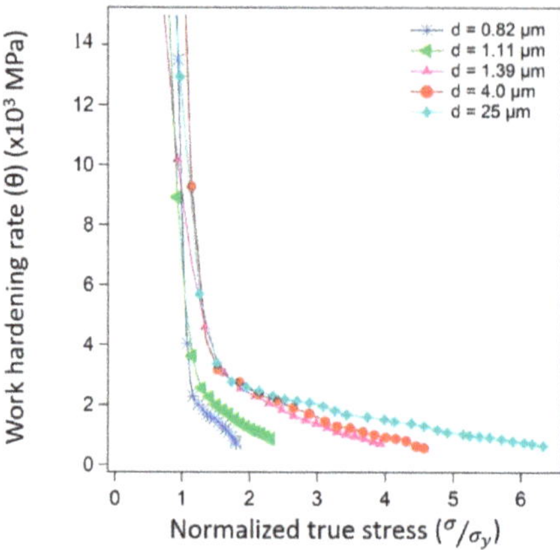

Figure 3. Evolution of the work hardening rate ($\theta = \left(\frac{d\sigma}{d\varepsilon}\right)$), determined from tensile testing to failure at a strain rate of 10^{-3} s^{-1}, with true stress normalized by yield strength for Ni samples prepared by SPS with different grain sizes. The average grain size of the samples is indicated in a color scale for each figure.

The first term of the right-hand side of Equation (2) concerns the accumulation of mobile dislocations, where β represents the ratio between the dislocation mean free path (Λ) and the near neighbor dislocation spacing (l_d): $\beta = \frac{\Lambda}{l_d}$, and the other parameters have their usual meaning. The dislocation mean free path is the distance travelled by a dislocation before being stopped by crystalline defects (i.e., other dislocations, precipitates and grain boundaries) and becomes stored [47]. The second term accounts for the generation of geometrically necessary dislocations (GNDs) at grain boundaries [46,48], where k_g represents a geometric factor that depends on the morphology of grains. Finally, the last term corresponds to dislocation annihilation that results from generalized cross-slip. The probability of annihilation is represented by P and y_a symbolizes the distance of annihilation of two dislocations by cross-slip [46,48]. If Equations (1) and (2) are combined the following expression is obtained:

$$\sigma \cdot \theta = \sigma \frac{d\sigma}{d\varepsilon} = \frac{\alpha \mu M^2}{2\beta}\sigma + \frac{\alpha^2 \mu^2 b M^3 k_g}{2d} - \frac{\alpha \mu M P y_a \sigma^2}{b}. \tag{3}$$

The low probability of dislocation annihilation in the beginning of strain hardening makes the last term in Equation (3) negligible throughout stages I and II. Thus, the study of the transition between hardening stages can be performed from the evolution of the product $\sigma \cdot \theta$ as it displays clear slope changes. Stage I is characterized by a steep decrease in $\sigma \cdot \theta$ and is characterized by non-homogeneous deformation with a gradual elastic to plastic transition, which depends on grain orientation and size. Planar/single slip is the main deformation mechanism and the elevated hardening rate is due to deformation incompatibility between grains. Then, a linear part of increasing $\sigma \cdot \theta$ represents stage II, which is characterized by multiple slip, beginning of cross-slip and the formation of dense dislocation structures, including tangles, dislocation walls and cells [49]. Assuming the Mecking–Kocks model and the approximation of low probability of dislocation annihilation, stage II can be investigated as dependent on two terms [50]:

$$\sigma \cdot \theta = \frac{\alpha \mu M^2}{2\beta}\sigma + \frac{\alpha^2 \mu^2 b M^3 k_g}{2d} = \Delta_{II}\sigma + (\sigma\theta)_0. \tag{4}$$

The first one (Δ_{II}), known as the latent hardening rate, is related to the accumulation and interaction between dislocations. The second one ($(\sigma\theta)_0$) corresponds, for a polycrystal of a pure single-phase metal, to the contribution of grain boundaries, through GNDs, to strain hardening. The contribution of $(\sigma\theta)_0$ to strain hardening increases when the microstructure is refined. Both terms can be obtained from the fraction of the plots of $\sigma\cdot\theta$ as a function of stress situated between the transition stress from stage I to stage III ($\sigma_{I/II}$) and the one from stage II to III ($\sigma_{II/III}$). A linear fit of this section gives Δ_{II} as the slope and $(\sigma\theta)_0$ as the origin intercept (Figure 4). Once linearity is lost, stage III of strain hardening begins, where cross-slip is generally occurring and dislocation annihilation predominantly takes place.

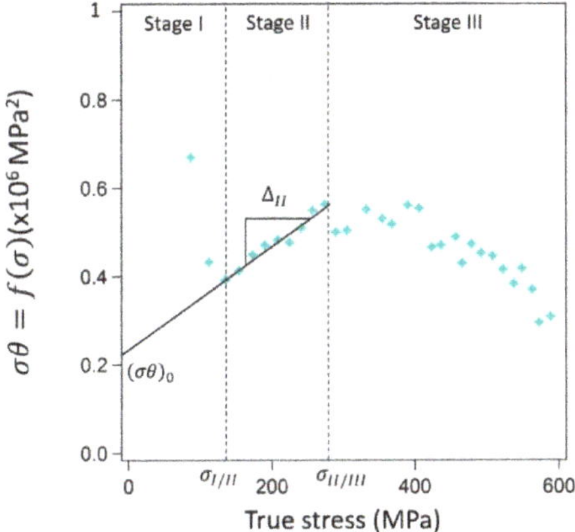

Figure 4. Example of the evolution of the $\sigma\cdot\theta$ product with stress for polycrystalline Ni ($d = 25\ \mu m$), displaying the three stages of strain hardening, the transition stresses $\sigma_{I/II}$ and $\sigma_{II/III}$, Δ_{II} and $(\sigma\theta)_0$.

Figure 5a displays the evolution of $\sigma\cdot\theta$ with stress for the sintered samples. A strong effect of grain size on $\sigma_{I/II}$ is observed, arising from the difference in σ_y between individual samples. When normalizing stress by σ_y (Figure 5b), the transition took place at a value of σ/σ_y close to 1.5 independently of the grain size.

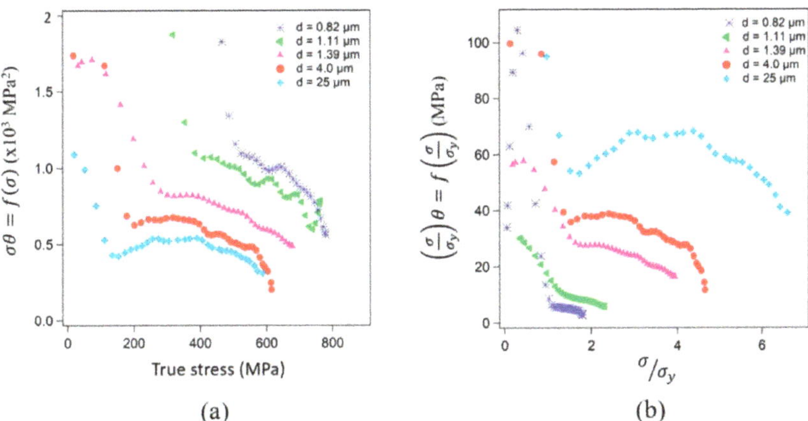

Figure 5. Evolution of the $\sigma\cdot\theta$ product with stress for the Ni samples processed by SPS with different grain sizes (**a**) and the same curves after normalization of stress by yield strength σ_y (**b**). The average grain size of the samples is indicated in a color scale for each figure.

The transition stress from stage II to stage III ($\sigma_{II/III}$), normalized by yield strength decreased with decreasing grain size (Figure 6). The stress necessary for the transition to stage III for CG nickel (sample 5) was 3.5 times its yield strength, whereas for the two UFG samples (framed in blue) the transition took place shortly after the initial yield stress. The faster transition from stage II to stage III in UFG samples suggests different hardening mechanisms compared to CG specimens.

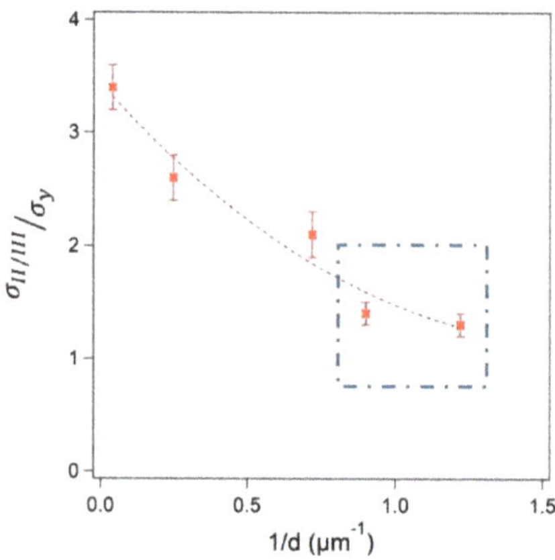

Figure 6. Evolution of the transition stress from work hardening stage II to stage III ($\sigma_{II/III}$) normalized by yield strength as a function of 1/d for the five sintered samples. The two UFG samples are framed in blue.

The evolution of Δ_{II} with grain size is plotted in Figure 7a. Despite the important scattering of experimental values, resulting from noisy data, a decrease of the latent hardening rate with decreasing grain size can be observed, approaching zero for the smallest grain sizes. The latent hardening rate is related to the rate in which mobile dislocations are stored within the grain from the interaction with the dislocation density that is already stored [43]. In the second work hardening stage, dislocations store by rearranging into low energy structures such as dislocation cells. The faster transition to stage III in UFG specimens suggests that generalized cross-slip and annihilation of dislocations takes place in the beginning of plastic deformation, reducing the possibility of dislocation interaction and reorganization. A decrease in Δ_{II} implies a decrease of the interaction coefficient, α, and/or an increase of the parameter β (Equation (4)). A lower α suggests low dislocation interaction and organization. A higher β entails an increase in Λ and/or a decrease in l_d, thus, a lower possibility of interaction between dislocations. The evolution of $(\sigma\theta)_0$ as a function of the inverse value of grain size is displayed in Figure 7b. A linear trend can be depicted, where the parameter $(\sigma\theta)_0$ increases with decreasing grain size. The value of k_g can be determined from the slope:

$$(\sigma\theta)_0 = \frac{\alpha^2\mu^2 bM^3 k_g}{2}\cdot\frac{1}{d}. \tag{5}$$

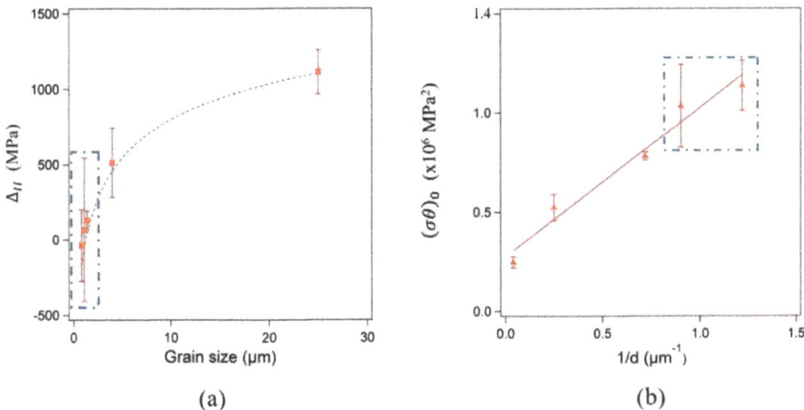

Figure 7. Evolution of the latent hardening rate (Δ_{II}) in stage II of strain hardening as a function of grain size (**a**) for the five sintered samples. Evolution of $(\sigma\theta)_0$ as a function of 1/d for the five sintered samples (**b**). The two UFG samples are framed in blue.

Considering the value of α estimated for polycrystals in [45], $\alpha = 0.45$, the obtained value of k_g in this study was 0.15 ± 0.02, which was in good agreement with the experimental value obtained in [45] for polycrystalline nickel, $k_g = 0.20$. Thus, the Mecking–Kocks model applies to the study of the plasticity mechanisms in this work [51]. Given that $(\sigma\theta)_0$ is related to the contribution of grain boundaries to strain hardening, these results suggest that for UFG samples, the hardening mechanisms mainly take place at grain boundaries. In consequence, when grain size decreases, dislocation interaction with grain boundaries becomes the main hardening mechanism.

3.2. Investigation of the Deformed State at Fracture by EBSD and TEM Observations

Previous conclusions issued from the Mecking–Kocks model suggest a change in strain hardening mechanisms with grain refinement. Dislocation accumulation through cell formation would be present in the core of the grain for coarse grain specimens. Whereas in UFG samples, hardening would be mainly related to interactions between dislocations and grain boundaries. To explore this possibility, the deformed state at fracture was investigated by means of EBSD.

Figure 8 displays a global outlook of the deformed microstructures from the inverse pole figure (IPF) taking the tensile stress direction as reference. The good quality of the data was ensured by an indexation rate above 88% for all samples. A preferred crystallographic orientation is observed where grains rotate so that a <111> or <100> axis tends to become parallel to the tensile axis, which is the expected behavior for highly deformed face centered cubic (FCC) metals [52,53]. In addition, elongation of grains following the tensile stress axis can be observed, leading to an increase in the grain boundary area. This phenomenon is more present for samples 4 and 5 depicting higher grain size (Figure 8c,d) and almost negligible for sample 1, with the smallest grain size (Figure 8a). The area increase requires a high fraction of dislocations storing in the grain boundaries, which accounts for a significant part of the strain hardening energy [54]. The difference in elongation to fracture (ε_f) could explain such a difference as $\varepsilon_f = 49\% \pm 3\%$ for sample 4 (d = 4 µm) and $\varepsilon_f = 39\% \pm 3\%$ for sample 1 [28]. Nevertheless, Q.H. Bui et al. [33] reported a similar result for samples of CG and UFG Ni deformed in compression at the same level of deformation.

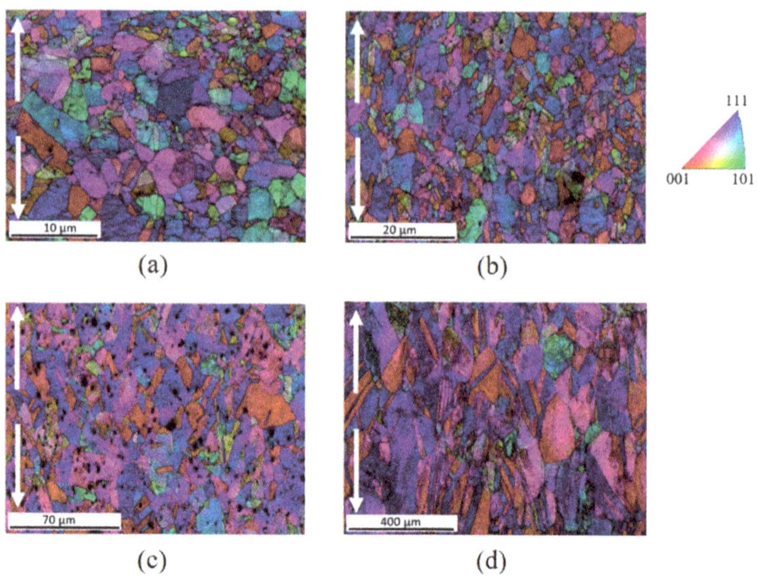

Figure 8. Superposition of inverse pole figure (IPF) + image quality (IQ) cartography for samples 1 (d = 0.82 μm) (**a**), 3 (d = 1.39 μm) (**b**), 4 (d = 4.0 μm) (**c**) and 5 (d = 25 μm) (**d**) with the tensile stress direction as reference, for fractured samples. White arrows indicate the tensile direction.

A change in the GBCD is also observed after deformation (Table 2), where a decrease in the Σ3 grain boundary density was accompanied by an important increase in the LAGB density. The lower fraction of Σ3 grain boundaries can be explained as the outcome of gliding dislocations interacting with pre-existing twin boundaries (a particular case of Σ3 grain boundaries), resulting in the accumulation of the slip dislocations in such boundaries. This can be achieved by dissociation of slip dislocations into partials under high stress [55,56]. Gubicza et al. [57] observed the same phenomenon for UFG Ni prepared by SPS from nanopowders deformed by the compression test. The higher fraction of LAGBs is the outcome of the arrangement of dislocations into subgrain boundaries at high levels of deformation [58].

Table 2. Grain boundary character distribution (GBCD) of samples in the initial state and deformed at fracture.

Sample	Grain Size (μm)	Initial State [28]				Deformed at Fracture			
		LAGBs	Σ3	CSL	HABs	LAGBs	Σ3	CSL	HABs
1	0.82 ± 0.67	0.036	0.495	0.056	0.413	0.057	0.388	0.047	0.508
2	1.11 ± 0.84	0.04	0.456	0.044	0.46	0.133	0.246	0.056	0.565
3	1.39 ± 1.11	0.023	0.535	0.057	0.385	0.093	0.336	0.044	0.527
4	4.0 ± 2.4	0.02	0.526	0.046	0.408	0.383	0.202	0.036	0.379
5	25 ± 17	0.02	0.616	0.037	0.327	0.73	0.088	0.022	0.16

Table 2 shows that bigger grain size entailed a sharper decrease in Σ3 grain boundaries and a stronger increase of LAGBs. The probability of reducing the fraction of Σ3 grain boundaries was thus higher for samples with grain sizes in the CG regime. The considerable difference in the fraction of LAGBs for fractured samples 4 and 5, compared to samples 1–3, can be related to the short strain hardening stage II in UFG samples. In this second stage, dislocations reorganize into lower energy structures to accommodate deformation. They evolve with increasing strain from tangled dislocations to dislocation cells due to a multiple slip, which enables the multiplication of dislocations. In stage III, cross-slip is generalized and dislocation cells are the predominant dislocation arrangement. For higher deformation levels, the misorientation across dislocation walls increases, which results in the development of subgrains from the original dislocation cells [59], ultimately forming LAGBs [54]. The fast occurrence of stage III in the UFG samples hinders the formation dislocation cells and their evolution to LAGBs.

EBSD is also useful in the investigation of the misorientations within grains, as this can be related to the presence of dislocations [29,38]. Figure 9 displays GOS cartographies of samples 1 (d = 0.82 μm) (a), 3 (d = 1.39 μm) (b) and 4 (d = 4 μm) (c), calculated for the initial state (left) and the deformed state at fracture (right). The average GOS value increased for all samples from 0.6°, in the initial state, to around 2°, in fractured samples, with the smallest value corresponding to sample 1 (GOS = 1.7° ± 0.9°). Nevertheless, the distribution of the GOS value highly depended on grain size. For instance, the fraction of grains with a GOS < 2° for sample 5 (d = 25 μm) was 2.9%, whereas sample 1 (d = 0.82 μm) exhibited a fraction of 29.8%.

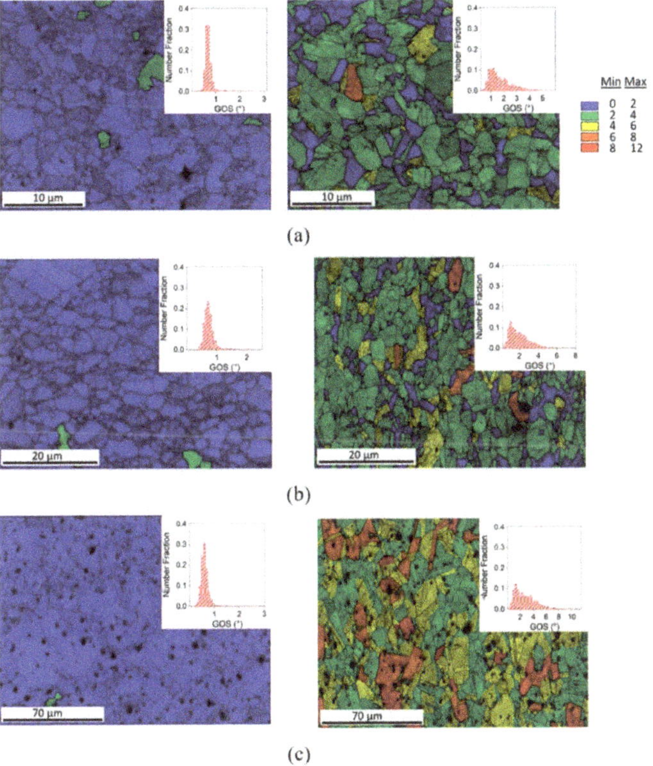

Figure 9. Grain orientation spread (GOS) + IQ map of samples 1 (d = 0.82 μm) (**a**), 3 (d = 1.39 μm) (**b**) and 4 (d = 4.0 μm) (**c**) at the initial state (left) and deformed state at fracture (right). Color legend represents the orientation spread in degrees. Inserts display the distribution of GOS values for each state and sample.

To investigate the correlation between grain size and orientation spread, the EBSD data for sample 1 at fracture (d = 0.82 μm) was divided into two groups, based on the GOS value. The first group included grains that display a GOS < 2° and the second group grains with GOS > 2°. The measured average grain sizes of each partition were d = 0.54 ± 0.38 μm and d = 1.39 ± 0.84 μm, respectively. These results indicate that deformed grains in the UFG regime display less orientation spread and thus exhibited a lower level of deformation [60].

This difference in deformation capability offers an explanation to the heterogeneous deformation (Lüders-type plateau) at the beginning of plastic deformation. In addition, Figure 9 shows that at higher deformation levels, despite the macroscopic homogeneous deformation displayed in Figure 2, the deformation at a microscopic scale is heterogeneous and dependent on grain size. The same analysis was performed in samples 2 and 3, and a similar trend was observed concerning the grains in the UFG regime. The fraction of grains that displayed a GOS < 1° (considered as non-deformed grains) was less than 3% for all samples.

GOS cartographies offer a general view of deformation considering averaged values for each grain. Nevertheless, different deformation levels are present on a microscale inside the grains, where dislocations are frequently distributed in a heterogeneous manner, alternating regions of high and low local dislocation density. KAM cartographies can be employed for a local description of the presence of misorientations within a grain [61]. The first neighbor KAM cartographies of samples 1 (d = 0.82 μm) (a), 3 (d = 1.39 μm) (b) and 4 (d = 4 μm) (c) are displayed in Figure 10. In the deformed state in sample 4 (Figure 10c right), it is noticeable that high misorientations (above 1°) are present at grain boundaries and in the bulk of grains. In addition, the average KAM value in this sample increased from 0.7° in the initial state to 1.4° deformed at fracture, as expected for deformed samples [62]. With grain refinement, lower misorientations are observed and are mostly located at grain boundaries (Figure 10a,b right). The generalized higher dislocation density at grain boundaries is in agreement with the model developed by Meyers and Ashworth based on a different deformation behavior between the inside of grains and their boundaries [63]. Deformation incompatibility between grains develops a stress concentration at grain boundaries, which entails the activation of supplementary slip systems as compared to the core of the grain [7]. Hence, regions closer to the grain boundaries will strain harden faster. This effect is particularly noticeable in the UFG grains, which display low misorientations in the core of grains. These observations are in agreement with the results obtained in the previous section and suggest hardening mechanisms mainly related to interactions between dislocations and grain boundaries. Samples 1–3 display no significant change in the average KAM value between the initial state and the deformed state.

To complete these observations, TEM observations were performed at fracture for two samples, representing the FG and the UFG range. Figure 11 displays bright field TEM images of sample 4 (d = 4.0 μm). Grains divided by geometrical dislocation cells (or subgrains) can be observed with different misorientations as depicted in Figure 11a for a single grain observed using a diffraction vector $\vec{g} = [1\bar{1}1]$. The high density of dislocations in the cell walls and cores can be observed in Figure 11b, where a grain boundary is displayed edge-on. The average size of the dislocation cells was estimated as 0.53 ± 0.18 μm, from 28 cells in different grains. This value, rather independent of grain size, is in agreement with the value of 0.35–0.55 μm observed in CG nickel [64] for a high plastic strain. The features of the deformed microstructure correspond to those expected for highly deformed nickel, characterized by high dislocation density and dislocation cells [65].

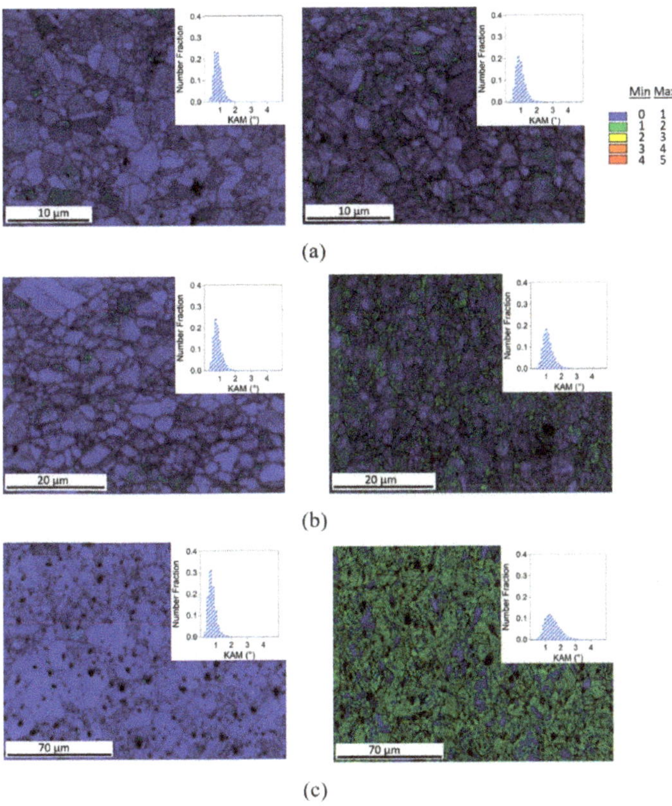

Figure 10. First neighbor kernel average misorientation (KAM)+IQ cartographies of samples 1 (d = 0.82 μm) (**a**), 3 (d = 1.39 μm) (**b**) and 4 (d = 4 μm) (**c**) in the initial state (**left**) and for fractured samples (**right**). Color legend represents the kernel average misorientation in degrees. Inserts display the distribution of first neighbor KAM values for each state and sample.

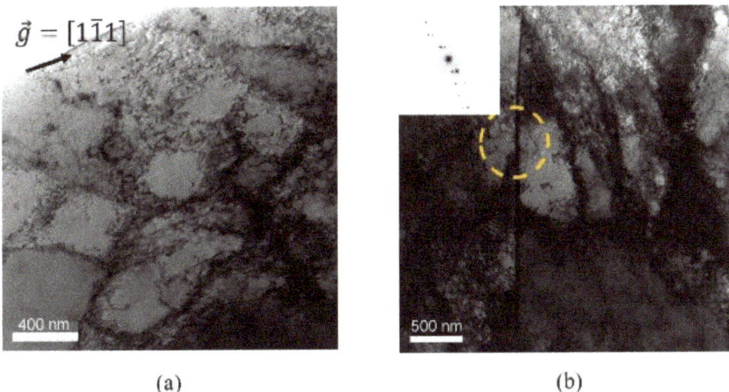

Figure 11. TEM images of grains of fractured sample 4 (d = 4.0 μm), containing dislocation cells. Accumulation of geometric dislocation cells of similar size in a single grain ($\vec{g} = [1\bar{1}1]$) (**a**) and contiguous grains displaying dislocation cells with a high density of dislocations (**b**).

Concerning the UFG regime, the deformed state of sample 1 (d = 0.82 μm) displays a much different appearance (Figure 12). Figure 12a shows an overview of small equiaxed grains with a lower dislocation density inside grains, compared with sample 4 (d = 4.0 μm) (Figure 11). Moreover, dislocations tend to accumulate preferably close to grain boundaries. The lack of dislocation walls, tangles or cells indicates low misorientation (as seen in EBSD) and hardening mechanisms related mainly to the interaction between individual dislocations. A decrease in the interactions between dislocations and dense dislocation structures implies a decrease in the value of α and thus of Δ_{II}. Small size dislocation cells were rarely found in the deformed microstructure (Figure 12b). Given the grain size (d = 0.82 μm), which approaches the cell size in CG Ni, the few dislocation cells observed display in all cases a smaller size than those observed in FG specimens. Nevertheless, TEM observations show that dislocation cells are not a characteristic feature of the deformed state of UFG Ni. These results are in agreement with the EBSD analyses, and with the strain hardening behavior of UFG samples. The lower density of dislocations and the lack of dislocation substructures hinder the hardening capability of UFG specimens and has a major impact in their strain hardening mechanisms.

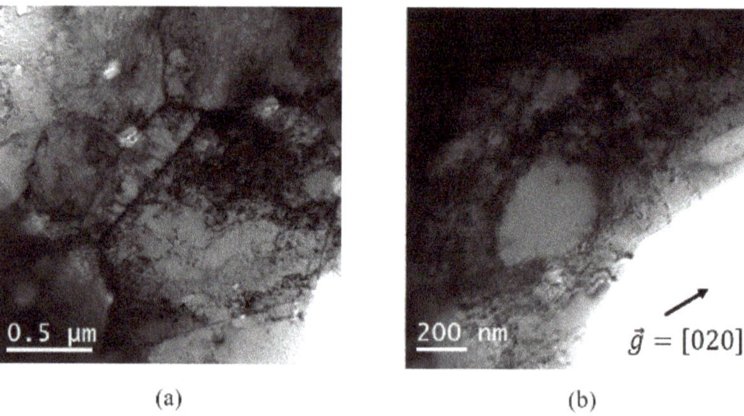

(a)   (b)

Figure 12. TEM images of sample 1 (d = 0.82 μm) deformed up to fracture. Grains with different contrasts possibly due to a high density of dislocations (**a**) and a nanometric dislocation cell ($\vec{g} = [020]$) (**b**).

4. Conclusions

Changes in the plasticity mechanisms were investigated in Ni samples obtained by SPS with grain sizes ranging from 0.82 to 25 μm, through tensile testing up to fracture and microstructural observations. The main results obtained in this study can be summarized as follows:

- Considerable strain hardening is displayed by all samples, including those with grain sizes in the UFG regime, although a decrease in the strain hardening capacity is observed with decreasing grain size.
- The three stages characteristic of plastic deformation in CG specimens were identified in the UFG samples. Considering the normalized stress σ/σ_y, the transition stress from stage I to II was similar in all samples, whereas a very short stage II was displayed by the UFG samples, entering in stage III just after initial yielding.
- A decrease of Δ_{II} with decreasing grain size was observed suggesting low accumulation and interaction between dislocations in UFG specimens. In addition, the increase of $(\sigma\theta)_0$ with 1/d indicate a high contribution of dislocation interaction with grain boundaries to the strain hardening of such samples.

- Microstructural observations performed by EBSD showed fewer changes in the microstructure, such as changes in morphology or low formation of LAGBs, in samples with grain sizes in the UFG range.
- Misorientation cartographies (GOS and KAM), suggest low deformation of small size grains with low values of GOS and KAM, and preferred accumulation of dislocations close to grain boundaries.
- TEM observations of the deformed state of UFG samples (d = 0.82 μm) showed low dislocation density heterogeneously distributed, and lack of well-defined dislocation cells.

From these results, we could conclude that with grain refinement in the UFG regime, dislocation storage was lessened by generalized cross-slip mechanisms and dislocation annihilation starting just after yielding, hindering the formation of organized dislocation structures in the grain core. Hardening is thus mainly mediated by dislocation interactions with grain boundaries, for grain sizes below d = 1.11 μm, which enable good ductility contrary to deformation mechanism such as grain sliding, observed in smaller size UFG and NsM metals.

Author Contributions: Conceptualization, L.G.d.l.C. and E.H.; methodology, L.G.d.l.C., M.M.C. and E.H.; validation, E.H. and C.K.; formal analysis, L.G.d.l.C., C.K. and E.H.; investigation, all authors; writing—original draft preparation, L.G.d.l.C.; writing—review and editing, all authors; supervision, E.H.; project administration, E.H. All authors have read and agreed to the published version of the manuscript.

Funding: This research received no external funding.

Institutional Review Board Statement: Not applicable.

Informed Consent Statement: Not applicable.

Data Availability Statement: The data presented in this study are available on request from the corresponding author.

Conflicts of Interest: The authors declare no conflict of interest.

Glossary

Abbreviations

BM	Ball milling
CG	Coarse grained
CSL	Coincidence site lattice
EBSD	Electron backscatter diffraction
FCC	Face centered cubic
FG	Fine grained
GBCD	Grain boundary character distribution
GNDs	Geometrically necessary dislocations
GOS	Grain orientation spread
HABs	High angle boundaries
IQ	Image quality
IPF	Inverse pole figure
KAM	Kernel average misorientation
LAGBs	Low angle grain boundaries
NsM	Nanostructured materials
SPS	Spark plasma sintering
TEM	Transmission electron microscopy
UFG	Ultrafine grained

Abbreviations
Symbols

α	Parameter of dislocation interaction and configuration
b	Norm of the Burgers vector
β	Parameter that relates Λ and l_d.
d	Grain size
Δ_{II}	Latent hardening rate
ε	True strain
ε_f	Elongation to failure
$\vec{g}$	Diffraction vector
θ	Work hardening rate
k_g	Geometric factor related to grain morphology
l_d	Average distance between dislocations
Λ	Average distance traveled by a dislocation
M	Taylor factor
P	Probability of dislocation annihilation
ρ	Total dislocation density
ρ_{rel}	Relative density
σ	True stress
σ_y	Yield strength
$\sigma_{I/II}$	Stress of transition between hardening stages I and II.
$\sigma_{II/II}$	Stress of transition between hardening stages II and III.
y_a	Dislocation annihilation distance
$(\sigma\theta)_0$	Parameter of contribution of grain boundaries to strain hardening

References

1. Lu, L.; Yongfeng, S.; Chen, X.; Qian, L.; Lu, K. Ultrahigh Strength and High Electrical Conductivity in Copper. *Science* **2004**, *304*, 422–426. [CrossRef] [PubMed]
2. Zhu, K.N.; Godfrey, A.; Hansen, N.; Zhang, X.D. Microstructure and Mechanical Strength of Near- and Sub-Micrometre Grain Size Copper Prepared by Spark Plasma Sintering. *Mater. Des.* **2017**, *117*, 95–103. [CrossRef]
3. Gu, C.; Lian, J.; Jiang, Z.; Jiang, Q. Enhanced Tensile Ductility in an Electrodeposited Nanocrystalline Ni. *Scr. Mater.* **2006**, *54*, 579–584. [CrossRef]
4. Dhal, A.; Panigrahi, S.K.; Shunmugam, M.S. Achieving Excellent Microformability in Aluminum by Engineering a Unique Ultrafine-Grained Microstructure. *Sci. Rep.* **2019**, *9*, 10683. [CrossRef] [PubMed]
5. Le, G.M.; Godfrey, A.; Hansen, N.; Liu, W.; Winther, G.; Huang, X. Influence of Grain Size in the Near-Micrometre Regime on the Deformation Microstructure in Aluminium. *Acta Mater.* **2013**, *61*, 7072–7086. [CrossRef]
6. Zhu, Y.T.; Wu, X.L. Ductility and Plasticity of Nanostructured Metals: Differences and Issues. *Mater. Today Nano* **2018**, *2*, 15–20. [CrossRef]
7. Ashby, M.F. The Deformation of Plastically Non-Homogeneous Materials. *Philos. Mag.* **1970**, *21*, 399–424. [CrossRef]
8. Cheng, S.; Spencer, J.A.; Milligan, W.W. Strength and Tension/Compression Asymmetry in Nanostructured and Ultrafine-Grain Metals. *Acta Mater.* **2003**, *51*, 4505–4518. [CrossRef]
9. Hayes, R.W.; Witkin, D.; Zhou, F.; Lavernia, E.J. Deformation and Activation Volumes of Cryomilled Ultrafine-Grained Aluminum. *Acta Mater.* **2004**, *52*, 4259–4271. [CrossRef]
10. Kumar, K.S.; Suresh, S.; Chisholm, M.F.; Horton, J.A.; Wang, P. Deformation of Electrodeposited Nanocrystalline Nickel. *Acta Mater.* **2003**, *51*, 387–405. [CrossRef]
11. Ma, E. Instabilities and Ductility of Nanocrystalline and Ultrafine-Grained Metals. *Scr. Mater.* **2003**, *49*, 663–668. [CrossRef]
12. Krasilnikov, N.; Lojkowski, W.; Pakiela, Z.; Valiev, R. Tensile Strength and Ductility of Ultra-Fine-Grained Nickel Processed by Severe Plastic Deformation. *Mater. Sci. Eng. A* **2005**, *397*, 330–337. [CrossRef]
13. Samigullina, A.A.; Nazarov, A.A.; Mulyukov, R.R.; Tsarenko, Y.V.; Rubanik, V.V. Effect of Ultrasonic Treatment on the Strength and Ductility of Bulk Nanostructured Nickel Processed by Equal-Channel Angular Pressing. *Rev. Adv. Mater. Sci.* **2014**, *39*, 48–53. [CrossRef]
14. Dalla Torre, F.; Van Swygenhoven, H.; Victoria, M. Nanocrystalline Electrodeposited Ni: Microstructure and Tensile Properties. *Acta Mater.* **2002**, *50*, 3957–3970. [CrossRef]
15. Dalla Torre, F.; Spatig, P.; Schaublin, R.; Victoria, M. Deformation Behaviour and Microstructure of Nanocrystalline Electrodeposited and High Pressure Torsioned Nickel. *Acta Mater.* **2005**, *53*, 2337–2349. [CrossRef]
16. Meyers, M.A.; Mishra, A.; Benson, D.J. Mechanical Properties of Nanocrystalline Materials. *Prog. Mater. Sci.* **2006**, *51*, 427–556. [CrossRef]
17. Erb, U. Electrodeposited Nanocrystals: Synthesis, Properties and Industrial Applications. *Nanostruct. Mater.* **1995**, *6*, 533–538. [CrossRef]

18. Valiev, R.Z.; Islamgaliev, R.K.; Alexandrov, I.V. Bulk Nanostructured Materials from Severe Plastic Deformation. *Prog. Mater. Sci.* **2000**, *45*, 103–189. [CrossRef]
19. Barry, A.H.; Dirras, G.; Schoenstein, F.; Tétard, F.; Jouini, N. Microstructure and Mechanical Properties of Bulk Highly Faulted Fcc/Hcp Nanostructured Cobalt Microstructures. *Mater. Charact.* **2014**, *91*, 26–33. [CrossRef]
20. Ebrahimi, F.; Bourne, G.R.; Kelly, M.S.; Matthews, T.E. Mechanical Properties of Nanocrystalline Nickel Produced by Electrodeposition. *Nanostruct. Mater.* **1999**, *11*, 343–350. [CrossRef]
21. Valiev, R.Z.; Langdon, T.G. The Art and Science of Tailoring Materials by Nanostructuring for Advanced Properties Using SPD Techniques. *Adv. Eng. Mater.* **2010**, *12*, 677–691. [CrossRef]
22. Gleiter, H. Materials with Ultrafine Microstructures: Retrospectives and Perspectives. *Nanostruct. Mater.* **1992**, *1*, 1–19. [CrossRef]
23. Gleiter, H. Nanostructured Materials: Basic Concepts and Microstructure. *Acta Mater.* **2000**, *48*, 1–29. [CrossRef]
24. Estrin, Y.; Vinogradov, A. Extreme Grain Refinement by Severe Plastic Deformation: A Wealth of Challenging Science. *Acta Mater.* **2013**, *61*, 782–817. [CrossRef]
25. Tian, Y.Z.; Gao, S.; Zhao, L.J.; Lu, S.; Pippan, R.; Zhang, Z.F.; Tsuji, N. Remarkable Transitions of Yield Behavior and Lüders Deformation in Pure Cu by Changing Grain Sizes. *Scr. Mater.* **2018**, *142*, 88–91. [CrossRef]
26. Yu, C.Y.; Kao, P.W.; Chang, C.P. Transition of Tensile Deformation Behaviors in Ultrafine-Grained Aluminum. *Acta Mater.* **2005**, *53*, 4019–4028. [CrossRef]
27. Gao, S.; Bai, Y.; Zheng, R.; Tian, Y.; Mao, W.; Shibata, A.; Tsuji, N. Mechanism of Huge Lüders-Type Deformation in Ultrafine Grained Austenitic Stainless Steel. *Scr. Mater.* **2019**, *159*, 28–32. [CrossRef]
28. García de la Cruz, L.; Martinez, M.; Keller, C.; Hug, E. Achieving Good Tensile Properties in Ultrafine Grained Nickel by Spark Plasma Sintering. *Mater. Sci. Eng. A* **2020**, *772*, 138770. [CrossRef]
29. Jin, Y.; Lin, B.; Bernacki, M.; Rohrer, G.S.; Rollett, A.D.; Bozzolo, N. Annealing Twin Development during Recrystallization and Grain Growth in Pure Nickel. *Mater. Sci. Eng. A* **2014**, *597*, 295–303. [CrossRef]
30. García de la Cruz, L.; Domenges, B.; Divinski, S.V.; Wilde, G.; Hug, E. Ultrafast Atomic Diffusion Paths in Fine-Grained Nickel Obtained by Spark Plasma Sintering. *Met. Mat. Trans. A* **2020**, *51*, 3425–3434. [CrossRef]
31. Song, J.-M.; Lee, J.-S. Self-Assembled Nanostructured Resistive Switching Memory Devices Fabricated by Templated Bottom-up Growth. *Sci. Rep.* **2016**, *6*, 18967. [CrossRef] [PubMed]
32. Cabibbo, M. Grain Refinement and Hardness Saturation in Pure Nickel Subjected to a Sequence of ECAP and HPT. *Metall. Ital.* **2015**, *9*, 37–48.
33. Bui, Q.H.; Dirras, G.; Ramtani, S.; Gubicza, J. On the Strengthening Behavior of Ultrafine-Grained Nickel Processed from Nanopowders. *Mater. Sci. Eng. A* **2010**, *527*, 3227–3235. [CrossRef]
34. Randle, V. The Coincidence Site Lattice and the 'Sigma Enigma'. *Mater. Charact.* **2001**, *47*, 411–416. [CrossRef]
35. King, A.H.; Shekhar, S. What Does It Mean to Be Special? The Significance and Application of the Brandon Criterion. *J. Mater. Sci.* **2006**, *41*, 7675–7682. [CrossRef]
36. de la Cruz, L.G.; Flipon, B.; Keller, C.; Martinez, M.; Hug, E. Nanostructuration of Metals via Spark Plasma Sintering Using Activated Powder Obtained by Ball-Milling: Impact on the Strain-Hardening Mechanisms. In Proceedings of the AIP Conference Proceedings 1896, Dublin, Ireland, 26–28 April 2017; Volume 1896, p. 200002. [CrossRef]
37. García de la Cruz, L. Ultrafine Grained Nickel Processed by Powder Metallurgy: Microstructure, Mechanical Properties and Thermal Stability. Ph.D. Thesis, Normandie Université, Caen, France, 2019.
38. Sitarama Raju, K.; Ghanashyam Krishna, M.; Padmanabhan, K.A.; Subramanya Sarma, V.; Gurao, N.P.; Wilde, G. Microstructure Evolution and Hardness Variation during Annealing of Equal Channel Angular Pressed Ultra-Fine Grained Nickel Subjected to 12 Passes. *J. Mater. Sci.* **2011**, *46*, 2662–2671. [CrossRef]
39. Feaugas, X.; Haddou, H. Grain-Size Effects on Tensile Behavior of Nickel and AISI 316L Stainless Steel. *Metall. Mater. Trans. A* **2003**, *34*, 2329–2340. [CrossRef]
40. Kocks, U.F.; Mecking, H. Physics and Phenomenology of Strain Hardening: The FCC Case. *Prog. Mater. Sci.* **2003**, *48*, 171–273. [CrossRef]
41. Mecking, H.; Kocks, U.F. Kinetics of Flow and Strain-Hardening. *Acta Metall.* **1981**, *29*, 1865–1875. [CrossRef]
42. Kuhlmann-Wilsdorf, D. Theory of Plastic Deformation:-Properties of Low Energy Dislocation Structures. *Mater. Sci. Eng. A* **1989**, *113*, 1–41. [CrossRef]
43. Kubin, L.; Devincre, B.; Hoc, T. Modeling Dislocation Storage Rates and Mean Free Paths in Face-Centered Cubic Crystals. *Acta Mater.* **2008**, *56*, 6040–6049. [CrossRef]
44. Jordan, L.; Swanger, W.H. The Properties of Pure Nickel. Bur. Standards. *J. Res.* **1930**, *10*, 6028.
45. Narutani, T.; Takamura, J. Grain-Size Strengthening in Terms of Dislocation Density Measured by Resistivity. *Acta Metall. Mater.* **1991**, *39*, 2037–2049. [CrossRef]
46. Keller, C. Etude Expérimentale Des Transitions Volume/Surface Des Propriétés Mécaniques Du Nickel Polycristallin de Haute Pureté. Ph.D. Thesis, Université de Caen Basse-Normandie, Caen, France, 2009.
47. Devincre, B.; Hoc, T.; Kubin, L. Dislocation Mean Free Paths and Strain Hardening of Crystals. *Science* **2008**, *320*, 1745–1748. [CrossRef]
48. Dubos, P.A. Influence de La Témperature et Du Trajet de Chargement Sur Les Transitions Volume/Surafec Des Méteaux Cubiques à Faces Centrées. Ph.D. Thesis, Université de Caen Basse-Normandie, Caen, France, 2013.

49. Hansen, N.; Huang, X. Microstructure and Flow Stress of Polycrystals and Single Crystals. *Acta Mater.* **1998**, *46*, 1827–1836. [CrossRef]
50. Hug, E.; Keller, C. Intrinsic Effects Due to the Reduction of Thickness on the Mechanical Behavior of Nickel Polycrystals. *Metall. Mater. Trans. A* **2010**, *41A*, 2498–2506. [CrossRef]
51. Estrin, Y.; Mecking, H. A Unified Phenomenological Description of Work Hardening and Creep Based on One-Parameter Models. *Acta Metall.* **1984**, *32*, 57–70. [CrossRef]
52. Jensen, D.J.; Thompson, A.W.; Hansen, N. The Role of Grain Size and Strain in Work Hardening and Texture Development. *Metall. Trans. A* **1989**, *20*, 2803–2810. [CrossRef]
53. Taylor, G.I. Plastic Strain in Metals. *J. Inst. Met.* **1938**, *62*, 307–324.
54. Humphreys, F.J.; Hatherly, M. *Recrystallization and Related Annealing Phenomena*, 2nd ed.; Elsevier: Oxford, UK, 2004.
55. Rémy, L. The Interaction between Slip and Twinning Systems and the Influence of Twinning on the Mechanical Behavior of Fcc Metals and Alloys. *Metall. Trans. A* **1981**, *12*, 387–408. [CrossRef]
56. Yamakov, V.; Wolf, D.; Phillpot, S.R.; Gleiter, H. Dislocation–Dislocation and Dislocation–Twin Reactions in Nanocrystalline Al by Molecular Dynamics Simulation. *Acta Mater.* **2003**, *51*, 4135–4147. [CrossRef]
57. Gubicza, J.; Bui, H.-Q.; Fellah, F.; Dirras, G.F. Microstructure and Mechanical Behavior of Ultrafine-Grained Ni Processed by Different Powder Metallurgy Methods. *J. Mater. Res.* **2009**, *24*, 217–226. [CrossRef]
58. Brewer, L.N.; Field, D.P.; Merriman, C.C. Mapping and Assessing Plastic Deformation Using EBSD. In *Electron Backscatter Diffraction in Materials Science*; Springer Science & Business Media: Berlin/Heidelberg, Germany, 2009.
59. Kuhlmann-Wilsdorf, D.; Hansen, N. Geometrically Necessary, Incidental and Subgrain Boundaries. *Scr. Metall. Mater.* **1991**, *25*, 1557–1562. [CrossRef]
60. Field, D.P.; Bradford, L.T.; Nowell, M.; Lillo, T.M. The Role of Annealing Twins during Recrystallization of Cu. *Acta Mater.* **2007**, *55*, 4233–4241. [CrossRef]
61. Schwartz, A.J.; Kumar, M.; Adams, E. *Backscatter Diffraction in Materials Science*; Field, D.P., Adams, B.L., Kumar, M., Schwartz, A.J., Eds.; Springer US: Boston, MA, USA, 2000.
62. Sakakibara, Y.; Kubushiro, K. Stress Evaluation at the Maximum Strained State by EBSD and Several Residual Stress Measurements for Plastic Deformed Austenitic. *World J. Mech.* **2017**, *7*, 195–210. [CrossRef]
63. Fu, H.-H.; Benson, D.J.; Meyers, M.A. Analytical and Computational Description of Effect of Grain Size on Yield Stress of Metals. *Acta Mater.* **2001**, *49*, 2567–2582. [CrossRef]
64. Feaugas, X.; Haddou, H. Effects of Grain Size on Dislocation Organization and Internal Stresses Developed under Tensile Loading in Fcc Metals. *Philos. Mag.* **2007**, *87*, 989–1018. [CrossRef]
65. Keller, C.; Hug, E.; Retoux, R.; Feaugas, X. TEM Study of Dislocation Patterns in Near-Surface and Core Regions of Deformed Nickel Polycrystals with Few Grains across the Cross Section. *Mech. Mater.* **2010**, *42*, 44–54. [CrossRef]

Review

Elaboration of Metallic Materials by SPS: Processing, Microstructures, Properties, and Shaping

Jean-Philippe Monchoux [1,2,*], Alain Couret [1,2], Lise Durand [1,2], Thomas Voisin [3], Zofia Trzaska [4,5] and Marc Thomas [6]

1   Centre d'Elaboration de Matériaux et d'Etudes Structurales, Centre National de la Recherche Scientifique, Unité Propre de Recherche 8011, 29 rue J. Marvig, BP 94347, CEDEX 4, 31055 Toulouse, France; alain.couret@cemes.fr (A.C.); lise.durand@cemes.fr (L.D.)

2   Université Toulouse III—Paul Sabatier, 118 Route de Narbonne, CEDEX 9, 31062 Toulouse, France

3   Lawrence Livermore National Laboratory, 7000 East Av., Livermore, CA 94550, USA; voisin2@llnl.gov

4   Institut de Chimie et des Matériaux Paris-Est, Unité Mixte de Recherche 7182, 2-8, rue Henri Dunant, 94320 Thiais, France; zofia.trzaska@univ-paris13.fr

5   Université Sorbonne Paris Nord, 99, Avenue Jean-Baptiste Clément, 93430 Villetaneuse, France

6   ONERA—The French Aerospace Lab, Department of Materials and Structures, Université Paris-Saclay, 29 Avenue de la Division Leclerc, BP 72, 92322 Châtillon CEDEX, France; marc.thomas@onera.fr

*   Correspondence: monchoux@cemes.fr

Abstract: After a few decades of increasing interest, spark plasma sintering (SPS) has now become a mature powder metallurgy technique, which allows assessing its performances toward fabricating enhanced materials. Here, the case of metals and alloys will be presented. The main advantage of SPS lies in its rapid heating capability enabled by the application of high intensity electric currents to a metallic powder. This presents numerous advantages balanced by some limitations that will be addressed in this review. The first section will be devoted to sintering issues, with an emphasis on the effect of the electric current on the densification mechanisms. Then, typical as-SPS microstructures and properties will be presented. In some cases, they will be compared with that of materials processed by conventional techniques. As such, examples of nanostructured materials, intermetallics, metallic glasses, and high entropy alloys, will be presented. Finally, the implementation of SPS as a technique to manufacture complex, near-net shape industrial parts will be discussed.

Keywords: SPS; metals; TiAl; microstructure; mechanical properties; shaping

Citation: Monchoux, J.-P.; Couret, A.; Durand, L.; Voisin, T.; Trzaska, Z.; Thomas, M. Elaboration of Metallic Materials by SPS: Processing, Microstructures, Properties, and Shaping. *Metals* **2021**, *11*, 322. https://doi.org/10.3390/met11020322

Academic Editor: Eric Hug

Received: 18 January 2021
Accepted: 7 February 2021
Published: 12 February 2021

Publisher's Note: MDPI stays neutral with regard to jurisdictional claims in published maps and institutional affiliations.

1. Introduction

Spark plasma sintering (SPS) has become a widely used processing technique for a broad class of materials, including ceramics, metals, and polymers. Consequently, significant improvements were achieved since the review conducted by Orru et al. [1] in 2009, allowing SPS to reach pre-industrialization. Early on, the ability of SPS to rapidly process many different materials with improved microstructures and properties was attributed to electricity-induced phenomena, such as the occurrence of sparks or plasma at the powder particle interfaces (which the process was named after), the involvement of electromigration, electroplasticiy, and so on. However, after two decades of investigations, it is now possible to discuss more accurately these early hypotheses, and to establish more firmly its physical principles.

In this chapter, we will concentrate on the case of the metallic materials. Regarding potential electrical effects, these materials exhibit the particularity of a high electric conductivity. Therefore, they have been particularly the subject of investigation of electrically-induced effects. We will present studies devoted to these questions, but also, more generally, to the densification mechanisms and kinetics. Then, we will present the original microstructures, which can be achieved, characterized often by nanostructuration, and we will present results showing the relation between the microstructures obtained by

93

SPS and those predicted by the equilibrium phase diagram. This will lead us to make a review of materials, which have been the subject of particular interest regarding their processing by SPS: intermetallics, nanostructured materials, metallic glasses, and high entropy alloys. Then, because processing of parts in complex geometries is becoming a major issue, we will deal with the necessary steps involved to achieve this goal: thermo-mechanical modeling, upscaling, and shaping.

2. Overview of the SPS Process

SPS technology consists in densifying powders by applying high electric current pulses under mechanical pressure. This results in rapid heating and compaction of the powder into fully dense materials. As shown in the reviews of Orru et al. [1] and of Munir et al. [2], the principle of the SPS originates in 1922. The ideas of local heating at the particle contacts was proposed in 1945, and the hypothesis of vaporization of superficial oxide layers by dielectric breakdown was formulated in the 1950s. The term *spark sintering* appeared in 1965, and Inoue enriched these concepts by describing phenomena, such as electric discharge, ionization, or local melting, and developed devices based on the use of high frequency electric currents. The works of Inoue lead to the development of Be-based materials for missiles at the Lockheed Corporation at the end of the 1960s. However, the extensive spreading of the current-assisted sintering technologies occurred at the beginning of the 1990s, when the Sumitomo Coal Mining company developed the so-called *spark plasma sintering* technology, based on patents of the 1960s. Since then, the number of works devoted to the SPS approach has increased very rapidly. For example, an analysis of the number of articles published per year containing the keywords "spark plasma sintering" is given in Figure 1. It shows an impressive increase in publications since the first time this term was encountered in literature, in 1994. Among this production, an analysis of the review of Orru et al. in 2009 [1] shows that the studies on metallic systems represent $\approx 15\%$ of the total, a proportion probably stable over time.

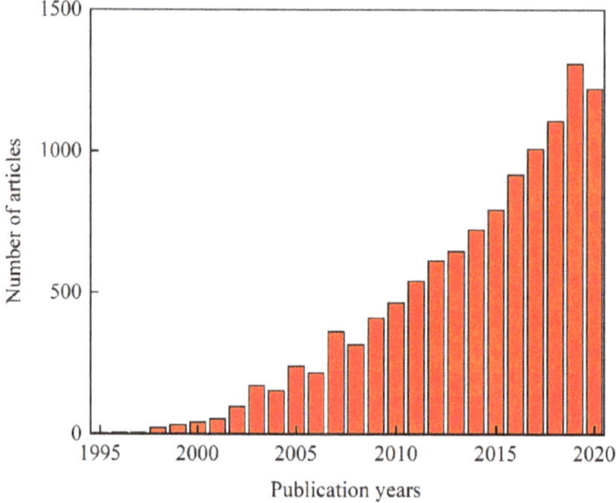

Figure 1. Number of articles per year related to spark plasma sintering (SPS) technology (source: Web of Science).

A simplified schematic of the powder sintering setup as found in most commercial SPS machines is depicted in Figure 2. The powder is filled inside a graphite mold, generally made of graphite, consisting in a cylindrical container closed by two compression electrodes, and this assembly is introduced into an evacuated chamber. As a high pressure (typically around 100 MPa) and a high intensity current are applied through the pistons, the entire setup, including the powder if conductive, rapidly heats up by Joule effect. In fact, because the heat production occurs internally, high heating rates can be achieved (typically 50–1000 °C/min, depending on the setup, material, and intensities). As a result, short processing times (5–60 min) can be achieved. Figure 3 gives an example of a typical processing cycle for TiAl disks 36 mm in diameter and 8 mm thick. It can be seen, in this example, that the current intensity reached 3000 A. Because the temperature is usually not measured at the sample but either on the external surface of the die or inside one of the electrodes, a few cm away from the sample, a temperature gradient between the measure and the sample exists. In this case, corrections can be applied based on experimental calibrations or simulations. This will be discussed in Section 5.

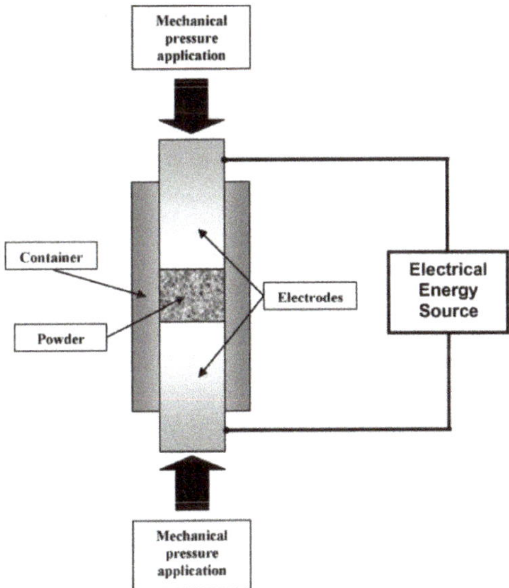

Figure 2. Schematic representation of the SPS process [1]. Reproduced with permission from *Materials Science and Engineering: R*, published by Elsevier, 2009.

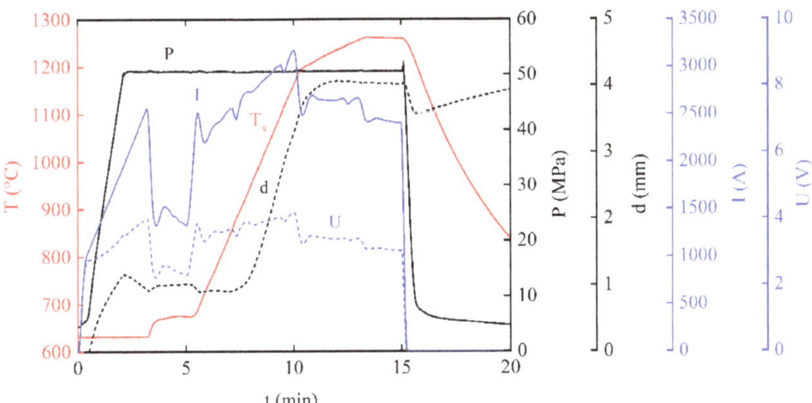

Figure 3. Evolution of the SPS parameters during the processing of a metallic TiAl alloy, in the case of a 36 mm sample (T$_s$: corrected sample temperature, P: applied mechanical pressure, I: intensity, U: voltage and d: sample thickness reduction).

3. Densification Mechanisms and Kinetics

It has been suggested that the high intensity electric current applied upon processing could intrinsically influence the sintering kinetics and the densification mechanisms [1–3]. We present here the main efforts dedicated to verifying this hypothesis.

3.1. Microscopic Densification Mechanisms

Several mechanisms that can be induced by the high-intensity electrical current during SPS can be classified as follow: (i) dielectric breakdown of the oxide layers at the surface of the powder particles. (ii) Arcs and plasma generation between powder particles. (iii) Local overheating due to a current density increase at the contact between powder particles. (iv) Activation of electromigration and electroplasticity.

Mechanisms (i) and (ii) have been first proposed by Tokita [3]. Breakdowns of superficial oxide layers, mechanism (i), were reported in many studies [4–7], but without providing convincing experimental evidences. Arcing, as in mechanism (ii), has been on the contrary more clearly observed between adjacent Cu powder particles 550 μm in diameter (Figure 4) [8]. However, in this work the applied pressure was very low, which did not reproduce the conditions encountered during an actual SPS cycle. Detection of arcs by emission spectroscopy has been attempted by two teams. Plasma was reported in a W powder for applied voltages >50 V by Saunders et al. [9]. Conversely, Hulbert et al. [10] concluded to the absence of plasma in various metallic powders (Al, Mg, Zn) for voltages ≈ 1–2 V, which are closer to that observed in typical SPS experiments. Even though spark and/or plasma events may still happen at the onset of sintering, the heavy plastic deformation of the interparticle contact regions due to the applied high pressure may too quickly remove the geometrical singularities at the origin of electric current intensification and arcing.

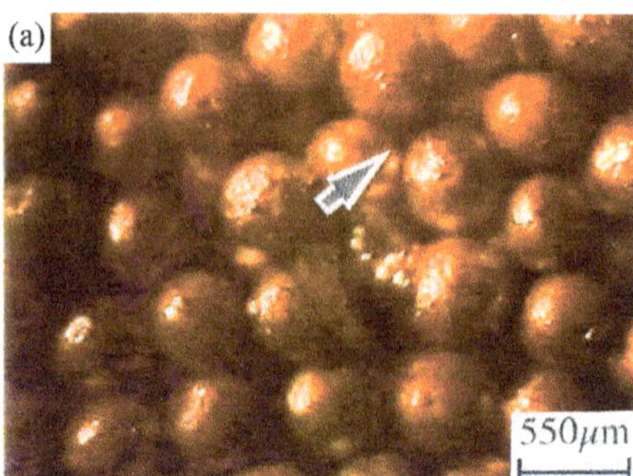

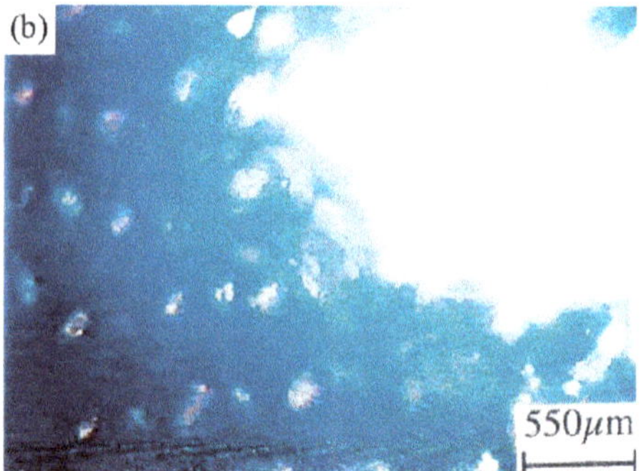

Figure 4. (**a**) Cu powder particles 550 μm in diameter. (**b**) Generation of an electric arc by a current pulse of 6700 A/cm^2 during 100 ms, at the location indicated by an arrow in Figure (**a**) [8]. Reproduced with permission from *Materials Science and Engineering: A*, published by Elsevier, 2003.

Local Joule overheating at the contacts between powder particles, mechanism (iii) has mainly been investigated theoretically. In the early works, very elevated local current densities, leading to local high overheating at the contacts (up to 10,000 °C), were computationally calculated [8,11]. The main limitation of these simulations is that they did not consider heat conduction within the particles, which was later shown to play a major role [12–16]. Even though more recent simulations predict local current densities as high as ≈ 5×10^4 A/cm^2 at the contacts between particles [15–17], it has been shown that, for regular-size powder particles (≈ 100 μm), the temperature homogenization by heat conduction takes only milliseconds, which was not compatible with stabilization of "hot spots" at the particle contacts [15,16]. Few experimental works with FeAl [18] and metallic glasses [19] concluded local overheating, but clear and direct evidence could not be provided. Therefore, the formation of "hot spots" at the regions of contact between powder particles due to local electrical current concentrations is still debated in the literature.

Finally, the activation of metallurgical effects by electric currents, mechanisms (iv), has been considered, because the high current densities in the particles contact regions ($\approx 5 \times 10^4$ A/cm^2) are near or above the activation threshold for electroplasticity (10^3–10^6 A/cm^2) [20–22] and electromigration (10^2–10^3 A/cm^2) [2]. For example, model experiments with 3 mm Cu spheres in contact with Cu plates showed enhanced growth rates of the sphere-plate contact zones with currents $\approx 5 \times 10^2$ A/cm^2 [17]. This effect, which could account for increase in densification rate of metallic powders by the current, was interpreted by electromigration phenomena. However, another interpretation can also be proposed. When the size of the powder particle increases, and reaches millimeters and above, the thermal equilibration of the powder particles takes longer time [12,13], and hot spots can therefore be stabilized in the contact regions. Consequently, the local increase of the temperature can lead to acceleration of plastic deformation by creep. Other experimental investigations have been performed by focused ion beam and transmission electron microscopy (TEM) in the contact regions between powder particles in TiAl [23] and Ni [24]. It has been shown that the elementary deformation mechanisms were dislocation glide and climb, followed by dislocations recovery and recrystallization. For example, Figure 5 shows TEM images of a thin foil extracted in the contact region between two Ni powder particles [24]. High dislocation densities, and recovery walls, can be observed. This suggests that the observed mechanisms, which are those classically involved in power-law creep, are not significantly modified by the electric current.

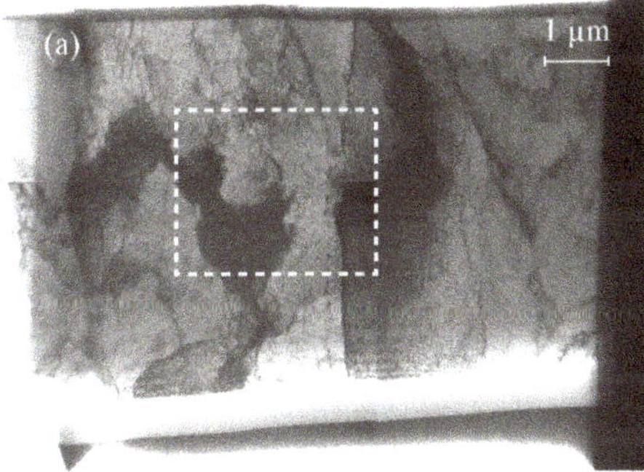

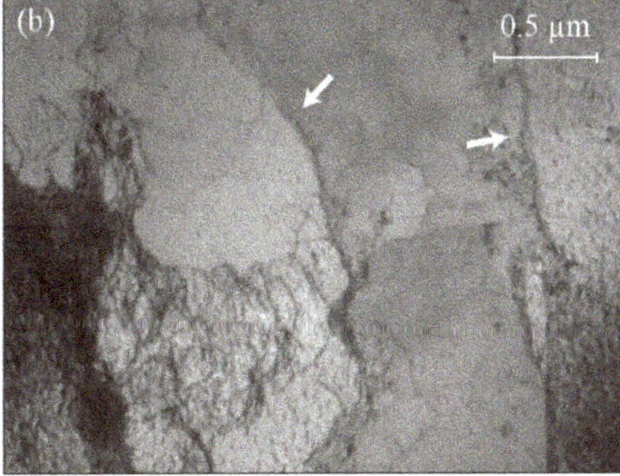

Figure 5. (**a**) TEM thin foil extracted from the contact region between two particles during densification of a Ni powder by SPS. (**b**) Zoom-in corresponding to the white dashed rectangle in (**a**) showing the presence of dislocations wall associated to dislocation recovery mechanisms [24]. Reproduced with permission from *Metallurgical and Materials Transactions: A*, published by Springer Nature, 2018.

3.2. Influence of High Intensity Currents on Powder Densification Kinetics

The effect of high intensity currents on the metallic powder densification at macroscopic scale has been the subject of several studies that gave contradictory results. For example, Figure 6 shows that when increased currents are applied to a Ni powder, by use of devices forcing all of the current to flow through the sample (setup 2), the final density of the sample is about 5% higher than that obtained in current-insulated experiments (setup 3), keeping the other parameters (temperature, pressure, dwell time, etc.) identical [25]. Similarly, Aleksandrova et al. [26] observed a density increase of Cu specimens when applying currents of 1000 A/cm^2, compared with experiments without current. On the contrary, a comparison between SPS and hot pressing with Cu [27] and TiAl [28] powders highlighted similar densification kinetics (density as a function of time) independently of the process employed.

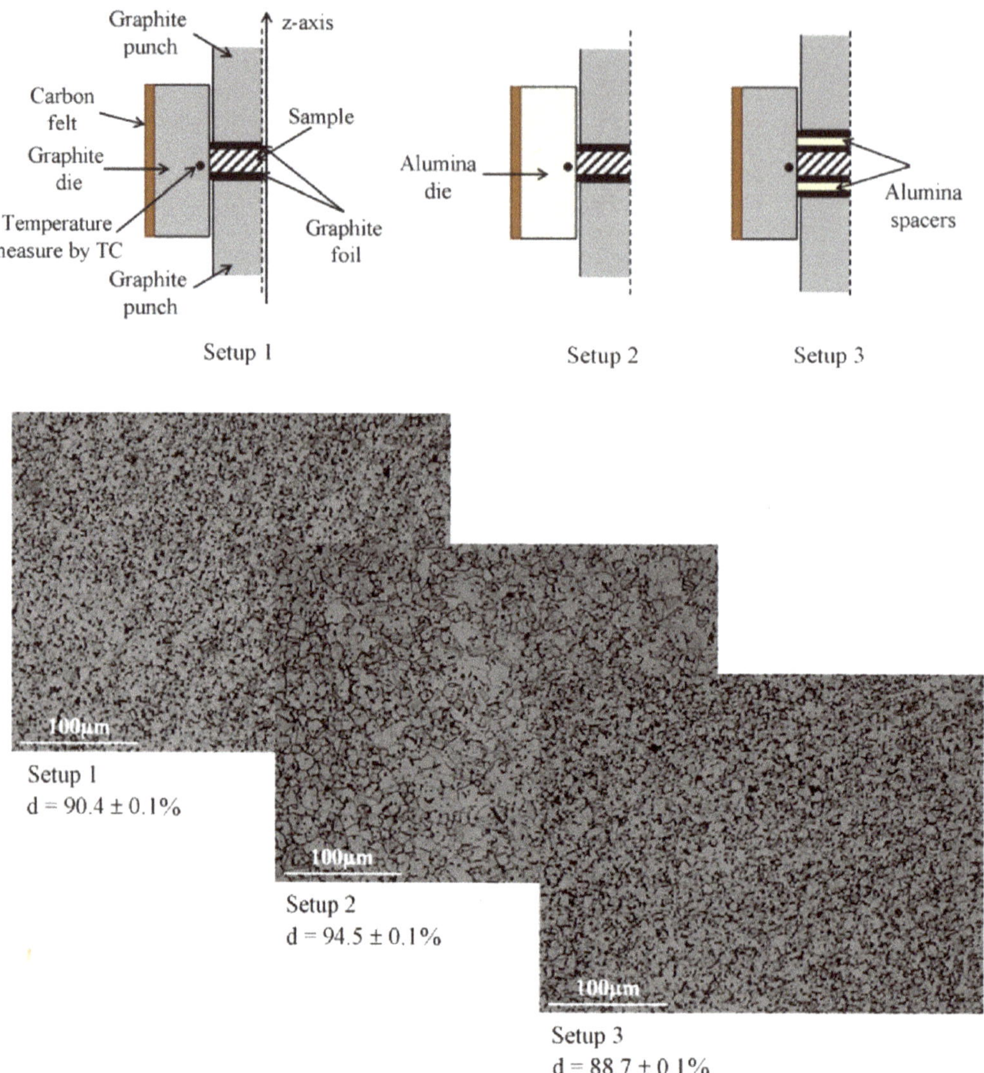

Figure 6. Relative density *d* of a Ni powder submitted to normal SPS conditions (setup 1), and to conditions where all of the SPS current flows through the sample (setup 2) and where the sample is electrically insulated (setup 3). Note in particular the difference in relative density *d* between samples processed using setup 2 (*d* = 94.5%) and setup 3 (*d* = 88.7%), showing an increase of sample relative density of about 5% due to the electric current [25]. Reproduced with permission from *Journal of Alloys and Compounds*, published by Elsevier, 2010.

3.3. Conclusions

What the investigations reported in this section agree on is that it is not possible to definitively conclude on the role of high intensity electric currents on the densification mechanisms and kinetics besides providing an effective heat source. Macroscopic effects are at most moderate [26], limited [25] or inexistent [27,28]. Moreover, at the microscopic scale, with the noticeable exception of the study of Yanagisawa et al. [8], no evident phenomena have been experimentally detected. The densification mechanisms are consequently similar to those classically involved in densification theories: deformation of the powder particles by power-law creep involving dislocations glide and climb, followed by recovery and recrystallization [23,24]. Therefore, it can reasonably be stated that the densification of metallic powders by SPS experiments is not governed by electrical effects. As such, the process can be adequately modeled using Norton law constitutive relations, as will be detailed in Section 5.

4. As-SPS Microstructure and Mechanical Properties

As a powder metallurgy technique, the SPS primarily purpose is to densify powders to obtain bulk metallic materials of controlled density. However, because the sintering takes place at high temperatures, it is also possible to optimize the microstructures. In many cases, this results in improved mechanical properties. Moreover, the rapid densification capability of the SPS technique allows obtain nanostructured or metastable materials by using very short thermal cycles. Another important advantage of the SPS technique is the microstructure homogeneity, which is generally more satisfactory than with other techniques, such as casting for example. However, the distribution of the electric current during SPS is often heterogeneous throughout the samples, which causes temperature gradients, especially for large samples, resulting in possible microstructure variations. Therefore, a significant effort has been dedicated to model the temperature fields in samples and molds toward optimizing the temperature distribution in large specimens, as will be discussed in Section 5. These microstructures heterogeneities are nevertheless less pronounced than with casting, because in this latter case solidification leads to many structural defects: cracks, porosities, contaminations, segregations and so on. Moreover, the problem of temperature heterogeneity in large parts tends to be mastered, by adapting the geometry of the molds, or by developing the so-called "hybrid" SPS technology, which provides additional heat sources leading to decrease of the temperature gradients within the parts. Therefore, the SPS shows now a high potential for developing advanced materials, for parts of increasing dimensions. This section is thus dedicated to discuss selected studies that best illustrate the potential of SPS to tune microstructures and improve the resulting mechanical properties.

4.1. Equilibrium Microstructures

Because of the rapidity of the SPS technique, it is anticipated that the phases and microstructures of metallic alloys can differ from those at equilibrium as predicted by the phase diagrams. This point has been investigated in the case of TiAl alloys, for which processing by conventional techniques in suitable fields of the binary phase diagram lead to well referenced near γ, duplex, and lamellar microstructures [29]. It has been shown that, in the case of the simplest TiAl-based GE alloy (Ti-Al$_{48}$-Cr$_2$-Nb$_2$), manufacturing by SPS leads to an accurate correspondence with the binary phase diagram (Figure 7 [30]). Each microstructure exhibits a specific mechanical response to uniaxial tension at room temperature and uniaxial tensile creep testing at high temperature. This provides the opportunity to find a microstructure with an optimized balance between room temperature ductility and high temperature properties. Unfortunately, such a compromise cannot be reached with the Ti-Al$_{48}$-Cr$_2$-Nb$_2$ composition as show in Figure 7. The duplex microstructure offering the best ductility exhibits high creep deformation rate and limited lifetime (Figure 7d,f). However, because the obtained microstructures were those predicted by the phase diagrams, it was possible to design materials of more complicated chemistries and microstructures, based on theoretically predicted phase diagrams, to improve the mechanical properties [30,31]. Thus, the beneficial roles of W on the creep resistance, through reduction of the diffusion rates, and that of B on the limitation of grain growth, through grain boundary pinning on TiB$_2$ precipitates, have been investigated. This resulted in the so-called IRIS alloy, of Ti-48Al-2W-0.1B composition, which strength at high temperature was higher than alloys elaborated by other techniques (Figure 8 [32,33]).

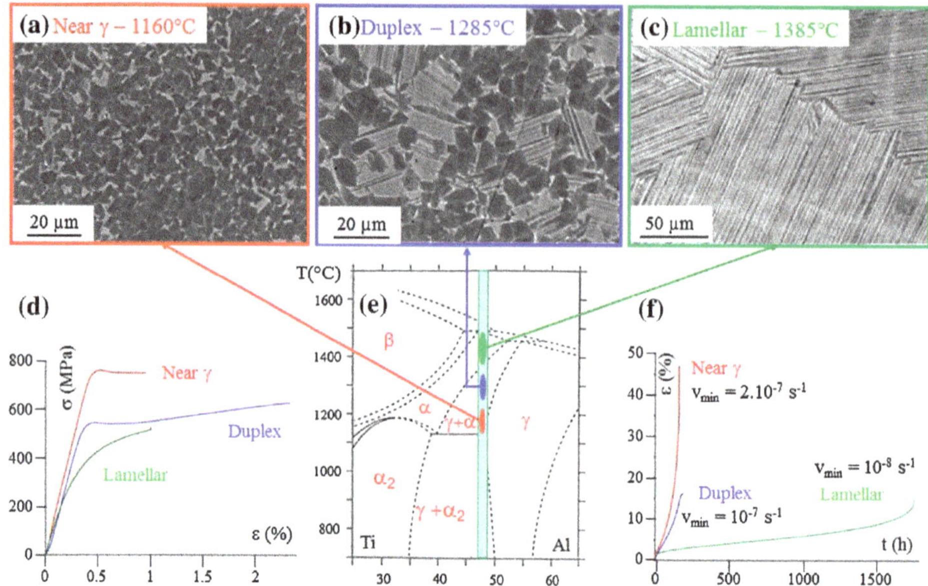

Figure 7. (**a**–**c**) Near γ, duplex, and lamellar microstructure of the Ti-Al$_{48}$-Cr$_2$-Nb$_2$ alloy obtained by elaboration by SPS, compared with the binary phase diagram of this alloy (**e**). Mechanical properties in tension at room temperature (**d**) and in creep (700 °C—300 MPa) (**f**) of the corresponding materials [30]. Reproduced with permission from *Journal of the Minerals, Metals and Materials Society*, published by Springer Nature, 2017.

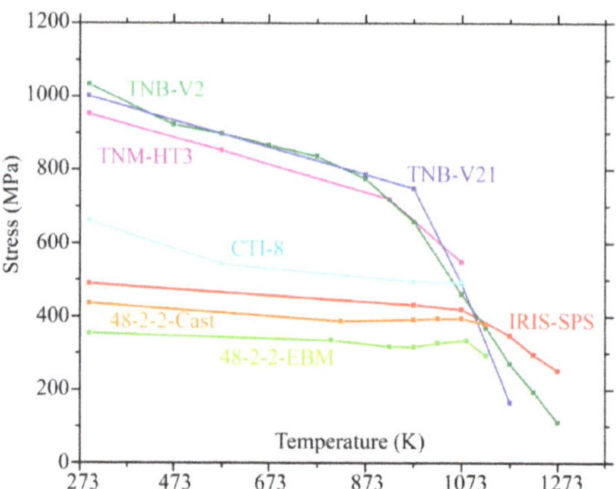

Figure 8. Variation of the yield stress as a function of temperature of the TiAl IRIS alloy elaborated by SPS, compared with that of alloys elaborated by other techniques (TNM-HT3: forging, TNB-V2, TNB-V21 and CTI-8: extrusion, 48-2-2: casting, 48-2-2-EBM: electron beam melting) [33]. Reproduced with permission from *Metallurgical and Materials Transactions: A*, published by Springer Nature, 2016.

4.2. Nanostructured Materials

Limiting the grain coarsening during manufacturing is one of the main advantages of SPS. Densification can occur fast enough for samples to be held at high temperature over short period of time, typically in the order of minutes. Therefore, nanostructured materials can be obtained with potentially original mechanical properties.

Various nanostructured metals and alloys were obtained by SPS using nanopowders such as Cu [34], TiNi [35], Al [7,36], and W [37]. For example, in the case of Al, the use of appropriate SPS parameters allows to keep an extremely fine microstructure, the average grain size being in the order of 50 nm [7]. In addition, using nanopowders [38–42], or micrometric mechanically activated powders [43–45], nanostructured Ni materials have successfully been obtained. In particular, these routes allowed elaboration of nanomaterials exhibiting very high-accumulated strains during cyclic tests [38], as well as significantly enhanced strength (Figure 9) [43]. However, in the case of nanopowders, oxides at the surface of the powder particles were found to affect the cohesion of the material [42], leading to low ductility in compression [39] and tension [41]. Incidentally, these results indicate that a removal of the superficial oxide layers by electric arcs and/or plasma, as proposed in several studies [4–7], does not seem to operate. Finally, original harmonic microstructures consisting in nanograins surrounding large grains could be stabilized with β-Ti due to the rapidity of the SPS process (Figure 10) [46–49]. These materials exhibited enhanced resistance under monotonic solicitation [46,47] and fatigue [48,49].

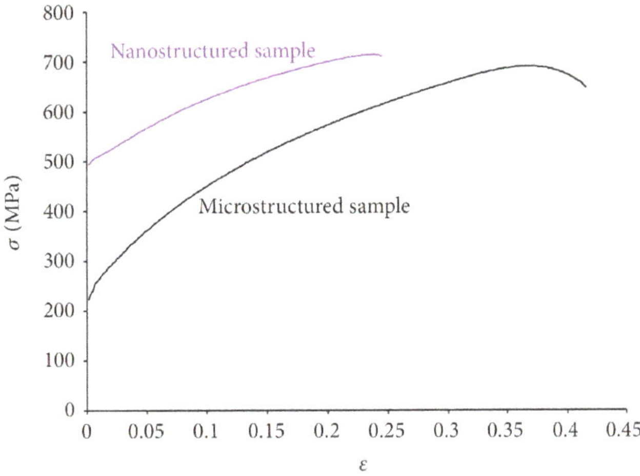

Figure 9. Tensile curves of micro- and nanostructured Ni materials elaborated by SPS [43]. Reproduced with permission from *Journal of Nanomaterials*, published by Hindawi, 2013.

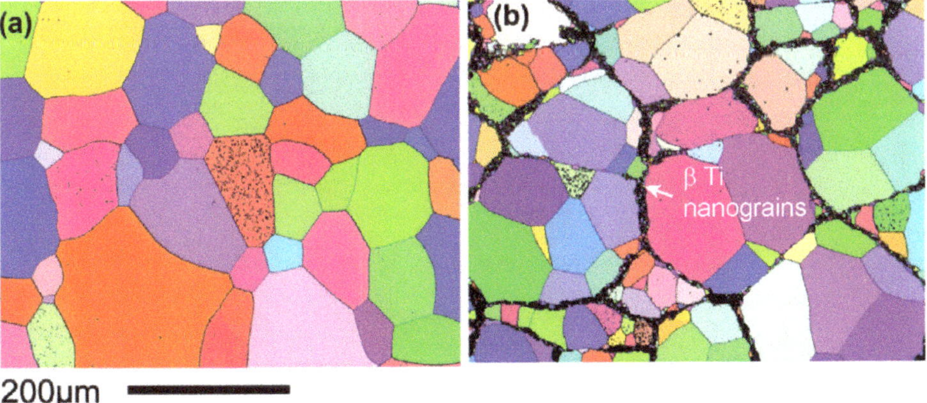

Figure 10. SEM electron back-scattered diffraction inverse pole figure maps of the β-Ti phase comparing the coarse grain microstructure of a material conventionally processed (**a**), to the harmonic microstructure, consisting of large grains surrounded by nanograins, of a material processed by SPS (**b**) [48]. Reproduced with permission from *Scripta Materialia*, published by Elsevier, 2017.

4.3. Metallic Glasses

The SPS technique has shown a high potential for developing amorphous metallic materials. This is first the consequence that the pre-alloyed powders can be prepared by atomization, the very fast cooling rates of the liquid alloy droplets (10^5–10^8 °C/s) [50] during the atomization process, favoring solidification in amorphous state. Therefore, the atomization route has been successfully employed since the late 1970s [51]. Then, to avoid devitrification during processing of the amorphous powder at high temperature, the SPS method is particularly interesting in reason of its rapidity. As such, many metallic glasses have been developed as recently reviewed by Perrière et al. [52]. However, in some cases with Zr-based metallic glasses, devitrification was observed at the powder particle contacts at the early stages of densification, a phenomenon that was interpreted by local overheating effects [19,53]. Such devitrification effects can be detrimental for the development of metallic glasses by SPS, and need now to be better controlled.

4.4. High Entropy Alloys

The use of the SPS for developing high entropy alloys (HEA) results from the better homogeneity obtained by this technique as compared to casting, and also from the very fine microstructures, which can be achieved. However, sintering of elemental powders is most of the time not possible, because of incomplete reaction at the end of the elaboration. Therefore, mechanical alloying (MA) is frequently employed before elaboration by SPS. Surprisingly, there were only few attempts so far to investigate the gas atomization technique [54,55] despite the wide use of this technique to produce pre-alloyed powders. Therefore, the works reported below only refer to materials obtained by SPS after MA.

For example, the MA + SPS route allowed to obtain $Co_{25}Ni_{25}Fe_{25}Al_{7.5}Cu_{17.5}$ alloys exhibiting strength more than eight times higher than materials elaborated by arc melting [56], $Ni_{1.5}Co_{1.5}CrFeTi_{0.5}$ alloys exhibiting very good combination of strength and ductility [57], and other high-strength alloys (e.g., AlCoCrCuFe and NiCoCrCuFe [58], $Al_{0.75}FeNiCrCo$ [59], $Ni_{1.5}CoCuFeCr_{0.5}$ [60]). In all of these cases, the interest of the SPS was the preservation of the nanostructuration. In addition, the possibility to mix different powders by SPS allowed obtaining self-lubricating CoCrFeNi-(Ag, BaF2/CaF2) composites that can reach friction coefficients below 0.26 [61].

In these studies, the rapid densification by SPS allowed retaining nanostructures, which partially explains the high strength. Note that the potential powder contamination during MA by the milling medium or the environment can be overcome by developing suitable methodologies [62]. The main disadvantage of MA remains the long process time. For that, powder gas atomization appears more suited as it has demonstrated successful obtention of several pre-alloyed powders with metals of very different melting points, such as Al, Ti, Re, and W [63].

4.5. Conclusions

The examples shown in this part demonstrate that the SPS allows elaborating materials with elevated mechanical properties. Moreover, the microstructures can be accurately predicted provided that the theoretical phase diagram is known. Therefore, it has been attempted, in some cases, to manufacture parts of larger sizes and of increased geometrical complexity. This is the subject of the next section.

5. Towards Elaboration of Complex Parts

Beside manufacturing materials, it has been demonstrated that the SPS technique was also efficient for shaping. However, it has to be underlined that, if some achievements look impressive, the development of shaping by SPS is not straightforward. Because the processing occurs by submitting a powder to a uniaxial pressure at high temperature, many constraints have to be considered: the electrodes move in only one direction, the volume occupied by the parts evolves due to densification of the powder, the temperature is not homogeneous within the material, and so on. Therefore, it has been proved useful to first develop predictive modeling strategies, based on multi-physics finite element methods coupling the various physical parameters of the technique (electric fields, temperature gradients, stress distribution). After these modeling studies, it is generally necessary to develop suitable SPS cycles for parts of large dimensions. Due to energy limitations, most laboratory SPS equipment have a limit in maximum sample size that is, even though material dependent, usually around 5 cm in diameter for a couple cm in thickness. Finally, shaping of small or large specimen is attempted, using specially designed molds. This is this last part, which is by far the most critical, and different approaches have been proposed.

5.1. Modeling of the SPS Process

Modeling the SPS process is a complex task that requires accounting for several physics at once. The case of conducting metallic materials has been considered in the following references: [64–79]. As depicted by Achenani et al. [78], the ongoing modeling effort can be divided in two categories: thermoelectric coupling to simulate temperature gradients in fully-dense, static conditions [64–67,70,74,76,78,79], and stress distribution calculations to model the densification kinetics [28,68,69,71–73,75,77].

5.1.1. Thermoelectric Coupling

Knowing the temperature distribution in SPS samples during processing is key to control the temperature gradients and ensure full densification and homogeneous microstructures. However, the temperature fields within the samples and molds are complex, because they result from two main contributions. The first one is the heat generated by the Joule effect under the influence of the high intensity electric current. The current distribution being non-uniform, the heat production exhibits also high variations within the sample/mold assembly. The second is the cooling, by heat radiation from the external surfaces and by heat transfer towards the water-cooled electrodes (note that exothermic or endothermic reactions within the samples are here not considered.) Because the influence of these contributions on the resulting temperature fields can be difficult to anticipate, modeling has proved very useful. Thus, the finite element method (FEM) has been widely employed to predict the temperature fields within the sample/molds assembly. Equations coupling the contributions of heat production by Joule effect and of heat conduction are solved, which are generally of the following forms:

$$\nabla \times \nabla(V) = 0, \tag{1}$$

$$\nabla \times (-k\nabla T) + \rho c_p \left(\frac{\partial T}{\partial t} \right) = Q, \tag{2}$$

$$Q = \sigma(\nabla V)^2, \tag{3}$$

where V: electric potential, k: thermal conductivity, ρ: density, c_p: thermal capacity, σ: electrical conductivity, T: temperature and t: time. Using this approach, Anselmi-Tamburini et al. [80] showed that the current density profile across the sample and mold is significantly non-uniform and highly depending on the electrical conductivity of the sample (Figure 11a,b). Consequently, the temperature field is also non-uniform within the molds and the samples (Figure 11c). However, the shape of the thermal gradient cannot be deduced in a straightforward manner from the current density profile, and modeling proves here quite useful. By comparing insulating and conducting materials, Vanmeensel et al. [67] concluded that the amplitude of the temperature gradient was more elevated in the latter case (Figure 12). Similar simulations were developed and experimentally validated for TiAl alloys by Voisin et al. [74]. As discussed in Section 4.1, Ti-Al$_{48}$-Cr$_2$-Nb$_2$ undergoes a microstructure transition from duplex to lamellar at 1335 °C. This was used as an indicator of the local temperature to validate FEM models predictions. It was thus shown that the predicted temperature gradients in TiAl samples was depending on the geometry of the setup, in quantitative agreement with experimental metallographic observations (Figure 13). In particular, the use of elongated molds allowed to move away from each other the hot spots created at the extremities of the molds, and consequently to lower the temperature gradients in the parts.

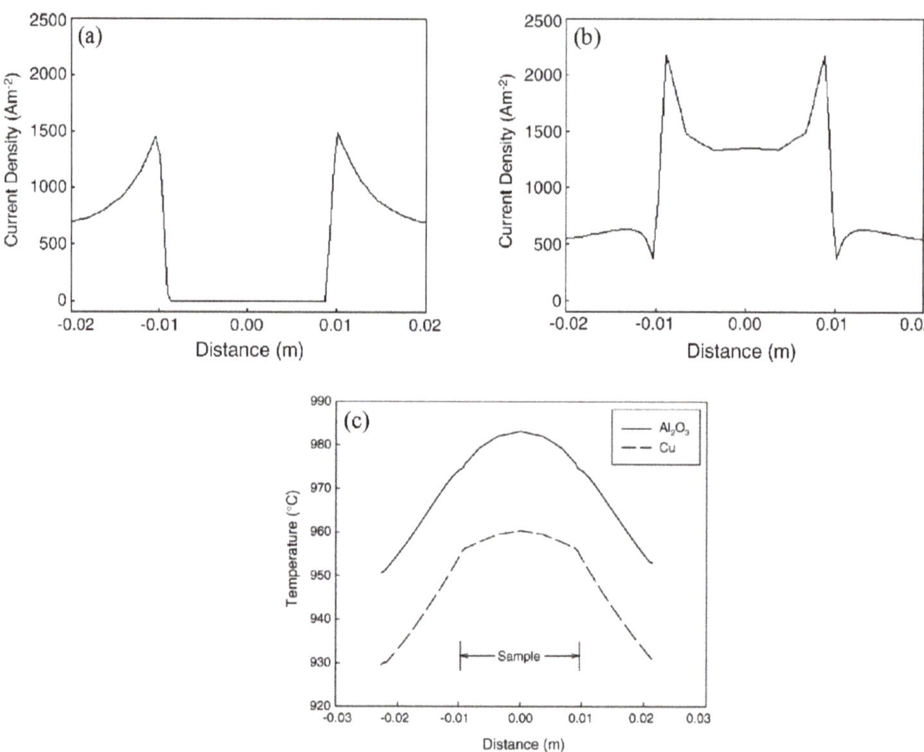

Figure 11. Radial current density distribution for non-conducting and conducting materials (applied voltage: 5V): Al_2O_3 (**a**) and Cu (**b**). Temperature distribution for Al_2O_3 and Cu (constant current applied: 1000 A) (**c**). In all cases, the profile is passing through the center of the samples and though the graphite molds surrounding them [80]. Reproduced with permission from *Materials Science and Engineering: A*, published by Elsevier, 2005.

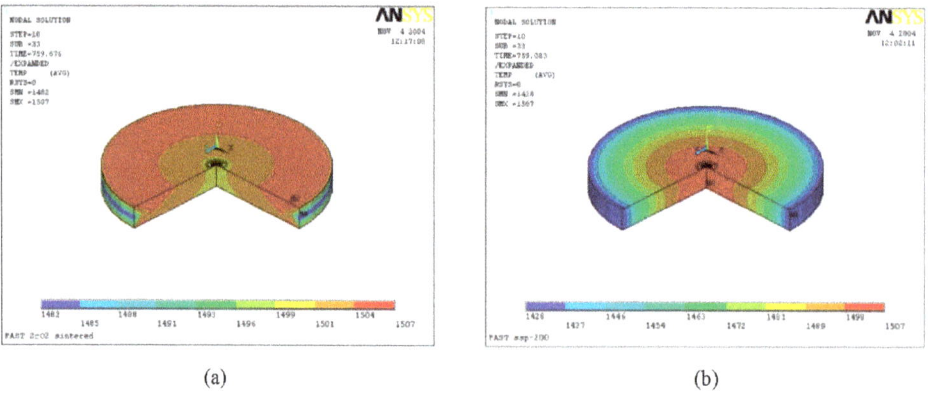

Figure 12. Temperature distribution in insulating ZrO_2 (**a**) and conducting TiN (**b**) samples during holding at 1500 °C [67]. Reproduced with permission from *Acta Materialia*, published by Elsevier, 2005.

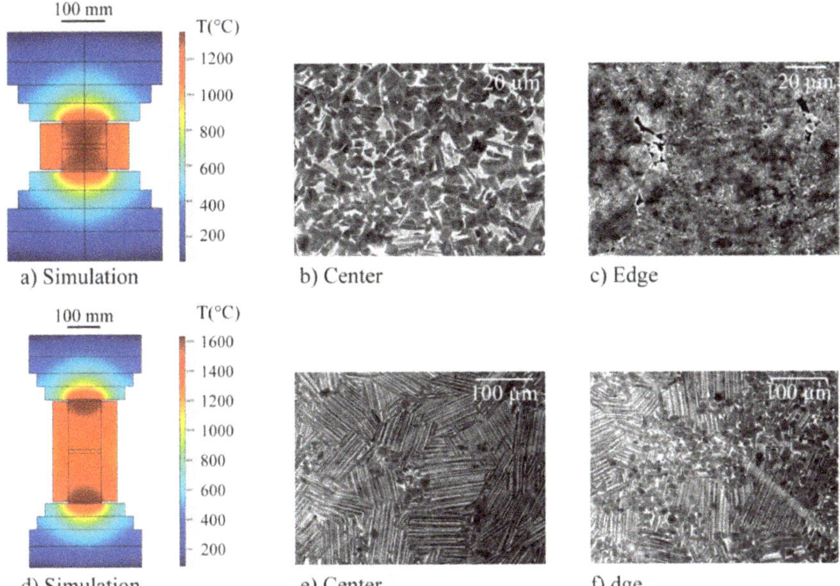

Figure 13. FEM simulations of two SPS setups, compared with metallographic observations of the center and edge of TiAl samples. For setup (**a**), the microstructure of the center is duplex (**b**), and that of the edge is near-γ with some remaining porosities (**c**). This indicates a temperature gradient between the center and the edge of the sample of about 100 °C, in reasonable agreement with the FEM modeling (**a**), which predict a gradient of 125 °C. For setup (**d**), the FEM simulation predicts a temperature gradient between center and edge of the samples of 30 °C. The corresponding microstructures are near lamellar at the center (**e**) and duplex at the edge (**f**), indicating a thermal gradient of 25–50 °C, close to the simulations [74]. Reproduced with permission from *Journal of Materials Processing Technology*, published by Elsevier, 2013.

In summary, the temperature distribution throughout samples during SPS can be accurately predicted with FEM. Indeed, it is a necessary tool for designing molds that minimize thermal gradients and increase the microstructure homogeneity while limiting more expensive and time-consuming experimental testing.

5.1.2. Modeling of Densification Kinetics

There are two ways to model the densification of porous materials: a micromechanical approach based on analytical calculations [81,82] or continuum mechanics derivations [83,84]. The interest of the micromechanical approach is its simplicity, but its limitation is that the model is highly idealized (random dense packing of equal spheres [81]), and can therefore describe too roughly the actual geometrical features of the problem (e.g., powder particles sizes distributed over granulometry ranges). Moreover, the expressions are discontinuous for a density D of 0.9. Despite these shortcomings, it proved to be reasonably accurate. However, to increase results precision, more complex models based on continuum mechanics have been developed notably by Abouaf and Olevsky.

One common input to these calculations is the material constitutive law. For metals, the constitutive laws are most of the time of the following form (Norton relation):

$$\dot{\varepsilon} = A_0 exp\left[-\frac{Q}{RT}\right]\sigma^n, \tag{4}$$

where $\dot{\varepsilon}$: deformation rate, σ: stress, Q: activation energy, n: stress exponent, A_0: material constant, R: gas constant and T: temperature. Using micromechanical or continuum mechanics models, simulations of the densification kinetics have been successfully performed for TiAl [28,73,85], ZrC [86], Ni [77], Fe and Al [87], and Ti-6Al-4V [88]. An example of such calculations applied to TiAl is shown in Figure 14, in the case of a continuum mechanics model. It can be seen that the calculated densification kinetics reproduces the experimental one with an accuracy better that 5%. Therefore, the models seem to be reasonably accurate.

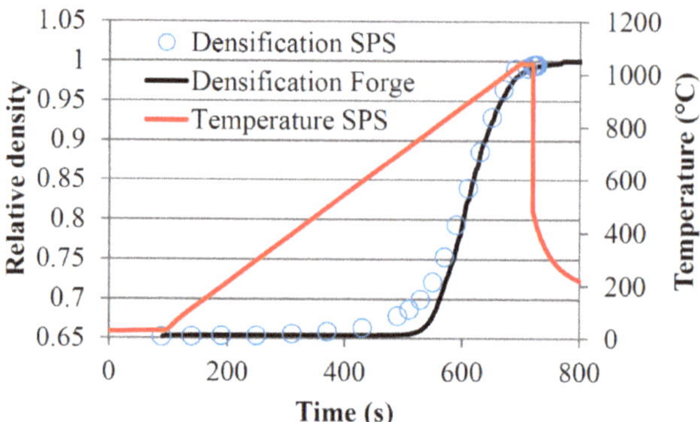

Figure 14. Densification kinetics by SPS of a TiAl alloy, compared with modeling by Forge software and continuum mechanics, using constitutive relations of bulk TiAl [85]. Reproduced with permission from *Intermetallics*, published by Elsevier, 2017.

However, these calculations are very sensitive to the material stress exponent [28]. For example, Figure 15a shows constitutive relations for a recrystallized TiAl powder, in which the activation parameters have been experimentally determined: $n = 1.9$, and $Q = 308$ kJ/mol [28]. This law agrees with literature data for bulk TiAl and well captures the SPS kinetics (Figure 15b). However, if n is changed to 3 for a same Q (= 308 kJ/mol) for example, the constitutive relations obtained are still in reasonable agreement with the literature data (Figure 15a), but the model predication is no longer in phase with the experiment (Figure 15b). Therefore, it appears that the models used to predict the SPS densification behavior are very sensitive to the material parameters, which are difficult to accurately determine. Moreover, as can be seen in Figure 15a, the constitutive relations from literature are significantly scattered, and making proper choices of n and Q can be difficult. Therefore, the determination of the material constants has to be carefully carried out.

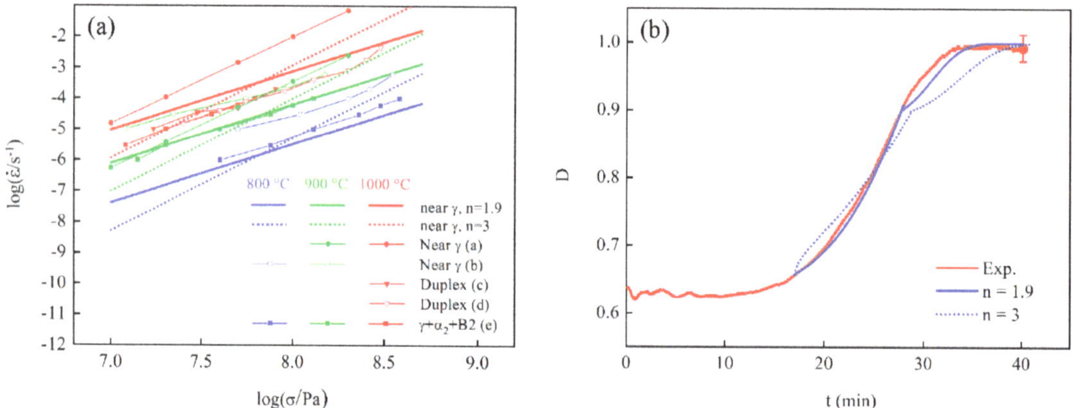

Figure 15. (a) Norton laws of a near γ TiAl alloy, taking $n = 1.9$ and $n = 3$, the activation energy being identical ($Q = 308$ kJ/mol). Comparison with literature data of alloys of similar near-γ and duplex microstructures ((a): [85], (b): [89], (c): [90], (d): [91], (e): [92]). (b) Densification curves of TiAl (D: relative density, t: time) calculated using $n = 1.9$ [28], and $n = 3$, compared with experiment.

5.2. Up-Scaling

Manufacturing big parts is an important issue, to obtain pieces of dimensions suitable for industrial applications. Sintering of metallic disks up to 350 mm in diameter has been reported [93], but a limited number of studies have been published to date. Early works report on processing of large specimens constituted of the following materials: Al [94], Ti (Figure 16) [95], TiAl [74,96]. One of the main difficulties, when processing parts of large dimensions is the presence of thermal and hence microstructure gradients in the materials, which are amplified with respect to those existing in small size samples. Therefore, the geometry of the molds needs to be optimized to homogenize the temperature distribution as previously discussed in Section 5.1.1 by using predictive FEM simulations. Another method is emerging with hybrid SPS machines. Since the temperature gradients are caused by a difference in current density between the center and the edges of the samples, a complimentary external furnace is added to heat the entire mold, thus limiting the temperature gradients. Unfortunately, it seems that no work on this promising approach has been published to date.

Figure 16. Ti composite part of large dimensions (20 cm in diameter), elaborated by SPS [95]. Reproduced with permission from *Materials Science and Engineering: A*, published by Elsevier, 2016.

5.3. Complex Shaping

Finally, shaping of parts of complex geometries is also an important issue, because this would allow to avoid or to limit machining, which can be difficult or even impossible, depending on the material. An early study reported the sintering of WC/Co/Ni screws for extrusion machines by SPS [93]. Unfortunately, little information is provided concerning the methodology. Later on, a method based on the deformation of an interface between a metallic powder and using sacrificial powders transmitting the pressure to the component has allowed forming 3 cm long CoNiCrAlY turbine blades (Figure 17) [97]. Finally, studies have reported on the development of parts in TiAl and NbSi reproducing the shape of turbine blades of aircraft engines, constituted of elongated and massive portions, as long as 10 cm (Figure 18) [30,32,98]. Here, the main challenge lied in compacting regions of different thicknesses such as encountered in this case between the elongated and massive portions. This was overcome by using separate plungers [98].

5.4. Conclusions

The development of simulations of thermoelectric coupling and of densification kinetics proved to be key to optimizing the processing of complex parts. This enabled up-scaling laboratory proof-of-concept to full-size components. Near-net shaping complex, full-scale parts has been successfully demonstrated, opening the doors to a wide range of possibilities, notably for materials hard to obtain with conventional manufacturing techniques.

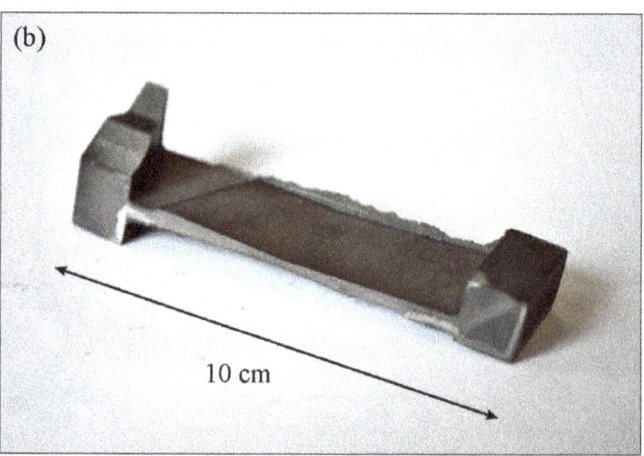

Figure 17. Example of turbine blade in CoNiCrAlY obtained by the deformed interfaces method. (**a**) Target shape. (**b**) Complex part obtained by SPS [97]. Reproduced with permission from *Powder Technology*, published by Elsevier, 2017.

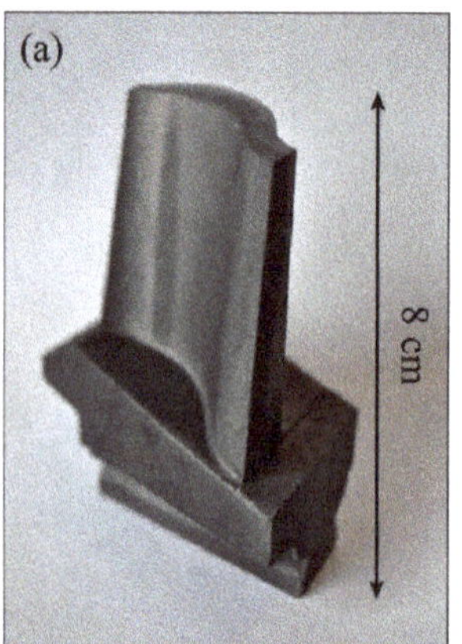

Figure 18. Turbine blade preforms in TiAl (**a**) 8 cm in height [32], and (**b**) 10 cm long [98], shaped by SPS. Reproduced with permission from *Advanced Engineering Materials*, published by John Wiley and Sons, 2015, and from *Springer eBook*, published by Springer Nature, 2019.

However, the challenge is now to implement these approaches in industrial processes, and to develop integrated production sectors. For this purpose, two critical steps can be identified: production of high quality powders, and automation of the SPS process. If satisfactory answers are given to these questions, the SPS technology could potentially experience a large development, because of the elevated properties reached in many cases by the elaborated materials. Therefore, the development of SPS-based processes deserves now increased industrial efforts.

6. General Conclusions

Over the past two decades, SPS has emerged as a technique of choice for producing innovative materials with optimized microstructures and properties, and for shaping complex parts. Despite the absence of demonstrated current-induced effects, the rapid heating by Joule effect allowing short densification cycles appears very efficient at retaining fine, homogeneous microstructures. Design of equilibrium or metastable microstructures can be controlled, depending on the alloy and on the processing conditions. As a result, a wide range of materials can be obtained with improved mechanical properties such as nanostructured materials, intermetallics, metallic glasses, and high entropy alloys. Moving forward, introducing SPS to new industrial chains will require stepping toward more automated processes, allowing production of parts in larger quantities.

Author Contributions: Methodology, J.-P.M., Z.T., T.V.; software, L.D., T.V.; investigation, J.-P.M., Z.T., T.V.; writing—original draft preparation, J.-P.M.; writing—review and editing, J.-P.M., A.C., L.D., T.V., Z.T., M.T.; supervision, A.C., M.T.; project administration, A.C.; funding acquisition, A.C. All authors have read and agreed to the published version of the manuscript.

Funding: This research was funded by Agence Nationale de la Recherche, grant numbers MF2-ANR-2011-PBS09-020 and IRIS-ANR-08-MAPR-0018. T. Voisin was supported by the Laboratory Directed Research and Development (LDRD) program (21-LW-027) at Lawrence Livermore National Laboratory. T. Voisin's work was performed under the auspices of the U.S. Department of Energy by Lawrence Livermore National Laboratory under contract no. DE-AC52-07NA27344.

Institutional Review Board Statement: Not applicable.

Informed Consent Statement: Not applicable.

Data Availability Statement: Data is contained within the article.

Acknowledgments: SPS experiments were performed using the Sumitomo 2080 SPS machine of the Plateforme Nationale de Frittage Flash du CNRS, Paul Sabatier University, Toulouse, France.

Conflicts of Interest: The authors declare no conflict of interest. The funders had no role in the design of the study; in the collection, analyses, or interpretation of data; in the writing of the manuscript, or in the decision to publish the results.

References

1. Orru, R.; Licheri, R.; Locci, A.M.; Cincotti, A.; Cao, G. Consolidation/synthesis of materials by electric current activated/assisted sintering. *Mater. Sci. Eng. R Rep.* **2009**, *63*, 127–287. [CrossRef]
2. Munir, Z.A.; Anselmi-Tamburini, U.; Ohyanagi, M. The effect of electric field and pres-sure on the synthesis and consolidation of materials: A review of the spark plasma sintering method. *J. Mater. Sci.* **2006**, *41*, 763–777. [CrossRef]
3. Tokita, M. Development of large-size ceramic/metal bulk FGM fabricated by spark plasma sintering. *Mater. Sci. Forum* **1999**, *308–311*, 83–88. [CrossRef]
4. Omori, M. Sintering, consolidation, reaction and crystal growth by the spark plasma system (SPS). *Mater. Sci. Eng. A* **2000**, *287*, 183–188. [CrossRef]
5. Okazaki, K. Electro-discharge consolidation applied to nanocrystalline and RSP/MA powders. *Mater. Sci. Eng. A* **2000**, *287*, 189–197. [CrossRef]
6. Groza, J.R.; Zavaliangos, A. Sintering activation by external electrical field. *Mater. Sci. Eng. A* **2000**, *287*, 171–177. [CrossRef]
7. Ye, J.; Ajdelsztajn, L.; Schoenung, J.M. Bulk nanocrystalline aluminum 5083 alloy fabricated by a novel technique: Cryomilling and spark plasma sintering. *Metall. Mater. Trans. A* **2006**, *37*, 2569–2579. [CrossRef]
8. Yanagisawa, O.; Kuramoto, H.; Matsugi, K.; Komatsu, M. Observation of particle behavior in copper powder compact during pulsed electric discharge. *Mat. Sci. Eng. A* **2003**, *350*, 184–189. [CrossRef]
9. Saunders, T.; Grasso, S.; Reece, M.J. Plasma formation during electric discharge (50 V) through conductive powder compacts. *J. Eur. Ceram. Soc.* **2015**, *35*, 871–877. [CrossRef]
10. Hulbert, D.M.; Anders, A.; Andersson, J.; Lavernia, E.J.; Mukherjee, A.K. A discussion on the absence of plasma in spark plasma sintering. *Scripta Mater.* **2009**, *60*, 835–838. [CrossRef]
11. Song, X.; Liu, X.; Zhang, J. Neck Formation and Self-Adjusting Mechanism of Neck Growth of Conducting Powders in Spark Plasma Sintering. *J. Am. Ceram. Soc.* **2006**, *89*, 494–500. [CrossRef]
12. Xiong, Y.; Liu, D.; Li, Y.; Zheng, B.; Haines, C.; Paras, J.; Martin, D.; Kapoor, D.; Lavernia, E.J.; Schoenung, J.M. Spark plasma sintering of cryomilled nanocrystalline Al alloy—Part I: Microstructure evolution. *Metall. Mater. Trans. A* **2012**, *43*, 327–339. [CrossRef]
13. Diouf, J.S.; Fedrizzi, A.; Molinari, A. A fractographic and microstructural analysis of the neck regions of coarse copper particles consolidated by SPS. *Powder Tech.* **2013**, *221*, 220–227. [CrossRef]
14. Trapp, J.; Kieback, B. Temperature distribution in metallic powder particles during initial stage of field-activated sintering. *J. Am. Ceram. Soc.* **2015**, *98*, 3547–3552. [CrossRef]
15. Collard, C.; Trzaska, Z.; Durand, L.; Chaix, J.M.; Monchoux, J.P. Theoretical and experimental investigations of local overheating at particle contacts in spark plasma sintering. *Powder Tech.* **2017**, *321*, 458–470. [CrossRef]
16. Trzaska, Z.; Collard, C.; Durand, L.; Couret, A.; Chaix, J.M.; Fantozzi, G.; Monchoux, J.P. SPS microscopic mechanisms of metallic systems: Experiments and simulations. *J. Am. Ceram. Soc.* **2019**, *102*, 654–661.

17. Frei, J.M.; Anselmi-Tamburini, U.; Munir, Z.A. Current effects on neck growth in the sintering of copper spheres to copper plates by the pulsed electric current method. *J. Appl. Phys.* **2007**, *101*, 114914. [CrossRef]
18. Ji, G.; Grosdidier, T.; Bozzolo, N.; Launois, S. The mechanisms of microstructure formation in a nanostructured oxide dispersion strengthened FeAl alloy obtained by spark plasma sintering. *Intermetallics* **2007**, *15*, 108–118. [CrossRef]
19. Nowak, S.; Perrière, L.; Dembinski, L.; Tusseau-Nenez, S.; Champion, Y. Approach of the spark plasma sintering mechanism in Zr57Cu20Al10Ni8Ti5 metallic glass. *J. Alloy. Compd.* **2011**, *509*, 1011–1019. [CrossRef]
20. Sprecher, A.F.; Mannan, S.L.; Conrad, H. On the mechanisms for the electroplastic effect in metals. *Acta Metall.* **1986**, *34*, 1145–1162. [CrossRef]
21. Conrad, H.; Sprecher, A.F.; Cao, W.D.; Lu, X.P. Electroplasticity—The effect of electricity on the mechanical properties of metals. *JOM* **1990**, *42*, 28–33. [CrossRef]
22. Guan, L.; Tang, G.; Chu, P.K. Recent advances and challenges in electroplastic manufacturing processing of metals. *J. Mater. Res.* **2010**, *25*, 1215–1224. [CrossRef]
23. Trzaska, Z.; Couret, A.; Monchoux, J.P. Spark plasma sintering mechanisms at the necks between TiAl powder particles. *Acta Mater.* **2016**, *118*, 100–108. [CrossRef]
24. Trzaska, Z.; Cours, R.; Monchoux, J.P. Densification of Ni and TiAl by SPS: Kinetics and microscopic mechanisms. *Metall. Mater. Trans. A* **2018**, *49*, 4849–4859. [CrossRef]
25. Minier, L.; Le Gallet, S.; Grin, Y.; Bernard, F. Influence of the current flow on the SPS sintering of a Ni powder. *J. Alloys. Compd.* **2010**, *508*, 412–418. [CrossRef]
26. Aleksandrova, E.V.; Ilyina, A.M.; Grigoryev, E.G.; Olevsky, E.A. Contribution of electric current into densification kinetics during spark plasma sintering of conductive powder. *J. Am. Ceram. Soc.* **2015**, *98*, 3509–3517. [CrossRef]
27. Collet, R.; Le Gallet, S.; Naimi, F.; Charlot, F.; Bonnefont, G.; Fantozzi, G.; Chaix, J.M.; Bernard, F. Effect of current on the sintering of pre-oxidized copper powders by SPS. *J. Alloy. Compd.* **2017**, *692*, 478–484.
28. Trzaska, Z.; Bonnefont, G.; Fantozzi, G.; Monchoux, J.P. Comparison of densification kinetics of a TiAl powder by spark plasma sintering and hot pressing. *Acta Mater.* **2017**, *135*, 1–13. [CrossRef]
29. Bewlay, B.P.; Nag, S.; Suzuki, A.; Weimer, M.J. TiAl alloys in commercial aircraft engines. *Mater. High Temp.* **2016**, *33*, 549–559. [CrossRef]
30. Couret, A.; Voisin, T.; Thomas, M.; Monchoux, J.P. Development of a TiAl alloy by spark plasma sintering. *JOM* **2017**, *69*, 2576–2582. [CrossRef]
31. Voisin, T.; Monchoux, J.P.; Perrut, M.; Couret, A. Obtaining of a fine near-lamellar microstructure in TiAl alloys by Spark Plasma Sintering. *Intermetallics* **2016**, *71*, 88–97. [CrossRef]
32. Voisin, T.; Monchoux, J.P.; Durand, L.; Karnatak, N.; Thomas, M.; Couret, A. An innovative way to produce-TiAl blades: Spark plasma sintering. *Adv. Eng. Mater.* **2015**, *17*, 1408–1413. [CrossRef]
33. Voisin, T.; Monchoux, J.P.; Thomas, M.; Deshayes, C.; Couret, A. Mechanical properties of the TiAl IRIS alloy. *Metall. Mater. Trans. A* **2016**, *47*, 6097–6108. [CrossRef]
34. Zhang, Z.; Wang, F.; Lee, S.; Liu, Y.; Cheng, J.; Liang, Y. Microstructure characteristic, mechanical properties and sintering mechanism of nanocrystalline copper obtained by SPS process. *Mat. Sci. Eng. A* **2009**, *523*, 134–138. [CrossRef]
35. Shearwood, C.; Fu, Y.; Yu, L.; Khor, K. Spark plasma sintering of TiNi nano-powder. *Scripta Mater.* **2005**, *52*, 455–460. [CrossRef]
36. Kellogg, F.; McWilliams, B.; Sietins, J.; Giri, A.; Cho, K. Comparison of SPS processing behavior between as atomized and cryomilled aluminum alloy 5083 powder. *Metall. Mater. Trans A* **2017**, *48*, 5492–5499. [CrossRef]
37. El-Atwania, O.; Quach, D.V.; Efe, M.; Cantwell, P.R.; Heim, B.; Schultz, B.; Stach, E.A.; Groza, J.R.; Allain, J.P. Multimodal grain size distribution and high hardness in fine grained tungsten fabricated by spark plasma sintering. *Mat. Sci. Eng. A* **2011**, *528*, 5670–5677. [CrossRef]
38. Dirras, G.; Bouvier, S.; Gubicza, J.; Hasni, B.; Szilagyi, T. Mechanical characteristics under monotonic and cyclic simple shear of spark plasma sintered ultrafine-grained nickel. *Mat. Sci. Eng. A* **2009**, *526*, 201–210. [CrossRef]
39. Bui, Q.H.; Dirras, G.; Ramtani, S.; Gubicza, J. On the strengthening behavior of ultrafine-grained nickel processed from nanopowders. *Mat. Sci. Eng. A* **2010**, *527*, 3227–3235. [CrossRef]
40. Dutel, G.D.; Tingaud, D.; Langlois, P.; Dirras, G. Nickel with multimodal grain size distribution achieved by SPS: Microstructure and mechanical properties. *J. Mater. Sci.* **2012**, *47*, 7926–7931. [CrossRef]
41. Farbaniec, L.; Dirras, G.; Krawczynska, A.; Mompiou, F.; Couque, H.; Naimi, F.; Bernard, F.; Tingaud, D. Powder metallurgy processing and deformation characteristics of bulk multimodal nickel. *Mater. Charac.* **2014**, *94*, 126–137. [CrossRef]
42. Tingaud, D.; Jenei, P.; Krawczynska, A.; Mompiou, F.; Gubicza, J.; Dirras, G. Investigation of deformation micro-mechanisms in nickel consolidated from a bimodal powder by spark plasma sintering. *Mater. Charac.* **2015**, *99*, 118–127. [CrossRef]
43. Naimi, F.; Minier, L.; Le Gallet, S.; Couque, H.; Bernard, F. Dense nanostructured nickel produced by SPS from mechanically activated powders: Enhancement of mechanical properties. *J. Nanomater.* **2013**, *2013*, 674843. [CrossRef]
44. García de la Cruz, L.; Flipon, B.; Keller, C.; Martinez, M.; Hug, E. Nanostructuration of metals via spark plasma sintering using activated powder obtained by ball-milling: Impact on the strain-hardening mechanisms. *AIP Conf. Proc.* **2017**, *1896*, 200002.
45. García de la Cruz, L.; Martinez, M.; Keller, C.; Hug, E. Achieving good tensile properties in ultrafine grained nickel by spark plasma sintering. *Mat. Sci. Eng. A* **2020**, *772*, 138770. [CrossRef]

46. Vajpai, S.K.; Yu, H.; Ota, M.; Watanabe, I.; Dirras, G.; Ameyama, K. Three-dimensionally gradient and periodic harmonic structure for high performance advanced structural materials. *Mater. Trans.* **2016**, *57*, 1424–1432. [CrossRef]

47. Dirras, G.; Tingaud, D.; Ueda, D.; Hocini, A.; Ameyama, K. Dynamic Hall-Petch versus grain-size gradient effects on the mechanical behavior under simple shear loading of β-titanium Ti-25Nb-25Zr alloys. *Mater. Lett.* **2017**, *206*, 214–216. [CrossRef]

48. Dirras, G.; Ueda, D.; Hocini, A.; Tingaud, D.; Ameyama, K. Cyclic shear behavior of conventional and harmonic structure-designed Ti-25Nb-25Zr β-titanium alloy: Back-stress hardening and twinning inhibition. *Scripta Mater.* **2017**, *138*, 44–47. [CrossRef]

49. Ueda, D.; Dirras, G.; Hocini, A.; Tingaud, D.; Ameyama, K.; Langlois, P.; Vrel, D.; Trzaska, Z. Data on processing of Ti-25Nb-25Zr β-titanium alloys via powder metallurgy route: Methodology, microstructure and mechanical properties. *Data Brief* **2018**, *17*, 703–708. [CrossRef]

50. Charpentier, M. Hétérogénéités Héritées de la Solidification et Formation des Microstructures Dans L'alliage Ti48Al2Cr2Nb. Contribution au Développement des Alliages Intermétalliques de Base γ-TiAl. Ph.D. Thesis, Ecole des Mines, Nancy, France, 2003.

51. Miller, S.A.; Murphy, R.J. A gas-water atomization process for producing amorphous powders. *Scripta Metall.* **1979**, *13*, 673–676. [CrossRef]

52. Perrière, L.; Champion, Y.; Bernard, F. Spark plasma sintering of metallic glasses. In *Spark Plasma Sintering of Materials*; Cavaliere, P., Ed.; Springer: Cham, Switzerland, 2019; pp. 291–335.

53. Perrière, L.; Thai, M.T.; Tusseau-Nenez, S.; Blétry, M.; Champion, Y. Spark plasma sintering of a Zr-based metallic glass. *Adv. Eng. Mater.* **2011**, *13*, 581–586. [CrossRef]

54. Liu, Y.; Wang, J.S.; Fang, Q.H.; Liu, B.; Wu, Y.; Chen, S.Q. Preparation of superfine-grained high entropy alloy by spark plasma sintering gas atomized powder. *Intermetallics* **2016**, *68*, 16–22. [CrossRef]

55. Zhou, P.F.; Xiao, D.H.; Wu, Z.; Song, M. Microstructure and mechanical properties of AlCoCrFeNi high entropy alloys produced by spark plasma sintering. *Mater. Res. Express* **2019**, *6*, 0865e7. [CrossRef]

56. Fu, Z.; Chen, W.; Wen, H.; Zhang, D.; Chen, Z.; Zheng, B.; Zhou, Y.; Lavernia, E.J. Microstructure and strengthening mechanism in an fcc structured single-phase nanocrystalline Co25Ni25Fe25Al7.5Cu17.5 high-entropy alloy. *Acta Mater.* **2016**, *107*, 59–71. [CrossRef]

57. Moravcik, I.; Cizek, J.; Zapletal, J.; Kovacova, Z.; Vesely, J.; Minarik, P.; Kitzmantel, M.; Neubauer, E.; Dlouhy, I. Microstructure and mechanical properties of Ni1,5Co1,5CrFeTi0,5 high entropy alloy fabricated by mechanical alloying and spark plasma sintering. *Mater. Des.* **2017**, *119*, 141–150. [CrossRef]

58. Praveen, S.; Murty, B.S.; Kottada, R.S. Alloying behavior in multi-component AlCoCrCuFe and NiCoCrCuFe high entropy alloys. *Mater. Sci. Eng. A* **2012**, *534*, 83–89. [CrossRef]

59. Chen, Z.; Chen, W.; Wu, B.; Cao, X.; Liu, L.; Fu, Z. Effects of Co and Ti on microstructure and mechanical behaviour of Al0.75FeNiCrCo high entropy alloy prepared by mechanical alloying and spark plasma sintering. *Mater. Sci. Eng. A* **2015**, *648*, 217–224. [CrossRef]

60. Wang, P.; Cai, H.; Cheng, X. Effect of Ni/Cr ratio on phase, microstructure and mechanical properties of $Ni_xCoCuFeCr_{2-x}$ (x = 1.0, 1.2, 1.5, 1.8 moL) high entropy alloys. *J. Alloy. Compd.* **2016**, *662*, 20–31. [CrossRef]

61. Zhang, A.; Han, J.; Su, B.; Meng, J. A novel CoCrFeNi high entropy alloy matrix selflubricating composite. *J. Alloy. Compd.* **2017**, *725*, 700–710. [CrossRef]

62. Suryanarayana, C. Mechanical alloying and milling. *Prog. Mater. Sci.* **2001**, *46*, 1–184. [CrossRef]

63. Jabbar, H.; Couret, A.; Durand, L.; Monchoux, J.P. Identification of microstructural mechanisms during densification of a TiAl alloy by spark plasma sintering. *J. Alloy. Compd.* **2011**, *509*, 9826–9835. [CrossRef]

64. Yucheng, W.; Zhengyi, F. Study of temperature field in spark plasma sintering. *Mater. Sci. Eng. B* **2002**, *90*, 34–37. [CrossRef]

65. Matsugi, K.; Kuramoto, H.; Hatayama, T.; Yanagisawa, O. Temperature distribution at steady state under constant current discharge in spark sintering process of Ti and Al2O3 powders. *J. Mater. Process. Tech.* **2004**, *146*, 274–281. [CrossRef]

66. Zavaliangos, A.; Zhang, J.; Krammer, M.; Groza, J.R. Temperature evolution during field activated sintering. *Mater. Sci. Eng. A* **2004**, *379*, 218–228. [CrossRef]

67. Vanmeensel, K.; Laptev, A.; Hennicke, J.; Vleugels, J.; Van der Biest, O. Modelling of the temperature distribution during field assisted sintering. *Acta Mater.* **2005**, *53*, 4379–4388. [CrossRef]

68. Wang, X.; Casolco, S.R.; Xu, G.; Garay, J.E. Finite element modeling of electric current activated sintering: The effect of coupled electrical potential, temperature and stress. *Acta Mater.* **2007**, *55*, 3611–3622. [CrossRef]

69. McWilliams, B.; Zavaliangos, A. Multi-phenomena simulation of electric field assisted sintering. *J. Mater. Sci.* **2008**, *43*, 5031–5035. [CrossRef]

70. Molénat, G.; Durand, L.; Galy, J.; Couret, A. Temperature control in spark plasma sintering: An FEM approach. *J. Metallurgy* **2010**, *145431*. [CrossRef]

71. Muñoz, S.; Anselmi-Tamburini, U. Temperature and stress fields evolution during spark plasma sintering processes. *J. Mater. Sci.* **2010**, *45*, 6528–6539. [CrossRef]

72. Song, Y.; Li, Y.; Zhou, Z.; Lai, Y.; Ye, Y. A multi-field, coupled FEM model for one-step forming process of spark plasma sintering considering local densification of powder material. *J. Mater. Sci.* **2011**, *46*, 5645–5656. [CrossRef]

73. Mondalek, P.; Silva, L.; Bellet, M. A numerical model for powder densification by SPS technique. *Adv. Eng. Mater.* **2011**, *13*, 587–593. [CrossRef]

74. Voisin, T.; Durand, L.; Karnatak, N.; Le Gallet, S.; Thomas, M.; Le Berre, Y.; Castagné, J.F.; Couret, A. Temperature control during Spark Plasma Sintering and application to up-scaling and complex shaping. *J. Mater. Proc. Tech.* **2013**, *213*, 269–278. [CrossRef]

75. Wolff, C.; Mercier, S.; Couque, H.; Molinari, A. Modeling of conventional hot compaction and spark plasma sintering based on modified micromechanical models of porous materials. *Mech. Mater.* **2012**, *49*, 72–91. [CrossRef]

76. Pavia, A.; Durand, L.; Ajustron, F.; Bley, V.; Chevallier, G.; Peigney, A.; Estournès, C. Electro-thermal measurements and finite element method simulations of a spark plasma sintering device. *J. Mater. Process. Tech.* **2013**, *213*, 1327–1336. [CrossRef]

77. Wolff, C.; Mercier, S.; Couque, H.; Molinari, A.; Bernard, F.; Naimi, F. Thermal-electrical-mechanical simulation of the nickel densification by Spark Plasma Sintering. Comparison with experiments. *Mech. Mater.* **2016**, *100*, 126–147. [CrossRef]

78. Achenani, Y.; Saâdaoui, M.; Cheddadi, A.; Bonnefont, G.; Fantozzi, G. Finite element modeling of spark plasma sintering: Application to the reduction of temperature inhomogeneities, case of alumina. *Mater. Des.* **2017**, *116*, 504–514. [CrossRef]

79. Manière, C.; Pavia, A.; Durand, L.; Chevallier, G.; Afanga, K.; Estournès, C. Finite-element modeling of the electro-thermal contacts in the spark plasma sintering process. *J. Eur. Ceram. Soc.* **2016**, *36*, 741–748. [CrossRef]

80. Anselmi-Tamburini, U.; Gennari, S.; Garay, J.E.; Munir, Z.A. Fundamental investigations on the spark plasma sintering/synthesis process: II. Modeling of current and temperature distributions. *Mater. Sci. Eng. A* **2005**, *394*, 139–148. [CrossRef]

81. Arzt, E.; Ashby, M.F.; Easterling, K.E. Practical applications of hot-isostatic pressing dia-grams: Four case studies. *Metall. Trans. A* **1983**, *14*, 211–221. [CrossRef]

82. Helle, A.S.; Easterling, K.E.; Ashby, M.F. Hot-isostatic pressing diagrams: New develop-ments. *Acta Metall.* **1985**, *33*, 2163–2174. [CrossRef]

83. Abouaf, M. Modélisation de la Compaction de Poudres Métalliques Frittées—Approche par la Mécanique des Milieux Continus. Ph.D. Thesis, Institut National Polytechnique, Grenoble, France, 1985.

84. Olevsky, E.A. Theory of sintering: From discrete to continuum. *Mat. Sci. Eng. R* **1998**, *23*, 41–100. [CrossRef]

85. Martins, D.; Grumbach, F.; Manière, C.; Sallot, P.; Mocellin, K.; Bellet, M.; Estournès, C. In-situ creep law determination for modeling spark plasma sintering of TiAl 48-2-2 powder. *Intermetallics* **2017**, *86*, 147–155. [CrossRef]

86. Wei, X.; Back, C.; Izhvanov, O.; Khasanov, O.L.; Haines, C.D.; Olevsky, E.A. Spark plasma sintering of commercial zirconium carbide powders: Densification behavior and mechanical properties. *Materials* **2015**, *8*, 6043–6061. [CrossRef]

87. Manière, C.; Durand, L.; Chevallier, G.; Estournès, C. A spark plasma sintering densification modeling approach: From polymer, metals to ceramics. *J. Mater. Sci.* **2018**, *53*, 7869–7876. [CrossRef]

88. Manière, C.; Kus, U.; Durand, L.; Mainguy, R.; Huez, J.; Delagnes, D.; Estournès, C. Identification of the Norton-Green compaction model for the prediction of the Ti-6Al-4V densification during the spark plasma sintering process. *Adv. Eng. Mater.* **2016**, *18*, 1720–1727. [CrossRef]

89. Sun, J.; He, Y.H.; Wu, J.S. Characterization of low-temperature superplasticity in a thermomechanically processed TiAl based alloy. *Mat. Sci. Eng. A* **2002**, *329–331*, 885–890. [CrossRef]

90. Sun, J.; Wu, J.S.; He, Y.H. Superplastic properties of a TiAl based alloy with a duplex microstructure. *Mater. Sci.* **2000**, *35*, 4919–4922. [CrossRef]

91. Cheng, S.C.; Wolfenstine, J.; Sherby, O.D. Superplastic behavior of 2-phase titanium aluminides. *Metall. Trans. A* **1992**, *23*, 1509–1513. [CrossRef]

92. Nieh, T.G.; Hsiung, L.M.; Wadsworth, J. Superplastic behavior of a powder metallurgy TiAl alloy with a metastable microstructure. *Intermetallics* **1999**, *7*, 163–170. [CrossRef]

93. Tokita, M. The potential of spark plasma sintering (SPS) method for the fabrication on an industrial scale of functionally graded materials. *Adv. Sci. Tech.* **2010**, *63*, 322–331. [CrossRef]

94. Sweet, G.A.; Brochu, M.; Hexemer, R.L.; Donaldson, I.W.; Bishop, D.P. Microstructure and mechanical properties of air atomized aluminum powder consolidated via spark plasma sintering. *Mater. Sci. Eng. A* **2014**, *608*, 273–282. [CrossRef]

95. Lagos, M.A.; Agote, V.; Atxaga, G.; Adarraga, O.; Pambaguian, L. Fabrication and characterisation of Titanium Matrix Composites obtained using a combination of Self propagating High temperature Synthesis and Spark Plasma Sintering. *Mater. Sci. Eng. A* **2016**, *655*, 44–49. [CrossRef]

96. Naanani, S.; Hantcherli, M.; Hor, A.; Mabru, C.; Monchoux, J.P.; Couret, A. Study of the low cyclic behaviour of the IRIS alloy at high temperature. *MATEC Web Conf.* **2018**, *165*, 06007. [CrossRef]

97. Manière, C.; Nigito, E.; Durand, L.; Weibel, A.; Beynet, Y.; Estournes, C. Spark plasma sintering and complex shapes: The deformed interfaces approach. *Powder Tech.* **2017**, *320*, 340–345. [CrossRef]

98. Voisin, T.; Monchoux, J.P.; Couret, A. Near-net shaping of titanium-aluminum jet engine turbine blades by SPS. In *Spark Plasma Sintering of Materials*; Cavaliere, P., Ed.; Springer: Cham, Switzerland, 2019; pp. 713–737.

 *metals*

MDPI

Article

SHS Synthesis, SPS Densification and Mechanical Properties of Nanometric Tungsten

Sarah Dine [1], Elodie Bernard [2], Nathalie Herlin [3], Christian Grisolia [2], David Tingaud [1] and Dominique Vrel [1,*

[1]   Laboratoire des Sciences des Procédés et des Matériaux (LSPM), Université Sorbonne Paris Nord, UPR 3407 CNRS, 93430 Villetaneuse, France; sarah.dine@lspm.cnrs.fr (S.D.); david.tingaud@lspm.cnrs.fr (D.T.)
[2]   CEA, Institut de Recherche sur la Fusion par Confinement Magnétique (IRFM), 13108 Saint Paul lez Durance, France; Elodie.BERNARD@cea.fr (E.B.); Christian.GRISOLIA@cea.fr (C.G.)
[3]   CEA, Institut Rayonnement Matière de Saclay (IRAMIS) UMR Nanosciences et Innovation pour les Matériaux, la Biomédecine et l'Énergie (NIMBE), Université Paris Saclay, 91191 Gif-sur-Yvette, France; nathalie.herlin@cea.fr
*   Correspondence: dominique.vrel@lspm.cnrs.fr; Tel.: +33-1-4940-3411

Abstract: Recent studies have shown that low grain sizes are favorable to improve ductility and machinability in tungsten, as well as a resistance to ablation and spallation, which are key properties for the use of this material in a thermonuclear fusion environment (Tokamaks such as ITER). However, as one of the possible incidents during Tokamak operation is the leakage of air or water from the cooling system inside the chamber, resulting in the so-called loss of vacuum accident (LOVA), extensive oxidation may arise on tungsten components, and the use of an alloy with improved oxidation resistance is therefore highly desirable. As current production routes are not suitable for the fabrication of bulk nanostructured tungsten or tungsten alloys samples, we have proposed a new methodology based on powder metallurgy, including the powder synthesis, the densification procedure, and preliminary mechanical testing, which was successfully applied to pure tungsten. A similar study is hereby presented on tungsten-chromium alloys with up to 6 wt.% Cr. Results show that full tungsten densification may be obtained by SPS at a temperature lower than 1600 °C. The resulting morphology strongly depends on the amount of the alloying element, presenting a possible second phase of chromium oxide, but always keeps a partial nanostructure inherited from the synthesized powders. Such microstructure had previously been identified as being favorable to the use of these materials in fusion environments and for improved mechanical properties, including hardness, yield strength and ductility, all of which is confirmed by the present study.

Keywords: tungsten alloys; ductility; nanostructure; mechanical properties

Citation: Dine, S.; Bernard, E.; Herlin, N.; Grisolia, C.; Tingaud, D.; Vrel, D. SHS Synthesis, SPS Densification and Mechanical Properties of Nanometric Tungsten. *Metals* **2021**, *11*, 252. https://doi.org/10.3390/met11020252

Academic Editor: Javier S. Blázquez Gámez

Received: 2 December 2020
Accepted: 30 January 2021
Published: 2 February 2021

Publisher's Note: MDPI stays neutral with regard to jurisdictional claims in published maps and institutional affiliations.

1. Introduction

Tungsten has been chosen as one of the principal candidates for plasma-facing components (PFC) in the construction of thermonuclear fusion tokamaks [1–6]. Indeed, among all the pure elements, it has the highest melting point while presenting a good thermal conductivity, which makes it interesting for any high-temperature application. Moreover, considering the specific conditions in tokamaks, it also presents a low activation under neutronic irradiation and low plasma sputtering. However, tungsten presents a strong drawback at room temperature for any structural application, as it is usually brittle. As a function of the preparation method used, and specifically of the presence of some impurities which may coalesce at grain boundaries [7], tungsten usually exhibit a ductile-to-brittle-transition-temperature (DBTT) within the 200–400 °C range. Indeed, due to this brittleness, cracking may pre-exist before operation due to machining, but more important, thermal cycling of PFC may also favor their formation and thus drastically reduce their life span [8,9].

In order to induce and improve ductility in tungsten, different pathways are currently being considered [6]. While in metallurgy, the use of alloying elements is usually the first solution investigated to improve one specific property in metal, this problem is hard to circumvent for tungsten ductility, as only rhenium (Re) has such a capacity [10]. This costly element is unfortunately inadequate for the construction of large-scale pieces such as in ITER plasma-facing components, and especially because in the specific operational conditions of thermonuclear fusion, with high neutron bombardment, rhenium is transmuted into long half-life radioactive osmium, inducing waste management and safety issues [11,12]. For similar reasons, the use of niobium and molybdenum should be avoided [13]. Neutron-induced embrittlement, reducing thermal conductivity, is also a strong drawback for which these elements should not be considered.

With these considerations in mind, the list of acceptable alloying elements in tungsten is reduced to only a few, including Ta, V, Cr, Ti, Si [14–19]. Among these possible alloying elements, Cr should be specifically mentioned for its possible contribution to the improvement of tungsten oxidation resistance [20–23]. It is indeed well-known that chromium has a natural tendency to form a thin protective oxide layer that passivates the material, which is the basic principle of its use in stainless steel.

However, tungsten wires in old-fashioned light bulbs are ductile enough to be shaped as small spirals, and this is due to the fine microstructure of the material resulting from the extrusion process used to produce these wires. Unfortunately, due to the high-temperature reached during the light bulb operation, recrystallization occurs, and this ductility is lost. To avoid or limit this phenomenon, some studies have investigated the addition of dispersed phases, such as La_2O_3, Y_2O_3, CeO_2 [6,24,25], TiC [26,27], ZrC [28] or HfC [11]. As these phases are insoluble in W, they precipitate at the grain boundaries and thus limit grain growth. As a result, the overall tungsten ductility after thermal cycling is improved, and the machinability may be preserved. Although these oxides and carbides are thermodynamically stable, their use in plasma-facing components that will be subjected to erosion should be considered cautiously. Indeed, the potential release of oxygen or carbon in the hydrogen plasma would result in detrimental pollution, with the possible formation of water or hydrocarbons. In addition, some of the metals in these compounds may form metal hydrides, resulting in large tritium retention.

For larger samples, obtaining these fine microstructures is more complex and may require processes such as equal channel angular pressing [29,30], cold rolling [31–33], surface attrition [34], or even high-pressure torsion [35]. Due to the high forces required in these processes, and the high elastic limit of tungsten, they are unfortunately not appropriate in order to obtain nanostructured plasma-facing components as large as the design of the ITER tokamak requires. Using these fine microstructures, recent investigations have shown that the DBTT can be reduced to a value as low as 77 K [31].

The use of chromium as a possible alloying element in tungsten, studied here mainly as a possible solution to increase tungsten oxidation resistance, has been previously investigated by other authors. It has been reported that the synthesis of dense W-Cr alloys is difficult due to low interdiffusion coefficients in this system [36] and that the use of a third element such as Pd would be necessary to improve interdiffusion [36–38]. However, the presence of palladium dissolves rather large amounts of tungsten and chromium, making the whole system more complex [39]. Moreover, the existence of a relatively low Cr-Pd eutectic may have a detrimental effect on the high-temperature properties of the resulting material [38,40], especially because this phase has been shown to form at grain boundaries [41]. Nanometric W-Cr alloys were also synthesized using high-energy milling from micrometric powders [22,42]. This technique, however, needing long milling times, implies a high level of energy consumption and leads inevitably to the contamination of the powders from the milling media.

2. Materials and Methods

In our previous works on pure W [43] and MoNbW alloys [44,45], we developed a new methodology to synthesize refractory metals nanopowders using self-propagating high-temperature synthesis (SHS). To preserve the nanostructure after densification, these powders were sintered by spark plasma sintering (SPS) [46], which allows a full or nearly full densification in short times. A similar study is conducted here on W-Cr alloys, with up to 6 wt.% Cr. Samples will be characterized using XRD (Equinox 1000 diffractometer, Cu $k\alpha_1$ radiation, INEL, Artenay, France), SEM observations for the powders and the densified samples (ZEISS FEG-SEM Supra 40VP, Oberkochen, Germany), and these characterizations will also include electron backscattered diffraction (EBSD), and mechanical testing will be performed (hardness and compressive tests, DURAMIN 20 apparatus, STRUERS, Ballerup, Denmark) for the latter.

2.1. Nanopowders Synthesis

The powders were synthesized using the self-propagating high-temperature synthesis process, using the reduction of metal oxides by magnesium in thermite-like reactions, according to the protocol described in [43–45]. WO_3 (Alfa Aesar (Thermo Fisher GmbH, Kandel, Germany), 99.8%, ref. 11828 batch Q12F015), Cr_2O_3 (Alfa Aesar, 99.8%, ref. 12285 batch D26Z027), Mg (Sigma-Aldrich (Merck KGaA, Darmstadt, Germany), 97.9%, ref. 8.18506.500 batch S7714006-901) and NaCl (VWR (Avantor, Radnor, PA, USA), 100%, ref. 27800.360 batch 17C144128), acting, respectively as an oxidizer, reducer, and moderator were first carefully weighted in order to retrieve a 20 g final mass of alloy with 2, 4 or 6 wt.% of Cr (respectively mentioned to hereunder as WCr2, WCr4 and WCr6, Table 1), then thoroughly mixed in the powder form using a 3D Turbula mixer (WAB, Muttenz, Switzerland) for 24 h, and placed in a sealed reactor, surrounded by NaCl, before the following thermitic reaction was started by Joule-heating a tungsten wire placed at the bottom of the sample:

$$WO_3 + x\, Cr_2O_3 + (3 + 3x + y)\, Mg \rightarrow W + 2x\, Cr + 3\, MgO + y\, Mg$$

SEM micrographs of the reactants are presented in Figure 1. In order to make sure that the reaction is complete, 50% excess Mg was added from the stoichiometric proportions. As for NaCl, its introduction provides an increase of the global heat capacity of the mixture, and its amount was calculated to decrease the overall product's final temperature to 1800 °C. However, as the placement of a thermocouple proved to be difficult considering the temperatures reached, and as pyrometric measurements were not an option in our pressure sealed reactor, no temperature measurements were performed. The value of 1800 °C should therefore be considered as an estimate, allowing us to synthesize all our different samples in similar conditions. On the other hand, the introduction of NaCl in the SHS reaction is known to favor nanometric grain sizes in the resulting powders [43–45]. As magnesium has a low boiling point, these reactions were performed under confinement, for most of the gaseous magnesium to stay available for the reaction. For reference, the WCr2 experiment was also performed without confinement.

Table 1. Weighing of the reactants for the synthesis of 20 g of the different alloys. WCr2, WCr4 and WCr6 stand for 2, 4 and 6 weight percent chromium in the resulting tungsten alloy.

Alloy	Cr wt.%	Cr at.%	WO_3 (g)	Cr_2O_3 (g)	Mg (g)	NaCl (g)
WCr2	2	6.73	24.717	0.585	12.082	16.13
WCr4	4	12.84	24.213	1.169	12.264	15.76
WCr6	6	18.41	23.708	1.754	12.447	15.39

Figure 1. SEM observations of the reactants: (**a**) WO_3; (**b**) Cr_2O_3; (**c**) Mg; (**d**) NaCl. Note that tungsten trioxide particles seem to be made of large grains covered with submicronic particles (insert).

As the reaction scheme yields to the presence of MgO, residual Mg, and NaCl in the final powder mixture, these were lixiviated using 2 N hydrochloric acid for 2 h under stirring at 80 °C. The powders were then retrieved using a 0.22 µm polyethersulfone membrane filtration (MB Express Millipore®, MERCK, Damrstadt, Germany). To ensure complete dissolution of the contaminants, this procedure was performed twice, and the retrieved powders were then rinsed before a last filtration, and finally dried in an autoclave at 50 °C for 24 h. A similar reaction was previously investigated to synthesize stoichiometric W-Cr alloys [47] using aluminum as a reducer and without a reaction moderator. However, in this case, the separation between molten alumina and the molten alloy is performed using a high gravity field, inducing a composition gradient within the alloy.

After these steps are completed, the crystal structure of the obtained powders was analyzed using X-ray diffraction. This characterization was then followed by Rietveld refinement using the Materials analysis using diffraction (MAUD) software to determine the lattice parameters of the different alloys.

2.2. SPS Densification

Spark plasma sintering (SPS) was performed in the "plateforme IdF de frittage flash" (Thiais, France), on a Syntec 515 S machine (Fuji Electronic Industrial Co., Ltd., Saitama, Japan). ~5 g of the as-synthesized powders were inserted in graphite mold with a 10 mm diameter. The temperature is followed with an infrared pyrometer pointing to a hole allowing temperature measurement in the direct vicinity of the sample; consequently, temperature measurements were not possible under 600 °C. Pressure and piston movement measurements were also recorded, providing all the necessary information to analyze the different steps of the sintering.

Densification experiments were performed at 1600, 1800 and 2000 °C with a plateau time of 1 to 15 min. Most cycles were made of a single ramp with a heating rate of $50 °C \cdot min^{-1}$. However, to minimize the risk of a possible overshoot, the temperature ramp is set to 20 °C·min^{-1} when the temperature is less than 50 °C under the maximum temperature. Some more complex temperature cycles were also tested, but with no significant differences in the final density and will not be presented. The compaction pressure is always 100 MPa, set at the beginning of the heating cycle, a pre-compaction at 42 MPa being set before. This pressure is released during the quenching of the sample.

2.3. Densified Samples

Densified samples were polished with 68 μm, 9 μm and 3 μm cloths (P220—Struers MD-Piano 220; P2400—Struers MD-Allegro; and P4000—Struers MD-Dac, respectively, STRUERS, Ballerup, Denmark), with an OPS finishing (colloidal silica, 40 nm—Struers MD-Chem). Densities have then been measured using Archimedes' method in orthoxylene.

Bulk samples were analyzed using X-ray diffraction then by Rietveld refinement. SEM observations were conducted, together with EDX local analysis and mapping. Grain sizes, preferential orientations and nature of grain boundaries were analyzed using electron backscattered diffraction (EBSD).

As for mechanical properties, Vickers hardness was measured on the plane perpendicular to the densification pressing direction using a 1.96 N force applied for 10 s. In order to increase the statistical precision of the measure, this measure was repeated 10 times on each sample. Due to the small sizes of the samples (10 mm diameter and 3–4 mm in thickness), we were unable to cut tensile test specimens. Instead, compression tests were performed on parallelepiped samples cut by electro-erosion, with dimensions 2.4 × 2.4 × 5.0 mm^3, by pressing along the main axis of the sample.

3. Results and Discussion

3.1. Nanopowders Synthesis

SEM observations, Figure 2, reveal mostly rounded particle shapes, with a large polydispersity, typically ranging from 10 to 200 nm. However, in some areas, particles seem bounded together as if local melting had occurred during the reaction, whatever the composition. This morphology is significantly different from what had been obtained with pure W [46], where particle shapes were mostly in the form of platelets. Moreover, we can observe that, while keeping a wide size distribution, the average particle size seems to decrease when the amount of Cr is increased.

Figure 2. SEM micrographs of the submicrometric powders synthesized by SHS (×20,000). (**a**) W-2 wt.% Cr (WCr2), (**b**) W-4 wt.% Cr (WCr4), and (**c**) W-6 wt.% Cr (WCr6).

X-ray diffraction patterns of the resulting powders are presented in Figure 3, and the results of the Rietveld analysis performed on these diffraction patterns are summarized in Table 2. In these patterns, the five main peaks come from the diffraction of tungsten, revealing the α, Body Centered Cubic (BCC) crystal structure, as expected. However, specifically, when the reaction chamber is not sealed, some additional peaks are also observed and are the signature of the presence of WO_3 and WO_2, which remain despite the excess Mg used. These residual tungsten oxide peaks are significant for the WCr2 sample synthesized without confinement and much less, but still clearly visible when confinement was used. Indeed, the reaction temperature is ~700 °C above the boiling point of magnesium, and, despite using a sealed reactor, this magnesium might escape away from the reaction zone and condense on the salt surrounding the sample, yielding an incompletely converted reaction product. However, we have shown that when increasing the size of the sample, the amount of oxidized tungsten present in the final powder decreases [45]. Nevertheless, it seems that the total amount of residual oxide phases decreases when the Cr amount is increased. Although the presence of chromium inside the alloy was intended to increase its oxidation resistance, it is unlikely that the residual oxides observed on WCr2 and WCr4 result from oxidation of the nanopowders; indeed, although this oxidation most likely takes place during air exposure, it should produce amorphous oxides, which would be undetected by XRD. We rather suspect that the presence of Cr_2O_3 together with WO_3 modifies the reaction kinetics and improves its final yield. Finally, there is a distinct shift of the diffraction peaks when the Cr amount is increased, which could be attributed to a greater amount of Cr within the W lattice, as the Cr lattice is smaller than the one for W. Moreover, an imperfect mixing of the two elements is suspected for WCr4 and WCr6, as distinct shoulders are visible to the right of the peaks (Figure 3b).

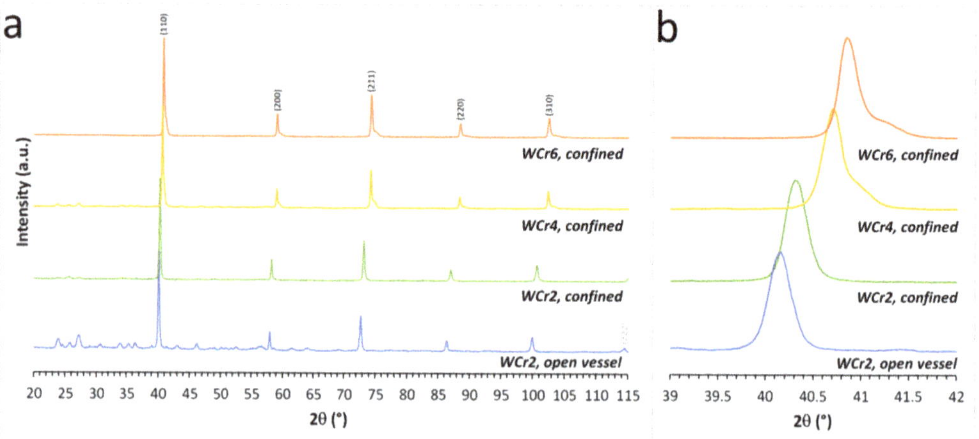

Figure 3. (**a**) X-ray diffraction pattern of the powders after synthesis, lixiviation, rinsing and drying. Miller indices represented correspond to the expected tungsten BCC crystal structure. The other peaks observed on the XRD patterns are attributed to tungsten oxides WO_3 and WO_2. (**b**) Magnification of the (110) peak.

Table 2. Rietveld refinement of powders XRD patterns. Results shown correspond to the main BCC phase only.

Alloy	Lattice Parameter (nm)
WCr2, open vessel	0.31814
WCr2, confined	0.31642
WCr4, confined	0.31241
WCr6, confined	0.31220

3.2. SPS Densification

For the sake of clarity, as no significant differences are observed between the different cycles, we will present here the results concerning the 3 different alloys compacted at 1800 °C, Figure 4. As the mass and thicknesses of the three samples are different, the curves representing the kinetics of densification dz/dt are normalized, so their integrals are equal to 1.

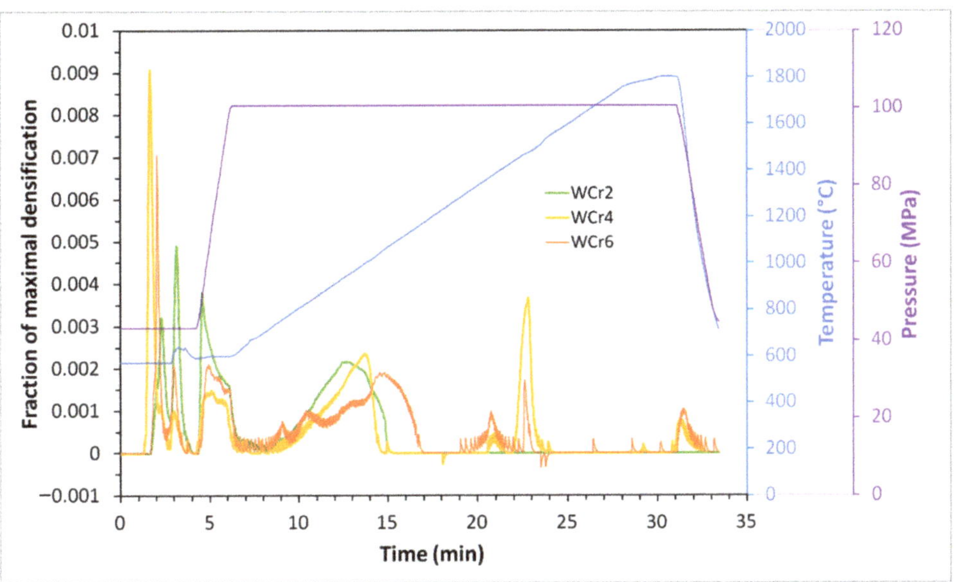

Figure 4. Spark plasma sintering (SPS) cycle (1800 °C—1′). Purple: pressure; blue: temperature; and green, yellow and orange: kinetics of densification as a fraction of total densification (integral of each curve is equal to 1).

A first densification peak is observed at low-temperature, below the lower pyrometer limit at 600 °C, and is due to outgassing, possibly of residual water if the drying step was incomplete. From 4 to 6.5 min, while the temperature is constant just above the detection limit of the pyrometer, the pressure on the sample is increased, and a second densification peak is observed on all three samples. After this first step, densification stops temporarily and then resumes progressively as the temperature is increased, with maximal densification kinetic at 12.4, 13.6 and 14.8 min for WCr2, WCr4 and WCr6, respectively (with corresponding temperatures of 930, 995 and 1055 °C, respectively). These low temperatures thus show the influence of the nanometric grain sizes on the sintering of tungsten. It is worth mentioning that despite chromium has a lower melting temperature, and despite the fact that powders particle sizes decrease with the increasing amount of chromium, the sintering nevertheless arises with increasing temperatures. A third, abrupt peak is observed when the temperature is in the 1465–1475 °C range. As this temperature corresponds to the tungsten trioxide melting temperature (1473 °C), we believe that a liquid phase is then present, and densification may proceed almost freely. No significant evolution in the density can be observed when the temperature is further increased to 1800 °C, nor during the 5 min plateau. Although the WCr2 yielded an incomplete reading of the densification process (saturation of the detector after 15′), it should be noted that the intensity of the shrinkage of the sample decreases when the Cr amount increases at that temperature. Indeed, the WCr6 powders had fewer residual oxides (no crystalized oxides detectable on the XRD pattern, Figure 2) and most certainly less amorphous oxides at the surface of the nanopowders after air exposure due to the Cr-induced passivation of the powders. Finally, at the end of the cycle, the temperature is abruptly decreased, yielding a shrinking of the sample due to natural thermal expansion, and an apparent densification step is observed.

A temperature of 1600 °C seems therefore sufficient to fully densify the W nanopowders produced. However, due to the melting of the oxides at or around 1473 °C, and their subsequent flowing outside of the sample, a small mass loss was observed. It is, however, considerably smaller than what had been observed when pure tungsten samples were sintered [46].

3.3. Densified Samples

Measured densities using Archimedes' method were greater than 90% of the theoretical density for all samples and up to 99.91% (Table 3). Comparatively to previously published results, these densities extremely promising: Yao et al. [48] studied the densification of commercial 50 nm tungsten nanometer at a similar temperature, only limiting the pressure to 50 MPa, for a final relative density of 97.8%, and, to obtain a density of 99%, similar to the results presented here, the use of resistance sintering under ultra high-pressure (RSUHP) at 9 GPa was necessary. Comparatively, the densification of micrometric commercial tungsten powder (2 μm) using SPS could only provide a density of 84.3% (50 MPa, 1600 °C) [18], and 92.16% using hot pressing (1800 °C, 2 h, 20 MPa) [49]. The specific granulometry and its polydispersity, the specific shape of the initial powder grains seem, therefore, very favorable to the densification.

Table 3. Relative density, Vickers hardness and Rietveld refinement results on SPS-densified samples. In bulk are the samples described more specifically in the text.

Alloy	Sintering Temperature (°C)	Plateau Duration (min)	Relative Density (%)	Vickers Hardness	Lattice Parameter (nm)	Crystallite Size (nm)
WCr2	1600	1	90.86	335.7	0.31645	75.9
WCr2	1800	1	98.10	314.8	0.31672	68.4
WCr4	1800	1	99.91	349.7	0.31682	68.5
WCr4	1800	5	99.78	244.1	0.31669	78.0
WCr4	2000	15	98.85	264.5	0.31666	78.3
WCr6	1800	1	99.12	420.6	0.31676	79.6

Bulk samples were analyzed using X-ray diffraction, Figure 5, then by Rietveld refinement, Figure 5 presents the X-ray diffraction patterns obtained on densified samples, and Table 3 summarizes the results of the Rietveld analysis. These diffraction patterns only display the characteristic peaks of the tungsten BCC crystal structure, and no residual (crystalized) oxide phases can be detected, accordingly to the hypothesis formulated when analyzing densification curves, attributing the third densification peak at 1473 °C to the melting of the tungsten oxides and their subsequent flowing outside of the sample. This, unfortunately, implies a small mass loss, although significantly lower than what was observed on pure W [46], but raises the possibility of obtaining pure nanostructured bulk tungsten alloys.

Contrary to what could be expected from the XRD analysis performed previously on the powders, no peak shift is observed on the different samples, and the calculated lattice parameters are very similar. Moreover, crystallites sizes are comparable, in the 68–80 nm range, despite a significant difference observed on the powder SEM images. To analyze further these diffraction patterns, we performed a magnification of the (211) peak. Whereas the positions of this peak do not seem to shift significantly, its asymmetry seems to slightly vary with the composition due to the presence of different compositional gradients as a function of the Cr content, which Rietveld analysis seems to not take correctly into account. However, this seems strongly insufficient to explain the variation of the lattice parameters from the powders to the bulk samples. As backscattered images of the three samples (not presented) show a strong increase of the Cr-rich phase (in black on Figure 6) with the Cr content, a Cr depletion cannot be the cause of the newly gained homogeneity of the lattice parameters. On the contrary, we suspect that the strong shift observed on the powders do not provide only from the addition of Cr but also from some unaccounted form of out-of-equilibria state of these powders, which disappears after densification.

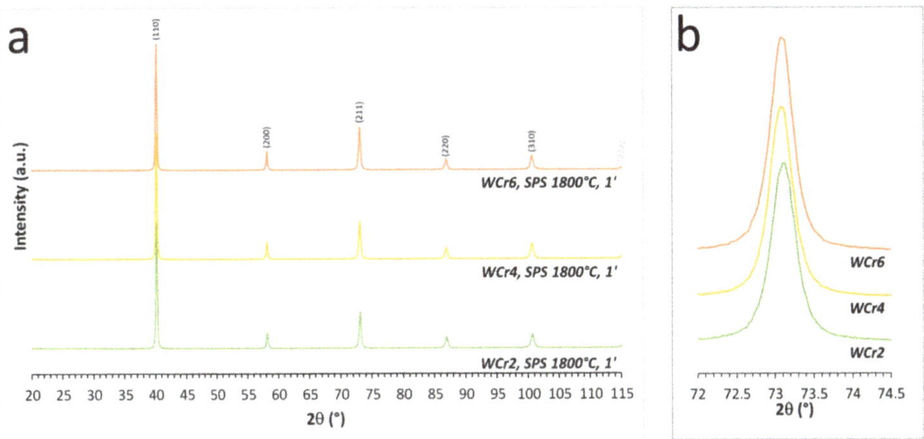

Figure 5. (**a**) XRD patterns of SPS sintered samples. Miller indices represented correspond to the expected tungsten BCC crystal structure. (**b**) magnification of the (211) peak.

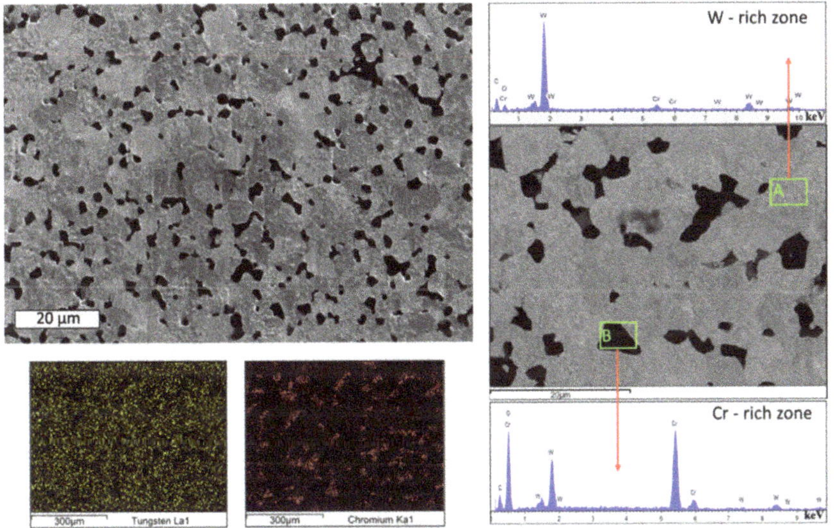

Figure 6. Upper left: SEM observation of the WCr6 sintered sample (2000 °C—5′). Right: EDX analysis of W-rich and Cr-rich regions; bottom left: EDX mapping on a large-scale of the sample.

Figure 6 shows an SEM observation of the WCr6 samples sintered at 2000 °C for 5′. On this image, the microstructure of the densified samples seems to be within a grain size range of 5 to 10 µm, and this apparent crystal growth seems more pronounced for higher W contents. Moreover, when the Cr content is 4 or 6 wt.%, dark areas appear and are naturally in greater proportions for WCr6 than for WCr4.

EDS local analysis and mapping of the sample proved these regions to also be a W-Cr alloy but significantly richer in Cr than the average of the sample. However, when performing EBSD analysis to study this apparent loss of the nanostructure, Figure 7, these micrometric grains turn out to be, for the most part, nanostructured or with a high concentration of low-angle grain boundaries (LAGB), with grain sizes consistent to the measured value obtained by Rietveld refinement (Table 3). Surprisingly, the presence of a real nanosubstructure is the strongest for WCr2 and WCr6 and significantly less pronounced for WCr4. This nanostructuration is also of utmost importance for Plasma Facing Components as several experimental studies on metals [50,51], intermetallics [52], and ceramic materials [53,54] have shown that it would improve irradiation resistance. Dark areas are these images correspond to unresolved Kikuchi line patterns, which could be due to a different crystal structure, e.g., the presence of an oxide, or to a local loss of flatness of the sample such as a grain tear-off during polishing. In order to resolve this issue, Figure 8 presents the same magnified area of the sample observed on EBSD and SEM, where it can clearly be seen that the dark areas observed on the SEM images are properly indexed within the BCC structure, which confirms that they are made of a W-Cr alloy (and are therefore not recrystallized chromium oxide, which crystallizes with the corundum, hexagonal structure). Simultaneously, black, unindexed regions on the EBSD image due to unresolved Kikuchi lines patterns correspond to regions on the SEM image where significant irregularities are observed, seemingly from grain tear-off during polishing. This result is in good agreement with the XRD pattern, where no other phase was detected besides the BCC structure.

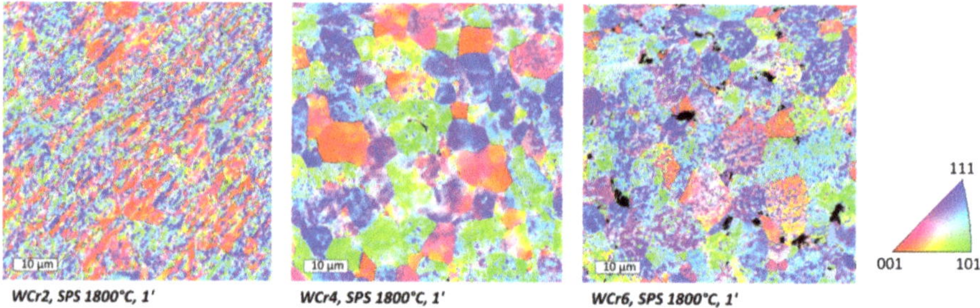

Figure 7. Electron backscattered diffraction (EBSD) images representing the grain orientations on $60 \times 60\ \mu m^2$ areas.

Figure 8. SEM (**left**) and EBSD (**center**) and (**right**) images of the same area, showing the dark SEM areas to be correctly indexed with the BCC structure, and the black EBSD areas to correspond to surface irregularities (marked grain is for the eye's guidance only); the right image shows the positions of the High-Angle Grain Boundaries (HAGB) (in black) and Low-Angle Grain Boundaries (LAGB) (2–5° rotation angle in blue, 5–15° in green).

From the EBSD patterns in Figure 7, it is possible to retrieve the (110), (200) and (211) pole figures as shown for the WCr4 sample in Figure 9. While the symmetry of these patterns is poor, inverse pole figures calculated from these and presented in Figure 10 show a significant trend for the material towards a 111 preferential orientation. However, this analysis is performed on a small, $60 \times 60\ \mu m^2$ area, and is most certainly not statistically significant. A similar result was obtained on pure W, although in that case, grains were in the form of platelets, and a preferential orientation could be explained from an anisotropic compression behavior of the powders during SPS sintering [46]. It is much harder to explain the origin of the preferential orientation here, as the powders were, if not spherical, at least mainly rounded (Figure 2), with only a few sharp-edged particles. A more precise analysis on a larger area, e.g., using quantitative texture analysis using XRD, would be necessary to confirm this result, but was unfortunately not performed before the samples were destroyed during mechanical testing.

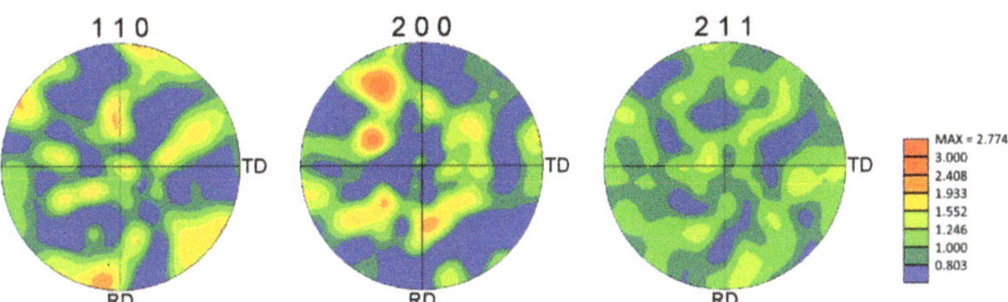

Figure 9. 110, 200 and 211 pole figures calculated from the EBSD mapping of the WCr4 sample presented in Figure 7.

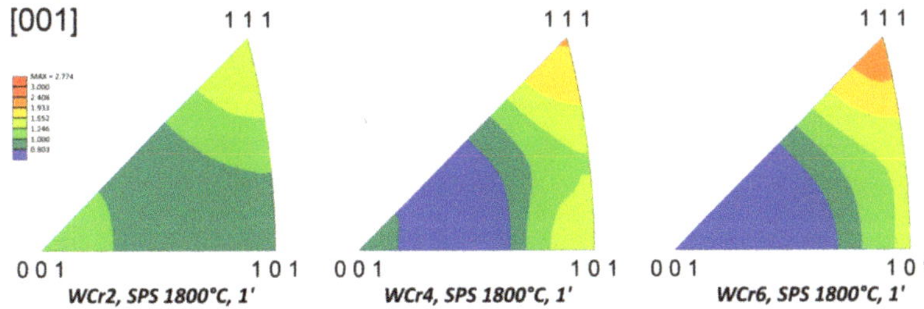

Figure 10. Inverse pole figures calculated from the three EBSD mappings (Figure 7).

Vickers hardness measurements are summarized in Table 3. The theoretical hardness for pure tungsten, as provided in the literature, is 343 HV, and, as the nanostructuration of a material is known to increase the hardness [48], increased hardnesses were expected for the samples presented here. Moreover, the hardness of 1142 HV was found in the literature for stoichiometric W-Cr alloy [47], adding another reason for an increase in hardness to be expected in our alloys. Indeed, results presented in Table 3 show an increase in hardness related to the composition: the three alloys sintered at 1800 °C for 1′ have hardness values from 315 Hv for WCr2 to 420 Hv for WCr6. In the meantime, comparing the WCr2 samples sintered at 1600 °C and 1800 °C, respectively, a drop in hardness value is observed, as partial recrystallization may occur at high-temperature. A similar result is observed on the WCr4 samples, simply by increasing the plateau duration to 5′. This drop in hardness is, however, limited, and the value measured for the sample sintered at 2000 °C with a plateau duration of 15′ is almost identical to the previous one. It is, however, difficult, in these conditions, to clearly identify if this increase in hardness is due to the presence of an alloying element or to the nanostructuration of the material

Results of compression tests are presented in Figure 11. They depict a ductility at room temperature in the range of 9 to 10%. Apparent mechanical properties deduced from these curves are summarized in Table 4. These results show a slight increase in the elastic limit, compared to pure, micrometric, tungsten, but is still slightly lower than the values obtained with pure, "cold rolled" (at temperatures ranging from 600 °C to 1000 °C) pure tungsten [32]. Indeed, the well-known Hall–Petch effect associates the decrease of the grain sizes to improved mechanical properties, and this result was therefore expected. However, the main improvement concerns the obtained ductility of the resulting materials, as the same study only reached a maximal engineering strain of 3–4% before rupture and <1% before the maximum stress is reached ([34], Figure 2). With values reaching 9–10% for the strain at the maximum stress and 16–18% strain at failure at room temperature, our results are therefore very promising for the development of new refractory ductile components.

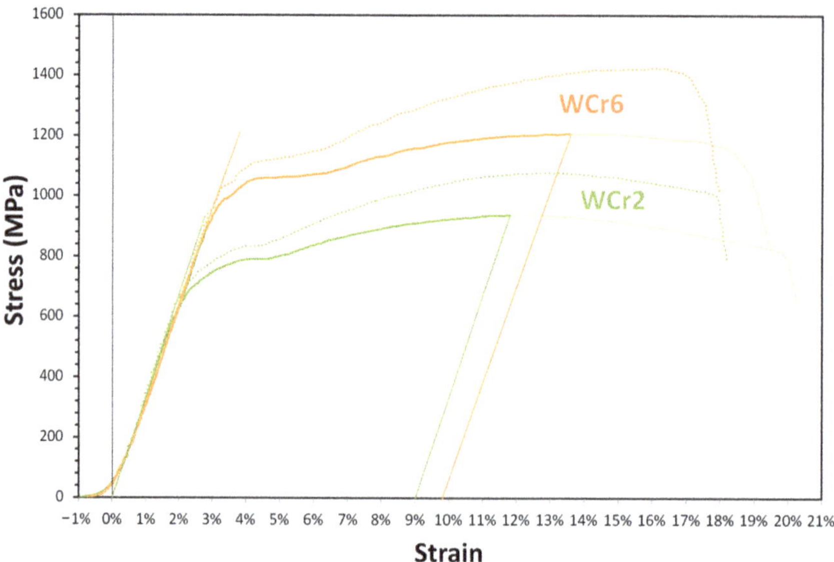

Figure 11. Results from the compression tests. Dotted lines: engineering strain/stress curves; full lines: true stress/strain curves, showing a ductility of 9.00% and 9.74%.

Table 4. Results of compression tests. Ductility is the plastic deformation only at the maximum strength.

Alloy	Elastic Limit (MPa)	Apparent Young's Modulus (GPa)	Yield Strength (MPa)	Ductility (%)
WCr2	600	33.8	937	9.00
WCr6	950	31.6	1208	9.74

4. Conclusions

Magnesio-thermitic reactions were used to synthesize tungsten-chromium alloys with a morphology of nanometric to submicronic powders. Densification was then performed using SPS, with a final density of up to 99.9% of the theoretical value while preserving the nanometric substructure of the grains. Such a microstructure is thought to be favorable to increase resistance to ablation and spallation. On the other hand, the main aim of our effort to obtain this microstructure was to increase mechanical properties, and specifically ductility. Compressive testing of the samples presented an elastic limit ranging from 700 to 1000 MPa, depending on the alloy composition, while displaying a ductility greater than 9%. Future investigations will focus on slightly more complex alloys, with 2–3 alloying elements, in order not only to improve further mechanical properties but also to study the resistance of the materials to recrystallization at high-temperature.

Author Contributions: S.D. was the principal investigator of this study, performed unser the supervision and in collaboration with D.V.; C.G., working with E.B., was the coordinator of a broader topic on Plasma Facing Materials, and brought valuable insight all along this study; N.H. provided clear information and inerpretation of various analyses in this study, expecially SEM. D.T. provided help durind g SPS synthesis and performed EBDS analyses. The paper has been mainly written by D.V. All authors have read and agreed to the published version of the manuscript.

Funding: This research received no external funding.

Data Availability Statement: No significant data in this study were created or analyzed in this study, aside from the one presented here. Additional data are available on request from the corresponding author.

Conflicts of Interest: The authors declare no conflict of interest.

References

1. Joachim, R.; Klaus, S. Hydrogen in tungsten as plasma-facing material. *Phys. Scr.* **2011**, *T145*, 014031.
2. Bolt, H.; Barabash, V.; Krauss, W.; Linke, J.; Neu, R.; Suzuki, S.; Yoshida, N.; Team, A.U. Materials for the plasma-facing components of fusion reactors. *J. Nucl. Mater.* **2004**, *329–333*, 66–73. [CrossRef]
3. Rieth, M.; Boutard, J.L.; Dudarev, S.L.; Ahlgren, T.; Antusch, S.; Baluc, N.; Barthe, M.-F.; Becquart, C.S.; Ciupinskie, L.; Correia, J.B.; et al. Review on the EFDA programme on tungsten materials technology and science. *J. Nucl. Mater.* **2011**, *417*, 463–467. [CrossRef]
4. Zhou, Z.; Ma, Y.; Du, J.; Linke, J. Fabrication and characterization of ultra-fine grained tungsten by resistance sintering under ultra-high pressure. *Mater. Sci. Eng. A* **2009**, *505*, 131–135. [CrossRef]
5. Cottrell, G.A. Sigma phase formation in irradiated tungsten, tantalum and molybdenum in a fusion power plant. *J. Nucl. Mater.* **2004**, *334*, 166–168. [CrossRef]
6. Pintsuk, G. Tungsten as a Plasma-Facing Material, Reference Module in Materials Science and Materials Engineering—Comprehensive. *Nucl. Mater.* **2012**, *4*, 551–581.
7. El-Kharbachi, A.; Chene, J.; Garcia-Argote, S.; Marchetti, L.; Martin, F.; Miserque, F.; Vrel, D.; Redolfi, M.; Malard, V.; Grisolia, C.; et al. Tritium absorption/desorption in ITER-like tungsten particles. *Int. J. Hydrogen Energy* **2014**, *39*, 10525–10536. [CrossRef]
8. Pestchanyi, S.E.; Linke, J. Simulation of cracks in tungsten under ITER specific transient heat loads. *Fusion Eng. Des.* **2007**, *82*, 1657–1663. [CrossRef]
9. Wittlich, K.; Hirai, T.; Compan, J.; Klimov, N.; Linke, J.; Loarte, A.; Merola, M.; Pintsuk, G.; Podkovyrov, V.; Singheiser, L.; et al. Damage structure in divertor armor materials exposed to multiple ITER relevant ELM loads. *Fusion Eng. Des.* **2009**, *84*, 1982–1986. [CrossRef]
10. Mutoh, Y.; Ichikawa, K.; Nagata, K.; Takeuchi, M. Effect of rhenium addition on fracture toughness of tungsten at elevated temperatures. *J. Mater. Sci.* **1995**, *30*, 770–775. [CrossRef]
11. Barabash, V.; Akiba, M.; Mazul, I.; Ulrickson, M.; Vieider, G. Selection, development and charaterisation of plasma facing materials for ITER. *J. Nucl. Mater.* **1996**, *233–237*, 718–723. [CrossRef]
12. El-Guebaly, L.; Kurtz, R.; Rieth, M.; Kurishita, H.; Robinson, A.; Team, A. W-based alloys for advanced divertor designs: Options and environmental impact of state-of-the-art alloys. *Fusion Sci. Technol.* **2011**, *60*, 185–189. [CrossRef]
13. Wurster, S.; Baluc, N.; Battabyal, M.; Crosby, T.; Du, J.; García-Rosales, C.; Hasegawa, A.; Hoffmann, A.; Kimura, A.; Kurishita, H.; et al. Recent progress in R&D on tungsten alloys for divertor structural and plasma facing materials. *J. Nucl. Mater.* **2013**, *442*, S181–S189.
14. Grisolia, C.; Hodille, E.; Chene, J.; Garcia-Argote, S.; Pieters, G.; El-Kharbachi, A.; Marchetti, L.; Martin, F.; Miserque, F.; Vrel, D.; et al. Tritium absorption and desorption in ITER relevant materials: Comparative study of tungsten dust and massive samples. *J. Nucl. Mater.* **2015**, *463*, 885–888. [CrossRef]
15. Coenen, J.W.; Philipps, V.; Brezinsek, S.; Pintsuk, G.; Uytdenhouwen, I.; Wirtz, M.; Kreter, A.; Sugiyama, K.; Kurishita, H.; Torikai, Y.; et al. Melt-layer ejection and material changes of three different tungsten materials under high heat-flux conditions in the tokamak edge plasma of TEXTOR. *Nucl. Fusion* **2011**, *51*, 113020. [CrossRef]
16. Aguirre, M.V.; Martín, A.; Pastor, J.Y.; Llorca, J.; Monge, M.A.; Pareja, R. Mechanical behavior of W-Y$_2$O$_3$ and W-Ti alloys from 25 °C to 1000 °C. *Metall. Mater. Trans. A* **2009**, *40*, 2283–2290. [CrossRef]
17. Aguirre, M.V.; Martín, A.; Pastor, J.Y.; Llorca, J.; Monge, M.A.; Pareja, R. Mechanical properties of Y$_2$O$_3$-doped W-Ti alloys. *J. Nucl. Mater.* **2010**, *404*, 203–209. [CrossRef]
18. Arshad, K.; Guo, W.; Wang, J.; Zhao, M.Y.; Yuan, Y.; Zhang, Y.; Wang, B.; Zhou, Z.-J.; Lu, G.-H. Influence of vanadium precursor powder size on microstructures and properties of W-V alloy. *Int. J. Refract. Met. Hard Mater.* **2015**, *50*, 59–64. [CrossRef]
19. Arshad, K.; Yuan, Y.; Cheng, L.; Wang, J.; Zhou, Z.J.; De Temmerman, G.; Lu, G.-H. Deuterium blistering in tungsten and tungsten vanadium alloys. *Fusion Eng. Des.* **2016**, *107*, 25–31. [CrossRef]
20. López-Ruiz, P.; Koch, F.; Ordás, N.; Lindig, S.; García-Rosales, C. Manufacturing of self-passivating W-Cr-Si alloys by mechanical alloying and HIP. *Fusion Eng. Des.* **2011**, *86*, 1719–1723. [CrossRef]
21. Telu, S.; Mitra, R.; Pabi, S.K. High temperature oxidation behavior of W–Cr–Nb Alloys in the Temperature Range of 800–1200 °C. *Int. J. Refract. Met. Hard Mater.* **2013**, *38*, 47–59. [CrossRef]
22. Telu, S.; Patra, A.; Sankaranarayana, M.; Mitra, R.; Pabi, S.K. Microstructure and cyclic oxidation behavior of W-Cr alloys prepared by sintering of mechanically alloyed nanocrystalline powders. *Int. J. Refract. Met. Hard Mater.* **2013**, *36*, 191–203. [CrossRef]
23. Zhou, Y.; Sun, Q.X.; Xie, Z.M.; Liu, R.; Wang, X.P.; Fang, Q.F.; Liu, C.S. The microstructure and microhardness of W–5wt% Cr alloy fabricated by spark plasma sintering. *J. Alloys Compd.* **2014**, *585*, 771–775. [CrossRef]

24. Veleva, L.; Oksiuta, Z.; Vogt, U.; Baluc, N. Sintering and characterization of W–Y and W–Y$_2$O$_3$ materials. *Fusion Eng. Des.* **2009**, *84*, 1920–1924. [CrossRef]
25. Liu, R.; Xie, Z.M.; Fang, Q.F.; Zhang, T.; Wang, X.P.; Haoa, T.; Liua, C.S.; Daic, Y. Nanostructured yttria dispersion-strengthened tungsten synthesized by sol-gel method. *J. Alloys Compd.* **2016**, *657*, 73–80. [CrossRef]
26. Kurishita, H.; Matsuo, S.; Arakawa, H.; Narui, M.; Yamazaki, M.; Sakamotob, T.; Kobayashib, S.; Nakaib, K.; Takidac, T.; Takebec, K.; et al. High temperature tensile properties and their application to toughness enhancement in ultra-fine grained W-(0-1.5)wt% TiC. *J. Nucl. Mater.* **2009**, *386–388*, 579–582. [CrossRef]
27. Kurishita, H.; Matsuo, S.; Arakawa, H.; Sakamoto, T.; Kobayashi, S.; Nakai, K.; Takida, T.; Kato, M.; Kawai, M.; Yoshida, N. Development of re-crystallized W–1.1%TiC with enhanced room-temperature ductility and radiation performance. *J. Nucl. Mater.* **2010**, *398*, 87–92. [CrossRef]
28. Zhang, T.; Wang, Y.; Zhou, Y.; Song, G. Effect of heat treatment on microstructure and mechanical properties of ZrC particles reinforced tungsten-matrix composites. *Mater. Sci. Eng. A* **2009**, *512*, 19–25. [CrossRef]
29. Hao, T.; Fan, Z.Q.; Zhang, T.; Luo, G.N.; Wang, X.P.; Liu, C.; Fang, Q. Strength and ductility improvement of ultrafine-grained tungsten produced by equal-channel angular pressing. *J. Nucl. Mater.* **2014**, *455*, 595–599. [CrossRef]
30. Levin, Z.S.; Hartwig, K.T. Strong ductile bulk tungsten. *Mater. Sci. Eng. A* **2017**, *707*, 602–611. [CrossRef]
31. Németh, A.A.N.; Reiser, J.; Armstrong, D.E.J.; Rieth, M. The nature of the brittle-to-ductile transition of ultrafine grained tungsten (W) foil. *Int. J. Refract. Met. Hard Mater.* **2015**, *50*, 9–15. [CrossRef]
32. Reiser, J.; Hoffmann, J.; Jäntsch, U.; Klimenkov, M.; Bonk, S.; Bonnekoh, C.; Rieth, M.; Hoffmann, A.; Mrotzek, T. Ductilisation of tungsten (W): On the shift of the brittle-to-ductile transition (BDT) to lower temperatures through cold rolling. *Int. J. Refract. Met. Hard Mater.* **2016**, *54*, 351–369. [CrossRef]
33. Reiser, J.; Hoffmann, J.; Jäntsch, U.; Klimenkov, M.; Bonk, S.; Bonnekoh, C.; Hoffmann, A.; Mrotzek, T.; Rieth, M. Ductilisation of tungsten (W): On the increase of strength AND room-temperature tensile ductility through cold-rolling. *Int. J. Refract. Met. Hard Mater.* **2017**, *64*, 261–278. [CrossRef]
34. Guo, H.Y.; Xia, M.; Wu, Z.T.; Chan, L.C.; Dai, Y.; Wang, K.; Yan, Q.-Z.; He, M.-C.; Ge, C.; Lu, J. Nanocrystalline-grained tungsten prepared by surface mechanical attrition treatment: Microstructure and mechanical properties. *J. Nucl. Mater.* **2016**, *480*, 281–288. [CrossRef]
35. Chen, L.; Ping, L.; Ye, T.; Lingfeng, L.; Kemin, X.; Meng, Z. Observations on the Ductility and Thermostability of Tungsten Processed from Micropowder by Improved High-pressure Torsion. *Rare Met. Mater. Eng.* **2016**, *45*, 3089–3094. [CrossRef]
36. Dzykovich, I.Y.; Panichkina, V.V.; Skorokhod, V.V.; Shaiderman, L.I. Effect of palladium on diffusion processes in the system tungsten-chromium. *Sov. Powder Met. Met. Ceram.* **1976**, *115*, 151–153. [CrossRef]
37. Itagaki, T.; Yoda, R. The Effect of Palladium on Oxidation Behavior of Sintered Tungsten-Chromium-Palladium Alloys. *J. Jpn. Inst. Met.* **1974**, *38*, 486–492. [CrossRef]
38. Zhang, Q.; Zhao, J.C. Impurity and interdiffusion coefficients of the Cr–X (X = Co, Fe, Mo, Nb, Ni, Pd, Pt, Ta) binary systems. *J. Alloys Compd.* **2014**, *604*, 142–150. [CrossRef]
39. Lee, D.B.; Simkovich, G. Oxidation of Mo-W-Cr-Pd alloys. *J. Less Common Met.* **1990**, *163*, 1–62. [CrossRef]
40. Waterstrat, R.M. Analysis of selected alloys in the systems Cr-Pd, Cr-Ru, V-Pd and Ta-Pt. *J. Less Common Met.* **1981**, *80*, P31–P36. [CrossRef]
41. Li, Y.; Li, P.; Bian, H.; Tang, N.; Koizumi, Y.; Chiba, A. The hot forging behaviour and its effects on the oxidation behaviour of W–Cr alloy. *Corros. Sci.* **2014**, *83*, 367–374. [CrossRef]
42. Malewar, R.; Kumar, K.S.; Murty, B.S.; Sarma, B.; Pabi, S.K. On sinterability of nanostructured W produced by high-energy ball milling. *J. Mater. Res.* **2007**, *22*, 1200–1206. [CrossRef]
43. Dine, S.; Aïd, S.; Ouaras, K.; Malard, V.; Odorico, M.; Herlin-Boime, N.; Habert, A.; Gerbil-Margueron, A.; Grisolia, C.; Chêne, J.; et al. Synthesis of tungsten nanopowders: Comparison of milling, SHS, MASHS and milling-induced chemical processes. *Adv. Powder Technol.* **2015**, *26*, 1300–1305. [CrossRef]
44. Kentheswaran, V.; Dine, S.; Vrel, D.; Couzinié, J.P.; Dirras, G. Synthesis of nanometric refractory alloys powders in the MoNbW system. *J. Alloys Compd.* **2016**, *679*, 80–87. [CrossRef]
45. Dine, S.; Kentheswaran, V.; Vrel, D.; Couzinié, J.P.; Dirras, G. Synthesis of nanometric MoNbW alloy using self-propagating high-temperature synthesis. *Adv. Powder Technol.* **2017**, *28*, 1739–1744. [CrossRef]
46. Dine, S.; Bernard, E.; Herlin, N.; Grisolia, C.; Tingaud, D.; Vrel, D. SHS synthesis, SPS densification and mechanical properties of nanometric tungsten. *Adv. Eng. Mater.* **2018**. [CrossRef]
47. Liu, G.; Li, J.; Chen, K.; He, G.; Yang, Z.; Guo, S. High-gravity combustion synthesis of W-Cr alloys with improved hardness. *Mater. Chem. Phys.* **2016**, *182*, 6–9. [CrossRef]
48. Yao, M.; Zhangjian, Z.; Jun, T.; Ming, L. Fabrication of Ultra-fine Grain Tungsten by Combining Spark Plasma Sintering with Resistance Sintering under Ultra High Pressure. *Rare Met. Mater. Eng.* **2011**, *40*, 0004–0008. [CrossRef]
49. Guo, W.; Arshad, K.; Yuan, Y.; Zhao, M.Y.; Shu, X.L.; Zhou, Z.-J.; Zhang, Y.; Lu, G.-H. Effects of vanadium alloying on the microstructures and mechanical properties of hot-pressed tungsten material. *Mod. Phys. Lett. B* **2016**, *30*, 1650216. [CrossRef]
50. Wurster, S.; Pippan, R. Nanostructured metals under irradiation. *Scr. Mater.* **2009**, *60*, 1083–1087. [CrossRef]
51. Nita, N.; Schaeublin, R.; Victoria, M. Impact of irradiation on the microstructure of nanocrystalline materials. *J. Nucl. Mater.* **2004**, *329–333*, 953–957. [CrossRef]

52. Kilmametov, A.R.; Gunderov, D.V.; Valiev, R.Z.; Balogh, A.G.; Hahn, H. Enhanced ion irradiation resistance of bulk nanocrystalline TiNi alloy. *Scr. Mater.* **2008**, *59*, 1027–1030. [CrossRef]
53. Shen, T.D.; Feng, S.; Tang, M.; Valdez, J.A.; Wang, Y.; Sickafus, K.E. Enhanced radiation tolerance in nanocrystalline MgGa$_2$O$_4$. *Appl. Phys. Lett.* **2007**, *90*, 263115. [CrossRef]
54. Yang, T.; Huang, X.; Wang, C.; Zhang, Y.; Xue, J.; Yan, S.; Wang, Y. Enhanced structural stability of nanoporous zirconia under irradiation of He. *J. Nucl. Mater.* **2012**, *427*, 225–232. [CrossRef]

Article

Influence of Carbon Diffusion and the Presence of Oxygen on the Microstructure of Molybdenum Powders Densified by SPS

Mathias Moser [1,2], Sylvain Lorand [1], Florian Bussiere [1], Frédéric Demoisson [1], Hervé Couque [2] and Frédéric Bernard [1,*]

[1] Laboratoire Interdisciplinaire Carnot de Bourgogne, UMR6303 CNRS-Université de Bourgogne-Franche-Comté, BP 47870, 21078 Dijon, France; Mathias_Moser@etu.u-bourgogne.fr (M.M.); sylvain.lorand@live.fr (S.L.); florian.bussiere@u-bourgogne.fr (F.B.); Frederic.Demoisson@u-bourgogne.fr (F.D.)

[2] Nexter Munitions, 18 route de Guerry, 18000 Bourges, France; h.couque@nexter-group.fr

[*] Correspondence: fbernard@u-bourgogne.fr; Tel.: +33-380-396-125

Received: 5 June 2020; Accepted: 8 July 2020; Published: 14 July 2020

Abstract: Due to molybdenum's Body-Centered Cubic (BCC) crystalline structure, its ductile–brittle transition temperature is sensitive to shaping, purity and microstructure. Dense molybdenum parts are usually shaped by the powder metallurgy process. The aim of this work concerns the spark plasma sintering of high-purity powders prepared by inductively coupled plasma. The influence of carbon diffusion and its interaction with oxygen on the density (i.e., the densification stage) and on the microstructure (i.e., the grain growth stage) during spark plasma sintering was investigated. The formation of carbide is usually expected for a sintering temperature above 1500 °C leading to grain growth (e.g., more than 10 times larger than the initial powder grain size after sintering at 1900 °C for 10 min). The brittleness was also affected by the segregation of molybdenum carbides at the grain boundaries (i.e., intergranular brittle fracture). Consequently, to reduce the sintering temperature to below 1500 °C, mechanically activated powders were used. From these milled powders, a dense molybdenum disc (60 mm in diameter and 10 mm in thickness) sintered at 1450 °C under a pressure of 70 MPa for 30 min was obtained. It is composed of a fine microstructure without carbide and oxide, its ductility is close to 13% with a maximum resistance of 550 MPa.

Keywords: molybdenum; grain growth; carbon diffusion; ball milling; spark plasma sintering

---

1. Introduction

Molybdenum is a refractory metal with a high-temperature resistance, due to its melting point of 2620 °C. Such a characteristic, along with its good mechnical properties, has motivated its use in defense [1] as well as civilian applications, such as the production of ribbons and wires for the lighting industry, electrodes for glass melting and thermal zones for high-temperature ovens. However, its high melting temperature makes the implementation of the material by traditional metallurgy (casting or forging) a complex and expensive process. That is why much research has been conducted into the powder metallurgy of molybdenum [2–4]. However, the sintering process remains complex and long. Indeed, conventional sintering of molybdenum requires temperatures between 1800 and 2000 °C for several hours. In addition, the density obtained from such a sintering process is less than 95% of the theoretical maximum density (TMD) [5,6], which is often not sufficient for target applications such as in defense. Many studies have been performed to increase the density of the end-product by using: (i) different sizes of molybdenum powder [7], (ii) heat treatment of the initial molybdenum powders before sintering [6], (iii) specific additives as densification aids [8], or (iv) alternative methods of sintering such as explosive consolidation [9,10], microwave energy [11] or spark plasma sintering (SPS) [3,4,7,12,13].

It is well known that the presence of some impurities in molybdenum powder can modify the mechanical behavior of the sintered product. The presence of oxygen influences the mechanical properties by segregating at the grain boundaries [14–16], while other impurities, such as carbon in a reasonable proportion, increase the ductility [17,18], but have a negative effect in excess. Some studies have also investigated the relationship between impurities and the final density. For example, the addition of nickel [19,20] or silicon [21] increases the sinterability of molybdenum leading to a better density of the final product.

Apart from the direct effect of such impurities on the ductility or the density, there is a grain growth phenomenon during the sintering of molybdenum powder at high temperature with a non-negligible effect on the density of the sintered product and, consequently, on the mechanical, electrical and thermal properties. In fact, Lee et al. [22] reported that the concentration of oxygen leading to molybdenum oxide formation is an accelerator factor for grain growth during sintering.

The main objective of this study was to investigate the fabrication by SPS of dense parts starting from powders prepared using inductively coupled plasma technology (ICP). Clearly, it is essential to understand the grain growth and densification stages of such powders in order to control them. Consequently, the present work evaluated several powders and sintering conditions to observe and understand the grain growth and the effects of the presence of oxygen and carbon.

2. Experimental Procedure

A 99.95% pure molybdenum powder with a narrow particle size distribution (i.e., D10: 11.04 μm, D50: 16.33 μm and D90: 28.88 μm) made by the Tekna Company (Macon, France) (i.e., hereafter named Mo-45) was used. This powder was prepared using ICP, which has the advantage of having a high level of sphericity (Figure 1) and a high density (10.2 g · cm^{-3} obtained by He pycnometry) without any oxide either on the surface of or inside the particle. Scanning Electron Microscope (SEM) images highlight the spherical morphology with regularity in size and shape (Figure 1a). The presence of agglomerated clusters is certainly due to some precursor powder particles not having entirely undergone the spheroidization step because they were expelled too early from the plasma by reflux movements. However, the majority of the grains have smooth surfaces. Despite these few irregularities, two populations of particles were mainly formed: particles with smooth surfaces (Figure 1b) and particles having large facets (Figure 1c). A higher magnification observation shows that the majority of smooth particles are actually composed of large faceted grains covered with a very thin layer on the surface. The smoothing of these particles may be due to faster migration of the surface species during spheroidization than occurred on particles with well-marked facets.

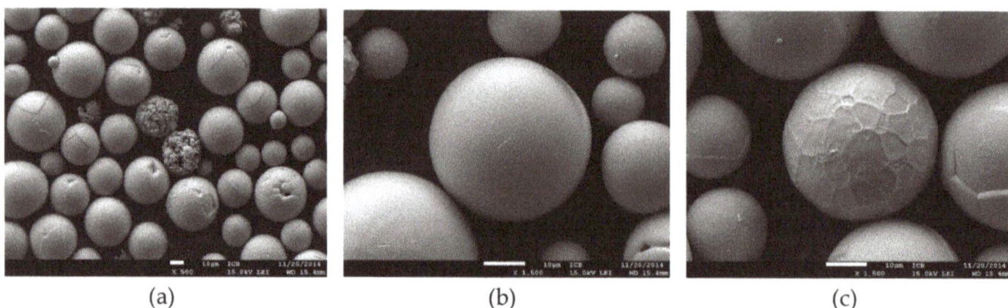

(a)   (b)   (c)

Figure 1. (a) SEM image of molybdenum powder Mo-45 (Tekna); (b) and (c) SEM images with a higher magnification showing different powder morphologies.

However, in addition, it is essential to verify the presence on the surface of a native molybdenum oxide. Indeed, some molybdenum oxides, MoO_3 or MoO_2, are usually observed on the very surface of the powder, due to humidity, but this thin layer can be removed by a thermal treatment at 800 °C under vacuum as proved by XPS (X-ray photoelectron spectroscopy) analysis (Figure 2) performed on Mo-45 powder. XPS spectra are presented of molybdenum powder before (green) and after (red) thermal treatment under vacuum at 800 °C. This study shows that it is not necessary to work under a reducing atmosphere since heating under vacuum at 800 °C is sufficient to remove the thin layer of oxide present on the surface of the molybdenum particles. Indeed, the Mo_{ox} $3d^{3/2}$ peak at 235 eV and the Mo_{ox} $3d^{5/2}$ peak at 232 eV associated with the presence of oxidized molybdenum disappeared in favor of metallic molybdenum after a heat treatment at 800 °C under vacuum (Figure 2). Consequently, as the SPS sintering was also carried out under vacuum, these oxides will be reduced during the heating stage, the rate of which was limited to 50 °C/min for all SPS tests.

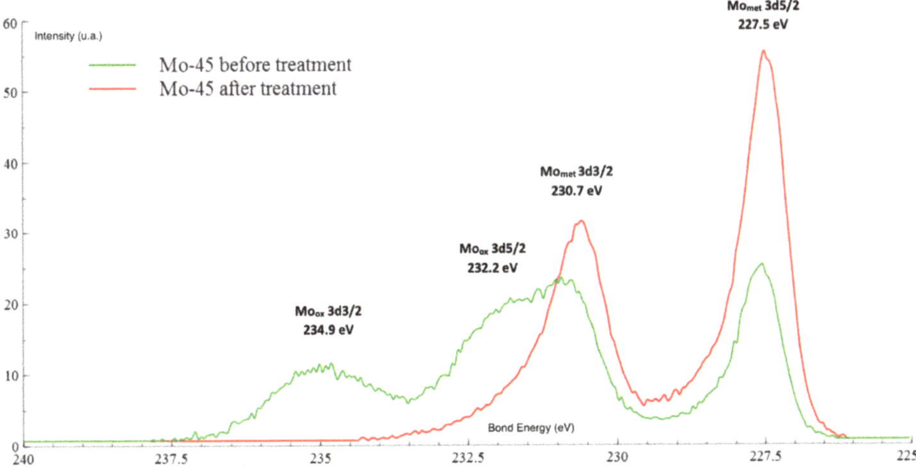

Figure 2. XPS spectra of molybdenum powder before (green) and after (red) thermal treatment performed under vacuum at 800 °C.

The SPS assembly used in this work is described in Figure 3a. As molybdenum is a conductive material, the current go through the sample and, consequently, this latter is heated by Joule effect. To favor this "internal heating", a spray of boron nitride (BN, an electrical insulating) has been deposited all around the sample (Figure 3c).

A graphite die of 60 mm inner diameter lined with a 0.35 mm thick layer of graphite foil (named Papyex®) (Mersen, Gennevilliers, France) was filled with the powder and coated with a boron nitride (BN) spray to prevent radial carbon diffusion (compare Figure 3c with the classical configuration shown in Figure 3b). An additional limit to axial carbon diffusion was provided by 3 mm thick tantalum discs placed between the punches and the powder (Figure 3d). A 3 cm thick layer of graphite felt was wrapped around the die to prevent and limit thermal gradients inside the sample during heating by reducing heat losses by radiation. Sintered samples of 60 mm in diameter and 10 mm in height were then sintered in an FCT HPD-125 spark plasma sintering machine (FCT Systeme, Rauenstein, Germany).

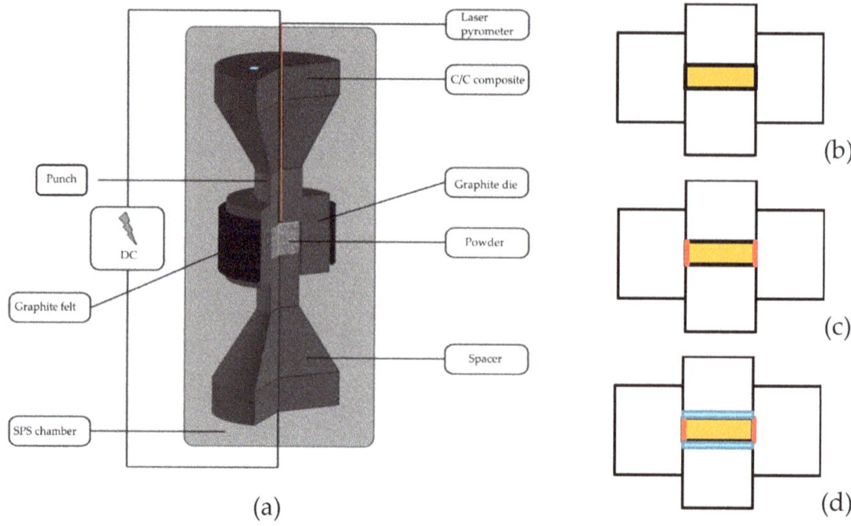

Figure 3. (**a**) General view of spark plasma sintering (SPS) assembly; (**b**), (**c**) and (**d**) represent different sample configurations: (**b**) classical assembly using graphite foil all around the sample, (**c**) assembly using a boron nitride (BN) spray on graphite foil around the sample to limit carbon radial diffusion, and (**d**) assembly with a BN barrier as previously and a tantalum disc (3 mm) on the top and bottom of the powder to limit axial carbon diffusion.

Sintering was carried out under vacuum using a DC electric current (~4000 A) and a heating rate of 50 °C · min^{-1}. Above 400 °C, the temperature was measured and regulated using an optical pyrometer looking at a spot located at 7 mm on the sample top. The temperature was varied from 1750 to 2000 °C with a hold time at the maximum temperature of between 0 to 30 min. The uniaxial pressure was varied from 28 MPa (the minimum value to ensure sufficient electrical contact) to 90 MPa (the maximum value for graphite tools). The pressure was applied from RT to 450 °C, just before the activation of the laser pyrometer. To reduce thermal stresses inside the sintered part, it was cooled in a controlled manner at a rate of 50 °C/min. Simultaneously, during the cooling, the pressure was slowly decreased at a rate of 6 KN/min. In experiment 10 (see Table 1), two massive discs of tantalum (3 mm thick) located top and bottom of the powder and a BN barrier around the powder (Figure 3d) were used to avoid any carbon diffusion from the graphite into the molybdenum. The sintering conditions for each sample are listed in Table 1.

Table 1. Table of sintering conditions of Mo-45 powder.

Experiment No.	Pressure (MPa)	Temperature (°C)	Holding Time (min)	Protection
1	90	1750	0	BN
2	90	1900	0	BN
3	90	2000	0	BN
4	70	1900	0	BN
5	70	1900	10	BN
6	70	1900	30	BN
7	70	1750	10	BN
8	28	1900	0	BN
9	70 (delayed at 900 °C)	1900	0	BN
10	70	1900	30	BN + Ta

After sintering, the graphite foil on the sample surface was removed by a sanding operation. Two types of preparation for analysis were performed on the sintered parts. First, a transverse cut was made, usually using a circular saw. The relative densities of the sintered molybdenum samples were measured using Archimedes' technique and found to have a value of 10.2. Then, the surface of this slice was mechanically polished, starting with SiC papers down to 3 μm diamond abrasive and, finished with an alumina super-finishing solution. After ultrasonic cleaning, two types of chemical etching were then performed on the cross sections. For observation of the microstructure by optical microscopy or scanning electron microscopy, a Murakami reagent ($K_3Fe(CN)_6$ /KOH/H_2O (1:1:1)) was applied for 3 min while for the observation of carbides, Hasson's reagent (C_2H_5OH/HCl (32%)/$FeCl_3^-$ $6H_2O$ (1300 g·L^{-1}) (3:1:2)) was applied for 2 min. In both cases, the samples were cleaned using distilled water and dried with ethanol.

A standard optical microscope (ZEISS) (Carl Zeiss, Jena, Germany) was used to characterize the microstructure of the SPS sintered samples. The microstructural analysis of the material was carried out using a field emission gun (FEG) scanning electron microscope (SEM JEOL JSM-7600F, Musashino, Akishima, Tokyo, Japan) coupled with a LINK OXFORD energy dispersive X-ray spectrometer (EDXS) (High Wycombe, London, UK) Tensile tests are a suitable method to determine mechanical properties such as the yield strength, the maximum tensile strength and the ductility. Three specimens were extracted from each sintered sample using electrical discharge machining (EDM) as shown in Figure 4. The gauge length and diameter of the gauge section of the samples were 16 and 4 mm, respectively. Uniaxial tensile tests were performed at a constant strain rate of 10^{-3} s^{-1} following the ASTM E8/E8M standard test method using a Testwell machine with a 1 MN load cell and a 10 mm long clip gage.

Figure 4. Location of the three tensile test pieces machined from within sintered discs (60 mm in diameter and 10 mm in thickness).

3. Experimental Results

3.1. Density, Microstructure and Grain Size

Chemical etching on cross-sections of the molybdenum powder consolidated by SPS allows a description of the sintered molybdenum microstructure. Particular attention was paid to following both the evolution of the grain size and the porosity during sintering. An investigation was performed of the growth of the molybdenum grains as a function of the temperature, the holding time and the value of the uniaxial pressure. Figure 5 shows the effect of the sintering temperature (1750, 1900 and 2000 °C) on the average molybdenum grain size (corresponding to experiments 1, 2, 3 in Table 1, respectively). Thus, a sintered sample at 1750 °C under 90 MPa pressure and without holding time (Figure 5a) has a microstructure with a grain size close to the initial size of the powder particles (i.e., 25 μm determined by laser granulometry). At this temperature, any sign of grain growth is highlighted. By contrast at 1900 °C, grain growth starts but is not homogeneous since two different microstructures are observed: a central part composed of large grains (150 μm, Figure 5c, right hand side) whereas the sample edge is made up of grains having a similar size as initially (i.e., 25 μm, Figure 5c, left hand side). Such an observation is probably the result of a thermal gradient during SPS sintering. The existence of this thermal gradient is due to the absence of a holding time at the sintering temperature which does not allow thermal homogeneization. Effectively, a thermal gradient exists (i.e., as Mo is a conductive material, the center is hotter that the periphery), this latter may be reduced by a modification of SPS tools dimensions and by the addition of graphite felt which limits the thermal losses by radiation [23]. Finally, at 2000 °C, the grain growth occurs over the whole sample (Figure 5b) and the final density is better (98.4 ± 0.2% TMD Theoritical Maximum Density)) in comparison with those obtained at lower temperatures for which the density is not adequate, 96.5 ± 0.2% TMD at 1750 °C, 97.8 ± 0.2% TMD at 1900 °C without any holding time.

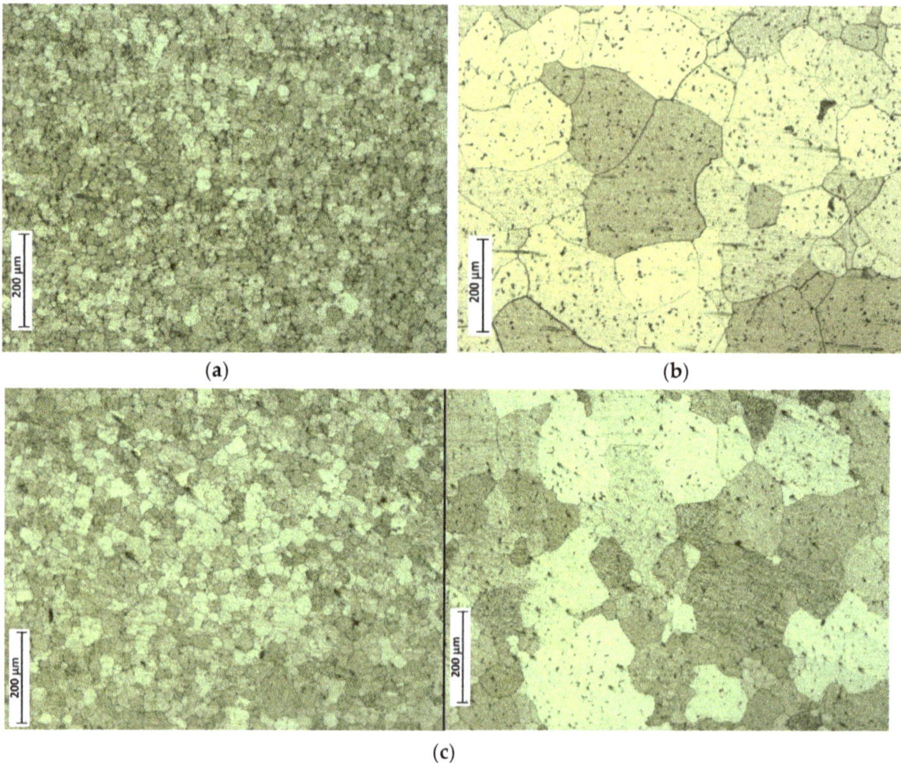

Figure 5. Microstructures after chemical etching of molybdenum powder (Mo-45) sintered under 90 MPa pressure without holding time at: (**a**) 1750, (**b**) 2000 and (**c**) 1900 °C (left: edge, right: core).

To evaluate the influence of holding time on molybdenum grain growth, studies were performed at 1900 °C for various holding times (0, 10 and 30 min) under 70 MPa pressure. These tests correspond to experiments 4, 5, 6, in Table 1. As shown in Figure 6b, after 10 min of holding time, the microstructure is different, with an irregular, but generalized grain growth compared to a sintered sample without holding time (Figure 6a). The latter has a fine microstructure on the shell but a beginning of grain growth in the core. After 30 min, the growth is total and regular in the whole sintered sample, as can be seen in Figure 6c. In reality, the grain growth starts after 2 or 3 min at temperature and it is difficult to obtain a microstructural difference between the core and shell. Because of this abrupt change in microstructure, it is possible to qualify this grain growth as exaggerated, in comparison with the phenomena usually observed in metal powders during sintering. The density is also affected by the holding time. While the density of the molybdenum powder after sintering without holding time is around 98.2 ± 0.2% TMD, after 10 min this becomes 99.2% ± 0.2% TMD and after 30 min, maximum densification is reached, with a density of 99.9% ± 0.2% TMD.

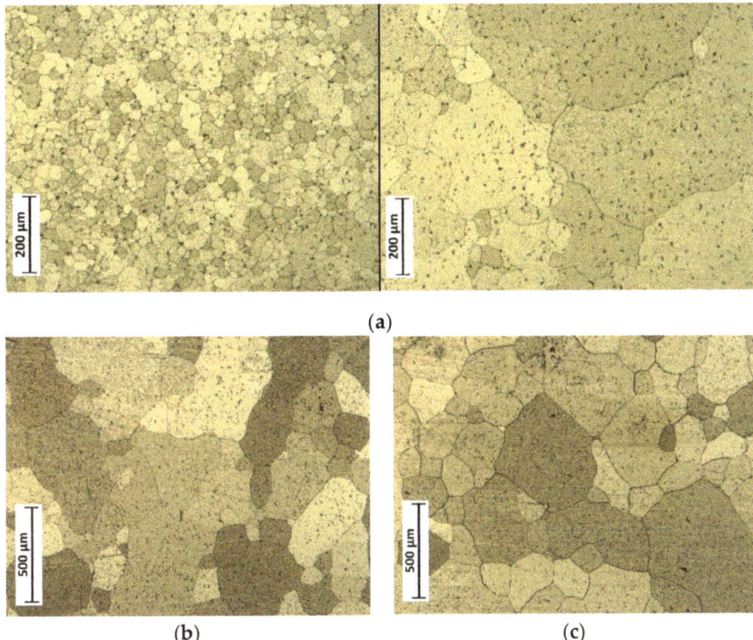

Figure 6. Microstructures after chemical etching of molybdenum powder (Mo-45) sintered at 1900 °C under 70 MPa pressure: (**a**) without holding (left: shell, right: core), (**b**) 10 min and (**c**) 30 min.

In addition (Figure 7), the Mo-45 powder, sintered at 1750 °C under 70 MPa pressure and held for 30 min (i.e., corresponding to experiment 7 in Table 1) shows a different microstructure than seen previously at 1900 °C for the same holding time (Figure 6c). The microstructure obtained is close to those obtained at 1900 °C without holding (Figure 5a). In fact, it is necessary to find a good compromise between temperature and holding time to obtain dense molybdenum with homogeneous microstructure while avoiding exaggerated grain growth.

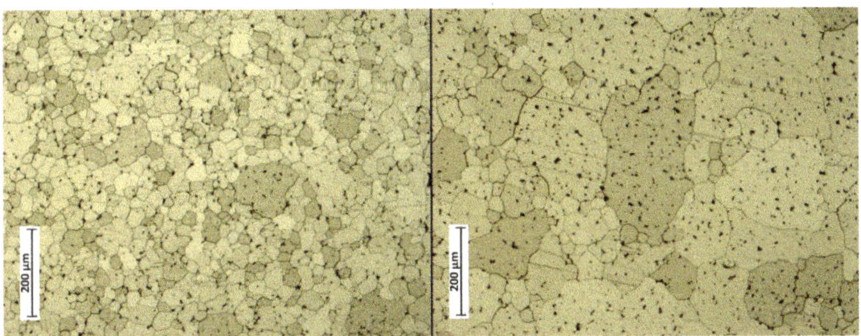

Figure 7. Microstructures after chemical etching of molybdenum powder (Mo-45) sintered at 1750 °C under 70 MPa pressure for 30 min (left: shell, right: core). Black dots are due colloidal silica used as reactant.

The influence of the mechanical pressure was also investigated for samples sintered at 1900 °C without holding time. Figures 5c, 6a and 8a show the evolution of molybdenum microstructure for three different pressures, 28, 70 and 90 MPa corresponding to experiments 8, 4 and 2 in Table 1, respectively.

There is no grain growth at 28 MPa even with 30 min of holding time. So at this pressure, the grain growth mechanisms are not activated and the density is limited to 90.7 ± 0.2% TMD. But, at and above 70 MPa pressure, the grain growth is similar and seems to be governed by the same driving forces as previously since grain growth and a non-homogeneous microstructure are observed. The density of this sample was close to 98.2 ± 0.2% TMD.

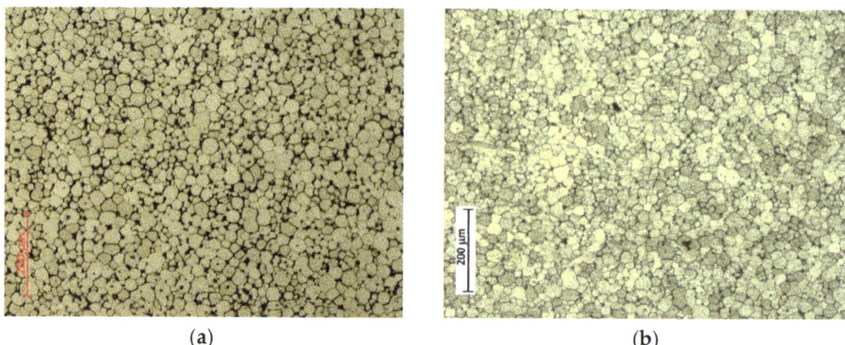

(a)   (b)

Figure 8. Microstructures after chemical etching of molybdenum powder (Mo-45) after sintering at 1900 °C without holding time: (**a**) under 28MPa and (**b**) under 70MPa applied at 900 °C.

However, if the pressure is not applied from the beginning of the sintering cycle but at a temperature of 900 °C, for example (i.e., corresponding to experiment 9 in Table 1), the microstructure is affected as shown in Figure 8b. Indeed, grain growth has not occurred and the microstructure is also homogeneous with a density of 98.1 ± 0.2% TMD. Several phenomena can explain those microstructural changes, such as the effect of the mechanical pressure which induces plastic deformation (including creep) and, consequently, changes the surface contact between particles leading to a modification of the local temperature.

3.2. Diffusion of Carbon

During sintering at 1900 °C under 70 MPa pressure for 30 min (experiment 6 in Table 1), the presence of a carbon layer on the molybdenum surface was demonstrated (Figure 9). This phenomenon has already been highlighted by Mouawad [12]. This layer with a thickness of 100 µm was composed of molybdenum carbides, as experimentally shown by X-Ray Diffraction (XRD). These molybdenum carbides can exist in two stable forms: the α-Mo_2C phase and the β-Mo_2C phase. XRD analysis confirmed that this layer was mainly composed of the hexagonal phase β-Mo_2C (the α-Mo_2C phase is orthorhombic). The thickness of the carbide layers depends on how far the carbon diffuses into the molybdenum, i.e., it depends on the sintering temperature, the holding time and the pressure. However, although this layer of molybdenum carbide leads to a hardness higher than a sample without it, it has proved to be harmful since it is very brittle, which does favor some mechanical properties such as ductility.

The formation of this carbide layer is mainly due to the presence of the graphite foil in contact with the sample and of the SPS chamber environment which is super-saturated with carbon (from the dies, spacers). In order to prevent carbon diffusion, the graphite foil located on the inner part of the die was coated with a boron nitride spray to create a chemical barrier. It also enhances the passage of the electric current inside the sample. However, as it is necessary to preserve the passage of the current within the sample, it is therefore not possible to deposit boron nitride on the surface of each punch. It is difficult to find a pertinent way to protect molybdenum sample surfaces from the formation of this layer, which can be removed easily by sandblasting, but this causes cracking and then the spalling of the molybdenum carbide layer.

Figure 9. Microstructures after chemical etching of molybdenum powder after sintering at 1900 °C under 70 MPa pressure for 30 min showing the carbide layer on the surface.

Unfortunately, the diffusion of carbon is not limited to the first 200 μm of the samples. Chemical etching shows the presence of molybdenum carbides both in the bulk and at the sample periphery. A study of those carbides shows that a relationship exists between the quantity, the location and the form of those carbides and the growth of molybdenum grains during sintering. Indeed, as can be seen in Figure 10a for the case of molybdenum with no grain growth during sintering, internal diffusion of carbon from the sample surface to the bulk and the formation of carbides are not highlighted whereas when grain growth occurs (Figure 10b), carbides are present at grain boundaries.

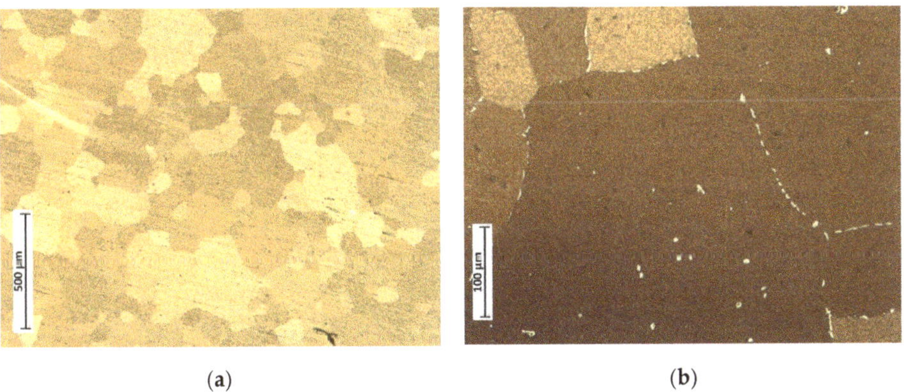

(a) (b)

Figure 10. Microstructures after chemical etching of molybdenum powder sintered at 1900 °C under 70 MPa pressure: (**a**) without a holding time and (**b**) with a 30 min holding time in both cases at the sample centre.

Such carbon diffusion depends not only on the temperature, the holding time and the mechanical charge but also on the sintering environment. To reduce the presence of such carbides, it is essential either to avoid the diffusion of carbon or, if diffusion occurs, the formation of those carbides.

Our observations show that molybdenum carbides form at a high temperature, above 1500 °C, in a carbon saturated environment. At 1900 °C and without a holding time, the diffusion is minimal as is the grain growth but, in that case, the density remains low (<98% TMD ± 0.2%). A first solution is to create a barrier to the diffusion of carbon into samples. Only two materials have a melting point higher than that of the molybdenum, namely tantalum and tungsten. Two 300 mm thick discs of tantalum were placed on either side of the powder to create a carbon diffusion barrier. This barrier enables the limitation of grain growth by reducing carbon diffusion into the molybdenum and, consequently, the formation of molybdenum carbides. This barrier is efficient because for sintering at 1900 °C under 70 MPa pressure for 30 min, there is no diffusion, no formation of carbide and no molybdenum grain growth (Figure 11).

Figure 11. Microstructure after chemical etching of molybdenum sintered at 1900 °C under 70 MPa pressure, for a holding time of 30 min with a 3 mm tantalum barrier.

This solution shows clearly that it is essential to avoid long distance carbon diffusion. Consequently, the presence of carbon or carbides changes the grain growth mechanisms. However, this solution is expensive and non-reusable because tantalum carbides are formed, and is difficult to adapt to the sintering of complex shapes.

As the relationship between carbides and grain growth was unclear, new experiments were performed using a commercial molybdenum powder containing a high concentration of carbon. This powder is also produced by the TEKNA company (Mâcon, France) (the commercial name is Mo-45 HC) and the carbon concentration is controlled during the ICP process). This powder, which has a similar particle size as the previously studied powder (Mo-45), is composed of two phases: Mo and Mo_2C. Sintering of the Mo-45HC powder at 1900 °C under 70MPa pressure for 30 min and without a tantalum diffusion barrier produces a fine microstructure (Figure 12), with a grain size close to the initial particle size of the powder grain. The molybdenum carbides do not seem to be responsible for the exaggerated grain growth but the "free" carbon plays a role in this phenomenon.

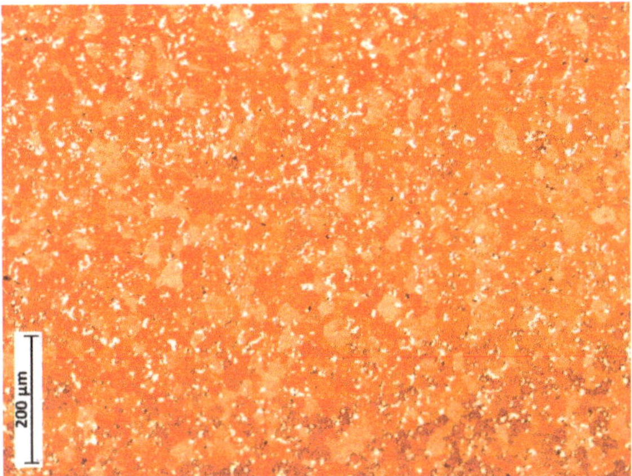

Figure 12. Microstructures after chemical etching of molybdenum powder with a high carbon content after sintering at 1900 °C under 70 MPa pressure for 30 min of holding.

3.3. Presence of Oxygen

XPS analysis performed on the molybdenum powder showed the presence of a thin layer (~4–5 nm) of molybdenum oxides at the grain surface. However, an exact measurement of the oxygen content in the bulk material after sintering remains difficult to obtain. In addition, the effect of this layer on the evolution of microstructure during sintering is uncertain. That is why, to amplify the effect of oxygen, a molybdenum powder was used with a particle size ranging from 3–7 µm (i.e., high surface area) and made without control of the oxygen concentration (produced by Alfa Aesar and named Mo-1). This powder has a smaller particle size than the previous ones, so the sintering temperature is lower. After a sintering cycle at 1750 °C under 70 MPa pressure (hold time 10 min) and using a tantalum diffusion barrier (disc of 3 mm thick), the microstructure did not show any grain growth or any molybdenum carbides. However, observation of the intergranular brittle fracture surfaces of this part (Figure 13a) shows the presence of a high concentration of spheres, composed of molybdenum oxides, i.e., MoO_2 and MoO_3.

In order to reduce the amount of molybdenum oxides, it was decided to add a reduction stage composed of a 650 °C temperature plateau lasting 30 min in vacuum before the sintering stage. After fracture of the sample, the surfaces (Figure 13b) also are characteristic of brittle fracture but there are no oxides at the grain surface. The reduction stage in vacuum is effective since an exaggerated grain growth, with grain sizes ten times larger than the initial grain size is observed. The spherical molybdenum oxides seem to block the grain boundaries, preventing exaggerated grain growth. The elimination of these oxides releases the grain boundaries and allows grain growth to occur.

Then, in order to limit the grain growth and to avoid the use of a tantalum barrier (i.e., limit the carbon diffusion), a sintering cycle composed of an oxide reduction stage at 650 °C and sintering at a temperature lower to that relative to carbide formation (i.e., 1450 °C) was carried out. These processing conditions produce a molybdenum disc without oxides and without grain growth (Figure 13c). Even if the oxide reduction stage is efficient, the effect of temperature remains non-negligible on the grain growth and also on the density (95.5 ± 0.2% TMD).

At high temperature, carbon strongly reduces molybdenum oxides. The diffusion of "free" carbon from the matrix to the center of the molybdenum grains induces the reduction of the molybdenum trioxides and dioxides if these have not been eliminated during the reduction stage performed in vacuum. These reductions release the grain boundaries. Several reactions occur, as described by Hegedus and Neugebauer [24], leading to the formation of some gaseous phases (CO, CO_2), which can be observed on the fracture surfaces of the powders sintered at high temperature with a finite holding time but without a reduction stage (Figure 14).

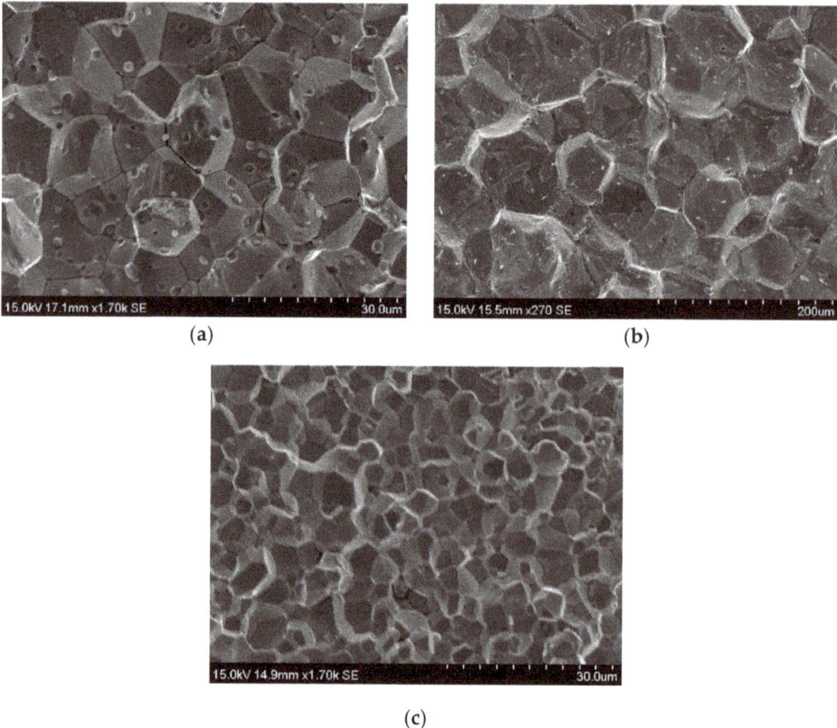

(c)

Figure 13. Fracture surfaces of high-oxidized sintered molybdenum: (**a**) at 1750 °C without reduction stage, (**b**) at 1750 °C and with an oxide reduction plateau at 650 °C and (**c**) at 1450 °C with an oxide reduction plateau at 650 °C.

Figure 14. Fracture surfaces of sintered molybdenum powder at 1450 °C under 70 MPa during 30 min without reduction stage at 650 °C, showing the effect of gaseous species.

After the reduction reaction between carbon and the molybdenum oxides, the excess carbon continues to diffuse into the sample and reacts with the molybdenum to form some carbides. These carbides are responsible for the high fragility of the material. The ideal sintered molybdenum is a dense solid without oxides and carbides. For the first species, an oxide reduction stage can be implemented in the sintering cycle, but at high temperature, which is a necessary condition to obtain high density, exaggerated grain growth is observed. This heterogeneous microstructure penalizes the mechanical properties of the final product. As for the carbides, the diffusion of carbon can be limited by the use of a tantalum barrier. However, this solution is not economically or technically feasible for the production of large and complex parts. For control of both of these impurities, sintering at a temperature lower than 1500 °C seems to be a good solution but these latter are not dense (~90% TMD ± 0.2%). Indeed, below this temperature, carbides do not form and if a reduction stage is implemented, the microstructure remains fine-grained without grain growth. However, it is still possible that some carbon diffusion occurs. However, the diffusion is slow, so even if it occurs, it remains limited.

In the case of the dense molybdenum produced from commercial powder made by Tekna, this sintering temperature is an interesting way of solving the carbide formation; however, its density is not sufficient (density close to 90% TMD ± 0.2%), meaning another solution has to be found. Several possibilities can be envisaged, such as a powder with smaller grains or a mechanically activated powder, but in both cases, the purity of the powder remains essential.

3.4. Mechanical Activation

As shown previously, a sintering temperature below 1500 °C is needed to limit grain growth and carbide formation. However, a lower temperature means a lower density and usually poor mechanical properties. Mechanical activation is a means of lowering the sintering temperature by reducing the sintering activation energy.

The ball milling stage (i.e., Fritsch pulverisette 4 planetary ball mill) was carried out using hardened steel balls which are introduced inside vials made of hardened steel. Planetary ball milling is characterized by two rotations expressed in rotations per minute (rpm): (i) rotation Ω of the plate to which the vials are fixed to obtain centrifugal acceleration, and (ii) rotation ω of the vials in the opposite direction. Depending on the selected rotation speeds, it is possible either to promote collisions or friction (energy and frequency of shocks) [25,26]. Moreover, the vials are filled in a glove box under an argon atmosphere in order to avoid oxidation. The molybdenum powders were ground according to the condition $\Omega/-\omega/h$ = 250 rpm/−250 rpm/4 h (i.e., friction mode). The size distribution of the commercial powder ranges from 10–30 to 10–75 μm after milling confirming the formation of nanostructured agglomerates as we have seen in the case of copper [25]. On the other hand, the agglomerates are smaller because of the more fragile character of molybdenum (Figure 15).

(a)   (b)   (c)

Figure 15. SEM microstructural analysis of commercial Mo-45 powders (**a**) and mechanically activated molybdenum powder agglomerates (**b**,**c**); (**b**) overview of the agglomerates showing that they can reach a size greater than 50 μm, and (**c**) observation of mechanically activated agglomerates.

X-ray diffraction patterns of powders that were mechanically activated by milling show a significant broadening of the XRD peaks corresponding to a decrease in the size of the crystallites and an increase in structural defects when high energy milling is applied. The influence of mechanical milling on the sintering behavior is confirmed by the curves in Figure 16 which compares the shrinkage curve of unmilled commercial powder shown as a solid green line (the solid red curve corresponds to the densification speed) and that for mechanically activated powder shown as a dotted green line (the dotted red curve corresponds to the densification speed). In fact, this figure shows that sintering begins at a lower temperature, that it is more active (higher slope) and that it finishes before 1500 °C, a temperature not to be exceeded to avoid the formation of molybdenum carbides and oxides. In addition, using the master curves method Lorand [27] has shown that high-energy mechanical grinding induces a decrease in the activation energy of sintering (Q = 195 kJ/mol instead of ~330 kJ/mol) due to the presence of structural defects.

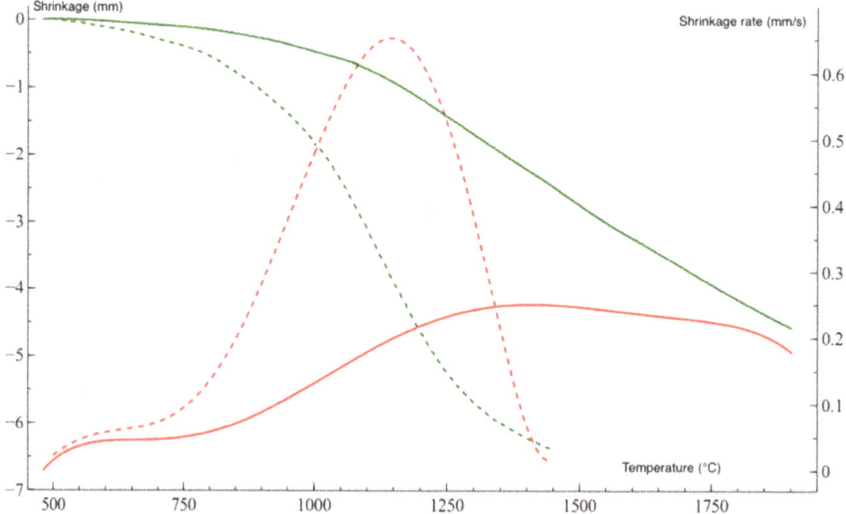

Figure 16. Shrinkage curves and densification speed for an unmilled Mo-45 powder (solid line, green shrinkage, red speed) and mechanically activated Mo powder (dotted line green shrinkage, speed in red). This figure clearly shows the advantage of milling to modify the sintering conditions of molybdenum powder.

SPS sintering of a mechanically activated molybdenum powder under the conditions 1450 °C, 70 MPa pressure for 30 min led to the formation of a dense molybdenum (98.7 ± 0.2% TMD) having a fine microstructure with no carbides or oxides. The formation of carbides is avoided if the sintering temperature is smaller than 1500 °C. However, the presence of free carbon may be interesting because this latter in low concentration enables the increasing of the Mo ductility [17,18].

Tensile tests have been carried out on samples sintered from unmilled and from mechanically activated powders (Figure 17a). The tensile test results highlight the improvement of ductility after ball milling, as a ductility of 13% was reached. We are in presence of a particular ductile failure process, encountered with BCC systems, with a cleavage fracture triggered after local plastic deformation. This cleavage process consists in plastic deformation of the crystal grains, the "ductile" aspect, followed by a transgranular failure of parallel dense atomic planes revealing facet with river appearances, the "cleavage" aspect of dense planes without dimples. This particular ductile failure process has been observed with tungsten particles, BCC as molybdenum, as reported by Lankford et al. [28]. In this work the ductile failure process has been called "transgranular shear cleavage".

However, some pores remain after sintering (Figure 17b). These results show the efficiency of the milling process since it reduces the sintering temperature and so avoids carbide formation. This is confirmed by the ductility of the sintered molybdenum. The large difference in the elastic limit of the unmilled powder (near 500 MPa) and milled powder (about 300 MPa) compacts comes from that the grain growth during sintering of the milled powder is greater than with as received powder. Precisely, the unmilled microstructure reveals a grain size of about 10 μm, while the grain size of the milled microstructure has been estimated at 7 μm. Such grain size data apply to the Hall Petch relation, along with the yield stress data of Figure 17a.

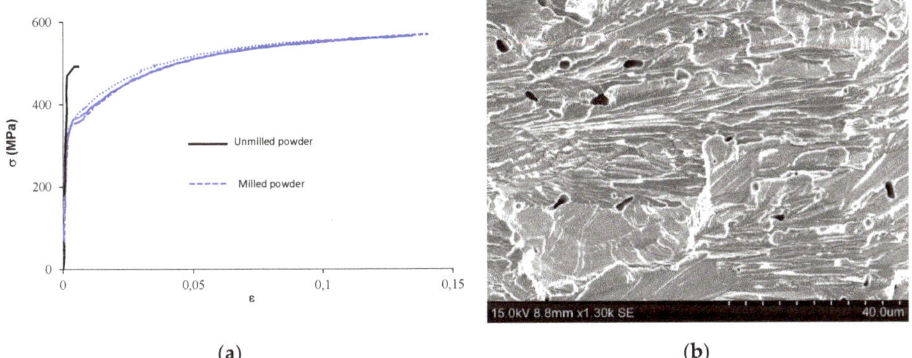

Figure 17. (**a**) True tensile stress strain curves of molybdenum samples sintered from unmilled and milled powder (**b**) SEM image of fracture surface of a sintered milled powder.

4. Conclusions

When oxygen is present as an oxide in the bulk material, the molybdenum grain boundaries are locked, thus preventing grain growth of the molybdenum grains, an undesirable effect during sintering. In addition, this pollution is prejudicial for the purity of the final product as well as for its mechanical properties. Therefore, it is necessary to remove any trace of oxygen during sintering by application of an oxide reduction stage. This can be done by the addition of carbon via the diffusion from the graphite present in the SPS tools. However, the formation of a carbide phase is also prejudicial to molybdenum for the same reasons as the oxide. It is therefore necessary to go through an oxide reduction stage and to add protection against carbon diffusion in order to obtain dense molybdenum with a pure chemical composition. It has been shown that the solution of introducing a barrier against diffusion of carbon is not viable for SPS sintering especially in the case of complex shapes. Moreover, by eliminating oxygen, densification will inevitably lead to exaggerated grain growth. The coarser microstructure of molybdenum so formed will then affect the mechanical properties of the final part.

Sintering temperature conditions with oxide reduction and no carbon diffusion are therefore required. This can be hypothetically performed if during sintering an oxide reduction stage at 650 °C is added, and if sintering is carried out below the carbide formation temperature, i.e., at a temperature lower than 1500 °C. However, for a temperature below 1500 °C, the final product is not dense enough. The solution adopted to meet this challenge was to implement high energy mechanical milling of the molybdenum powder. SPS sintering of a mechanically activated molybdenum powder according to the conditions 1450 °C/30 min/70 MPa has been performed. This mechanical activation of molybdenum powders led to the formation of a dense product (98.7% TMD) having a fine microstructure without either carbides or oxides. A ductility close to 13% could be reached via the observation of transgranular rupture still containing a few pores resulting from mechanical activation. Our results are very encouraging, because the tensile properties Rm = 550 MPa and A% = 13% of the raw SPS products are comparable to those obtained by Plansee for a sintered and annealed molybdenum (Rm = 560 MPa and

14%). On the other hand, they are much lower than those obtained in sintered, forged and annealed products for which a ductility of 40% has been obtained [1]. Consequently, it will be essential to perform further SPS sintering investigations so as to improve the ductility, while keeping in mind that the SPS approach provides a unique advantage compared to other processes with the ability to elaborate near shape parts.

Author Contributions: Methodology, M.M. and F.B. (Florian Bussiere); validation, S.L. and F.D.; formal analysis, F.B. (Frédéric Bernard) and H.C.; writing—original draft preparation, M.M. and S.L.; writing—review and editing, F.D., F.B. (Frédéric Bernard) and H.C.; supervision, F.B. (Frédéric Bernard) and H.C. All authors have read and agreed to the published version of the manuscript.

Funding: The authors thank the Regional Council of Bourgogne Franche-Comté for funding the Ph.D. grant of Sylvain Lorand Ph-D, and the French MOD for funding the Ph.D. grant of Mathias Moser.

Acknowledgments: Specific thanks to Stephen Walley from Cambridge University for his valuable comments.

Conflicts of Interest: The authors declare no conflict of interest.

References

1. Lichtenberger, A. Ductile behavior of some materials in the shaped charge jet. In *Metallrgical and Materials Applications of Shock-Wave and High-Strain-Rate Phenomena*; Murr, L.E., Staudhammer, K.P., Meyers, M.A., Eds.; Marcel Dekker: New York, NY, USA, 1995; pp. 463–470.
2. Kim, Y. Consolidation behavior and hardness of P/M molybdenum. *Powder Technol.* **2008**, *186*, 213–217. [CrossRef]
3. Park, H.-K.; Ryu, J.-H.; Youn, H.-J.; Yang, J.-M.; Oh, I.-H. Fabrication and property evaluation of mo compacts for sputtering target application by spark plasma sintering process. *Mater. Trans.* **2012**, *53*, 1056–1061. [CrossRef]
4. Sakamoto, T. Sintering behavior of Mo powder compact applied SPS spark plasma sintering method. *J. Jpn. Soc. Powder Powder Metall.* **1997**, *44*, 845–850. [CrossRef]
5. Garg, P.; Park, S.-J.; German, R.M. Effect of die compaction pressure on densification behavior of molybdenum powders. *Int. J. Refract. Met. Hard Mater.* **2007**, *25*, 16–24. [CrossRef]
6. Kim, G.-S.; Kim, H.G.; Kim, D.-G.; Oh, S.-T.; Suk, M.-J.; Kim, Y.D. Densification behavior of Mo nanopowders prepared by mechanochemical processing. *J. Alloy. Compd.* **2009**, *469*, 401–405. [CrossRef]
7. Srivatsan, T.S.; Ravi, B.G.; Petraroli, M.; Sudarshan, T.S. The microhardness and microstructural characteristics of bulk molybdenum samples obtained by consolidating nanopowders by plasma pressure compaction. *Int. J. Refract. Met. Hard Mater.* **2002**, *20*, 181–186. [CrossRef]
8. Kim, Y.; Lee, S.; Noh, J.-W.; Lee, S.H.; Jeong, I.-D.; Park, S.-J. Rheological and sintering behaviors of nanostructured molybdenum powder. *Int. J. Refract. Met. Hard Mater.* **2013**, *41*, 442–448. [CrossRef]
9. Murr, L.E.; Tuominen, S.M.; Hare, A.W.; Wang, S.-H. Structure and hardness of explosively consolidated molybdenum. *Mater. Sci. Eng. A* **1983**, *57*, 107–111. [CrossRef]
10. Shankar, S.; Murr, L.E. Heat treatment of explosively consolidated molybedenum: TEM studies. *J. Mater. Sci. Lett.* **1984**, *3*, 15–17. [CrossRef]
11. Chhillar, P.; Agrawal, D.; Adair, J.H. Sintering of molybdenum metal powder using microwave energy. *Powder Metall.* **2008**, *51*, 182–187. [CrossRef]
12. Mouawad, B.; Soueidan, M.; Fabregue, D.; Buttay, C.; Bley, V.; Allard, B.; Morel, H. Full densification of Molybdenum powders using spark plasma sintering. *Metall. Mater. Trans. A* **2012**, *43*, 1–8. [CrossRef]
13. Ohser-Wiedemann, R.; Martin, U.; Seifert, H.J.; Müller, A. Densification behaviour of pure molybdenum powder by spark plasma sintering. *Int. J. Refract. Met. Hard Mater.* **2010**, *28*, 550–557. [CrossRef]
14. Huang, H.S.; Hwang, K.S. Deoxidation of molybdenum during vacuum sintering. *Metall. Mater. Trans. A* **2002**, *33*, 657–664. [CrossRef]
15. Krajnikov, A.V.; Drachinskiy, A.S.; Slyunyaev, V.N. Grain boundary segregation in recrystallized molybdenum alloys and its effect on brittle intergranular fracture. *Int. J. Refract. Met. Hard Mater.* **1992**, *11*, 175–180. [CrossRef]
16. Kumar, A.; Eyre, B.L. Grain boundary segregation and intergranular fracture in molybdenum. *Proc. R. Soc. Lond. A Math. Phys. Eng. Sci.* **1980**, *370*, 431–458.

17. Hiraoka, Y. Significant effect of carbon content in the low-temperature fracture behavior of molybdenum. *Mater. Trans. JIM* **1990**, *31*, 861–864. [CrossRef]

18. Kadokura, T.; Hiraoka, Y.; Yamamoto, Y.; Okamoto, K. Change of mechanical property and fracture mode of molybdenum by carbon addition. *Mater. Trans.* **2010**, *51*, 1296–1301. [CrossRef]

19. Hofmann, H.; Grosskopf, M.; Hofmann-Amtenbrink, M.; Petzow, G. Sintering behaviour and mechanical properties of activated sintered molybdenum. *Powder Metall.* **1986**, *29*, 201–206. [CrossRef]

20. Hwang, K.S.; Huang, H.S. Identification of the segregation layer and its effects on the activated sintering and ductility of Ni-doped molybdenum. *Acta Mater.* **2003**, *51*, 3915–3926. [CrossRef]

21. Zovas, P.E.; German, R.M. Retarded grain boundary mobility in activated sintered molybdenum. *Mater. Trans. A* **1984**, *15*, 1103–1110. [CrossRef]

22. Lee, B.-K.; Oh, J.-M.; Shon, I.-J.; Kim, H.-S.; Lim, J.-W. Influence of oxygen concentration on mechanical properties of molybdenum powder during sintering. *J. Nanosci. Nanotechnol.* **2014**, *14*, 9029–9032. [CrossRef] [PubMed]

23. Naimi, F.; Ariane, M.; Barthelemy, F.; Cury, R.; Mahot, P.; Couque, H.; Bernard, F. The hybrid-SPS, an effective tool to produce large-sized parts. In *Spark Plasma Sintering: Current Status, New Developments and Challenge Dedicated to Prof. ZA. Munir*; Cao, G., Orru, R., Estournés, C., Garay, J., Eds.; Elsevier: Amsterdam, The Netherland, 2019; pp. 301–312.

24. Hegedüs, A.J.; Neugebauer, J. Zum mechanismus der reaktion von molybdäntrioxyd mit kohlenstoff. *Z. Anorg. Allg. Chem.* **1960**, *305*, 216–226. [CrossRef]

25. Gaffet, E.; Bernard, F. Mechanically activated powder metallurgy processing: A versatile way towards nanomaterials synthesis. *Ann. Chim. Sci. Mat.* **2002**, *27*, 45–59.

26. Boytsov, O.; Gaffet, E.; Bernard, F.; Ustinov, A. Correlation between milling parameters and microstructure characteristics of nanocrystalline copper powder prepared via a high energy planetary ball mill. *J. Alloy. Compd.* **2007**, *432*, 103–110. [CrossRef]

27. Lorand, S. Maitrise De La Microstructure Du Molybdène: De La Poudre Préparée Par ICP Au Massif Fritté Par SPS. Ph.D Thesis, Université de Bourgogne, Dijon, France, November 2017.

28. Lankford, J.; Couque, H.; Bose, A.; German, R. Dynamic deformation and failure of tungsten heavy alloys. In *Tungsten and Tungsten Alloys—Recent Advances TMS Conference*; Crownson, A., Chen, E.S., Eds.; The Minerals, Metals & Materials Society: Warrendale, PA, USA, 1991; pp. 151–159.

Article

Spark Plasma Sintering as an Effective Texturing Tool for Reprocessing Recycled HDDR Nd-Fe-B Magnets with Lossless Coercivity

Awais Ikram [1,2,3,*], Muhammad Awais [4], Richard Sheridan [4], Allan Walton [4], Spomenka Kobe [2,3], Franci Pušavec [1] and Kristina Žužek Rožman [2,3]

[1]   Faculty of Mechanical Engineering, University of Ljubljana, Aškerčeva cesta 6, SI-1000 Ljubljana, Slovenia; franci.pusavec@fs.uni-lj.si

[2]   Department for Nanostructured Materials, Jožef Stefan Institute, Jamova 39, SI-1000 Ljubljana, Slovenia; spomenka.kobe@ijs.si (S.K.); tina.zuzek@ijs.si (K.Ž.R.)

[3]   Jožef Stefan International Postgraduate School, Jamova 39, SI-1000 Ljubljana, Slovenia

[4]   School of Metallurgy and Materials, University of Birmingham, Edgbaston, Birmingham B15 2TT, UK; m.awais@bham.ac.uk (M.A.); r.s.sheridan.1@bham.ac.uk (R.S.); a.walton@bham.ac.uk (A.W.)

*   Correspondence: rana.awaisikram@yahoo.com; Tel.: +386-68-67-3099

Received: 23 December 2019; Accepted: 23 March 2020; Published: 24 March 2020

Abstract: The low-pressure hot-deformation methodology was applied to reprocess the nanocrystalline hydrogenation–disproportionation–desorption–recombination (HDDR) Nd-Fe-B powders from end-of-life (EOL) permanent magnets' waste to determine the mechanism of texture development and the resultant improvement in remanence (and BH_{max}) in the recycled material. Both the hot-pressed and hot-deformed magnets produced via spark plasma sintering (SPS) were compared in terms of their magnetic properties with respect to forging pressures. Also, a comparison was established with the microstructure to cite the effectiveness of texture development at low deformation rates and pressures which is pivotal for retaining high coercivity. The hot-pressed magnets maintain the high coercivity (better than 100%) of the original recycled powder due to the control of SPS conditions. The hot deformation pressure was varied from 100–150 MPa at 750 °C processing temperature to identify the optimal texture development in the sintered HDDR Nd-Fe-B magnets. The effect of post-hot-deformation thermal treatment was also investigated, which helped in boosting the overall magnetic properties and better than the recycled feedstock. This low-pressure hot deformation process improved the remanence of the hot-pressed magnet by 11% over the starting recycled powder. The M_r/M_S ratio which was 0.5 for the hot-pressed magnets increased to 0.64 for the magnets hot-deformed at 150 MPa. Also, a 55% reduction in height of the sample was achieved with the *c*-axis texture, indicating approximately 23% higher remanence over the isotropic hot-pressed magnets. After hot deformation, the intrinsic coercivity (H_{Ci}) of 960 kA/m and the remanence (B_r) value of 1.01 T at 150 MPa is indicative that the controlled SPS reprocessing technique can prevent microstructure related losses in the magnetic properties of the recycled materials. This route also suggests that the scrap Nd-Fe-B magnets can be treated with recoverable magnetic properties subsequently via HDDR technique and controlled hot deformation with a follow-up annealing.

Keywords: rare earth permanent magnets; anisotropic HDDR $Nd_2Fe_{14}B$; recycling; spark plasma sintering; texture development; hot deformation

---

1. Introduction

The rare-earth based Nd-Fe-B permanent magnets (REPMs) are vital components for modern electronics, energy, medical imaging, and automotive industries due to their high magnetization and magnetostatic energy confined in a small volume (BH_{max}). These excellent magnetic properties ascend from a combination *4f* rare earth (RE) sublattices contributing to high magnetocrystalline anisotropy and *3d* transition metal (TM) sublattices return high magnetization, energy product and Curie temperature [1]. The BH_{max} is approximated to value $J_r^2/4\mu_0$ (J_r is the remnant polarization and related to magnetic induction as: $J = B - \mu_0H$, such that μ_0 is the permeability of free space), in the case when the coercivity (H_{Ci}-resistance to demagnetization) is nominally $\frac{1}{2}J_r$ [2]. Nanocrystalline Nd-Fe-B alloys are developed to retain the high coercivity in bulk magnets by virtue of their ultrafine near single-domain sized grains. Coercivity will be very high as the reverse domains in a magnetically decoupled system will not spread easily and their movement at low external demagnetizing fields can be reduced by the pinning sites. Moreover, the probability of defects is lower in these smaller near single domain sized grains [3,4]. Nanocrystalline Nd-Fe-B alloys cannot be conventionally sintered which eludes their high coercivity exploitation because the ultrafine grain structure will undergo excessive grain coarsening at elevated sintering temperatures, resulting into coercivity losses. Due to this reason, either the bonded magnets are their primary application [5] or the rapid processing methodologies are applied, e.g., spark plasma sintering, hot deformation, and die-upsetting [1,6,7]. Previously we have demonstrated the control in SPS parameters resulted in high coercivity ($H_{Ci} \approx 1200$ kA/m) of the recycled HDDR Nd-Fe-B-type isotropic magnets [7] and that the particle size has a critical significance on the sinterability and oxygen uptake of the recycled powder [8]. These microstructure-magnetic properties relationships were derived only for the isotropic magnets ($M_r/M_S \leq 0.5$) processed by SPS, whereas the starting recycled HDDR Nd-Fe-B powder contains anisotropic powder particles. In order to enhance the remanence and the energy product, the hot deformation methodology is necessary [9] but unlike so far, the hot forging technique has not been adopted for the recycled HDDR Nd-Fe-B system. Besides oxygen content also plays an important role in the magnet reprocessing by hot deformation and since the recycled materials contain approximately two folds or higher oxygen content than the commercial grade materials, the processing route needs to be validated [8].

The deformation techniques produce a high degree of texture in the nanocrystalline Nd-Fe-B systems (melt-spun ribbons or hydrogenation–disproportionation–desorption–recombination (HDDR), making the anisotropic grains oriented towards the easy-axis of magnetization. To alter the direction of magnetization vector away from easy c-axis, a significantly larger external field is necessary in anisotropic magnets, such that the energy stored in the crystallites to work against the anisotropy increases as the elastic modulus becomes smaller in parallel direction to c-axis with more texture induced at high temperature pressure conditions [10]. The changes to the grain shape and aspect ratio alter the shape anisotropy, as the extent of magnetization due to an external field for a perfectly spherical ferromagnetic particle would be the same in any direction. However, for the non-spherical grains, the magnetization is easier in the longer direction (of higher aspect ratio) than the shorter axis. [4]. The high anisotropy is accredited mutually to the rotation and the preferred growth of $Nd_2Fe_{14}B$ grains under pressure at elevated temperatures. Güth et al. [11] studied the anisotropy profiles of HDDR powder particles and suggested that the system demonstrated a biaxial {001} (100) local texture (pre-dominant crystal field interactions along the c-axis [001] i.e., the easy-axis alignment and the original grains are confined along a-axis grains). This theory was later confirmed by EBSD pole figures analysis by Sepehri-Amin et al. [12] and Kim et al. [13].

In rare-earth lean systems, the stress induced crystallographic orientation of the nanocrystalline Nd-Fe-B powders by hot deformation becomes critically difficult to achieve [14]. Early work of hot deformation by Müller et al. [15] on the isotropic HDDR Nd-Fe-B powder resulted in $B_r = 1.1$ T and $H_{Ci} \approx 710$ kA/m. Similarly, Kirchner et al. [16] studied the effect of Nd addition during the hot deformation and reported for 14 at.% Nd in the HDDR powder, the remanence of 1.2 T and coercivity 550 kA/m can be achieved but sample cracking was still obvious after forging. Increasing the Nd content to 16 at.% prevented the HDDR samples from cracking and improved the coercivity to 750 kA/m but the B_r reduced slightly to 1 T with the higher rare-earth content. Concurrently Gopalan et al. [17] were able to increase the remanence after hot deformation from 1.07 T to 1.35 T in the precursor HDDR powder of composition $Nd_{14.1}Fe_{78.2}Co_{1.0}B_{6.1}Cu_{0.1}Al_{0.5}$. Likewise, Li et al. [18] obtained $B_r = 1.09$ T and BH_{max} = 114 kJ/m^3 after HDDR powder hot deformation through SPS, but the H_{Ci} was only 384 kA/m, due to grain shape factor and interlinked with the abnormal grain growth of Nd$_2$Fe$_{14}$B crystallites. The hot deformation temperatures were identified by Li et al. [19], achieving 69% height reduction, B_r = 1.22 T and yet again low H_{Ci} = 181 kA/m and BH_{max} = 121 kJ/m^3 only. This inferior coercivity in hot-deformed HDDR Nd-Fe-B magnets tends to significantly reduce the squareness of hysteresis and lowers the BH_{max} values, even after high deformation ratio; effectively rendering such magnets unsuitable for commercial applications [9,19]. Concurrently no previous report cited the hot deformation-based reprocessing of the recycled Nd-Fe-B alloys to the extent of our literature review, but the strategic task is to revitalize the recycled powders to novel magnets, retaining loop squareness, initially by gaining high H_{Ci}, and then by tackling B_r and BH_{max}.

Evidently, the hot deformation in the HDDR Nd-Fe-B system has yielded nominal improvement in texture but at the expense of coercivity. Previously, the strain rate and the degree of deformation have been considered as the controlling factors for texture development, without quantifying the terminal pressures [9]. So, in such cases, the control on resultant squareness of hysteresis loop is lost and coercivity decline is most certain [19]. The enhancement of coercivity along with B$_r$ in hot-deformed magnets is becoming a subject of interest due to successful application of grain boundary (GB) modification of the hot deformed melt-spun ribbons yielded high $B_r \approx 1.4$ T and $H_{Ci} > 2000$ kA/m [20–32]. Similarly, in case of the fresh Nd-Fe-B HDDR system, Song et al. [33] utilized Pr-Cu low melting alloys to improve the coercivity of the hot-deformed HDDR Nd-Fe-B-type magnets to 1230 kA/m. Similarly, with the addition of up to 2 wt.% Nd-Cu alloys were reported to improve the coercivity of hot-deformed magnets to 1042 kA/m [34]. A significant improvement of 640 kA/m was also indicated after the grain boundary modification treatment of the HDDR-based Nd-Fe-B magnets with milled powder blend of NdH$_X$ + Cu [35].

However, the primary argument in view of magnetic scrap reprocessing questioned the suitability of applying hot deformation techniques to the recycled HDDR Nd-Fe-B isotropic magnets. The existing literature on the commercial grade MF15P or high performance anisotropic HDDR Nd-Fe-B powders indicated a loss in coercivity and BH$_{max}$ after texturing/forging and possible recovery attained only with the additional grain boundary modification treatments [9,19]. Thus, we imply this requirement in the recycled material, that retention of the coercivity and the squareness of the hysteresis loop is sustained as the texture is introduced. The experimental reported work in this manuscript is an effort to validate the relationship of texturing in the microstructure with the dependent properties to match commercial counterpart magnets.

This study answers these two queries: (a) can the recycled HDDR powder (with ≈4800 ppm oxygen content) be forged at lower pressures in view of the reduced total rare earth (TRE) content (below 30 wt.%) [7] and (b) if the process can be controlled to obtain the magnetic properties comparable to the precursor recycled powder (with the mechanism of texturing justified with the microstructure). Low forging pressures are suitable to retain texturing process control and the degree of deformation in order to retain the squareness of hysteresis loop and lossless coercivity. The SPS system and forging pressure was controlled in the terminal stages using a single step deformation to reduce the thickness of hot-pressed (SPS-ed isotropic) magnets. This low forging pressure at the terminal stage (750 °C) prevents excessive cracking at the edges of the deformed magnet in rare-earth lean systems (like recycled HDDR nanocrystalline powder), rendering them useful for bulk applications after post-finishing and protective coating operations. Systematically, the effect of post-hot deformation thermal treatments on the magnetic properties was also established, which has not been investigated previously.

The development of textured recycled materials for the remanence enhancement is carried out in two steps: first, aligning the recycled powder in the applied magnetic field and then hot-pressing (SPS consolidation) to full density; secondly, hot-forging or deforming under higher pressures (100–150 MPa) with different dies and tooling to induce texture in the microstructure perpendicular to the pressing direction. Heating rate and pressure sequence during hot forging were optimized and the most suitable parameters are reported in this study. The pressure range was defined to factor the degree of deformation without compromising the coercivity, as the BH_{max} and remanence were enhanced by low-pressure controlled hot forging. Taking pressure as a regulating factor helped us obtain texture in the bulk magnet, without exceptionally high strain rate or operating at very high forging temperatures. Therefore, the reported method can be adapted to sequentially tailor the magnetic properties of forged bulk magnets, which are better than the starting recycled HDDR Nd-Fe-B powder.

2. Contribution to the Field Statement

The bulk of the literature on the nanocrystalline Nd-Fe-B system has the vast majority of publications reporting melt-spun materials subjected to hot deformation. However, in this manuscript, we dealt only with the recycled HDDR Nd-Fe-B material. These two nanocrystalline Nd-Fe-B systems differ significantly and cannot be intermixed to devise results or a technological strategy. The progress of recycling with hydrogen-based technologies, like hydrogen decrepitation (HD) or hydrogenation–disproportionation–desorption–recombination (HDDR) processes is a widely applicable method for developing anisotropic and high coercivity bonded magnets with nanostructured $Nd_2Fe_{14}B$ grains ($\leq$400 nm) and has drawn significant attention for the end-of-life scrap processing. The melt-spun ribbons as produced are never anisotropic and their grain size is less than 100 nm on average, such that hot-pressing results in exceptionally good magnetic properties. On the contrary, the remanence and maximum energy products are limited due to the isotropic nature of ribbons or pressed magnets. This makes a three-step process to develop anisotropy in melt-spun ribbons. Evidently, no method has ever been reported previously on the recycled HDDR Nd-Fe-B system to develop bulk anisotropic magnets.

The microstructure—i.e., nanocrystalline grain size, phase boundaries, and particle interface—is completely different in melt-spun ribbons and HDDR Nd-Fe-B and as a general comparison commercial, MQU-F grade melt-spun ribbons have $H_{Ci} \approx$ 1560 kA/m and B_r = 0.6 T (isotropic flakes) while the commercial HDDR MF-15P powder has $H_{Ci} \approx$ 1100 kA/m and B_r = 1.35 T (anisotropic powder). The mechanism of sintering, grain growth kinetics, grain morphology, grain boundary structure, and derivative magnetic properties thus are very different for the melt-spun ribbons and the HDDR Nd-Fe-B.

To the best of our knowledge, to date, there is not a single report on the hot deformation of the recycled HDDR Nd-Fe-B. Previously McGuiness et al. [9] and Li et al. [19] worked on the hot deformation of fresh HDDR Nd-Fe-B alloys, but their reported magnetic properties (BH_{max} and H_{Ci}) are comparably inferior to our recycled material system. When we compare these reports to our previous publications, the remanence values are in the range of 0.7–0.8 T, returning M_r/M_S ratio $\approx$ 0.5 which indicates the isotropic nature of high coercivity magnets. After hot forging and 55% deformation ratio, we improved the remanence to 1.01 T which is $\geq$ 23% better than these isotropic magnets and subsequently M_r/M_S ratio improved to 0.64, indicating the production of anisotropic magnets. Additionally, the impact of thermal treatments on the microstructure and resultant magnetic properties has never been studied after the hot deformation of either commercial or recycled HDDR Nd-Fe-B system. This study validated these principles to devise industrial recycling and reprocessing strategy of rare-earth permanent magnets. Most importantly, we proposed a model to define the mechanism behind the improvement in the magnetic properties (simultaneously) after the thermal treatment. Hence, we believe these state-of-the-art values of remanence, BH_{max} and high coercivities warrant the recycled HDDR Nd-Fe-B system as a feasible method to recycle; and controlled hot forging can be expanded to the industrial scale for developing the texture in bulk magnets.

3. Materials and Methods

The recycled Nd-Fe-B powder was produced from the end-of-life (EOL) scrap magnets of composition: $Nd_{13.4}Dy_{0.67}Fe_{78.6}B_{6.19}Nb_{0.43}Al_{0.72}$, pulverized and reprocessed via the well-known dynamic hydrogenation–disproportionation–desorption–recombination (d-HDDR) method (controlled hydrogen pressure for anisotropy development in nanocrystalline $Nd_2Fe_{14}B$ grains along c-axis by texture memory effect-TME) [36,37]. The oxygen content analysis was done with the Eltra ON 900 oxygen and nitrogen analyzer (Haan, Germany). Based on the density of elements and their weight fractions in this composition, the theoretical density of the recycled powder was calculated as 7.57 ± 0.01 g/cm^3. The magnetic properties of the recycled HDDR Nd-Fe-B powder were measured with the Lakeshore 7400 Series (Westerville, OH, USA) Vibrating Sample Magnetometer (VSM), capable of measurements up to 2 T applied fields.

The powder handling and sample preparation was done in controlled argon (Ar) atmosphere glove box with O_2 content < 5 ppm and H_2O content < 1 ppm. About 3 g of the recycled HDDR Nd-Fe-B powder was poured into a 10 mm graphite dies with graphite spacers in between the solid punches at the top and bottom of the same material as the die. These powder containing dies were sealed in a vacuum bag within the glove box to avoid oxidation during further sample handling before sintering. Prior magnetic alignment with a 1.5 T applied field was carried out in the direction parallel to the pressing direction. The rapid sintering or hot-pressing operation was performed under dynamic vacuum of 2×10^{-2} mbar, the graphite dies containing blends were put inside the Syntex 3000 (Fuji DR. SINTER, Saitama, Japan) SPS furnace. The sintering and hot forging pressure was regulated uniaxially with a pressure controller. The hot-pressing temperature was set to 750 °C with a holding time of 1 min and the heating rate of 700 °C/min with a final ramp of 50 °C/min was maintained for the hot pressing. The temperature measurements were controlled by a calibrated infrared pyrometer. After the hot pressing, the samples were ground with SiC papers to remove the graphite spacers and the thickness of the cylindrical sample was reduced to 3 mm.

Hot deformation (hot forging) experiments were done also at 700–800 °C for 1 min, with the 16 mm WC + Co sintered cermet dies under 200 mbar Ar (99.9% purity) atmosphere and the forging die pressure was varied from 100–150 MPa. A calibrated K-type thermocouple was inserted in the cermet die to control the hot forging processing temperatures to minimize overshooting to <5 °C at peak conditions. The SPS operation was performed under 2×10^{-2} mbar dynamic vacuum. A constant pressure of 40 MPa and relatively slow heating rate of 50 °C/min was maintained up to the peak forging temperature in order to avoid high current density in the cermet dies, which otherwise may induce undesirable microstructural changes (grain growth, Nd-rich GB, and oxide phase transitions) [38]. Under these optimal settings, the heating rate was kept constant until the holding temperature in the range of 700–800 °C was reached, but the pressure in the final 50 °C step was subsequently increased to peak conditions of 100–150 MPa within one minute. A total of 3 min were given at the maximum pressure, i.e., one minute to reach the peak temperature, another one-minute holding at this hot-deformation temperature and the final one minute when the SPS current was switched off and cooling started instantly. The z-axis movement (up–down plunge of the SPS cermet dies) was recorded (as a measure against excessive cracking). The change in initial and final thickness was measured with the Vernier calipers to calculate the degree of deformation w.r.t forging pressure and temperatures.

The demagnetization measurements of bulk samples were taken on a permeameter (Magnet-Physik Dr. Steingroever, Cologne, Germany) after initially magnetizing along the c-axis. The relative density was determined with (Exelia AG DENSITEC, Zurich, Switzerland) bulk density-meter based on Archimedes principle by submerging the SPS processed samples in silicone oil.

After the hot deformation, the thermal treatment (annealing) was carried out at 750 °C within the horizontal tube furnace (Carbolite Gero Limited, Hope Valley, UK) under > 10^{-5} mbar vacuum and a heating rate of 50 °C/min. The magnetic measurements were retaken after annealing. The samples were thermally demagnetized at 450 °C for 10 min in the vacuum furnace (>10^{-5} mbar). For the microscopic examination, the demagnetized samples were grinded by 2400 and 4000 grit size SiC papers. Final polishing was done with a 1/4 μm diamond paste on the velvet cloth. The microstructural analysis was performed with JEOL 7600F, a Field Emission Scanning Electron Microscope (JEOL Ltd., Tokyo Japan) with an electron energy dispersive X-ray spectroscopy (EDXS) analysis fitted with a 20 mm^2 Oxford X-Max detector (Oxford Instruments, High Wycombe, UK) for chemical and elemental identification, done at 20 keV.

4. Results and Discussion

Summarized in Table 1, the starting EOL magnet had H_{Ci} = 1170 kA/m and B_r = 1.19 T which translated to the recycled HDDR powder having H_{Ci} = 830 kA/m, J_r = 0.9 T, J_S = 1.59 T and BH_{max} = 124 kJ/m^3 measured using a VSM with zero self-demagnetizing factor. The VSM measurements on the EOL scrap batches processed by HDDR method are shown in Figure 1, indicate the reprocessed batches are quite similar in polarization (*J*) vs. applied field (*H*) response, which indicates good consistency of results and recycling repeatability.

Table 1. Magnetic properties of the starting Nd-Fe-B materials prior to hot deformation.

Material Class	Coercivity H_{Ci} (kA/m)	Remanence B_r (T)	BH_{max} (kJ/m^3)	M_r/M_S Ratio
End-of-Life (EOL) Scrap Magnet	1170	1.19	250	0.74
Recycled HDDR Powder	830	0.9	124	0.56
SPS Processed (Hot Pressed) Magnet	938	0.78	100	0.49
Post-SPS Annealed Magnets	1120–1160	0.79–0.8	100–110	0.5

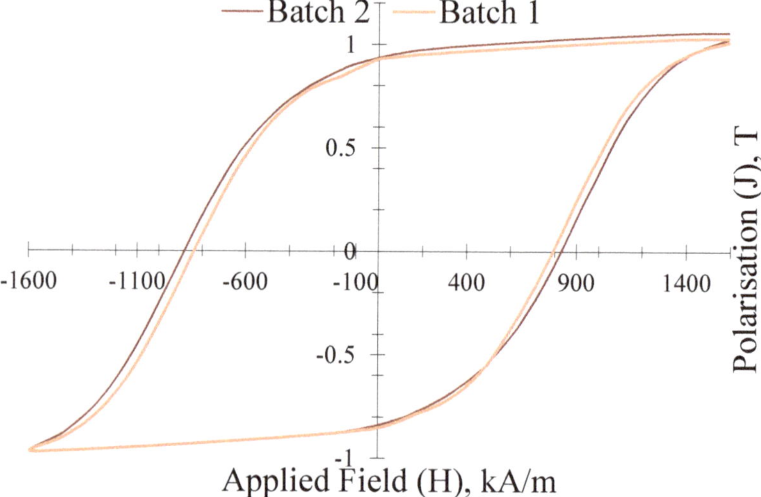

Figure 1. VSM analysis of recycled HDDR powder in two batches with approximately similar applied field vs. polarization response.

The oxygen content from the EOL magnet 2660 increased to 4760 ppm after the d-HDDR reprocessing in the recycled Nd-Fe-B powder [7]. Seemingly the oxygen content varies as the powder particle size is reduced below 100 μm and it was found oxygen content nearly doubled for < 30 μm sized fractions. Therefore, the powder was not milled to prevent additional oxide phases which degrade sinterability, density, remanence (reduction in ferromagnetic phases) [8], and subsequently the texture from hot deformation [1,19]. The hot pressing conditions were kept similar to the previously reported SPS work [8] in order to achieve high coercivity and the optimal microstructure, i.e., at 750 °C for 1 min, 100 MPa uniaxial pressure and the post-SPS thermal treatment was also done for 1 h at 750 °C (above ternary transition temperature of 665 °C) [7]. These annealed hot-pressed magnets had $H_{Ci} \geq$ 1120 kA/m and $B_r \geq$ 0.79 T, with $M_r/M_S \approx$ 0.5 indicating isotropic nature of the bulk magnets prior to hot deformation.

The average particle size of the recycled HDDR Nd-Fe-B powder (without any subsequent milling after d-HDDR) is 220 μm and low magnification image can be seen for particles' morphology in Figure 2a. Each individually d-HDDR treated particles contains 240–400 nm sized Nd$_2$Fe$_{14}$B grains separated by thin Nd-rich phase (as in Figure 2b), which is discontinuous and non-uniformly distributed in the microstructure of the recycled material as previously reported [7]. The effect of other elements in the composition of the HDDR powder (Dy, Nb, and Al) were also explained by Ikram et al. [7] to lower the eutectic transition temperature, such that Al is responsible for low temperature interactions (477 °C) with Nd/Dy at the Nd-rich grain boundary interface. The trace amount of Al in the composition stabilizes the $Nd_2Fe_{14}B$ grain morphology and increasing the coercivity. The addition of Al in the HDDR powder has also been suggested to reduce the melting temperature of the RE-rich grain boundary phase [1,3]. The heavy rare-earth elements—such as Dy, Tb, Ho, etc.—are added as dopants to increase the coercivity and high temperature stability of the Nd-Fe-B magnets [2]. Whereas Nb is reported to form the $NbFe_2$ *Laves* phase, which is finely dispersed in the nanocrystalline $Nd_2Fe_{14}B$ matrix. The *Laves* phase does not contribute detrimentally to the magnetic properties of the HDDR Nd-Fe-B during SPS consolidation or hot pressing, as they scavenge the additional soft ferromagnetic α-Fe during the recombination stage, which is necessary for retaining high coercivity in the HDDR system [7]. In a similar sense, the *Laves* phases are not anticipated to have deleterious effects during the hot deformation. The detailed phase analysis of the recycled HDDR powder is also presented in the preceding studies on explaining the oxygen content relationship to the particle size, microstructural changes, and sinterability [8].

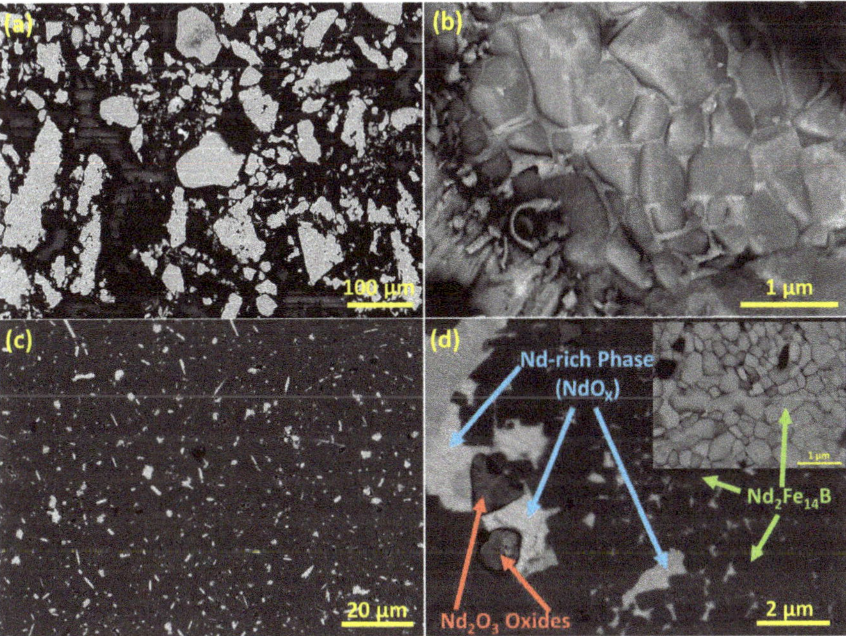

Figure 2. Backscattered electron (BSE) scanning electron microscope analysis of: (**a**) the recycled HDDR Nd-Fe-B powder particles of average 220 μm size and (**b**) fractal image at 20,000× magnification of approximately 400 nm sized $Nd_2Fe_{14}B$ grains within each powder particle, (**c**) optimally hot pressed and fully dense recycled powder processed at 750 °C, (**d**) high magnification image of the hot pressed annealed sample showing constituent phases of different contrasts and inset showing submicron $Nd_2Fe_{14}B$ grains exposed by etching with 3 M HCl and Cyphos ionic solution.

The microstructure of the 750 °C hot pressed sample is quite similar to the recycled HDDR powder (Figure 2c). It confirms that these conditions are suitable for attaining fully dense and high coercivity submicron sized $Nd_2Fe_{14}B$ grains by restricting their exaggerated grain growth and avoiding unnecessary transformation of the Nd-rich phase to Nd-oxides at elevated temperatures, as observed in the inset of Figure 2d. The backscattered electron imaging (presented in Figure 2d) provides the contrast to identify different phases in the complex microstructure of HDDR Nd-Fe-B system with grey matrix phase of submicron sized body centered tetragonal (bct) $Nd_2Fe_{14}B$ grains, Nd_2O_3 type oxides (cubic or hcp), and the bright intergranular Nd-rich phase (fcc-Nd or NdO_x type), which matches the previous reports [7,8]. The dark spots in the vicinity of matrix grains and triple pockets are pores.

The hot deformation at 100 MPa and temperature of 650 °C (below the ternary transition temperature of 665 °C [7], results not shown here) yielded H_{Ci} = 683 kA/m and J_r = 0.64 T (BH_{max} = 55 kJ/m^3) only. The remanence obtained in this case was significantly lower than the starting recycled powder. Therefore, in order to obtain better texturing in the hot-pressed samples, the temperature range was selected above this ternary transition temperature. Therefore, the hot deformation experiments were further performed at 100–150 MPa in the range of 700–800 °C. Further hot deformation temperatures higher than 800 °C, were not adapted in view of grain growth of the equiaxed and planer deformed grains which severely degrade the coercivity [14,19]. After hot deformation, these forged samples were subsequently annealed under vacuum at 750 °C for 1 h to recover the magnetic properties following strain relaxation and Nd-rich phase redistribution. Prolonging the hot forging to more than 1 min was found to have a detrimental effect on reprocessing of the HDDR Nd-Fe-B samples; therefore, further experiments were not evaluated beyond the optimized conditions.

For the hot deformation at 100 MPa pressure, the starting fully dense hot-pressed magnet after the thermal treatment had H_{Ci} ≈ 1120 kA/m, J_r = 0.8 T and BH_{max} = 110 kJ/m^3. Figure 3 illustrates the polarization (J) vs applied magnetic field (H) results of hot deformed samples at working pressures of 100 and 150 MPa in the range of 700–800 °C, with the hold time of 1 min only. In the first case when the hot deformation pressure was 100 MPa, holding at 700 °C yielded J_r = 0.83 T and H_{Ci} = 930 kA/m, which indicates weak texturing for this applied pressure and low temperatures. The holding temperature of 750 °C resulted in J_r = 0.87 T and H_{Ci} = 690 kA/m after hot forging. The J_r increased to 0.89 T and H_{Ci} was 754 kA/m when deformed at 800 °C under 100 MPa pressure. A slightly higher temperature promotes annihilation of pores and nominal grain growth, which indicates that the optimally forged microstructure is developed in 800 °C, 100 MPa conditions (above the ternary transition temperature).

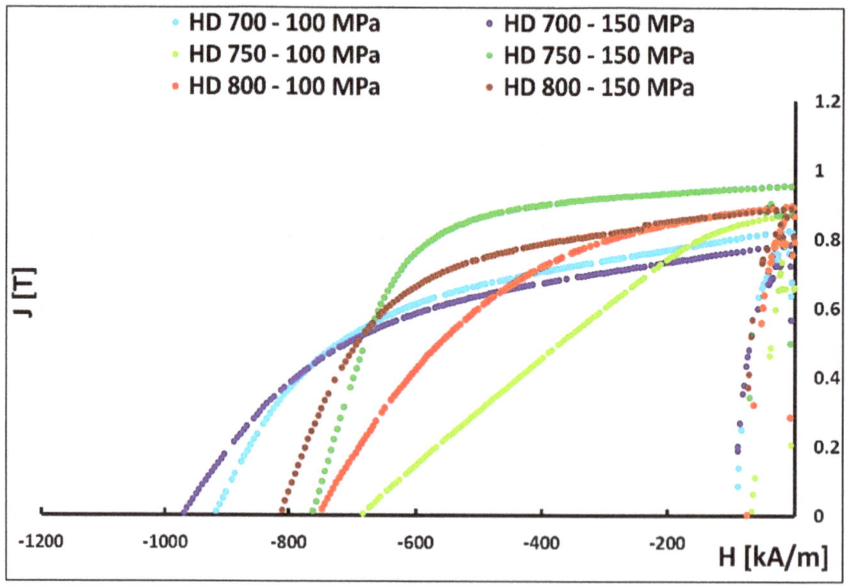

Figure 3. Comparison of the magnetic properties after hot deformation at 100 and 150 MPa.

At 700 °C, the onset of ternary transition is not complete yet, so the hot-forged samples retain the high coercivity of hot-pressed samples and the induced texture is very marginal in 100 and 150 MPa condition.

Comparing these findings with previous results on SPS processing, this slower ramp rate means low thermal stresses in the sample and H_{Ci} results are consistent with SPS reprocessing experiments [7].

As illustrated in Figure 3, by the application of higher hot deformation pressure at 150 MPa, the J_r increased in case of 750 °C sample to 0.96 T. On the contrary, at 800 °C, the remanence improvement over the 100 MPa forging was not observed with a similar value of J_r = 0.89 T, possibly due to edging over the $Nd_2Fe_{14}B$ grain growth regime at elevated temperature-pressure conditions [7,19,39]. The H_{Ci} in 150 MPa forging conditions followed a similar trend to 100 MPa samples, peaking at 990 kA/m at 700 °C to 767 kA/m at 750 °C and 817 kA/m at 800 °C. This improvement in the coercivity with temperature may be explained by the redistribution of the Nd-rich phase at the GBs and interfaces. The higher temperature and pressure conditions significantly reduce the porosity, so the Nd-rich phase achieves high wettability and transfuses between the nanocrystalline grains via capillary action and compressing forces, resulting in coercivity improvement with the temperature at 150 MPa. The low forging pressure condition are appropriate to attain control in texturing and the degree of deformation, which resulted into better squareness of hysteresis loop and high coercivity. The controlled forging pressure was applied in a single step deformation in the terminal stage to reduce the thickness of hot-pressed (SPS-ed isotropic) magnets. This low pressure in the final bulk deformation stage (750 °C) prevented excessive cracking at the edges of the deformed magnet. Clearly, this approach is more suitable in rare-earth lean systems (e.g., recycled HDDR nanocrystalline powder).

The magnetic properties with post-hot deformation thermal treatment at 750 °C for 1 h in a comparative plot for all temperature regime (700–800 °C) is shown in Figure 4 and indicates the optimal results (in H_{Ci} and B_r, better than starting recycled powder) were obtained at 750 °C hot forging conditions. For the thermally treated samples forged at 100 MPa, the H_{Ci} improved for 750 and 800 °C which can be attributed to redistribution of intergranular phase and relaxation of deformation stresses. At 800 °C, 100 MPa, the range of improvement is lower than 750 °C, which can be explained with a possible grain growth in parallel and perpendicular directions as not enough room for redistribution may be available in the former case. For hot deformation at 700 °C, 100 MPa, the sample was characterized with surface cracking which may generate voids and dislocations under deformation. The hot-pressed sample had H_{Ci} > 1120 kA/m which dropped to 930 kA/m after hot forging and slightly more to ≈900 kA/m with annealing. The slight decline in H_{Ci} after annealing can be attributed to inadequate redistribution of the Nd-rich phase below 746 °C (completion of ternary transition) into isolated voids that will not fully consolidate [7]. However, the remanence for all samples forged at 100 MPa pressure increased with the thermal treatments, as illustrated in Figure 4.

In the case of 150 MPa samples annealed at 750 °C, 1 h, the H_{Ci} increased subsequently for all the samples: 700 °C (H_{Ci} = 1007 kA/m), 750 °C (H_{Ci} = 960 kA/m), and very marginal for 800 °C sample (H_{Ci} = 820 kA/m). This is due to the Nd-rich phase experiencing prominent wettability and high compressing forces during the forging stage at 800 °C, 150 MPa. Therefore, the liquid phase has already diffused (redistributed) along the grain boundary of the particles and between the nanocrystalline grains via pressure-assisted capillary action. Whereas, the 750 °C in both pressure ranges gained 200–350 kA/m coercivity improvement by thermal treatment because the Nd-rich phase was not uniformly redistributed during the slower ramp rate of 50 °C/min during the forging stage. The pressure assisted coalescence of microstructure at 50 °C/min effectively diffused the Nd-rich phase to the grain boundary regions, but at higher temperatures, it assists in grain growth [7]. The J_r of 150 MPa forging condition increased subsequently for the lower temperature samples, i.e., 700 °C (0.9 T) and 750 °C (1.01 T) after the thermal treatment. However, the remanence improvement was insignificant after the thermal treatment for 800 °C sample (0.89 → 0.9 T), which as previously explained is due to the completion of pressure-assisted coalescence and Nd-rich phase diffusion during forging at elevated temperatures [19]. This mild increment in J_r can be related to the relaxation of the processing stresses [7].

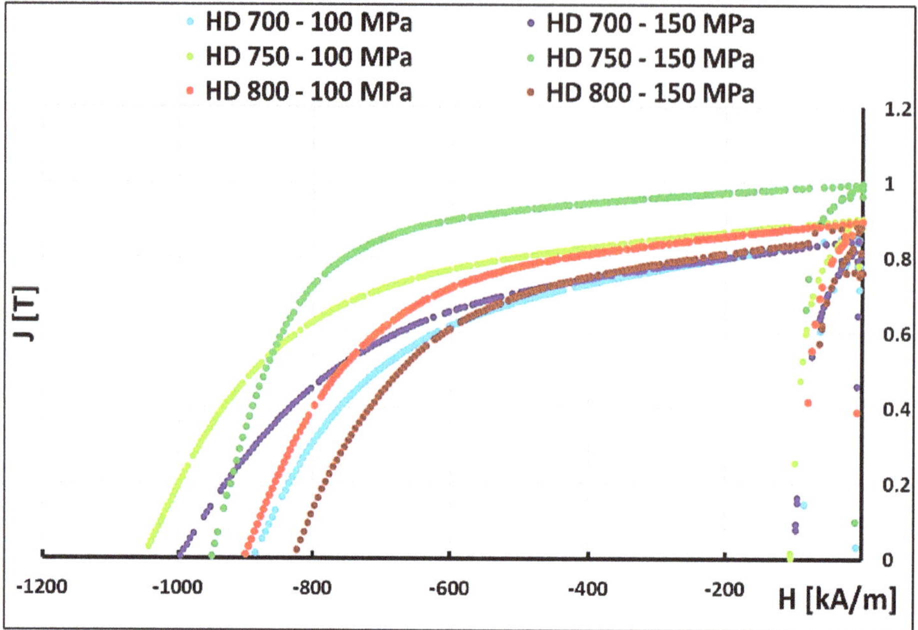

Figure 4. Comparative augmentation of magnetic properties of hot forged HDDR Nd-Fe-B magnets after vacuum post-processing thermal treatment at 750 °C for 1 h.

In short under low temperatures forging, the induced texture and J_r values are low, however the coercivity is high as the microstructure may correspond to hot-pressed magnets. As the temperature is increased at 100 MPa, texture development is better as Jr tends to increase and the annealing retains higher magnetic properties. With higher pressure—i.e., 150 MPa—the texture begins to develop at lower temperatures and the recovery of properties after the thermal treatment is mild since the hot deformation has already yielded the optimal microstructure. The thermal treatment for high temperature and pressure conditions correspond to stress relieving [7,8].

Moreover, under higher temperature–pressure conditions, the larger HDDR particles are prone to rupture and cracking, which could be associated with the 800 °C, 150 MPa sample not achieving high remanence as compared to 750 °C, 150 MPa [8]. These hot deformation-annealing results are in-line with the SPS study performed on the recycled HDDR Nd-Fe-B powders [8].

The 100 MPa hot-deformation process yielded ≈40% height reduction at 750 °C in the sample and the magnetic properties were: H_{Ci} = 690 kA/m, J_r = 0.87 T, and BH_{max} = 99 kJ/m^3. The hot deformation is a high strain process and therefore the stresses induced during the deformation can be considered to reduce the coercivity apart from the grain elongation [15,16,19,40]. The thermal treatment subsequently was performed at 750 °C for 1 h to release the microstructure of processing stresses and to also achieve the improvement in the magnetic properties due to the redistribution of the Nd-rich phase and the restructuring the surfaces of Nd$_2$Fe$_{14}$B matrix grain to accommodate the intergranular phase upon solidification [7]. Thus, after the heat treatment, the magnetic properties of 100 MPa hot forged sample improved to H_{Ci} ≈ 1060 kA/m, J_r = 0.91 T, and BH_{max} = 144 kJ/m^3, as shown in Figure 5a and indicates a 12% improvement in remanence over the hot-pressed magnet at 0.79 T.

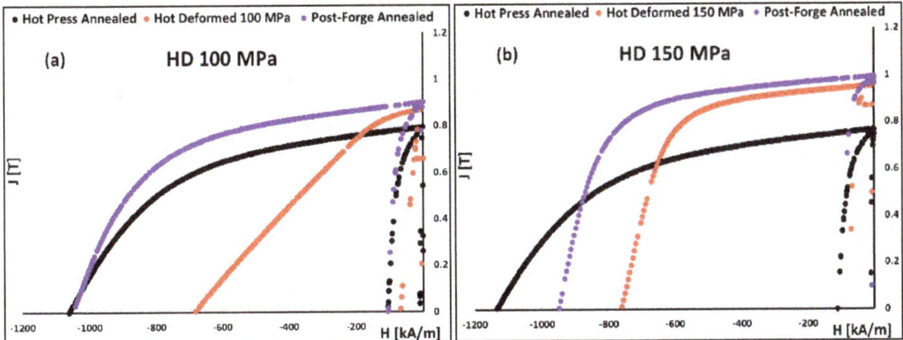

Figure 5. Comparison of the magnetic properties of recycled HDDR Nd-Fe-B based magnets optimally hot deformed at 750 °C under different forging pressures: (**a**) at 100 MPa and (**b**) 150 MPa. Plot legend: hot-pressing (black), hot-deformation (red), and post hot-deformation thermal treatment (purple).

Similarly, the hot-deformation experiments repeated at higher pressures resulted in better remanence than 100 MPa samples or hot-pressed magnets. The measured magnetic properties for the 750 °C, 150 MPa hot-deformed sample are illustrated in Figure 5b which gives ≈23% higher remanence with post-forging thermal treatment than isotropic hot-pressed magnets and 11% over the starting recycled anisotropic HDDR Nd-Fe-B powder. The heating rate was kept constant to 50 °C/min and during the final 3 min, the pressure was consequently maintained at 150 MPa in the second instance. The hot deformation pressure was increased rapidly from 40 MPa at 700 °C to 150 MPa at 750 °C within one minute to maximize the strain rate since a lower strain rate at high temperatures can cause grain coarsening as the deformation time is extended and the degree of alignment of the fine grains gets disturbed from the available Gibbs surface free (thermal) energy to undergo coarsening [14]. The magnetic properties of the starting hot-pressed magnet were: $H_{Ci} \approx 1157$ kA/m, $J_r = 0.79$ T, and $BH_{max} = 103$ kJ/m^3. The hot deformation at 150 MPa resulted in remanence enhancement and up to 55% height reduction. Therefore, the magnetic properties were measured to be: $H_{Ci} \approx 770$ kA/m, $J_r = 0.96$ T and $BH_{max} = 165$ kJ/m^3. Subsequently, the thermal treatment improved the overall magnetic properties and squareness of the loop in the hot-deformed magnet forged at 150 MPa, which were: $H_{Ci} = 960$ kA/m, $J_r = 1.01$ T, and $BH_{max} = 180$ kJ/m^3. In comparison, Li et al. [19] attained up to 69% height reduction with $B_r = 1.22$ T and $H_{Ci} = 181$ kA/m only from the fresh HDDR powder (possibly high degree of grain growth or hot forging stresses), measured parallel to the pressing direction.

In order to explain the behavior of remanence enhancement after annealing, we can systematically relate it according to the relationship of remanent magnetization [41]

$$M_r = A(1 - \beta)\frac{d}{d_0}\overline{\cos\theta}M_S \qquad (1)$$

In Equation (1), A denotes the positive parallel domains prior to the magnetic alignment, $(1 - \beta)$ represents the volume fraction of hard magnetic phase (Nd$_2$Fe$_{14}$B in this case), d_0 is theoretical density, d shows experimental density, the degree of orientation or texture is indicated by $\overline{\cos\theta}$ and M_S is the saturation magnetization. Therefore, a higher A factor indicates a low number of reverse magnetized domains in the system after annealing, as the H_{Ci} tends to increase as well due to microstructural factors (grain shape, morphology, stress relief, non-ferromagnetic grain boundary widening/decoupling effect, pinning field enhancement, and redistribution of Nd-rich phase). Moreover, no additional non-ferromagnetic additives were made part of hot-forging experiments, so A and β are not the influencing factors in hot deformation from Equation (1). Similarly, the densities of hot-forged and hot-pressed samples are nearly the same (99% of theoretical values, i.e., 7.57 g/cm^3), so they are not affecting the remanence after thermal treatments either. The saturation magnetization (M_S) remains largely unchanged before the HDDR reprocessing, hot pressing/deformation and subsequently annealing at 750 °C, 1 h. Therefore, a 5% improvement in remanence from 0.96 to 1.01 T is invariably linked to the degree of texturing perpendicular to the pressing direction which tends to get reduced subsequently after the thermal treatment. The microstructure composed of highly textured $Nd_2Fe_{14}B$ grains is beneficial for preferential grain growth during hot deformation and which will impart favorable remanence enhancement. In the case of nanocrystalline melt-spun ribbons, Wang et al. [41] identified with EBSD that the texture in platelet shaped layers on the basal plane was enhanced by increasing the annealing time. Moreover, the oriented platelet grains, although followed inhomogeneous grain growth pattern but suppressed the abnormal grain growth of non-aligned coarse grains and the number of mis-oriented grains decreased after annealing. Although EBSD was out of the scope of this research but the model from Equation (1) advocate the degree of orientation $\overline{\cos\theta}$ of $Nd_2Fe_{14}B$ hard phase grain tends to increase after the thermal treatment due to grain surface restructuring (both aligned and coarse non-aligned grains), besides relaxation of the forging stresses is also a contributing factor. Since the thermal treatments were performed above the ternary transition temperature of 665 °C, it may also be expected that the intergranular phases experience non-ferromagnetic transformations, which effectively improves the localized exchange decoupling behavior [41–44], however, such a study has not been proven on the HDDR Nd-Fe-B system.

In Nd-lean HDDR powder (e.g., recycled), the stress induced crystallographic orientation of the nanocrystalline grains by hot deformation is difficult and can only be achieved by critically controlling the forging parameters coupled with stress-relief thermal treatments. The applicable pressure of 150 MPa was in a sense limit for the operation mode and a further increase was not possible on this system. Therefore, from the comparison of 100 and 150 MPa range, the improvement (10%) is evident at 0.1 T approximately for 50 MPa increase. Thereby, if the system allowed hot deformation up to 500 MPa, the recycled HDDR powder is capable of reaching the remanence values much higher than these obtained but at a severe toll on coercivity as previous studies have shown for height reductions greater than 50% [17,45,46].

The cross sections of hot deformed magnets prepared at 750 °C were characterized for their microstructure by HR-SEM in backscattered electron imaging mode as shown in Figure 6. The pressing direction in each case is indicated by the arrows. As clearly seen from Figure 6a, when the hot deformation pressure is low at 100 MPa only, the microstructure is similar to the initial recycled HDDR Nd-Fe-B powder or the hot-pressed (optimally SPS-ed) samples in Figure 2. The equiaxed and the platelet-shaped grains both coexist with a poor degree of alignment due to inadequate deformation as marked with the red arrows in Figure 6a. Apart from only a few grains, the majority of the $Nd_2Fe_{14}B$ grains are not aligned after hot deformation with the pressing direction, as represented in Figure 6b. This suggests incomplete hot-deformation or the lack of uniform texture in the sample. The matrix grains are also not excessively deformed (Figure 6b), which is the reason for higher coercivity than the starting material. After the thermal treatment, we can expect the relaxation of deformation stresses, which improved the overall magnetic properties.

In the case of hot deformation at 150 MPa pressure, the microstructure in Figure 6c is more appropriately textured and several grains are deformed perpendicular to the pressing direction which resulted in higher B_r. In this case, the two-zone microstructure is not distinguishable as clearly the grains are suitably deformed when magnified in Figure 6d with the pressing direction. At 750 °C forging temperature, overall limited grain growth in equiaxed manner or perpendicular to the pressing direction was observed which is suitable to conserve the magnetic properties of the starting recycled powder. Higher forging pressure resulted in a larger degree of deformation and thereby improvement in the texture is apparent. The very high degree of deformation (above 80%) results in nanocrystalline grains elongated beyond > 1 μm size [14,15]. The elongation of matrix grains is also considered as grain growth and consequently, the demagnetizing field will be higher on severely deformed grain surfaces, so the coercivity starts to decrease as grains begin to grow, but more importantly with higher percentage reduction in height, the deformation stresses on the microstructure are higher, which degrade the H_{Ci} [15,16,19,40]. The in-homogeneities in the crystal structure, as well as the surface imperfections generated after hot deformation, are apparently the local sites for initiating demagnetization. The defects like irregularly shaped grains, non-ferromagnetic inclusions/grain boundaries, surface or geometric defects, etc. become the sites for local nucleation fields H_N (regions favoring magnetization reversal). These defects reduce the coercivity to

$$H_{Ci} = \alpha H_a - N_{eff} M_S \qquad (2)$$

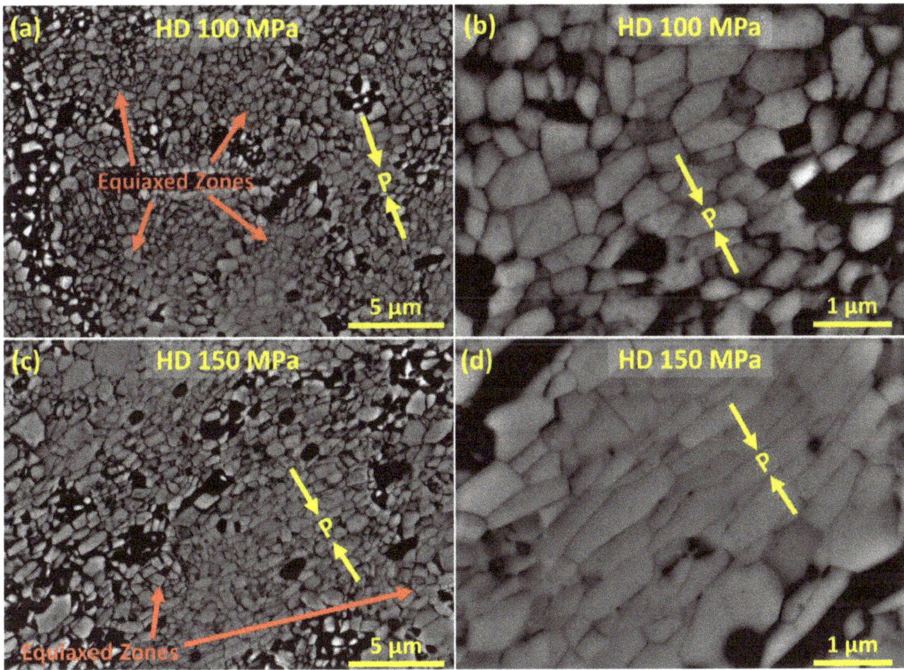

Figure 6. BSE micrographs of (non-annealed) hot deformed samples at (**a**) 100 MPa with two zone microstructure, (**b**) 100 MPa at high magnification indicating the pressing direction, and (**c**) 150 MPa with minimal equiaxed grains; and (**d**) 20,000× magnification of hot deformed $Nd_2Fe_{14}B$ grains; the arrows indicate the direction of pressing with "P" during SPS hot deformation.

Here in Equation (2), H_a is the anisotropy field, the Kronmüller's factor $\alpha \approx 0.2$–0.5 and the effective demagnetizing factor (N_{eff}) ≈ 0.8–1.7 [10]. The deformed platelet-shaped grained largely increase the N_{eff} value and the possible accelerated grain growth above 800 °C is reasoned to degrade the coercivity [14]. During the hot deformation, it has been advocated that the liquid grain boundary phase is squeezed out above the ternary transition temperature of 665 °C as it rotates and gets redistributed along with the smaller aspect ratio grain facets [47]. Therefore, the Nd-rich phase perpendicular to the pressing direction is forced out of the intergranular regions where textured parallel facing grains get into intermittent contact with each other. Therefore, the localized exchange coupling effects at these regions without the Nd-rich phase contribute to the lowering of the coercivity [17]. With the release of interfacial energy from the liquid Nd-rich phase during hot deformation, the stress concentration gets developed in the contact points and triple pockets which can lead to the formation of non-aligned grains in the vicinity thereby necessitates the annealing treatment [14]. In regions where the grain boundaries are absent, the nanocrystalline $Nd_2Fe_{14}B$ grains can rapidly grow above the ternary transition temperature and the lattice mismatch is gradually truncated, which will also release lattice from distortion under reprocessing (thermally induced) stresses [41]. This study verified that potential is very high for the HDDR Nd-Fe-B based commercial as well as recycled materials and provides the fundamental understanding of hot deformation application on the recycled HDDR Nd-Fe-B alloys, defining the limits for coercivity, processing conditions, annealing, and controlled texture development.

Henceforth, with the application of higher deformation pressures and the subsequent addition of low melting eutectic alloys or RE-rich nanoparticles blends, not only the remanence enhancement and BH_{max} are possible but coercivity can also be improved in low grade recycled materials. This scope of simultaneous improvement of H_{Ci} after hot deformation on the recycled HDDR Nd-Fe-B is currently under investigation and further results shall be explained in an alternate report.

5. Conclusions

This short study determines that the SPS hot pressed recycled HDDR Nd-Fe-B powder has coercivity values slightly higher than the commercial HDDR MF-15P powder because of the initial HDDR grain-size, microstructure preservation, and the SPS processing control. The achieved magnetic properties imply that the recycled HDDR Nd-Fe-B system can still achieve an optimal level of magnetic properties after hot deformation and without the addition of excessive Nd content, by controlling the forging parameters and supplementary thermal treatments. The application of higher pressures during hot deformation can develop texture in these originally isotropic magnets. The hot deformation pressure of 150 MPa resulted in 55% reduction in height and improvement in J_r to 0.96 T which is subsequently improved to 1.01 T (BH_{max} = 180 kJ/m^3) after the thermal treatment at 750 °C for 1 h, i.e., ≈11% improvement over the recycled magnetic powder. This reduced pressure optimal hot deformation suitably retains high coercivity and imparts texture, which evidentially yields magnetic properties better than the starting recycled HDDR Nd-Fe-B powder.

Author Contributions: Conceptualization, A.I.; Methodology, A.I.; Software, A.I.; Validation, A.I. and M.A.; Formal analysis, A.I.; Investigation, A.I.; Resources, M.A., R.S. and A.W.; Data curation, A.I.; Writing—original draft preparation, A.I.; Writing—review and editing, A.I., M.A. and F.P.; Visualization, A.I.; Supervision, S.K., A.W., K.Ž.R. and F.P.; Project administration, K.Ž.R., S.K. and A.W.; Funding acquisition, S.K., A.W. and K.Ž.R. All authors have read and agreed to the published version of the manuscript.

Funding: The study leading to these results has received the funding from the European Community's Horizon 2020 Program ([H2020/2014-2019]) under grant agreement no. 674973 (MSCA-ETN DEMETER). Project website: http://etn-demeter.eu/. This publication reflects only the authors' research findings, which are targeted to contribute to the betterment of the global community

Acknowledgments: The authors aptly acknowledge the Department for Nanostructured Materials (K7 Nano) for provisioning the magnet synthesis/measurement facilities and the Center for Electron Microscopy & Microanalysis (CEMM) for scanning electron microscopy support at the Jožef Stefan Institute, Slovenia.

References

1. Sugimoto, S. Current status and recent topics of rare-earth permanent magnets. *J. Phys. D Appl. Phys.* **2011**, *44*, 6. [CrossRef]
2. Gutfleisch, O. Controling the properties of high energy density permanent magnetic materials by different processing routes. *J. Phys. D Appl. Phys.* **2000**, *33*, 16. [CrossRef]
3. Morimoto, K.; Katayama, N.; Akamine, H.; Itakura, M. Coercivity enhancement of anisotropic Dy-free Nd–Fe–B powders by conventional HDDR process. *J. Magn. Magn. Mater.* **2012**, *324*, 3723–3726. [CrossRef]
4. Coey, J.M.D. *Magnetism and Magnetic Materials*; Cambridge University Press: New York, NY, USA, 2010.
5. Kimiabeigi, M.; Sheridan, R.S.; Widmer, J.D.; Walton, A.; Farr, M.; Scholes, B.; Harris, I.R. Production and Application of HPMS Recycled Bonded Permanent Magnets for a Traction Motor Application. *IEEE Trans. Ind. Electron.* **2018**, *65*, 3795–3804. [CrossRef]
6. Nakamura, H. The current and future status of rare earth permanent magnets. *Scr. Mater.* **2018**, *154*, 273–276. [CrossRef]

7. Ikram, A.; Mehmood, M.F.; Podlogar, M.; Eldosouky, A.; Sheridan, R.S.; Awais, M.; Walton, A.; Kržmanc, M.M.; Tomše, T.; Kobe, S.; et al. The sintering mechanism of fully dense and highly coercive Nd-Fe-B magnets from the recycled HDDR powders reprocessed by spark plasma sintering. *J. Alloys Compd.* **2019**, *774*, 1195–1206. [CrossRef]

8. Ikram, A.; Mehmood, F.; Sheridan, R.S.; Awais, M.; Walton, A.; Eldosouky, A.; Šturm, S.; Kobe, S.; Rožman, K.Ž. Particle size dependent sinterability and magnetic properties of recycled HDDR Nd–Fe–B powders consolidated with spark plasma sintering. *J. Rare Earths* **2020**, *38*, 90–99. [CrossRef]

9. McGuiness, P.J.; Short, C.; Wilson, A.F.; Harris, I.R. The production and characterization of bonded, hot-pressed and die-upset HDDR magnets. *J. Alloys Compd.* **1992**, *184*, 243–255. [CrossRef]

10. Cullity, B.D.; Graham, C.D. *Introduction to Magnetic Materials*, 2nd ed.; IEEE/Wiley: Hoboken, NJ, USA, 2009.

11. Güth, K.; Woodcock, T.G.; Thielsch, J.; Schultz, L.; Gutfleisch, O. Local orientation analysis by electron backscatter diffraction in highly textured sintered, die-upset, and hydrogenation disproportionation desorption and recombination Nd-Fe-B magnets. *J. Appl. Phys.* **2011**, *109*, 07A764. [CrossRef]

12. Sepehri-Amin, H.; Ohkubo, T.; Hono, K.; Güth, K.; Gutfleisch, O. Anisotropy inducement mechanism in hydrogen disproportionation (HDDR) processed Nd-Fe-B powders. *J. Magn. Soc. Jpn.* **2014**, *38*, 37–44.

13. Kim, T.-H.; Oh, J.-S.; Cha, H.-R.; Lee, J.-G.; Kwon, H.-W.; Yang, C.-W. Direct observation of texture memory in hydrogenation–disproportionation–desorption–recombination processed Nd-Fe-B magnets using electron backscatter diffraction. *Scr. Mater.* **2016**, *115*, 6–9. [CrossRef]

14. Chen, R.J.; Wang, Z.X.; Tang, X.; Yin, W.Z.; Jin, C.X.; Ju, J.Y.; Lee, D.; Yan, A.R. Rare earth permanent magnets prepared by hot deformation process. *Chin. Phys. B* **2018**, *27*, 117504. [CrossRef]

15. Müller, K.H.; Grünberger, W.; Hinz, D.; Gebel, B.; Eckert, D.; Handstein, A. Hot deformed HDDR Nd-Fe-B permanent magnets. *Mater. Lett.* **1998**, *34*, 50–54. [CrossRef]

16. Liesert, S.; Kirchner, A.; Grünberger, W.; Handstein, A.; de Rango, P.; Fruchart, D.; Schultz, L.; Müller, K.H. Preparation of anisotropic NdFeB magnets with different Nd contents by hot deformation (die-upsetting) using hot-pressed HDDR powders. *J. Alloys Compd.* **1998**, *266*, 260–265. [CrossRef]

17. Gopalan, R.; Sepehri-Amin, H.; Suresh, K.; Ohkubo, T.; Hono, K.; Nishiuchi, T.; Nozawa, N.; Hirosawa, S. Anisotropic Nd–Fe–B nanocrystalline magnets processed by spark plasma sintering and in situ hot pressing of hydrogenation–decomposition–desorption–recombination powder. *Scr. Mater.* **2009**, *61*, 978–981. [CrossRef]

18. Li, X.; Chen, Z.; Yang, C.; Qu, S. NdFeB Magnets Prepared by Spark Plasma Sintering and Hot Deformation, Xiyou Jinshu Cailiao Yu Gongcheng/Rare. *Met. Mater. Eng.* **2013**, *42*, 194–199.

19. Li, X.-Q.; Li, L.; Hu, K.; Chen, Z.-C.; Qu, S.-G.; Yang, C. Microstructure and magnetic properties of anisotropic Nd–Fe–B magnets prepared by spark plasma sintering and hot deformation. *Trans. Nonferrous Met. Soc. China* **2014**, *24*, 3142–3151. [CrossRef]

20. Xu, X.D.; Sepehri-Amin, H.; Sasaki, T.T.; Soderžnik, M.; Tang, X.; Ohkubo, T.; Hono, K. Comparison of coercivity and squareness in hot-deformed and sintered magnets produced from a Nd-Fe-B-Cu-Ga alloy. *Scr. Mater.* **2019**, *160*, 9–14. [CrossRef]

21. Song, J.; Guo, S.; Ding, G.F.; Chen, K.; Chen, R.J.; Lee, D.; Yan, A. Boundary structure modification and coercivity enhancement of the Nd-Fe-B magnets with TbCu doping by the process of pre-sintering and hot-pressing. *J. Magn. Magn. Mater.* **2019**, *469*, 613–617. [CrossRef]

22. Soderžnik, M.; Li, J.; Liu, L.H.; Sepehri-Amin, H.; Ohkubo, T.; Sakuma, N.; Shoji, T.; Kato, A.; Schrefl, T.; Hono, K. Magnetization reversal process of anisotropic hot-deformed magnets observed by magneto-optical Kerr effect microscopy. *J. Alloys Compd.* **2019**, *771*, 51–59. [CrossRef]

23. Li, J.; Liu, L.H.; Sepehri-Amin, H.; Tang, X.; Ohkubo, T.; Sakuma, N.; Shoji, T.; Kato, A.; Schrefl, T.; Hono, K. Coercivity and its thermal stability of Nd-Fe-B hot-deformed magnets enhanced by the eutectic grain boundary diffusion process. *Acta Mater.* **2018**, *161*, 171–181. [CrossRef]

24. Castle, E.; Sheridan, R.; Zhou, W.; Grasso, S.; Walton, A.; Reece, M.J. High coercivity, anisotropic, heavy rare earth-free Nd-Fe-B by Flash Spark Plasma Sintering. *Sci. Rep.* **2017**, *7*, 11134. [CrossRef] [PubMed]

25. Kim, T.-H.; Lee, S.-R.; Yun, S.J.; Lim, S.H.; Kim, H.-J.; Lee, M.-W.; Jang, T.-S. Anisotropic diffusion mechanism in grain boundary diffusion processed Nd–Fe–B sintered magnet. *Acta Mater.* **2016**, *112*, 59–66. [CrossRef]

26. Castle, E.; Sheridan, R.; Grasso, S.; Walton, A.; Reece, M. Rapid sintering of anisotropic, nanograined Nd–Fe–B by flash-spark plasma sintering. *J. Magn. Magn. Mater.* **2016**, *417*, 279–283. [CrossRef]

27. Sepehri-Amin, H.; Liu, L.H.; Ohkubo, T.; Yano, M.; Shoji, T.; Kato, A.; Schrefl, T.; Hono, K. Microstructure and temperature dependent of coercivity of hot-deformed Nd-Fe-B magnets diffusion processed with Pr-Cu alloy. *Acta Mater.* **2015**, *99*, 297–306. [CrossRef]

28. Liu, J.; Sepehri-Amin, H.; Ohkubo, T.; Hioki, K.; Hattori, A.; Schrefl, T.; Hono, K. Grain size dependence of coercivity of hot-deformed Nd–Fe–B anisotropic magnets. *Acta Mater.* **2015**, *82*, 336–343. [CrossRef]

29. Sepehri-Amin, H.; Liu, J.; Ohkubo, T.; Hioki, K.; Hattori, A.; Hono, K. Enhancement of coercivity of hot-deformed Nd-Fe-B anisotropic magnet by low-temperature grain boundary diffusion of $Nd_{60}Dy_{20}Cu_{20}$ eutectic alloy. *Scr. Mater.* **2013**, *69*, 647–650. [CrossRef]

30. Liu, L.H.; Sepehri-Amin, H.; Ohkubo, T.; Yano, M.; Kato, A.; Sakuma, N.; Shoji, T.; Hono, K. Coercivity enhancement of hot-deformed Nd-Fe-B magnets by the eutectic grain boundary diffusion process using $Nd_{62}Dy_{20}Al_{18}$ alloy. *Scr. Mater.* **2017**, *129*, 44–47. [CrossRef]

31. Seelam, U.M.R.; Liu, L.H.; Akiya, T.; Sepehri-Amin, H.; Ohkubo, T.; Sakuma, N.; Yano, M.; Kato, A.; Hono, K. Coercivity of the Nd-Fe-B hot-deformed magnets diffusion-processed with low melting temperature glass forming alloys. *J. Magn. Magn. Mater.* **2016**, *412*, 234–242. [CrossRef]

32. Akiya, T.; Liu, J.; Sepehri-Amin, H.; Ohkubo, T.; Hioki, K.; Hattori, A.; Hono, K. High-coercivity hot-deformed Nd-Fe-B permanent magnets processed by Nd-Cu eutectic diffusion under expansion constraint. *Scr. Mater.* **2014**, *81*, 48–51. [CrossRef]

33. Song, T.; Tang, X.; Yin, W.; Chen, R.; Yan, A. Coercivity enhancement of hot-pressed magnet prepared by HDDR Nd-Fe-B powders using Pr-Cu eutectic alloys diffusion. *J. Magn. Magn. Mater.* **2019**, *471*, 105–109. [CrossRef]

34. Wang, H.H.; Chen, R.J.; Yin, W.Z.; Zhu, M.Y.; Tang, X.; Wang, Z.X.; Jin, C.X.; Ju, J.Y.; Lee, D.; Yan, A. The effect of Nd-Cu diffusion during hot pressing and hot deformation on the coercivity and the deformation ability of Nd-Fe-B HDDR magnets. *J. Magn. Magn. Mater.* **2017**, *438*, 35–40. [CrossRef]

35. Cha, H.R.; Jeon, K.W.; Yu, J.H.; Kwon, H.W.; Kim, Y.D.; Lee, J.G. Coercivity enhancement of hot-deformed Nd-Fe-B magnet by grain boundary diffusion process using the reaction of NdHX and Cu nanopowders. *J. Alloys Compd.* **2017**, *693*, 744–748. [CrossRef]

36. Sheridan, R.S.; Sillitoe, R.; Zakotnik, M.; Harris, I.R.; Williams, A. Anisotropic powder from sintered NdFeB magnets by the HDDR processing route. *J. Magn. Magn. Mater.* **2012**, *324*, 63–67. [CrossRef]

37. Sheridan, R.S.; Williams, A.J.; Harris, I.R.; Walton, A. Improved HDDR processing route for production of anisotropic powder from sintered NdFeB type magnets. *J. Magn. Magn. Mater.* **2014**, *350*, 114–118. [CrossRef]

38. Ikram, A.; Mehmood, M.F.; Samardzija, Z.; Sheridan, R.S.; Awais, M.; Walton, A.; Šturm, S.; Kobe, S.; Žužek, R.K. Coercivity Increase of the Recycled HDDR Nd-Fe-B Powders Doped with DyF_3 and Processed via Spark Plasma Sintering & the Effect of Thermal Treatments. *Materials (Basel)* **2019**, *12*, 1498.

39. Seelam, U.M.R.; Ohkubo, T.; Abe, T.; Hirosawa, S.; Hono, K. Faceted shell structure in grain boundary diffusion-processed sintered Nd–Fe–B magnets. *J. Alloys Compd.* **2014**, *617*, 884–892. [CrossRef]

40. Kirchner, A.; Grünberger, W.; Gutfleisch, O.; Neu, V.; Müller, K.H.; Schultz, L. A comparison of the magnetic properties and deformation behaviour of Nd-Fe-B magnets made from melt-spun, mechanically alloyed and HDDR powders. *J. Phys. D Appl. Phys.* **1998**, *31*, 1660–1666. [CrossRef]

41. Wang, Z.; Ju, J.; Wang, J.; Yin, W.; Chen, R.; Li, M.; Jin, C.; Tang, X.; Lee, D.; Yan, A. Magnetic Properties Improvement of Die-upset Nd-Fe-B Magnets by Dy-Cu Press Injection and Subsequent Heat Treatment. *Sci. Rep.* **2016**, *6*, 38335. [CrossRef]

42. Hono, K.; Sepehri-Amin, H. Strategy for high-coercivity Nd–Fe–B magnets. *Scr. Mater.* **2012**, *67*, 530–535. [CrossRef]

43. Kim, T.-H.; Lee, S.-R.; Lee, M.-W.; Jang, T.-S.; Kim, J.W.; Kim, Y.D.; Kim, H.-J. Dependence of magnetic, phase-transformation and microstructural characteristics on the Cu content of Nd–Fe–B sintered magnet. *Acta Mater.* **2014**, *66*, 12–21. [CrossRef]

44. Mouri, T.; Kumano, M.; Yasuda, H.Y.; Nagase, T.; Kato, R.; Nakazawa, Y.; Shimizu, H. Effect of two-stage deformation on magnetic properties of hot-deformed Nd–Fe–B permanent magnets. *Scr. Mater.* **2014**, *78–79*, 37–40. [CrossRef]

45. Sadullahoğlu, G.; Altuncevahir, B.; Okan, A.A. Investigation of Grain Boundary Compositions and Magnetic Properties of Hot-Worked $Nd_{18}Tb_1Fe_{66.5}Co_5Al_{1.5}B_8$ Magnets. *Acta Phys. Pol. A* **2013**, *123*, 233–234. [CrossRef]

46. Liu, Z.W.; Huang, Y.L.; Hu, S.L.; Zhong, X.C.; Yu, H.Y.; Gao, X.X. Properties enhancement and recoil loop characteristics for hot deformed nanocrystalline NdFeB permanent magnets. *IOP Conf. Ser. Mater. Sci. Eng.* **2014**, *60*, 012013. [CrossRef]

47. Bance, S.; Seebacher, B.; Schrefl, T.; Exl, L.; Winklhofer, M.; Hrkac, G.; Zimanyi, G.; Shoji, T.; Yano, M.; Sakuma, N.; et al. Grain-size dependent demagnetizing factors in permanent magnets. *J. Appl. Phys.* **2014**, *116*, 233903. [CrossRef]

Review

Spark Plasma Sintering of Titanium Aluminides: A Progress Review on Processing, Structure-Property Relations, Alloy Development and Challenges

Ntebogeng F. Mogale * and Wallace R. Matizamhuka

Department of Metallurgical Engineering, Vaal University of Technology, Vanderbijlpark 1911, South Africa; wallace@vut.ac.za
* Correspondence: ntebogeng.mogale@yahoo.co.za

Received: 1 May 2020; Accepted: 5 June 2020; Published: 11 August 2020

Abstract: Titanium aluminides (TiAl) have the potential of substituting nickel-based superalloys (NBSAs) in the aerospace industries owing to their lightweight, good mechanical and oxidation properties. Functional simplicity, control of sintering parameters, exceptional sintering speeds, high reproducibility, consistency and safety are the main benefits of spark plasma sintering (SPS) over conventional methods. Though TiAl exhibit excellent high temperature properties, SPS has been employed to improve on the poor ductility at room temperature. Powder metallurgical processing techniques used to promote the formation of refined, homogeneous and contaminant-free structures, favouring improvements in ductility and other properties are discussed. This article further reviews published work on phase constituents, microstructures, alloy developments and mechanical properties of TiAl alloys produced by SPS. Finally, an overview of challenges in as far as the implementation of TiAl in industries of interest are highlighted.

Keywords: titanium aluminide; spark plasma sintering; microstructure; mechanical properties; alloy development

---

1. Introduction

Titanium-based intermetallics can be defined as metallic materials consisting of approximate stochiometric ratios in ordered crystal structures [1]. These have properties such as low densities and high melting points, good high-temperature strength, resistance to oxidation and creep [2]. The research interest in intermetallics for at least the past 30 years according to Muktinutalapati & Nageswara [3] has been due to the need to replace the previously used NBSAs (8–8.5 g/cm^3 in density) with lower density (4–7 g/cm^3) materials, saving about 55% of the weight gain of turbine engines [4].

Much attention was given to titanium (Ti) and nickel-based aluminides amongst many others. Of interest for this research work is titanium aluminides (TiAl). According to Muktinutalapati and Nageswara, intermetallics around TiAl can be classified into two-alpha (α_2)-Ti$_3$Al and gamma (γ)-TiAl phase. The greatest disadvantage of α_2-Ti$_3$Al is poor toughness and fatigue crack growth, shifting much research and development more on γ-TiAl. These possess properties which include good creep and oxidation properties at elevated temperature applications, low densities (3.9–4.2 g/cm^3 varying with composition), high stiffness and yield strength [5].

It has been over 20 years since the successful implementation of gamma titanium aluminide (γ-TiAl) alloys in aerospace components produced by companies such as General Electric Aircraft Engines, Pratt and Whitney and Rolls Royce [6–8]. The alloys have been employed in various aerospace components such as rotating and static engine components used in turbines, compressors, combustors, and nozzles. Research development of such alloys over the years primarily focused on the refinement of microstructure and improvement of properties, particularly ductility and formability, through compositional optimisations and application of various processing technologies.

Compositional variations including controlling gaseous impurities such as oxygen (O) and nitrogen (N), and the addition of chromium (Cr) and manganese (Mn) to TiAl alloys have been previously addressed with the aim of ductility improvement [9–11]. Furthermore, employing wrought processing techniques followed by post-treatments have also been extensively experimented [12–14]. However, microstructural inconsistencies resulting from solidification and phase transformations further deteriorate and scatter the mechanical properties of the alloys [15,16]. Spark plasma sintering (SPS) presents an opportunity to consolidate metallic powder materials without the deviations mentioned above. The process employs DC pulses of high intensity and pressure to achieve the required sintering temperature under a specified time. The SPS technique has been used in a large number of investigations and has advantages compared to traditional techniques such as shorter holding times, lower sintering temperatures and marked increases on the properties of materials [17–19]. This work is a summary of the progress, advances and challenges in the production and implementation of spark plasma sintered γ-TiAl alloys.

2. Powder Metallurgical Processing of γ-TiAl Alloys

2.1. Gas Atomization (GA)

The current preferred method for mass production of metallic powders consolidated using various methods including SPS is GA [20,21]. Although GA is said [21] to be more costly compared to water atomization, improved yields of spherical powders are achieved aiding flowability. The conventional basic operation of gas atomisation (GA) is illustrated in Figure 1. During GA, the metal or alloy is melted inside a crucible, followed by pouring the stream of hot metal liquid and finally pulverizing using a pressurized gas jet. The solidified metal droplets are then accumulated at the powder container for collection. According to Martín et al. [21], processing parameters such as the melt temperature and flow, nozzle type and gas purity and pressure affect the powder size and quality.

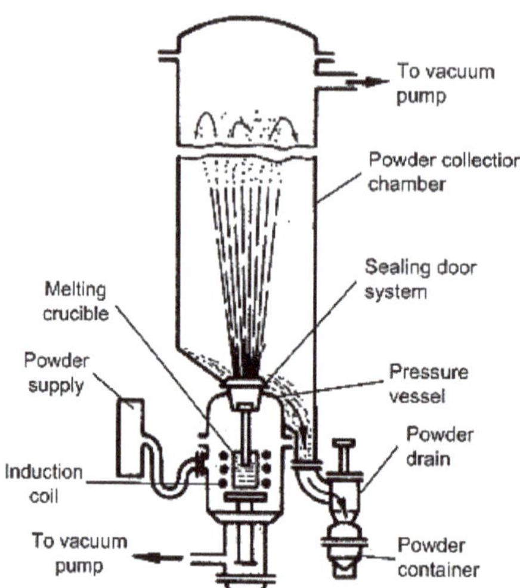

Figure 1. Gas atomisation (GA) schematic, reproduced from [22], with permission from Elsevier, 2019.

The production of γ-TiAl powders using GA is rather challenging. This can be ascribed to the high reactivity of the melt with the crucible lining material and the high impurity gas pick-up of the melt, atomized droplets and the hot powder particles [23]. As a result, cold-crucible or crucible-free techniques such as electrode induction melting gas atomization (EIGA) and plasma melting induction guiding gas atomization (PIGA) can be employed. For further reading on these techniques, the reader is referred to [24–26].

Powder characteristics of concern during the GA production of γ-TiAl affecting powder quality are particle size distributions, gaseous impurity levels and cooling rates [23]. The morphology of powders produced using GA, as shown in Figure 2, is spherical despite the particle size. Spherical powders are beneficial to subsequent PM processes, promoting flowability and packing density. The level of impurity gases such as O and N in GA produced powders can impair the mechanical properties after consolidation. Furthermore, oxide or nitride layers on powder surfaces may hinder compaction [23]. Minimal gas entrapment can be achieved by using an inert gas atmosphere during powder alloy atomization and handling. Finally, GA employs high cooling rates enhancing microstructural and compositional uniformity.

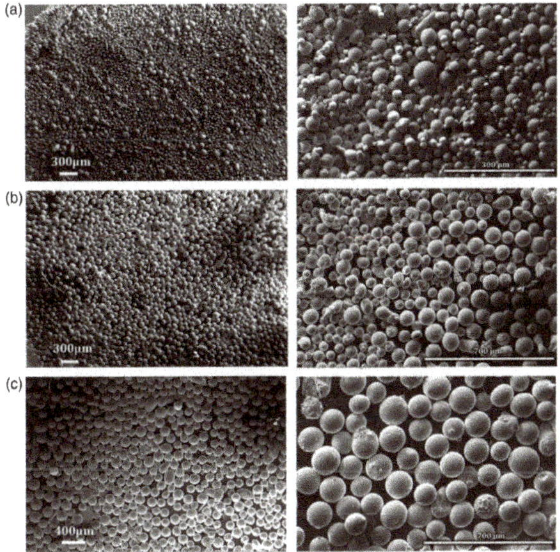

Figure 2. Scanning electron microscope (SEM) images of GA produced Ti4522XD powders with sizes: (**a**) d < 45 µm; (**b**) 45 µm < d < 100 µm; and (**c**) 100 µm < d < 150 µm, reproduced from [21], with permission from John Wiley and Sons, 2020.

Several studies involving the use of GA and SPS have been carried out to date [27–30]. Gu et al. [27] fabricated pre-alloyed (PA) Ti–43Al–5Nb–2V–Y (at. %) powders utilizing the PIGA technique and densified the powders at different temperatures by SPS. The morphology of the varying particle size (50–120 µm) PA powders was reported as spherical with satellite particles. Moreover, a cellular dendritic interior microstructure was observed, a characteristic microstructure of rapidly solidified metal droplets. In a related study, Liu and associates [28] studied the size-dependency on structural properties of a Ti–48Al–2Cr–8Nb (at. %) powder alloy produced using EIGA for PM and additive manufacturing technologies. It was found that decreasing the powder particle size decreased the content of the γ and increased that of α_2-phase.

2.2. Mechanical Milling (MM)

PA powder can be effectively manufactured using MM [31]. As stated by Xiao et al., MM aids solid diffusion and rapid compound formation. In work done by Hadef [32]; it is stated that milling includes grinding through impact, compression and attrition. During MM experiments, powders are introduced in required proportions into hardened steel (typically tungsten carbide, stainless steel and zirconia) vial along with milling balls of a selected size. Voluntarily, a process control agent (PCA) is also added to aid in minimising cold welding and lump formation. Collisions between the wall of the vial and the contents thereof exist at high frequencies and velocities, effectively milling the powder contained [33]. Milling time and speed, vial type, milling media, ball-to-powder weight ratio (BPR), milling atmosphere and the level of vial filling are parameters affecting the ultimate powder mix constituents.

Extensive research has been devoted over the years to study the effect of the various milling parameters on the constitution of the ultimate powder mix [34,35]. Oehring et al. [36], investigated the milling process of elemental Ti–Al powder blends using detailed X-ray diffraction (XRD) analysis and found that a dual mixture of hcp solid solution and an amorphous phase in alloy $Ti_{50}Al_{50}$ exist at lower milling intensities. Oehring et al. also found that energetic destabilisation of present phases occurring during milling results in the final observed intermetallic compounds. Zhang et al. [37] studied phase formation during milling of Ti–75Al powder mixtures using XRD analysis and concluded that a solid solution of Al(Ti) formed as a result diffusion of Ti into Al during the early milling stages. In addition to the study findings, an $L1_2$ ordered Al_3Ti phase with an average phase grain size of 18 nm was formed with extended milling time.

Wang et al. [38] studied the effects of SPS temperature and mechanical milling treatment parameters on the phase constitution and microstructure. The as-atomised powder following SPS at 1200 °C showed microstructural inhomogeneities consisting of α_2 and γ and some lamella colonies as compared to the homogeneous, fully dense MM powder. Additionally, it was found that extended milling times and speeds during MM attribute to the uniform distribution of the phases present.

Double mechanical milling (DMM) has recently gained attention [39,40] and involves two milling stages. The initial stage normally entails mixing of the elemental powders at low speeds, followed by high-energy ball milling at higher speeds for a shorter interval. Optionally, the powder mix is removed from the vial and heat-treated in a controlled atmosphere before the final milling stage at higher speeds and longer intervals. In a study employing DMM on TiAl alloys [40], the authors reported the formation of a regular shaped powder with a particle size decrease from a maximum of 80 to 40 μm after the final milling stage. Consequently, peak broadening was observed and can be attributed to the reactions and interdiffusion occurring between phases. In a similar study Shulong et al. [39] obtained an ultrafine regular powder morphology after DMM with sizes in the range of 20–40 μm.

2.3. Mechanical Alloying (MA)

The process of mechanical alloying (MA) involves loading of PA or elemental powders inside a high energy ball mill and the usage of grinding media to repeatedly cold weld, fracture and reweld powder particles [41]. Parameters including time, speed and BPR [42] are optimised to control and balance the resultant powder particle sizes. PCA are inorganic compounds [43] often used in MA of ductile metals to prevent agglomeration of powder particles. According to Suryanarayana [41], adding 1–2 wt. % of PCA can prevent excessive cold welding or agglomeration of the individual powders on the grinding media and the milling vessel. The working principles of MA involve the excessive deformation of powder particles through work hardening and flattening of colliding or contacting ductile particles using grinding media. In addition, intermetallic compounds are refined and fractured, while oxide dispersoids comminuted.

MA has proven to be effective in the production of non-equilibrium and nanostructured TiAl alloys [44–46]. Some of the successful developments in TiAl research through MA include the ability to produce fine-grained alloys [47], although some shortcomings, such as gas contamination [48] still need to be overcome. Sim et al. [46] found that alloys sintered from a 40 h MA powder showed promising yielding and ultimate strength in compression. The reported yielding and ultimate strength values from the previous study were as high as 1644 and 2542 MPa, respectively. Failure strains at both ambient temperatures and 650 °C were 31.3% and 55.3%. Forouzanmehr and colleagues [45] found that the milling of Ti–50Al (at. %) powder mixture formed a Ti(Al) solid solution earlier in milling, followed by a transformation of the amorphous structure to a Ti(Al) that is supersaturated with longer milling durations. Additionally, further annealing treatment for the amorphous and the supersaturated Ti(Al), formed a nanocrystalline TiAl intermetallic compound with an ultrafine grain size of ~50 nm and a microhardness value of ~11.67 GPa.

To limit elevated temperature grain growth in γ-TiAl and other PA powders, Bohn et al. [47] found that the addition of Si to Ti powders (resulting in precipitates of $Ti_5(Si,Al)_3$ embedding on grain boundaries of γ-TiAl matrix) gave refined microstructures with a grain size ranging from 160–480 nm for the matrix and 80–190 nm for the precipitates. To produce low contaminated TiAl powder, Bhattacharya et al. [48] found planetary ball billing to have to increase O contents. At the same time, insignificant impurity levels (O pick-up of $\approx$ 0.014 wt. % and N pick-up of 0.031 wt. %) were observed in attrition milled powders. However, in both cases, the powders used were handled in a glovebox with purified Ar atmosphere.

2.4. Cryomilling (CM)

Another interesting PM technique that evolved as a modification of the conventional high energy ball milling is cryomilling (CM). The process involves soaking grinding media in a cryogenic liquid (such as N) while optimising processing parameters to minimise recovery and recrystallisation, resulting in ultra-refined grain structures [49]. According to Lavernia et al., benefits of employing CM over traditional milling include the reduction of cold welding and powder agglomeration to the grinding media producing effective milling outcomes, limited powder oxidation occurrences and reduced milling times. Consolidation is required after CM processing, allowing the exploration of studies on phase and microstructural evolutions as a function of properties, comparable to those produced using conventional ball milling.

Figure 3 shows the basic setup used for CM. The apparatus involves thermocouples used to monitor the mill and ensure a constant cryogenic liquid level, thus maintaining a constant milling environment [50]. Additionally, the cryogenic liquid is in constant circulation into the mill with excess liquid drained through a particle filtering blower, enabling powder particle entrainment in the gas flow. Compatibly, the setup comprises of opening and closing valves to ease the flow of the cryogenic liquid and are monitored by temperature fluctuations above the slurry level. Some of the implemented configurations, according to Witkin & Lavernia, are 101 attritor mills used at the University of California in Davis, charging up to 1 kg of powder and the commercialised 35 kg capacity attritor mill sponsored by" Boeing's Rocketdyne division.

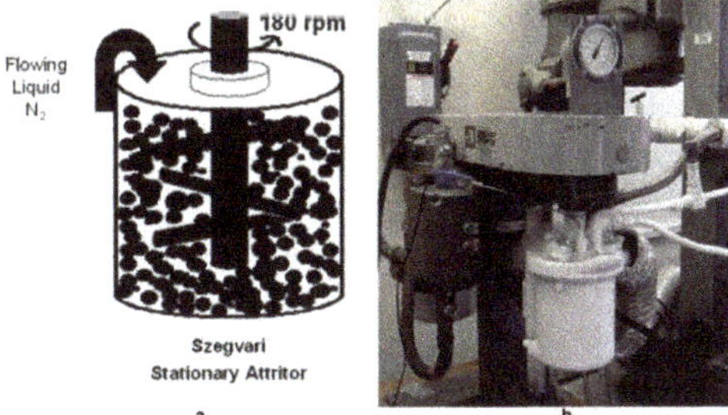

Figure 3. (**a**) Cryomilling (CM) apparatus representation; (**b**) a customised 500 g capacity mill in operation, reproduced from [50], with permission from Elsevier, 2006.

Several recent studies [51–54] have been devoted to understanding the processing and behaviour of nanostructured TiAl alloys produced using CM and SPS. Shanmugasundaram et al. [52], investigated the densification and microstructure of a γ-TiAl alloy produced using gas atomised CM powder and consolidated using SPS at 1050 and 1200 °C. CM resulted in a 100% powder yield, reduced particle

(2–50 µm) and crystalline sizes (40 nm). Although the alloy fully densified at 1200 °C, 8 h of CM resulted in a decrease in the sintering temperature. The obtained refined microstructure (with grain sizes of 0.9 and 0.6 µm at 1200 and 1050 °C) consisted of γ and α_2, with volume fraction of the later found to be dependent on the sintering temperature and O content. Deng et al. [51] obtained an ultrafine-grained FL and near γ microstructure promoted by CM and varying sintering temperatures of 900, 1000 and 1100 °C, respectively. The authors found that sintering at 1000 °C produced excellent compression properties, with yield strengths as high as 1575 MPa RT and 955 MPa at 850 °C.

3. Powder Consolidation by SPS

3.1. Basic Operating Principles

SPS, also referred to as field assisted sintering technique is a powder metallurgy (PM) method used to consolidate powders subjected to applied current and pressure [17]. As stated by Munir et al., the practice (schematically shown in Figure 4) employs uniaxial pressure and pulsed high direct current to consolidate powders. The science behind the successful consolidation of the powders is attributed to the sufficient generation of Joule heating [55], employing voltages below 10V and currents as high as 10 kA in conjunction with the electrically conductive tool materials used in the set-up. The widely used SPS technique has advantages compared to conventional manufacturing techniques such as hot-pressing, pressureless sintering, laser sintering and hot isostatic pressing (HIP). These advantages include shorter holding times, lower sintering temperatures and marked increases in the properties of materials [18,19,56].

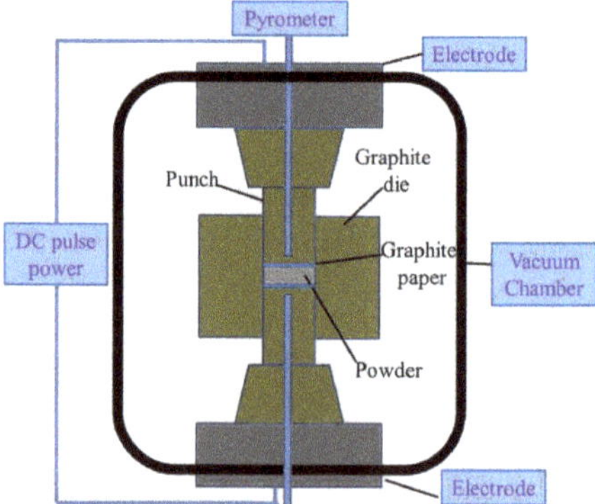

Figure 4. Spark plasma sintering (SPS) process representation, reproduced from [57], with permission from John Wiley and Sons, 2014.

3.2. Mechanisms of Sintering in SPS

The complex mechanisms and theories involved in SPS are as a result of electrical, thermal and mechanical effects [55]. Of the vast available models and theories, the standard and accepted involves effects due to joule heating, plasma generation and electroplasticity [58,59]. According to Matizamhuka [58], the electrical effects are related to the electrical properties of the powders. Furthermore, electrically conductive powders allow smooth current flow and are heated by the "Joule effect" and transferred to the bulk of the powder by conduction. The formation of necks between the powder particles through cleaning of powder surfaces, welding and vaporization is enhanced by the electric current flow through particles when pressure is applied [60]. Additionally, the electric current pulses produced during SPS are said to accelerate the densification kinetics existing between powder particle necks. The formation of arcs and plasma between powder particles promotes localized melting in amorphous material jet-form between spherical nanosized ceramic particles [61], through liquid wetting [62] and partial melting in W and ZrB_2 ceramic particles [63].

The mechanical effects promoting densification are related to the quasi-static compressive stress applied in SPS [55]. This results in better particle contact, altering the morphology and amount of contacting particles and further enhancing the existing densification kinetics related to viscous flow, lattice, and grain boundary diffusion. Alternatively, new mechanisms like grain boundary sliding or deformation plastically can be activated [64]. The most common thermal factor promoting densification during SPS is the heating rate. In the study conducted by Olevsky and colleagues [65], high heating rates reduced noncontributing surface diffusion, favoring sinterability and intensifying densification by grain boundary diffusion. Furthermore, high rates of heating were found to retard grain growth.

3.3. SPS Modeling for Complex-Net Shaping

Advanced nanomaterials can potentially be manufactured using SPS owing to the impressive characteristics offered by the process. These include pressures as high as 100 MPa, temperatures of up to 2500 °C, heating rates ranging up to 1000 K.min^{-1} and pulsed electric current readings of a few thousand amperes [66]. Despite these benefits, the challenge of manufacturing complex shapes beyond the traditional 2-D parts remains. This is due to the homogeneity variations experienced during the densification of components having high thickness regions, wherein the thicker regions require more shrinkage compared to the rest of the component [66,67]. To date, various models have been employed in SPS to overcome those, as mentioned earlier. These include modified punch designs [68], controlling punch displacement through the usage of sacrificial materials [69], deformed interface approach [70], finite element method (FEM) [67,71], and the controllable interface approach [66].

Voisin et al. [68], produced two GA γ-TiAl alloys to the nominal composition of $Ti_{49.92}Al_{48}W_2B_{0.08}$ and $Ti_{48}Al_{48}Cr_2Nb_2$ using SPS. Difficulty was experienced in as far as shaping a blade with a thick root and a thin foil, raising the need to design a mold of graphite (in Figure 5) consisting of a multiple punch assembly, the matrix and cylindrical pieces. The FEM was later employed to control the temperature at each point in the sample [72]. An 80 mm long blade was obtained by employing models in Figure 5 and FEM. The FEM was proven [71] to be reliable for qualitative predictions of powder sample grain growth and densification kinetics in SPS with a given temperature regime.

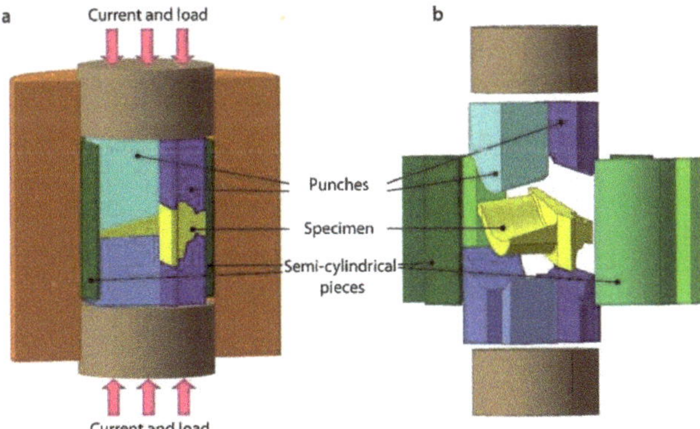

Figure 5. Representation of the graphite mold assembly with (**a**) punches in position and (**b**) side view of the mold, reproduced from [68], with permission from John Wiley and Sons, 2015.

Alternatively, a sacrificial material (such as in Figure 6) can be used to achieve complex part homogeneity. According to Manière and colleagues [69], this approach presents advantages such as possibilities of fully densifying parts of varying thickness complexities, even stress distributions as a result of shrinkage uniformity at all points on the sample and control over the final part shape.

Unfortunately, this approach has shortcomings related to material losses of the sacrificial component and the limitations of the number of thickness variations.

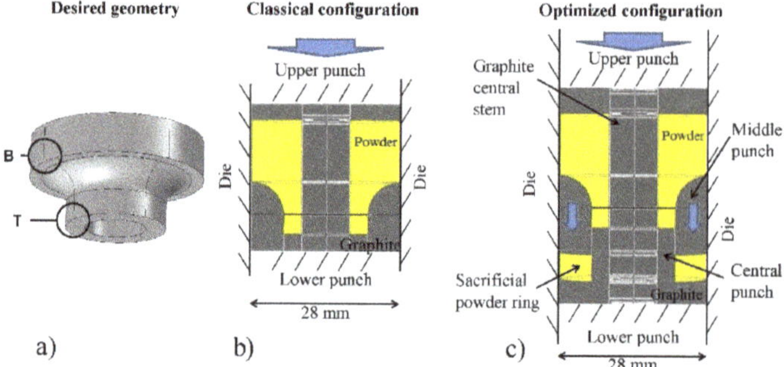

Figure 6. An illustration of the sacrificial material approach with (**a**) showing the desired geometry, (**b**) presenting the traditional configuration and (**c**) displaying the sacrificial material configuration reproduced from [69], with permission from Elsevier, 2016.

Commonly known as the "DEFORMINT" approach, the deformed interfaces method involves assembling a minimum of two porous materials (i.e., porous bodies or powder beds) separated by single or multiple interfaces [70]. Manière et al. described that the simple geometry on the outside (i.e., often cylindrical) is obtained by assembling a porous complex shape that is covered by a separation material and surrounded by an additional porous sacrificial material, that is the reverse of the initial shape. The detailed process steps for the deformed interfaces method are shown in Figure 7, and consists of the reproduction of porous assembly materials, densification of the assembly materials and sacrificial parts removal. Successful implementation of this method includes a 98% densification of a highly complex CoNiCrAlY alloy turbine blade developed by Manière et al. [70].

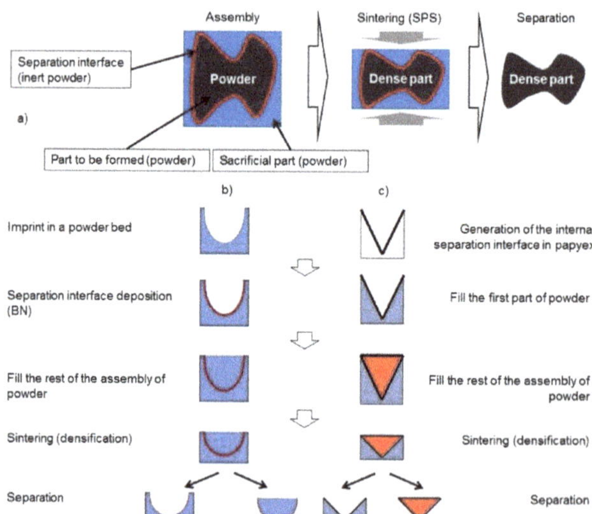

Figure 7. Deformed interfaces method process steps, (**a**) three main method steps comprising of: assembly, sintering, and the separation of the sacrificial tools, (**b**) imprint in powder bed approach, (**c**) graphite foil approach. reproduced from [70], with permission from Elsevier, 2017.

Finally, the recent controllable interface approach combines the adjustable interface thermal/electrical fluxes and the DEFORMINT approach. The controllable interface approach has been successfully used to manufacture 40 mm nickel gears [66], and the process steps (in Figure 8) for this consisted of powder loading into the assembly, SPS densification, removal of densified gear and polishing. SPS is modified by this approach through multiple die tooling, improving on the production number of complex shapes.

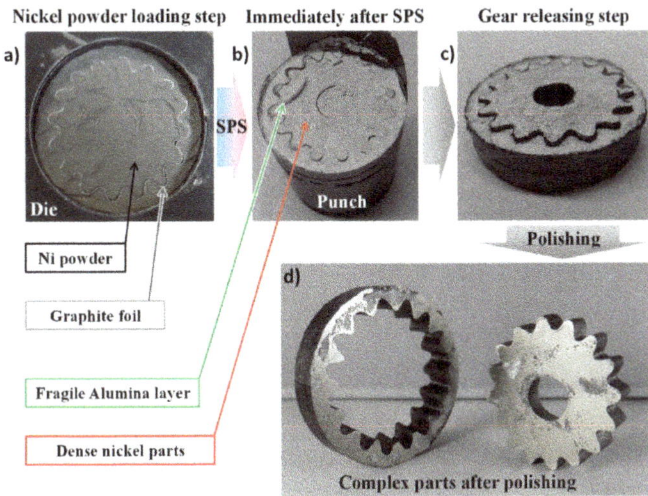

Figure 8. Controllable interface approach process steps for SPS densification of 40 mm nickel gears; where (**a**) shows loading; (**b**) following SPS; (**c**) gear releasing; and (**d**) polished parts [66].

3.4. Carburation at High Temperature during SPS

One of the major challenges during the processing of materials using SPS is carbon (C) contamination experienced at high temperatures, also known as carburation. For this reason, caution should be taken when selecting tooling materials. Pressing tools are generally covered with graphite foils/sheets to ease sample removal subsequent to SPS [73,74]. Therefore, at temperatures above 600 °C, reactions between the graphite covered tools and the O present in the sample may occur. Excess O, as stated by Franceschin et al. [73], can arise from the gaseous sintering atmosphere or moisture. The vigorous gaseous transportation occurring between the sample and the graphite mold may encourage the precipitation of C or carbides [55].

The effects of carburation on SPS manufactured materials are numerous. Waseem et al. [75] characterized a $Ti_xWTaVCr$ alloy synthesized using SPS and detected up to 0.83 wt. % of C during gas fusion analysis. The presence of C was ascribed to the diffusion of graphite into the samples during SPS. Furthermore, the presence of a TiC phase was observed during XRD analysis resulting in a ~30 μm thick layer on the surface of the samples. Moreover, the availability of TiC may be beneficial, improving on the resistance to irradiation and high heat fluxes. Meir et al. [76] showed that during the SPS densification of a magnesia–aluminate spinel, carbide precipitation was enhanced by the C containing atmosphere from the SPS apparatus. Carbide precipitation was found to impair light transmittance and even opacity in LiF-free samples. Additionally, volatile $(CF)n$ species were formed at high temperatures due to the interaction between C and LiF vapor.

Of much interest to this paper is carburation in TiAl alloys manufactured using SPS. Martin et al. [77] fully densified a commercial 48-2-2 TiAl powder by SPS and developed a two-fold protocol to obtain the desired microstructure and avoid TiC formation. Firstly, 8 and 10 mm samples were densified at 1200 °C below 50 MPa to obtain fully densified compacts with minimal (less than 1 μm) growth of the

TiC layer. Secondly, the sample was introduced into a 36 mm inner diameter graphite mold, avoiding any contact between the sample and mold walls. Lastly, an annealing treatment for 5 min at 1360 °C was carried out with minimal load application to promote current conduction. TiC layer formation was achieved through avoiding mutual contact between the mold walls and the sample.

To avoid carburation, graphite pressing tools can be reinforced with C fibers to improve on the mechanical strength [78]. Additionally, double-walled tools with inner ceramic die and an outer graphite covering may be employed. Tooling material made from refractories and steel such as silica, alumina, concrete, TNZ molybdenum and copper-beryllium alloys may be such employed at sintering temperatures below 1000 °C [79,80]. Finally, the graphite tools can be separated from the sample by using foil layers such as hexagonal boron nitride [81] or alumina [82].

4. Crystallographic Phases and Microstructures

4.1. Crystallographic Phases

The characteristic phases existing in SPS produced γ-TiAl alloys are TiAl (γ), Ti₃Al (α₂) or dual (γ + α₂). The γ-phase has an $L1_0$ type face-centred tetragonal structure [83] consisting of atomic layers at 90° to the c-axis [84] with lattice parameters a = 0.4005 nm, c = 0.4070 nm and a tetragonality ratio (c/a) of 1.02 [85,86]. Two peritectic reactions exist near the equiatomic composition [87]:

$L + \beta \rightarrow \alpha$ and $L + \alpha \rightarrow \gamma$.

The prime phase that crystallises in Ti-base alloys from the liquid is β. According to McCullough et al. [87] the solidification path is as follows:

$L \rightarrow [\beta + \alpha] \rightarrow [\alpha + \gamma] \rightarrow [\alpha_2 + \gamma]$.

On cooling, the phase becomes unstable and transforms into the γ-phase. With further solid-state cooling, the phase becomes unstable and orders into the two-phase around 1117 °C [88,89].

The α_2-Ti₃Al phase has a hexagonal symmetry of a $D0_{19}$ (Figure 9) structure with lattice parameters $a_{ord} = 2a_{disor}$ = 0.58 nm; c_{ord} = 0.48 nm. In a review by Djanarthany et al. [90], the last parameter is said to resemble that of an A3 type structure with a long-range ordering only in the direction perpendicular to the c-axis. Phase transformation according to $\beta \rightarrow \alpha \rightarrow \alpha_2$ resulting from cooling forms the α_2-phase, and orders between 1125 and 1150 °C [91]. As shown in Figure 10, the γ phase exists at a composition of ~48.5–60, α_2 at ~24–34 and (γ + α_2) at ~34–49 at. % Al. Although the α_2-phase shows good strength at high temperatures, it has been found to be very brittle, much related to the high absorption level of O and hydrogen [85,92]. Contrarily, the γ-phase exhibits excellent resistance to oxidation with low-level O and hydrogen absorption and poor ductility at room temperature (RT). In recent studies [51,52,93], TiAl powders were consolidated using SPS and the phases mentioned above were obtained.

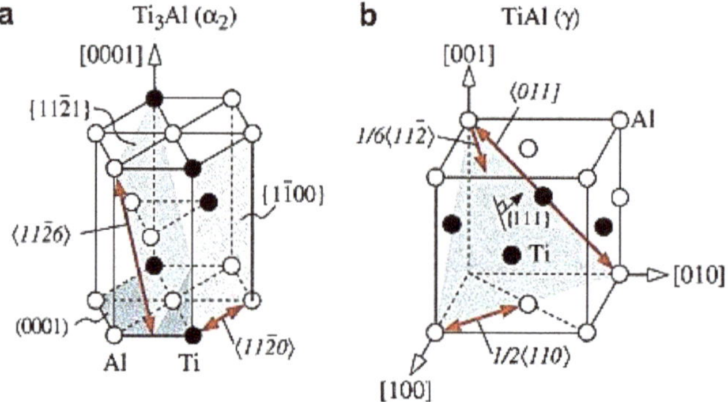

Figure 9. Titanium aluminide (TiAl) crystal structures, reproduced from [94], with permission from Elsevier, 2007.

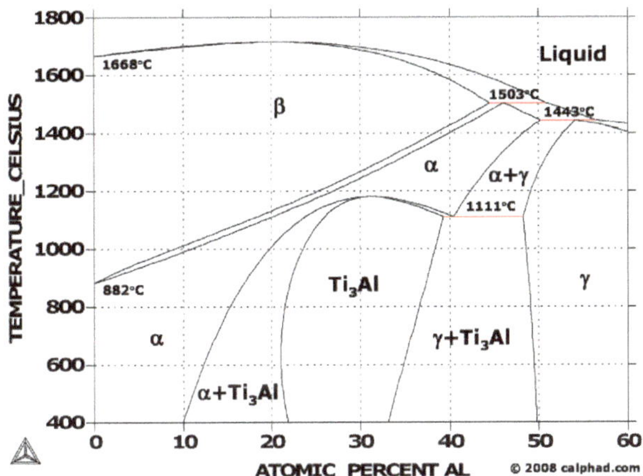

Figure 10. Ti-Al binary phase diagram.

A high performance Ti–48Al–2Cr–2Nb alloy was produced by employing SPS to consolidate spherical pre-alloyed (SP powder) and ball milled pre-alloyed powders (MP powder). In this work, Wang et al. [93] carried out an analysis of the variations in density, microstructure and mechanical properties at different sintering temperatures. Similar phases as mentioned above (γ-TiAl, α_2-Ti$_3$Al) were achieved with additional TiAl$_2$, TiAl$_3$ for both the SP and MP powders. Additionally, the MP powder showed the presence of a Ti$_2$Al phase. At sintering temperatures between 1200 and 1350 °C, the authors obtained γ-TiAl, TiAl$_3$ and TiAl$_2$ (see Figure 11).

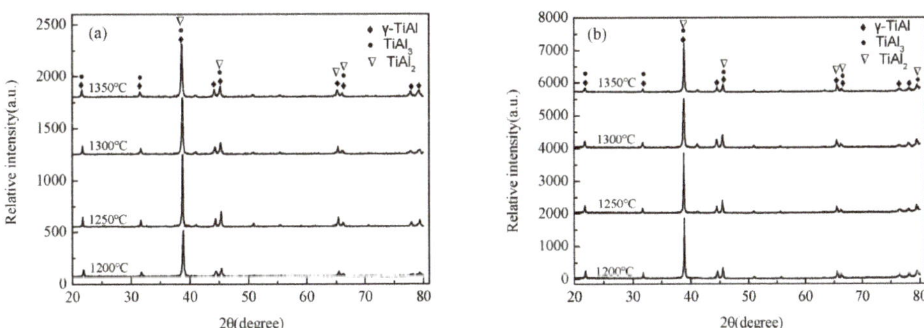

Figure 11. (**a**) XRD patterns of Ti-48Al-2Cr-8Nb alloys produced by SPS at different temperatures using spherical pre-alloyed (SP); and (**b**) milled ore-alloyed (MP) powders [93].

In a similar study, Deng et al. [51] demonstrated a novel avenue for the manufacturing of Ti–45Al–8Nb alloys using CM and SPS. To prevent cold welding and promote powder homogeneity, elemental powders were ball milled for 16 h at a rotating speed of 200 rpm. Subsequently, the blended powders were cryomilled for 40 h at a speed of 350 rpm. The alloys were sintered using a heating rate of 100 °C/min up to sintering temperatures of 900, 1000 and 1100 °C, respectively. XRD analysis showed the presence of the much dominant γ-TiAl phase with minor amounts of the α_2-Ti$_3$Al phase.

4.2. Microstructures of TiAl

With a composition ranging from ~46–52 at. % of Al; the general microstructure is either a dual α_2/γ or single pure γ-phase microstructure [95]. From the previously mentioned phases, with further processing and treatment characteristic microstructures referred to as equiaxed, duplex (DP), nearly lamellar (NL) and fully lamellar (FL) can be achieved [4]. The equiaxed structure contains a single phase γ-TiAl. The phase is stable in the region of the phase diagram above 50 at. %Al. Combined phase microstructure (commonly referred to as DP) made up of both α_2 and γ is formed in the dual-phase section where the at. % of Al is between 46 and 50, depending on the heat treatment employed.

The single-phase γ consists of a continuous built-up of smaller equiaxed grains [83], and is formed in the $\gamma + \alpha_2$ phase field resulting in grain coarsening of existing γ grains [96]. This microstructure possesses good properties at high temperatures. This microstructure is very brittle and, therefore, not desired in structural or engineering applications [96,97]. The disadvantages are its poor room-temperature properties such as fracture toughness and ductility [95].

Heating in the two-phase region, followed by cooling results in a DP microstructure consisting of γ and lamella grains [98]. Grain refined DP microstructures have ambient strength and ductility with poor response to creep and fatigue at elevated temperatures [99]. A NL microstructure is formed in the $\alpha_2 + \gamma$ phase region where the α/γ ratio is greater than 1 whereas a FL microstructure is obtained in the pure α-phase field, followed by air-cooling to ambient temperature [96]. Optimum tensile strengths with superior low-cycle fatigue performance are observed in fine-grained NL microstructures, while a coarse microstructure significantly affects the fatigue performance at low-cycles [100]. Improved hardness and toughness is mostly observed in FL structures [101,102].

5. Microstructure-Property Relations

Fully densified γ-TiAl alloys with less residual porosity can be successfully produced using SPS. It has been proved countlessly [27,103–106] that the superior properties of these alloys can also be achieved by optimising SPS parameters. A fully densified γ-TiAl by SPS will have three distinctive microstructures namely, equiaxed γ grains, DP and lamella colonies having alternating layers of α_2 and γ phase [9,103,107]. The lamella microstructure with $(\gamma + \alpha_2)$ phase is the preferred in structural applications [85,108] due to the ease of controlling the distributions and amounts of the α_2 and γ formed [109].

In a recent study, Gu and associates [27] produced a PA TiAl alloy with added yttrium using SPS. The microstructures (in Figure 12) obtained were DP, NL and FL by employing varying sintering temperatures ranging from 100 up to 1250 °C at a sintering pressure of 40 MPa. The compression results (Figure 13) obtained from the study showed that the compressive strength of the samples decreases with increasing sintering temperature. Furthermore, DP microstructure exhibited excellent yield, fracture strength and ultimate strain as compared to the samples with NL and FL microstructures. Wang et al. [103] investigated the effect of stresses and sintering temperature on the microstructural evolution and mechanical properties of a Ti–Al–Cr–Nb alloy produced by SPS. The authors reported that mechanical properties were reliant on the microstructures produced by the varied SPS conditions. An optimal true fracture strength of 1820 MPa and a plastic true strain of 32.6% were obtained for the alloy sintered at 1150 °C, consisting of a double phased and lamellar microstructure.

A NL microstructure has restricted RT ductility. According to a recent study, this effect can be defined by the difficult plastic deformation in γ/α_2 colonies due high-volume percent of α_2 present [106]. $\gamma + \alpha_2$ and DP microstructures offer enhanced yield strength, ductility and ultimate tensile strength at RT [18]. According to Couret and colleagues [18], these are governed by SPS parameter optimisation and addition of grain growth inhibitors such as boron. In the very same study, it was found that the poor strength and limited ductility in lamella microstructures can be attributed to the large grain size and lack of texture inhibiting propagation of dislocations in a preferred mode.

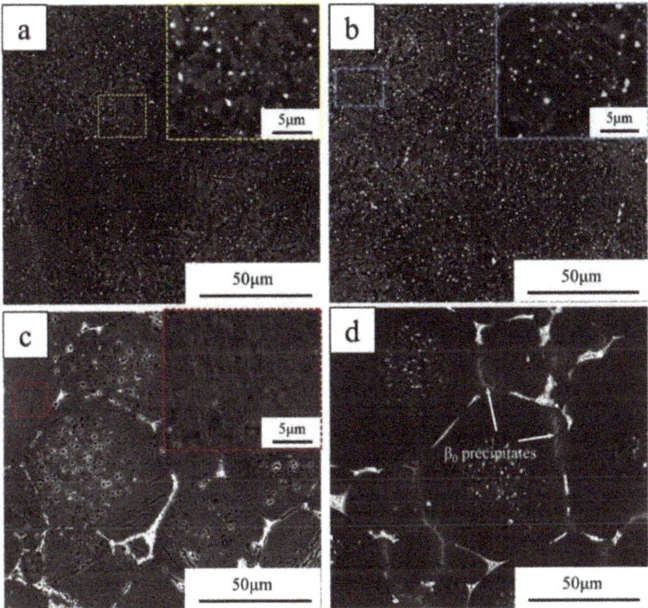

Figure 12. Scanning electron microscope (SEM) images of yttrium added TiAl alloys produced by SPS with (**a**) outer part of sample sintered at 1100 °C; (**b**) center of sample sintered at 1150 °C; (**c**) sample sintered at 1200 °C; and (**d**) 1250 °C (reproduced from [27], with permission from Elsevier, 2020).

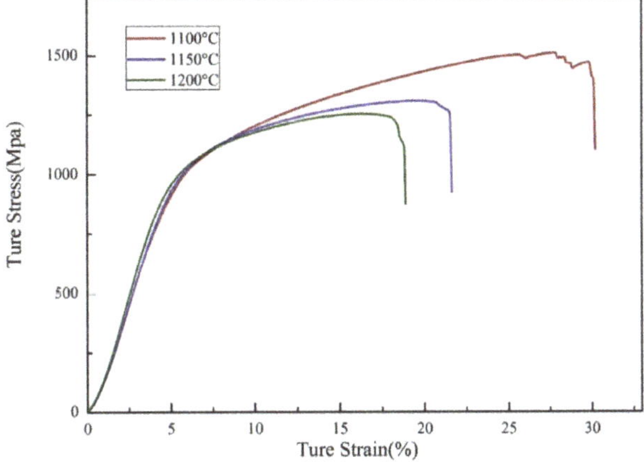

Figure 13. Stress-strain plots at room temperature (RT) of yttrium added TiAl alloys produced by SPS, reproduced from [27], with permission from Elsevier, 2020.

Voisin et al. [106], consolidated a TNM (γ-TiAl with Nb and Mo additions) alloy with chemical composition Ti–43.9Al–4Nb–0.95Mo-0.1B using SPS by varying sintering temperature of 1237–1429 °C. In addition, experiments on RT tensile deformation and creep at 700 °C employing stress of 300 MPa were conducted. A NL microstructure (in Figure 14a) was attained at sintering temperatures ranging from 1262 to 1336 °C. The highest ductility obtained from the results exhibited a fracture elongation of 0.86–0.9% for the specimen sintered at 1336 °C. TEM fractography also showed that the NL microstructure deformed using ordinary dislocation pileups and twinning (Figure 14c). Excellent creep resistance (Figure 14b) was obtained with lifetimes >2000 h at creep rates of $<10^{-8}$ s^{-1}, with the highest attained at a lifetime of ~4000 h for the alloy is that sintered at 1304 °C.

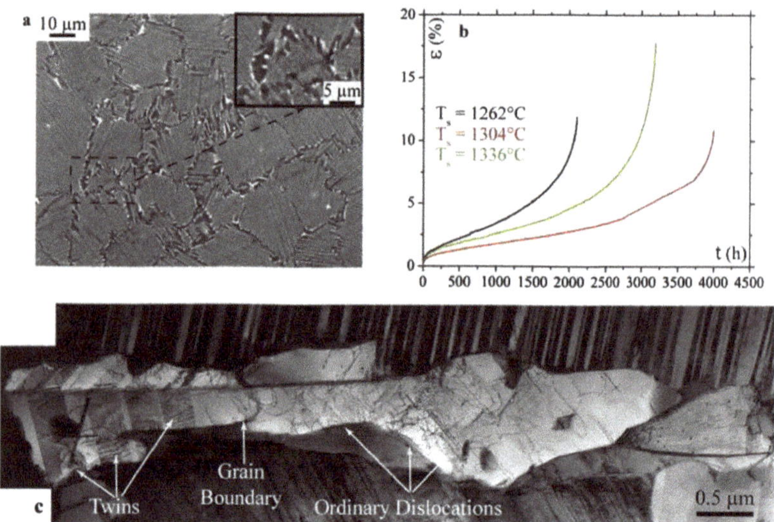

Figure 14. (**a**) Near lamella (NL) microstructure, (**b**) creep curves at 700 °C using stress of 300 MPa, (**c**) RT tensile deformation microstructure for SPS consolidated TNM alloy, reproduced from [106], with permission from Elsevier, 2014.

Another property of interest for TiAl alloys is oxidation. Resistance to oxidation, commonly known to as the resistance to corrosion at high temperatures, of γ-TiAl is said to be generally good at temperatures below 850 °C [110]. The kinetics of oxidation includes the formation of a non-protective mixed oxide scale of alumina (Al_2O_3) and titania (TiO_2) on the surface of bare γ-TiAl alloys in a favourable condition and environment [111]. Simultaneous formation of Al_2O_3 and TiO_2 occurs during the initial state of oxidation and thrive in the optimal growth direction [112]. Furthermore, TiO_2 grows at a higher velocity compared to Al_2O_3 attributable to the low growth activation energy of TiO_2 as compared to Al_2O_3. Additionally, the subsequent series of reactions are described as rapid and favour the formation of TiO_2. However, below the TiO_2 layer, an aluminium (Al)-rich layer exists because of Ti depletion. The oxide scale in TiAl consists of multi-layers of both TiO_2 and Al_2O_3. A decrease in the oxidation resistance of γ-TiAl as mentioned by Dai et al. [112] can be as a result of the thickening of the multi-layered oxide film due to mass transportation, weakening the bond existing between the oxide layer and the substrate.

A FL microstructure seems to be the most preferred as far as oxidation resistance is concerned. However, the average lamella grain size and compositional variations are to be considered. A lamella grain size of 25–20 μm with Nb additions was reported to suppress diffusion activity in TiAl alloys [113,114]. The kinetics of oxidation involves the simultaneous formation of a non-protective mixed oxide scale of alumina (Al_2O_3) and titania (TiO_2) on the surface of bare γ-TiAl alloys in a favourable condition and environment [111,112].

Several recent studies [113,115,116] aimed at improving the oxidation resistance of SPS produced TiAl alloys. Cobbinah and colleagues [115] investigated the effect of Ta additions (0.8, 4, and 8 at. %) on the isothermal oxidation resistance of a Ti–46.5Al (at. %) alloy produced by SPS. The results indicate that the superior resistance of the TiAl alloys is related to the formation of a non-porous and interconnected layer of Al_2O_3 at the interface of the metal-oxide. In works by Lu et al. [113], isothermal resistance at 1000 °C of SPS produced alloys of nominal composition Ti–45Al–8.5Nb–0.2B–0.2W–0.1Y and Ti–47.5Al–2.0V–1.0Cr were studied. A mixed oxide scale consisting of Al_2O_3/TiO_2 was formed next to the spalled TiO_2 layer for the Ti–47.5Al–2.5V–1.0Cr. Furthermore, the mixed scale produced proved to have no protective effect, resulting in high rates of oxidation (with mass gains of 51.06 mg/cm^2) as compared to alloy Ti–45Al–8.5Nb–0.2B–0.2W–0.1Y (with only 2.27 mg/cm^2 mass gain) tested under the same conditions. Relatively, the inner oxide scale of alloy Ti–45Al–8.5Nb–0.2B–0.2W–0.1Y exhibited an outer TiO_2-rich layer with traces of Al_2O_3 and TiN. The addition of 8.5% Nb promoted the formation of a Nb-rich diffusion layer between the substrate and the oxide layer, further enhancing the oxidation resistance of the alloy.

6. Alloy Development

During the initial advances in two-phase binary alloys, it was acknowledged that these alloys cannot be utilised owing to their poor oxidation and creep resistance [43]. Consequently, this resulted in a vast number of investigations aimed at understanding the effect of alloy additions in binary TiAl on the microstructure-property relations. To date, four generations [97,117] of TiAl alloys have been developed.

6.1. 1st and 2nd Generation of TiAl alloys

In this generation of alloys, elements such as Cr, V, Mn were added to Ti–(42–48)Al (at. %) to produce ternary alloys. These alloys were further processed using heat and thermomechanical treatments to improve ductility measures. In a previous study of interest [118], it was reported that additions of up to 4 at. % of Cr to binary Ti–(44–54)Al alloys consisting of DP microstructures led to partial ductilization due to the occupancy of Cr in Al lattice sites. Cr additions also modified the Al partitioning and the thermal stabilisation of transformed α_2 laths comparable to findings made when 0.4 at. % of V was added [119].

The 2nd generation of TiAl is based on the following composition [43]:

$$Ti\text{-}(45\text{-}48)Al\text{-}(1\text{-}3)X\text{-}(2\text{-}5)Y\text{-}(<1)Z \tag{1}$$

where X = Cr, Mn, Y = Nb, Ta, W, Mo; Z = Si, B, C.

The additions of the elements, as mentioned above, shift the position of the phase boundaries in the Ti–Al binary phase diagram [43,96]. Additions of element X improve the mechanical properties of TiAl alloys by increasing flow stress [120], reducing the stacking fault energy and thus enhancing the susceptibility of twinning [96]. Zhu et al. [121] studied the effect of Cr additions on the microstructure and nanohardness of Ti–48Al binary alloy produced using a single roller melt spinning and arc melting processing routes. With 2 at. % Cr, the microstructure of the alloy ribbons was lamella with equiaxed α_2 and small B2 phase particles. Additions of up to 4 at. % Cr increases the B2 phase grain sizes and the lamella structure disappears. The nanohardness of the alloys increased also with additions of up to 4 at. % of Cr. Additions of V and Mn have been reported [122] to increase the fracture and yield stresses much related to grain refinement and solid solution strengthening (particularly for V additions).

Adding elements like Ta, Mo and W (Y additions) improve the oxidation and creep properties at elevated temperatures. Recently [115], additions of 4 and 8 at. % of Ta to an SPS produced Ti-46.5Al alloy promoted the formation of an Al_2O_3 layer severing as a diffusion barrier on the metal-oxide interface, hence, resulting in outstanding resistance to oxidation. For further reading on the effect Ta on the oxidation performance of TiAl, the reader is referred to [123–126]. Remarkable creep properties were obtained in some research work [127–129] when Ta and W were added to TiAl alloys. Z additions such as B promote grain refinement utilising retarding grain growth in the α-phase field [96]. Moreover, the strength and resistance to creep is improved through the formation of Ti_3AlC perovskite precipitates when C is added. Numerous work has been conducted to date [130–134] on the effect B and C have on the microstructure and mechanical properties of TiAl, attesting to the above alloy addition benefits.

6.2. 3rd and 4th Generation of TiAl Alloys

This generation of alloys contains high Mo and or Nb contents to promote precipitation hardening and were developed for applications at elevated temperatures. The third generation of TiAl alloys follow the constitution below [43]:

$$\text{Ti-(42-48)Al-(0-10)X-(0-3)Y-(0-1)Z-(0-0.5)RE} \tag{2}$$

where X = Ta, Nb, Mn, Cr; Y = Zr, Hf, W, Mo; Z = C, B, Si and RE designates rare earth metals.

The additions of Zr is a known β-stabilizer, promoting occupancy of the Ti sites in the lattice of Ti–Al systems [135]. Furthermore, adding Zr to binary TiAl alloys improves compression strength. However, when added in the presence of Cr, the strength decreases with increasing elongation. Finally, phase evolutions occurring as a result of the existence of both Cr and Zr in Ti–43Al–4Nb–1Mo–0.1B (TNM-B1) alloy include the formation of ω, reduction of α_2 and increments in the amount of β phases. Y rare earth metal additions promote grain refinement, thus improving ductility [136,137], elevated temperature deformability [138] and heat resistance [139]. The oxidation rate of TNM alloys can be reduced by ~0.1 at. % RE metal additions such as La or Er [140]. Conversely, additions of ~0.2 at. % of aforementioned RE metals can impair the resistance to oxidation through the formation of hillocks in the oxide scale.

The fourth generation of TiAl, commonly referred to as TNM alloys, exhibit excellent oxidation resistance through Nb and or Ta additions and outstanding creep properties promoted by Mo additions. In addition, these alloys are said to have excellent workability making them universally applicable for various processing routes other than casting and hot forming such as additive manufacturing and SPS [141]. Some of the alloys developed to date include the β-solidifying Ti–43.5Al–4Nb–1Mo–0.1B [141], Ti–46Al–8Nb and Ti–46Al–8Ta [142]. Wimler et al. [141] studied the capabilities of SPS coupled with subsequent heat treatments to produce a Ti–43.5Al–4Nb–1Mo–0.1B alloy with strength and creep performance at 750 °C comparable to those produced by conventional processing techniques. After SPS consolidation of the gas atomised powder alloy at 1300 °C, the microstructure produced consisted of a refined and homogeneous NL γ and β_0. Subsequent two-step heat treatments resulted in the formation of NL γ, FL and NL β_0 structures (in Figure 15) decreasing the lamella spacing from 83 ± 7 in the as-sintered condition to 10 ± 3 nm. It was concluded that the strength and creep properties at 750 °C (in Figure 16) obtained can keep up with that of the conventional routes of processing.

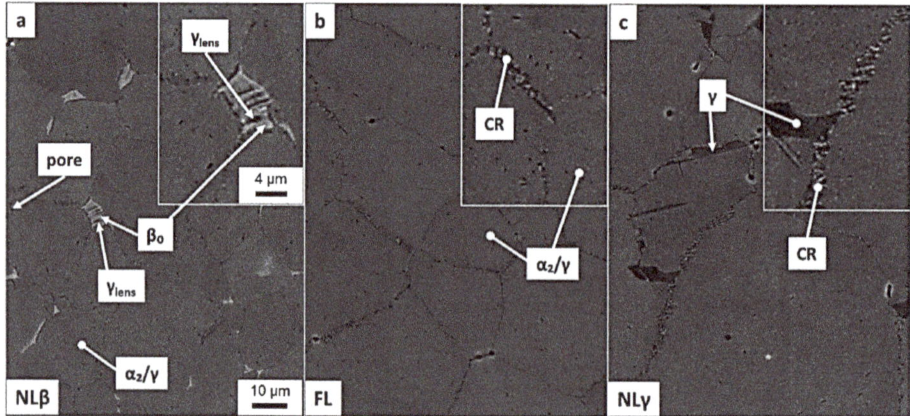

Figure 15. SEM images showing (**a**) NL beta (β); (**b**) fully lamella (FL); and (**c**) NL gamma (γ)microstructures after two-step heat treatment [141].

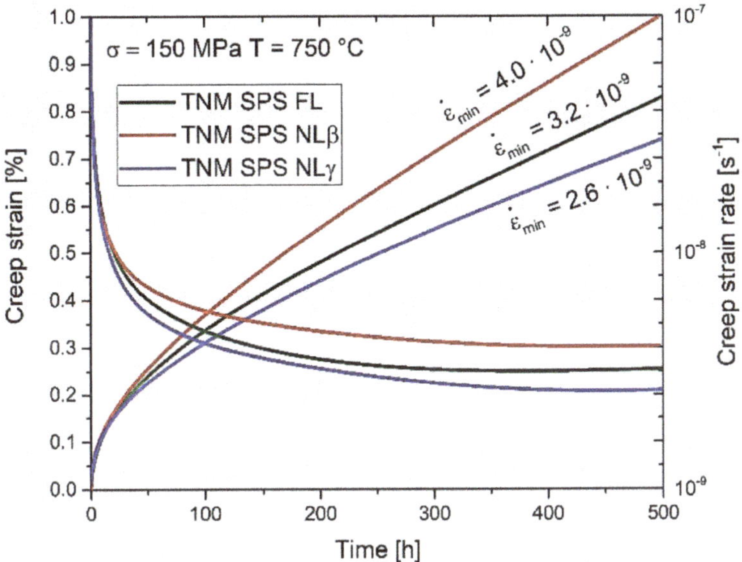

Figure 16. Plots of creep stress and strain versus time of the heat-treated microstructures tested at 750 °C and 150 MPa [141].

7. Perspectives and Conclusions

The realisation of the process has been highlighted by remarkable accomplishments in the production of a refined microstructure of γ-TiAl since it offers prompt rates of both heating and cooling. Currently, research and development have shifted attention to TiAl powder forming techniques such as DMM, GA, and, amongst others, pre-alloying. MA and SPS can be coupled in the production of nanocrystalline intermetallic TiAl owing to the advantages of obtaining non-equilibrium and amorphous phases, ultrafine grain sizes and, at the same time, retaining the grain size during consolidation. As an added advantage, SPS uses shorter holding times to obtain fully densified material as compared to conventional processes such as HIP. Consolidation methods aimed at balancing microstructure and properties have also been explored.

However, SPS shows an impressive balance of microstructure and mechanical properties.

A significant challenge in as far as TiAl alloys are concerned is commercialisation and costs. Although limited applications have been achieved, the reality is that implementation of TiAl alloys in both the automobile and aerospace industries is an on-going challenge. Although TiAl were designed as candidates to replace the Ni-based superalloys, the gap existing between the two is quite recognisable from both the properties and production costs perspective. Much research has been dedicated to improving the ductility at RT for TiAl alloys through compositional variations, opting for manufacturing technologies and thermomechanical treatments. These attempts aimed at refining and altering the microstructures and morphologies of the produced alloys. Advanced PM techniques present the opportunity to manufacture near-net-shape components with improved mechanical properties. However, challenges exist concerning methods that can be employed to characterise PM produced TiAl alloys effectively. Although the future seems exciting for TiAl alloys produced using PM, implementation on an industrial scale seems to lag far behind.

Author Contributions: Writing—original draft preparation, N.F.M.; writing—review and editing, W.R.M. All authors have read and agreed to the published version of the manuscript.

Funding: This research was funded by National Research Foundation of South Africa, grant number 116368.

Conflicts of Interest: The authors declare no conflict of interest.

References

1. Taub, A.I.; Fleischer, R.L. Intermetallic compounds for high-temperature structural use. *Science* **1989**, *243*, 616–621. [CrossRef] [PubMed]
2. Wolff, I.M.; Hill, P.J. Platinum metals-based intermetallics for high-temperature service. *Platin. Met. Rev.* **2000**, *44*, 158–166.
3. Muktinutalapati, N.R. Materials for gas turbines–an overview. In *Advances in Gas Turbine Technology*; IntechOpen: London, UK, 2011; Available online: https://www.intechopen.com/books/advances-in-gas-turbine-technology/materials-for-gas-turbines-an-overview (accessed on 1 May 2020).
4. Leyens, C.; Peters, M. *Titanium and Titanium Alloys: Fundamentals and Applications*; John Wiley & Sons: Hoboken, NJ, USA, 2003.
5. Clemens, H.; Smarsly, W. Light-weight intermetallic titanium aluminides–status of research and development. *Adv. Mater. Res.* **2011**, *278*, 551–556. [CrossRef]
6. Voice, W. The future use of gamma titanium aluminides by Rolls-Royce. *Aircr. Eng. Aerosp. Technol.* **1999**, *71*, 337–340. [CrossRef]
7. Bewlay, B.P.; Weimer, M.; Kelly, T.; Suzuki, A.; Subramanian, P.R. The science, technology, and implementation of TiAl alloys in commercial aircraft engines. *MRS Online Proc. Libr. Arch.* **2013**, *1516*, 49–58. [CrossRef]
8. Loria, E.A. Quo vadis gamma titanium aluminide. *Intermetallics* **2001**, *9*, 997–1001. [CrossRef]
9. Liu, B.; Liu, Y. 27-Powder metallurgy titanium aluminide alloys. In *Titanium Powder Metallurgy*; Elsevier Inc.: Amsterdam, The Netherlands, 2015; pp. 515–531.
10. Lamirand, M.; Bonnentien, J.L.; Ferrière, G.; Guérin, S.; Chevalier, J.P. Relative effects of chromium and niobium on microstructure and mechanical properties as a function of oxygen content in TiAl alloys. *Scr. Mater.* **2007**, *56*, 325–328. [CrossRef]
11. Wang, Q.; Ding, H.; Zhang, H.; Chen, R.; Guo, J.; Fu, H. Influence of Mn addition on the microstructure and mechanical properties of a directionally solidified γ-TiAl alloy. *Mater. Charact.* **2018**, *137*, 133–141. [CrossRef]
12. Imayev, V.; Oleneva, T.; Imayev, R.; Christ, H.J.; Fecht, H.J. Microstructure and mechanical properties of low and heavy alloyed γ-TiAl + α2-Ti3Al based alloys subjected to different treatments. *Intermetallics* **2012**, *26*, 91–97. [CrossRef]
13. Larsen, J.M.; Worth, B.D.; Balsone, S.J.; Jones, J.W. *An Overview of the Structural Capability of Available Gamma Titanium Aluminide Alloys*; U.S. Department of EnergyOffice of Scientific and Technical Informatio: Oak Ridge, TN, USA, 1995.
14. Tang, J.; Huang, B.; Liu, W.; He, Y.; Zhou, K.; Wu, A.; Peng, K.; Qin, W.; Du, Y. A high ductility TiAl alloy made by two-step heat treatment. *Mater. Res. Bull.* **2003**, *38*, 2019–2024. [CrossRef]
15. Lasalmonie, A. Intermetallics: Why is it so difficult to introduce them in gas turbine engines? *Intermetallics* **2006**, *14*, 1123–1129. [CrossRef]
16. Lagos, M.A.; Agote, I. SPS synthesis and consolidation of TiAl alloys from elemental powders: Microstructure evolution. *Intermetallics* **2013**, *36*, 51–56. [CrossRef]
17. Munir, Z.A.; Anselmi-Tamburini, U.; Ohyanagi, M. The effect of electric field and pressure on the synthesis and consolidation of materials: A review of the spark plasma sintering method. *J. Mater. Sci.* **2006**, *41*, 763–777. [CrossRef]
18. Couret, A.; Molénat, G.; Galy, J.; Thomas, M. Microstructures and mechanical properties of TiAl alloys consolidated by spark plasma sintering. *Intermetallics* **2008**, *16*, 1134–1141. [CrossRef]
19. Jaramillo, D.; Cuenca, R.; Juárez, F. Sintering comparison of NiCoCrAl-Ta powder processed by hot pressing and spark plasma. *Powder Technol.* **2012**, *221*, 264–270.
20. Chang, I.; Zhao, Y. *Advances in Powder Metallurgy: Properties, Processing and Applications*; Elsevier: Amsterdam, The Netherlands, 2013.
21. Martín, A.; Cepeda-Jiménez, C.M.; Pérez-Prado, M.T. Gas atomization of γ-TiAl Alloy Powder for Additive Manufacturing. *Adv. Eng. Mater.* **2020**, *22*, 1900594. [CrossRef]
22. Neikov, O.D. Chapter 4-Atomization and Granulation. In *Handbook of Non-Ferrous Metal Powders: Technologies and Applications*; Neikov, O.D., Naboychenko, S.S., Yefimov, N.A., Second, E., Eds.; Elsevier: Oxford, UK, 2019; pp. 125–185.
23. Gerling, R.; Clemens, H.; Schimansky, F.P. Powder Metallurgical Processing of Intermetallic Gamma Titanium Aluminides. *Adv. Eng. Mater.* **2004**, *6*, 23–38. [CrossRef]

24. Sun, P.; Fang, Z.Z.; Zhang, Y.; Xia, Y. Review of the methods for production of spherical Ti and Ti alloy powder. *JOM* **2017**, *69*, 1853–1860. [CrossRef]

25. Antony, L.V.M.; Reddy, R.G. Processes for production of high-purity metal powders. *JOM* **2003**, *55*, 14–18. [CrossRef]

26. Franz, H.; Plochl, L.; Schimansky, F.P. Recent advances of titanium alloy powder production by ceramic-free inert gas atomization. In Proceedings of the Titanium 2008, Las Vegas, NV, USA, 21–24 September 2008.

27. Gu, X.; Cao, F.; Liu, N.; Zhang, G.; Yang, D.; Shen, H.; Zhang, D.; Song, H.; Sun, J. Microstructural evolution and mechanical properties of a high yttrium containing TiAl based alloy densified by spark plasma sintering. *J. Alloy. Compd.* **2020**, *819*, 153264. [CrossRef]

28. Liu, B.; Wang, M.; Du, Y.; Li, J. Size-Dependent Structural Properties of a High-Nb TiAl Alloy Powder. *Materials* **2020**, *13*, 161. [CrossRef] [PubMed]

29. Xia, Y.; Zhao, J.L.; Qian, M. Spark Plasma Sintering of Ti-48Al-2Cr-2Nb Alloy Powder and Characterization of an Unexpected Phase. *JOM* **2019**, *71*, 2556–2563. [CrossRef]

30. Wang, J.; Wang, Y.; Liu, Y.; Li, J.; He, L.; Zhang, C. Densification and microstructural evolution of a high niobium containing TiAl alloy consolidated by spark plasma sintering. *Intermetallics* **2015**, *64*, 70–77. [CrossRef]

31. Xiao, S.L.; Jing, T.; Xu, L.J.; Chen, Y.Y.; Yu, H.B.; Han, J.C. Microstructures and mechanical properties of TiAl alloy prepared by spark plasma sintering. *Trans. Nonferrous Met. Soc. China* **2009**, *19*, 1423–1427. [CrossRef]

32. Hadef, F. Synthesis and disordering of B2 TM-Al (TM=Fe, Ni, Co) intermetallic alloys by high energy ball milling: A review. *Powder Technol.* **2017**, *311*, 556–578. [CrossRef]

33. Aliofkhazraei, M. *Handbook of Mechanical Nanostructuring*; John Wiley & Sons: Hoboken, NJ, USA, 2015.

34. Suryanarayana, C.; Ivanov, E.; Boldyrev, V.V. The science and technology of mechanical alloying. *Mater. Sci. Eng. A* **2001**, *304*, 151–158. [CrossRef]

35. Yadav, T.P.; Yadav, R.M.; Singh, D.P. Mechanical milling: A top down approach for the synthesis of nanomaterials and nanocomposites. *Nanosci. Nanotechnol.* **2012**, *2*, 22–48. [CrossRef]

36. Oehring, M.; Klassen, T.; Bormann, R. The formation of metastable Ti–Al solid solutions by mechanical alloying and ball milling. *J. Mater. Res.* **1993**, *8*, 2819–2829. [CrossRef]

37. Zhang, F.; Lu, L.; Lai, M.O. Study of thermal stability of mechanically alloyed Ti–75% Al powders. *J. Alloy. Compd.* **2000**, *297*, 211–218. [CrossRef]

38. Wang, Y.; Zhang, C.; Liu, Y.; Zhao, S.; Li, J. Microstructure characterization and mechanical properties of TiAl-based alloys prepared by mechanical milling and spark plasma sintering. *Mater. Charact.* **2017**, *128*, 75–84. [CrossRef]

39. Shulong, X.; Lijuan, X.; Hongbao, Y.; Yuyong, C. Effect of Y Addition on Microstructure and Mechanical Properties of TiAl-based Alloys Prepared by SPS. *Rare Met. Mater. Eng.* **2013**, *42*, 23–27. [CrossRef]

40. Xiao, S.; Xu, L.; Chen, Y.; Yu, H. Microstructure and mechanical properties of TiAl-based alloy prepared by double mechanical milling and spark plasma sintering. *Trans. Nonferrous Met. Soc. China* **2012**, *22*, 1086–1091. [CrossRef]

41. Suryanarayana, C. Mechanical Alloying: A Novel Technique to Synthesize Advanced Materials. *Research* **2019**, *2019*, 4219812. [CrossRef] [PubMed]

42. Suryanarayana, C. Mechanical alloying and milling. *Prog. Mater. Sci.* **2001**, *46*, 1–184. [CrossRef]

43. Clemens, H.; Mayer, S. Design, Processing, Microstructure, Properties, and Applications of Advanced Intermetallic TiAl Alloys. *Adv. Eng. Mater.* **2013**, *15*, 191–215. [CrossRef]

44. Lu, L.; Lai, M.O.; Froes, F.H. The mechanical alloying of titanium aluminides. *JOM* **2002**, *54*, 62–64. [CrossRef]

45. Forouzanmehr, N.; Karimzadeh, F.; Enayati, M.H. Study on solid-state reactions of nanocrystalline TiAl synthesized by mechanical alloying. *J. Alloy. Compd.* **2009**, *471*, 93–97. [CrossRef]

46. Sim, K.H.; Wang, G.; Son, R.C.; Choe, S.L. Influence of mechanical alloying on the microstructure and mechanical properties of powder metallurgy Ti2AlNb-based alloy. *Powder Technol.* **2017**, *317*, 133–141. [CrossRef]

47. Bohn, R.; Klassen, T.; Bormann, R. Mechanical behavior of submicron-grained γ-TiAl-based alloys at elevated temperatures. *Intermetallics* **2001**, *9*, 559–569. [CrossRef]

48. Bhattacharya, P.; Bellon, P.; Averback, R.S.; Hales, S.J. Nanocrystalline TiAl powders synthesized by high-energy ball milling: Effects of milling parameters on yield and contamination. *J. Alloy. Compd.* **2004**, *368*, 187–196. [CrossRef]

49. Lavernia, E.J.; Han, B.Q.; Schoenung, J.M. Cryomilled nanostructured materials: Processing and properties. *Mater. Sci. Eng.* **2008**, *493*, 207–214. [CrossRef]

50. Witkin, D.B.; Lavernia, E.J. Synthesis and mechanical behavior of nanostructured materials via cryomilling. *Prog. Mater. Sci.* **2006**, *51*, 1–60. [CrossRef]

51. Deng, H.; Chen, A.; Chen, L.; Wei, Y.; Xia, Z.; Tang, J. Bulk nanostructured Ti-45Al-8Nb alloy fabricated by cryomilling and spark plasma sintering. *J. Alloy. Compd.* **2019**, *772*, 140–149. [CrossRef]

52. Shanmugasundaram, T.; Guyon, J.; Monchoux, J.P.; Hazotte, A.; Bouzy, E. On grain refinement of a γ-TiAl alloy using cryo-milling followed by spark plasma sintering. *Intermetallics* **2015**, *66*, 141–148. [CrossRef]

53. Sun, Y.; Kulkarni, K.; Sachdev, A.K.; Lavernia, E.J. Synthesis of γ-TiAl by reactive spark plasma sintering of cryomilled Ti and Al powder blend, part I: Influence of processing and microstructural evolution. *Met. Mater. Trans. A* **2014**, *45*, 2750–2758. [CrossRef]

54. Sun, Y.; Kulkarni, K.; Sachdev, A.K.; Lavernia, E.J. Synthesis of γ-TiAl by reactive spark plasma sintering of cryomilled Ti and Al powder blend: Part II: Effects of electric field and microstructure on sintering kinetics. *Met. Mater. Trans. A* **2014**, *45*, 2759–2767. [CrossRef]

55. Guillon, O.; Gonzalez-Julian, J.; Dargatz, B.; Kessel, T.; Schierning, G.; Räthel, J.; Herrmann, M. Field-assisted sintering technology/spark plasma sintering: Mechanisms, materials, and technology developments. *Adv. Eng. Mater.* **2014**, *16*, 830–849. [CrossRef]

56. Guyon, J.; Hazotte, A.; Monchoux, J.P.; Bouzy, E. Effect of powder state on spark plasma sintering of TiAl alloys. *Intermetallics* **2013**, *34*, 94–100. [CrossRef]

57. Chen, F.; Yang, S.; Wu, J.; Galaviz Perez, J.A.; Shen, Q.; Schoenung, J.M.; Lavernia, E.J.; Zhang, L. Spark plasma sintering and densification mechanisms of conductive ceramics under coupled thermal/electric fields. *J. Am. Ceram. Soc.* **2015**, *98*, 732–740. [CrossRef]

58. Matizamhuka, W.R. Fabrication of Fine-Grained Functional Ceramics by Two-Step Sintering or Spark Plasma Sintering (SPS). In *Sintering [Working Title]*; IntechOpen: London, UK, 2019.

59. Matizamhuka, W.R. Spark plasma sintering (SPS)-an advanced sintering technique for structural nanocomposite materials. *J. South. African Inst. Min. Met.* **2016**, *116*, 1171–1180. [CrossRef]

60. Suárez, M.; Fernández, A.; Menéndez, J.L.; Torrecillas, R.; Kessel, H.U.; Hennicke, J.; Kirchner, R.; Kessel, T. Challenges and opportunities for spark plasma sintering: A key technology for a new generation of materials. *Sinter. Appl.* **2013**, *13*, 319–342.

61. Marder, R.; Estournès, C.; Chevallier, G.; Chaim, R. Spark and plasma in spark plasma sintering of rigid ceramic nanoparticles: A model system of YAG. *J. Eur. Ceram. Soc.* **2015**, *35*, 211–218. [CrossRef]

62. Marder, R.; Estournès, C.; Chevallier, G.; Chaim, R. Plasma in spark plasma sintering of ceramic particle compacts. *Scr. Mater.* **2014**, *82*, 57–60. [CrossRef]

63. Saunders, T.; Grasso, S.; Reece, M.J. Plasma formation during electric discharge (50V) through conductive powder compacts. *J. Eur. Ceram. Soc.* **2015**, *35*, 871–877. [CrossRef]

64. Rahaman, M.N. *Ceramic Processing and Sintering*; CRC press: Boca Raton, FL, USA, 2003.

65. Olevsky, E.A.; Kandukuri, S.; Froyen, L. Consolidation enhancement in spark-plasma sintering: Impact of high heating rates. *J. Appl. Phys.* **2007**, *102*, 114913. [CrossRef]

66. Manière, C.; Torresani, E.; Olevsky, E.A. Simultaneous spark plasma sintering of multiple complex shapes. *Materials* **2019**, *12*, 557. [CrossRef]

67. Manière, C.; Durand, L.; Weibel, A.; Estournès, C. Spark-plasma-sintering and finite element method: From the identification of the sintering parameters of a submicronic α-alumina powder to the development of complex shapes. *Acta Mater.* **2016**, *102*, 169–175. [CrossRef]

68. Voisin, T.; Monchoux, J.P.; Durand, L.; Karnatak, N.; Thomas, M.; Couret, A. An Innovative Way to Produce γ-TiAl Blades: Spark Plasma Sintering. *Adv. Eng. Mater.* **2015**, *17*, 1408–1413. [CrossRef]

69. Manière, C.; Durand, L.; Weibel, A.; Chevallier, G.; Estournès, C. A sacrificial material approach for spark plasma sintering of complex shapes. *Scr. Mater.* **2016**, *124*, 126–128. [CrossRef]

70. Manière, C.; Nigito, E.; Durand, L.; Weibel, A.; Beynet, Y.; Estournès, C. Spark plasma sintering and complex shapes: The deformed interfaces approach. *Powder Technol.* **2017**, *320*, 340–345. [CrossRef]

71. Olevsky, E.A.; Garcia-Cardona, C.; Bradbury, W.L.; Haines, C.D.; Martin, D.G.; Kapoor, D. Fundamental Aspects of Spark Plasma Sintering: II. Finite Element Analysis of Scalability. *J. Am. Ceram. Soc.* **2012**, *95*, 2414–2422. [CrossRef]

72. Voisin, T.; Durand, L.; Karnatak, N.; Le Gallet, S.; Thomas, M.; Le Berre, Y.; Castagné, J.F.; Couret, A. Temperature control during Spark Plasma Sintering and application to up-scaling and complex shaping. *J. Mater. Process. Technol.* **2013**, *213*, 269–278. [CrossRef]

73. Franceschin, G.; Flores-Martínez, N.; Vázquez-Victorio, G.; Ammar, S.; Valenzuela, R. Sintering and reactive sintering by spark plasma sintering (SPS). In *Sintering of Functional Materials*; IntechOpen: London, UK, 2017; Available online: https://www.intechopen.com/books/sintering-of-functional-materials/sintering-and-reactive-sintering-by-spark-plasma-sintering-sps- (accessed on 1 May 2020).

74. Isobe, T.; Daimon, K.; Sato, T.; Matsubara, T.; Hikichi, Y.; Ota, T. Spark plasma sintering technique for reaction sintering of Al2O3/Ni nanocomposite and its mechanical properties. *Ceram. Int.* **2008**, *34*, 213–217. [CrossRef]

75. Waseem, O.A.; Lee, J.; Lee, H.M.; Ryu, H.J. The effect of Ti on the sintering and mechanical properties of refractory high-entropy alloy TixWTaVCr fabricated via spark plasma sintering for fusion plasma-facing materials. *Mater. Chem. Phys.* **2018**, *210*, 87–94. [CrossRef]

76. Meir, S.; Kalabukhov, S.; Froumin, N.; Dariel, M.P.; Frage, N. Synthesis and densification of transparent magnesium aluminate spinel by SPS processing. *J. Am. Ceram. Soc.* **2009**, *92*, 358–364. [CrossRef]

77. Martins, D.; Grumbach, F.; Simoulin, A.; Sallot, P.; Mocellin, K.; Bellet, M.; Estournès, C. Spark plasma sintering of a commercial TiAl 48-2-2 powder: Densification and creep analysis. *Mater. Sci. Eng. A* **2018**, *711*, 313–316. [CrossRef]

78. Anselmi-Tamburini, U.; Garay, J.E.; Munir, Z.A. Fast low-temperature consolidation of bulk nanometric ceramic materials. *Scr. Mater.* **2006**, *54*, 823–828. [CrossRef]

79. Lenel, F. Resistance sintering under pressure. *JOM* **1955**, *7*, 158–167. [CrossRef]

80. Weissler, G.A. Resistance sintering with alumina dies. *Int. J. Powder Met. Powder Technol.* **1981**, *17*, 107–118.

81. Herrmann, M.; Schulz, I.; Bales, A.; Sempf, K.; Hoehn, S. "Snow flake" structures in silicon nitride ceramics–Reasons for large scale optical inhomogeneities. *J. Eur. Ceram. Soc.* **2008**, *28*, 1049–1056. [CrossRef]

82. Langer, J.; Hoffmann, M.J.; Guillon, O. Electric Field-Assisted Sintering in Comparison with the Hot Pressing of Yttria-Stabilized Zirconia. *J. Am. Ceram. Soc.* **2011**, *94*, 24–31. [CrossRef]

83. Westbrook, J.H.; Fleischer, R.L. *Crystal Structures of Intermetallic Compounds*; John Wiley & Sons: Hoboken, NJ, USA, 2000.

84. Paufler, P.J.H. *Westbrook, R.L. Intermetallic Compounds Principles and Practice*; John Wiley & Sons: Hoboken, NJ, USA, 1995.

85. Cobbinah, P.V.; Matizamhuka, W.R. Solid-State Processing Route, Mechanical Behaviour, and Oxidation Resistance of TiAl Alloys. *Adv. Mater. Sci. Eng.* **2019**, *2019*, 1–21. [CrossRef]

86. Froes, F.H.; Suryanarayana, C.; Eliezer, D. Synthesis, properties and applications of titanium aluminides. *J. Mater. Sci.* **1992**, *27*, 5113–5140. [CrossRef]

87. McCullough, C.; Valencia, J.J.; Levi, C.G.; Mehrabian, R. Phase equilibria and solidification in Ti-Al alloys. *Acta Met.* **1989**, *37*, 1321–1336. [CrossRef]

88. Jones, S.A.; Shull, R.D.; McAlister, A.J.; Kaufman, M.J. Microstructural studies of Ti-Al alloys in the vicinity of the "eutectoid" reaction ($\alpha \rightarrow \alpha 2\,\gamma$). *Scr. Met.* **1988**, *22*, 1235–1240. [CrossRef]

89. JH, P. Phase Reactions and Processing in the Ti-Al Based Intermetallics. *ISIJ Int.* **1991**, *31*, 1080–1087.

90. Djanarthany, S.; Viala, J.C.; Bouix, J. An overview of monolithic titanium aluminides based on Ti3Al and TiAl. *Mater. Chem. Phys.* **2001**, *72*, 301–319. [CrossRef]

91. Sastry, S.M.L.; Lipsitt, H.A. Ordering transformations and mechanical properties of Ti 3 Ai and Ti 3 Al-Nb alloys. *Met. Trans. A* **1977**, *8*, 1543–1552. [CrossRef]

92. Sarkar, S.; Datta, S.; Das, S.; Basu, D. Oxidation protection of gamma-titanium aluminide using glass–ceramic coatings. *Surf. Coat. Technol.* **2009**, *203*, 1797–1805. [CrossRef]

93. Wang, Z.; Sun, H.; Du, Y.; Yuan, J. Effects of Powder Preparation and Sintering Temperature on Properties of Spark Plasma Sintered Ti-48Al-2Cr-8Nb Alloy. *Metals* **2019**, *9*, 861. [CrossRef]

94. Werwer, M.; Kabir, R.; Cornec, A.; Schwalbe, K.H. Fracture in lamellar TiAl simulated with the cohesive model. *Eng. Fract. Mech.* **2007**, *74*, 2615–2638. [CrossRef]

95. Stoloff, N.S.; Sikka, V.K. *Physical Metallurgy and Processing of Intermetallic Compounds*; Springer Science & Business Media: Berlin/Heidelberg, Germany, 2012.

96. Kothari, K.; Radhakrishnan, R.; Wereley, N.M. Advances in gamma titanium aluminides and their manufacturing techniques. *Prog. Aerosp. Sci.* **2012**, *55*, 1–16. [CrossRef]

97. Kim, Y.W.; Dimiduk, D.M. Progress in the understanding of gamma titanium aluminides. *JOM* **1991**, *43*, 40–47. [CrossRef]
98. Wu, X. Review of alloy and process development of TiAl alloys. *Eur. Congr. Adv. Mater. Process.* **2006**, *14*, 1114–1122. [CrossRef]
99. Hernandez, J.; Murr, L.E.; Gaytan, S.M.; Martinez, E.; Medina, F.; Wicker, R.B. Microstructures for two-phase gamma titanium aluminide fabricated by electron beam melting. *Met. Microstruct Anal.* **2012**, *1*, 14–27. [CrossRef]
100. Recina, V. *Mechanical Properties of Gamma Titanium Aluminides*; Chalmers University of Technology: Gothenburg, Sweden, 2000.
101. Wang, G.X.; Dahms, M. Influence of heat treatment on microstructure of Ti-35wt%Al prepared by elemental powder metallurgy. *Scr. Met. Mater.* **1992**, *26*, 717–722. [CrossRef]
102. Gupta, R.K.; Pant, B.; Sinha, P.P. Theory and practice of γ α 2 Ti aluminide: A review. *Trans. Indian Inst. Met.* **2014**, *67*, 143–165. [CrossRef]
103. Wang, D.; Yuan, H.; Qiang, J. The microstructure evolution, mechanical properties and densification mechanism of TiAl-based alloys prepared by spark plasma sintering. *Metals* **2017**, *7*, 201. [CrossRef]
104. Wang, D.; Zhao, H.; Zheng, W. Effect of temperature-related factors on densification, microstructure and mechanical properties of powder metallurgy TiAl-based alloys. *Adv. Powder Technol.* **2019**, *30*, 2555–2563. [CrossRef]
105. Liu, J.; Li, Z.; Yan, H.; Jiang, K. Spark Plasma Sintering of Alumina Composites with Graphene Platelets and Silicon Carbide Nanoparticles. *Adv. Eng. Mater.* **2014**, *16*, 1111–1118. [CrossRef]
106. Voisin, T.; Monchoux, J.P.; Hantcherli, M.; Mayer, S.; Clemens, H.; Couret, A. Microstructures and mechanical properties of a multi-phase β-solidifying TiAl alloy densified by spark plasma sintering. *Acta Mater.* **2014**, *73*, 107–115. [CrossRef]
107. Zhang, C.; Zhang, K.; Wang, G. Dependence of heating rate in PCAS on microstructures and high temperature deformation properties of γ-TiAl intermetallic alloys. *Intermetallics* **2010**, *18*, 834–840. [CrossRef]
108. Mphahlele, M.R.; Olevsky, E.A.; Olubambi, P.A. Chapter 12-Spark plasma sintering of near net shape titanium aluminide: A review. In *Spark Plasma Sinter*; Elsevier Inc.: Amsterdam, The Netherlands, 2019; pp. 281–299.
109. Appel, F.; Clemens, H.; Fischer, F.D. Modeling concepts for intermetallic titanium aluminides. *Prog. Mater. Sci.* **2016**, *81*, 55–124. [CrossRef]
110. Bewlay, B.P.; Nag, S.; Suzuki, A.; Weimer, M.J. TiAl alloys in commercial aircraft engines. *Mater. High Temp.* **2016**, *33*, 549–559. [CrossRef]
111. Pflumm, R.; Friedle, S.; Schütze, M. Oxidation protection of γ-TiAl-based alloys–A review. *Intermetallics* **2015**, *56*, 1–14. [CrossRef]
112. Dai, J.; Zhu, J.; Chen, C.; Weng, F. High temperature oxidation behavior and research status of modifications on improving high temperature oxidation resistance of titanium alloys and titanium aluminides: A review. *J. Alloy. Compd.* **2016**, *685*, 784–798. [CrossRef]
113. Lu, X.; He, X.B.; Zhang, B.; Qu, X.H.; Zhang, L.; Guo, Z.X.; Tian, J.J. High-temperature oxidation behavior of TiAl-based alloys fabricated by spark plasma sintering. *J. Alloy. Compd.* **2009**, *478*, 220–225. [CrossRef]
114. Kumpfert, J.; Kim, Y.W.; Dimiduk, D.M. Effect of microstructure on fatigue and tensile properties of the gamma TiAl alloy Ti-46.5Al-3.0Nb-2.1Cr-0.2W. *Mater. Sci. Eng. A* **1995**, *192*, 465–473. [CrossRef]
115. Cobbinah, P.V.; Matizamhuka, W.; Machaka, R.; Shongwe, M.B.; Yamabe-Mitarai, Y. The effect of Ta additions on the oxidation resistance of SPS-produced TiAl alloys. *Int. J. Adv. Manuf. Technol.* **2020**, *106*, 3203–3215. [CrossRef]
116. Kenel, C.; Lis, A.; Dawson, K.; Stiefel, M.; Pecnik, C.; Barras, J.; Colella, A.; Hauser, C.; Tatlock, G.J.; Leinenbach, C.; et al. Mechanical performance and oxidation resistance of an ODS γ-TiAl alloy processed by spark plasma sintering and laser additive manufacturing. *Intermetallics* **2017**, *91*, 169–180. [CrossRef]
117. Kim, Y.W. Ordered intermetallic alloys, part III: Gamma titanium aluminides. *JOM* **1994**, *46*, 30–39. [CrossRef]
118. Huang, S.C.; Hall, E.L. The effects of Cr additions to binary TiAl-base alloys. *Met. Trans. A* **1991**, *22*, 2619–2627. [CrossRef]
119. Huang, S.C.; Hall, E.L. Characterization of the effect of vanadium additions to TiAl base alloys. *Acta Met. Mater.* **1991**, *39*, 1053–1060. [CrossRef]
120. Hashimoto, K. High-Temperature Tensile Properties of Ti-Al-X (X=Cr,W) Consisting of Î ± 2, Î² and Î³ in Three Phases. *Mater. Sci. Forum* **2012**, *706–709*, 1066–1070. [CrossRef]

121. Zhu, D.D.; Wang, H.W.; Qi, J.Q.; Zou, C.M.; Wei, Z.J. Effect of Cr addition on microstructures and nanohardness of rapidly solidified Ti–48Al alloy. *Mater. Sci. Technol.* **2012**, *28*, 1385–1390. [CrossRef]

122. Kawabata, T.; Fukai, H.; Izumi, O. Effect of ternary additions on mechanical properties of TiAl. *Acta Mater.* **1998**, *46*, 2185–2194. [CrossRef]

123. Pelachová, T.; Lapin, J. Cyclic oxidation behaviour of intermetallic Ti-46Al-8Ta alloy in air. *Kov. Mater* **2015**, *53*, 415–422. [CrossRef]

124. Yuanyuan, L.; Weidong, Z.; Zhengping, X.; Xiaonan, M.; Yingli, Y.; Jinping, W.; Hangbiao, S. Microstructure, mechanical properties and oxidation behavior of a hot-extruded TiAl containing Ta. *Rare Met. Mater. Eng.* **2015**, *44*, 282–287. [CrossRef]

125. Mitoraj, M.; Godlewska, E.M. Oxidation of Ti–46Al–8Ta in air at 700 °C and 800 °C under thermal cycling conditions. *Intermetallics* **2013**, *34*, 112–121. [CrossRef]

126. Popela, T.; Vojtěch, D.; Novák, P.; Knotek, V.; Průša, F.; Michalcová, A.; Novák, M.; Šerák, J. High-temperature oxidation of Ti-Al-Ta and Ti-Al-Nb alloys. *Metal* **2010**, *18*, 1–4.

127. Lapin, J.; Pelachová, T.; Dománková, M. Long-term creep behaviour of cast TiAl-Ta alloy. *Intermetallics* **2018**, *95*, 24–32. [CrossRef]

128. Lapin, J.; Pelachová, T.; Dománková, M. Creep behaviour of a new air-hardenable intermetallic Ti–46Al–8Ta alloy. *Intermetallics* **2011**, *19*, 814–819. [CrossRef]

129. Hodge, A.M.; Hsiung, L.M.; Nieh, T.G. Creep of nearly lamellar TiAl alloy containing W. *Scr. Mater.* **2004**, *51*, 411–415. [CrossRef]

130. Fang, H.; Chen, R.; Yang, Y.; Tan, Y.; Su, Y.; Ding, H.; Guo, J. Effects of Boron On Microstructure Evolution and Mechanical Properties in Ti46Al8Nb2.6C0.8Ta Alloys. *Adv. Eng. Mater.* **2019**, *21*, 1900143. [CrossRef]

131. Li, M.; Xiao, S.; Xiao, L.; Xu, L.; Tian, J.; Chen, Y. Effects of carbon and boron addition on microstructure and mechanical properties of TiAl alloys. *J. Alloy. Compd.* **2017**, *728*, 206–221. [CrossRef]

132. Čegan, T.; Szurman, I. Thermal stability and precipitation strengthening of fully lamellar Ti-45Al-5Nb-0.2 B-0.75 C alloy. *Met. Mater.* **2018**, *55*, 421–430.

133. Wang, Q.; Ding, H.; Zhang, H.; Chen, R.; Guo, J.; Fu, H. Variations of microstructure and tensile property of γ-TiAl alloys with 0–0.5at% C additives. *Mater. Sci. Eng. A* **2017**, *700*, 198–208. [CrossRef]

134. Li, M.; Xiao, S.; Chen, Y.; Xu, L.; Tian, J. The effect of boron addition on the high-temperature properties and microstructure evolution of high Nb containing TiAl alloys. *Mater. Sci. Eng. A* **2018**, *733*, 190–198. [CrossRef]

135. Bazhenov, V.E.; Kuprienko, V.S.; Fadeev, A.V.; Bazlov, A.I.; Belov, V.D.; Titov, A.Y.; Koltygin, A.V.; Komissarov, A.A.; Plisetskaya, I.V.; Logachev, I.A. Influence of Y and Zr on TiAl43Nb4Mo1B0.1 titanium aluminide microstructure and properties. *Mater. Sci. Technol.* **2020**, *36*, 548–555. [CrossRef]

136. Xu, W.; Huang, K.; Wu, S.; Zong, Y.; Shan, D. Influence of Mo content on microstructure and mechanical properties of β-containing TiAl alloy. *Trans. Nonferrous Met. Soc. China* **2017**, *27*, 820–828. [CrossRef]

137. Wu, Y.; Hwang, S.K. Microstructural refinement and improvement of mechanical properties and oxidation resistance in EPM TiAl-based intermetallics with yttrium addition. *Acta Mater.* **2002**, *50*, 1479–1493. [CrossRef]

138. Kong, F.T.; Chen, Y.Y.; Li, B.H. Influence of yttrium on the high temperature deformability of TiAl alloys. *Mater. Sci. Eng. A* **2009**, *499*, 53–57. [CrossRef]

139. Stringer, J. The reactive element effect in high-temperature corrosion. *Mater. Sci. Eng. A* **1989**, *120*, 129–137. [CrossRef]

140. Hadi, M.; Bayat, O.; Meratian, M.; Shafyei, A.; Ebrahimzadeh, I. Oxidation Properties of a Beta-Stabilized TiAl Alloy Modified by Rare Earth Elements. *Oxid. Met.* **2018**, *90*, 421–434. [CrossRef]

141. Wimler, D.; Lindemann, J.; Clemens, H.; Mayer, S. Microstructural Evolution and Mechanical Properties of an Advanced γ-TiAl Based Alloy Processed by Spark Plasma Sintering. *Materials* **2019**, *12*, 1523. [CrossRef]

142. Saage, H.; Huang, A.J.; Hu, D.; Loretto, M.H.; Wu, X. Microstructures and tensile properties of massively transformed and aged Ti46Al8Nb and Ti46Al8Ta alloys. *Intermetallics* **2009**, *17*, 32–38. [CrossRef]

Article

Joining of Oxide Dispersion-Strengthened Steel Using Spark Plasma Sintering

Foad Naimi [1], Jean-Claude Niepce [1], Mostapha Ariane [2], Cyril Cayron [3], José Calapez [3], Jean-Marie Gentzbittel [3] and Frédéric Bernard [2,*]

[1]   SINTERmat, 7 Avenue Maréchal Leclerc, 21500 Montbard, France; foad.naimi@sinter-mat.com (F.N.); jc.niepce@sinter-mat.com (J.-C.N.)

[2]   ICB UMR 6303 CNRS/ UBFC, 9 Av. A. Savary BP47078, 21078 Dijon, France; moustapha.ariane@u-bourgogne.fr

[3]   GRENOBLE-ALPES-CEA-LITEN, 38054 Grenoble CEDEX 9, France; cyril.cayron@epfl.ch (C.C.); jose.calapez@cea.fr (J.C.); jean-marie.gentzbittel@cea.fr (J.-M.G.)

*   Correspondence: fbernard@u-bourgogne.fr; Tel.: +33-3-80-39-61-25

Received: 5 June 2020; Accepted: 30 July 2020; Published: 2 August 2020

Abstract: Difficulties with joining oxide dispersion-strengthened (ODS) steels using classical welding processes have led to the development of alternative joining techniques such as spark plasma sintering (SPS). SPS, which is classically employed for performing sintering, may also be used to join relatively large components due to the simultaneous application of electrical pulsed current and uniaxial charge. SPS technology was tested by joining two ODS steel disks. The preliminary tests showed that it is necessary to control surface roughness before joining. Furthermore, the use of ground and lapped surfaces seemed to improve the quality of the interface. Tensile tests on two ODS cylinders joined using SPS were performed at 750 °C without any additives. Failure occurred away from the interface with a total elongation close to 50% and an ultimate stress of 110 MPa.

Keywords: ODS steel; SPS joining; interface; tensile tests

---

1. Introduction

Generation IV nuclear power plants need to increase efficiency and produce less radioactive material. Compared to current thermal-neutron reactors, the new generation of fast-neutron reactors can reduce the total radiotoxicity of nuclear waste by using all or almost all of the waste as fuel. Due to their excellent creep strength, corrosion resistance, and radiation resistance, ODS (oxide dispersion-strengthened) ferritic steels are good structural candidates for cladding and fuel applications in fast-neutron reactors [1–3]. The strengthening of these materials is due to the homogeneous distribution of nanosized oxide particles, which inhibits dislocation movement and stabilizes the grain boundary microstructure [4–6]. However, the joining of ODS alloys proves to be challenging. The classical fusion welding techniques such as electric arc welding or high energy fusion such as laser welding or electron beam welding, results in the agglomeration as well as coarsening of dispersed nanoparticles and a loss of strength in the joined material [7]. Solid-state welding processes such as inertia friction welding, friction stir welding, and pressure resistance welding have already been explored [8–10]. Mechanical testing and microstructure observation of ODS steels joined using solid-state diffusion bonding have also obtained encouraging results [11,12]. The SPS (spark plasma sintering) technique was employed in this study to join bulk ODS ferritic steel to itself without disrupting both the fine uniform dispersion of nanosized oxide particles and the grain structure. Joining was achieved by supplying an electrical pulsed current to the samples in order to cause rapid heating via the Joule effect, while simultaneously applying uniaxial pressure. The two cylindrical ODS steel samples were encapsulated in a graphite tool, formed by a die and two punches used to transmit the applied load. The electrical pulsed current flows through the setup formed by the graphite assembly and the bulk sample. This leads to direct heating via the Joule effect in a very short time [13,14]. Recently, SPS was used as an alternative consolidation technique of ODS steel powder [15,16]. A few studies have dealt with bulk metallic material joined using SPS [16] and only one of them has been performed on an extruded 20 wt.% chromium ODS PM2000 sample (PM2000 is a commercial product of the Plansee Company), which exhibits a fine microstructure [17]. The use of SPS technology as a means of joining two recrystallized large grain ODS PM2000 disks is the subject of the present article.

2. Materials and Methods

SPS joining was performed on an FCT HPD-25 spark plasma sintering machine (FCT Systeme, Rauenstein, Germany), located in the MATEIS laboratory at INSA Lyon using an on/off pulse sequence of 10:10 ms. Two joining configurations were tested: one with and one without a graphite die (Figure 1a,b), for which two thermal distributions were simulated (Figure 1c,d) using a thermo-electrical Abaqus model. In Figure 1a, the temperature was monitored using a radial pyrometer that determines the temperature at the centre of the graphite die wall 2 mm from the sample. SPS-joining was performed using a graphite foil of 0.2 mm thickness inserted between the various system components. Addition of the graphite foil helps to preserve the SPS tools as well as improve the electrical contact between each SPS component. In Figure 1b, the IR pyrometer is placed in contact with the joining area. This latter configuration was selected in order to eliminate any contamination of the material with graphite. Moreover, the temperature measurement via the radial pyrometer was performed at the contact zone between the two disks, which is un-joined at the beginning of the SPS cycle.

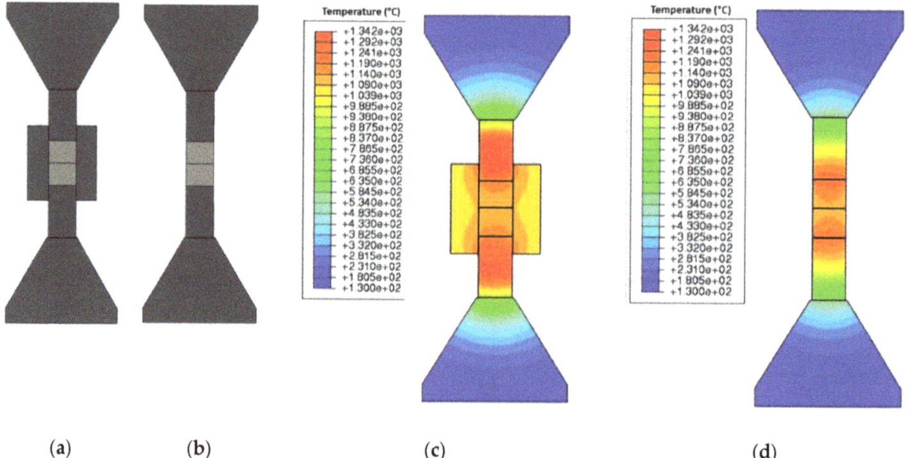

(a)   (b)   (c)   (d)

Figure 1. SPS configurations investigated in this study where (**a**) uses a die and (**b**) does not use a die. Numerical thermal mapping for both configurations where (**c**) represents the experimental setup with a die and (**d**) is without a die.

The thermal mappings for the two configurations previously described (Figure 1c,d) do not show a real difference regardless of the presence or absence of the SPS graphite die. Indeed, both thermal and electrical contact resistance between each component is assumed as perfect. Consequently, only the material properties can affect the flow of current. In our case, the sample and the die were both electrical conductors, therefore, a small proportion of current was diverted to the die, but the majority flowed through the sample (Figure 2).

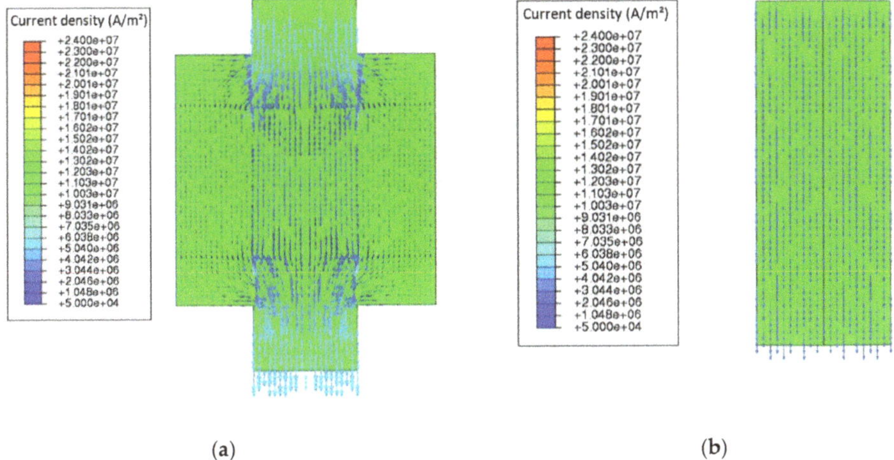

Figure 2. Current vector field for a sample (**a**) with a die and (**b**) without a die.

A recrystallized ODS PM2000 steel bar was delivered from Plansee AG, with a cylindrical diameter of 100 mm. PM2000 is a ferritic 19 wt.%, Cr–5.5 wt.%, Al–0.5 wt.%, Ti–0.5 wt.%, Y_2O_3 steel. The as-received microstructure was composed of large elongated grains (~few mm) parallel to the bar axis orientation, as shown in Figure 3.

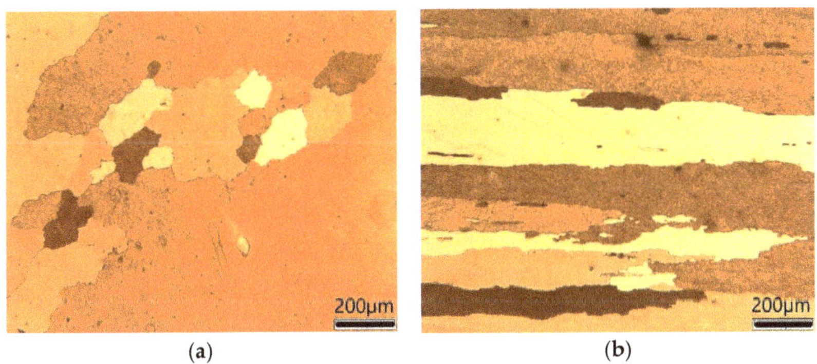

Figure 3. Optical micrographs of a PM2000 sample where (**a**) shows a transverse section and (**b**) depicts a longitudinal section.

Cylindrical samples of 10 mm diameter and 5 mm height were machined in the axis parallel to the bar axis. Only the orientation "transverse–transverse" of the join, shown in Figure 4, was tested.

In order to evaluate the SPS joining technology, a similar SPS procedure involving surface preparation, SPS environment, and thermal conditions was carried out without an additional interlayer (i.e., direct contact between parts to be assembled). SPS diffusion bonding of samples with varying surface roughness was investigated. Three levels of surface finish were tested corresponding to the following surface roughness values:

- Rough machined surface (Ra ~1.2 microns).
- Ground surface (Ra ~0.17 microns).
- Lapped surface (Ra ~0.04 microns).

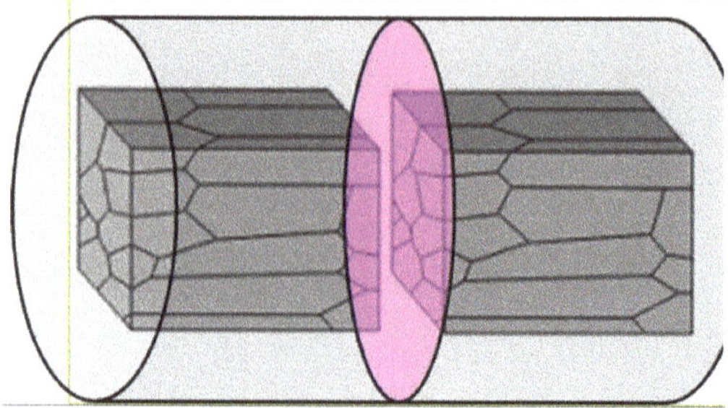

Figure 4. Diagram of the SPS joining orientation according to the sample orientation.

Several experimental SPS conditions, summarized in Table 1, were tested.

Table 1. Experimental conditions for SPS joining of ODS steel disks.

3 Sample Preparations	Heating Rate	Pressure	Temperature	Holding Time
Machined surface Ground surface Lapped surface	150 °C/min	20 MPa without a die to 64 MPa a with die	950 °C to 1100 °C	0 to 30 min

In order to evaluate the possibility of performing a SPS-joining on ODS steel, two sample dimensions were investigated:

(i) On small disks (two disks with a 10 mm diameter and 5 mm thickness) to determine the best SPS-joining conditions. In this case, the classical SPS environment (graphite die and graphite foil called papyex®) was used. The main objective of these preliminary tests was to study the influence of the surface roughness on the quality of the interface. The quality of SPS welding, especially the quality of the SPS interface, was evaluated by studying the metallographic micrographs obtained after cutting the samples in the longitudinal orientation. Then, the samples were polished and chemically etched with a solution (20 mL HF + 50 mL HCl + 10 mL HNO_3 + 30 mL water).

(ii) On larger samples (two cylinders with a 20 mm diameter and 15 mm length) for the purpose of performing tensile tests at room temperature and at 750 °C in order to evaluate the mechanical resistance of the interface. Moreover, to achieve a clean interface without graphite contamination, SPS-joining was performed without a die using the lowest possible applied load of 16 MPa.

3. Results and Discussion

3.1. Small Disks: Influence of Surface Roughness

The optical micrograph shown in Figure 5, relative to the sample prepared with a machined surface and joined using SPS at 975 °C for 10 min, revealed large voids corresponding to the initial surface roughness (Ra = 1.2 µm) and large unbounded areas measuring up to 150 µm. Additional experiments showed that time and temperature did not remove the pores or reduce the size of defects at the joining interface.

Metallographic observations showed that SPS-joining of samples with a lapped surface were more encouraging. Only some areas remained un-joined or contained pores (Figure 6a). In addition, some cracks were observed at lower temperatures (950 °C, Figure 6a) and remained after joining at higher temperatures (1025 °C and 1050 °C, Figure 6b). However, some joints seemed perfect, as shown in Figure 6b. The observed interface was clean, without cavities, and with few small grains at the interface.

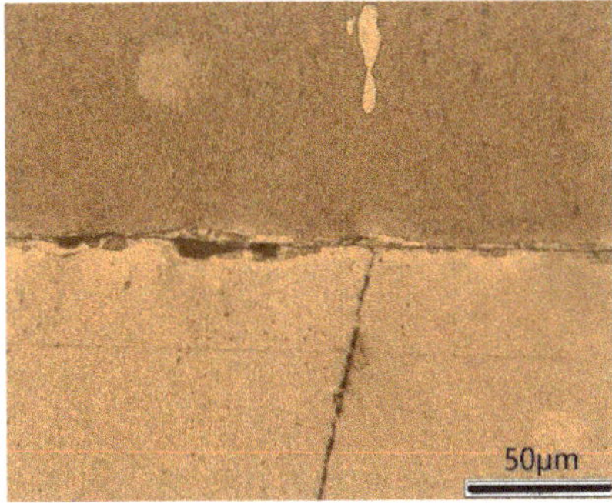

Figure 5. Micrograph of the bonding interface. SPS joined at 975 °C for 10 min from a sample with a rough machined surface.

(**a**)   (**b**)

Figure 6. Metallographic observations of the interface in a joint made from two pieces with lapped surfaces, showing: (**a**) a poor-quality joint at lower temperature (950 °C) and (**b**) a good quality joint at higher temperatures (1025 °C and 1050 °C).

Metallographic results of the samples prepared with a ground surface were quite similar to those obtained using the SPS of a lapped surface. A satisfactory joint was obtained at 975 °C for 10 min although some large defects appeared. The initially large elongated grains did not cross the interface. Moreover, a strong crystal lattice distortion in the interfacial zone was observed in a 10 μm wide area on both sides of the interface using backscattered electron (BSE) imaging under a scanning electron microscope (SEM) (Figure 7). The presence of micron-sized grains at the interface could be observed.

Electron backscatter diffraction (EBSD) images (Figure 8a–c) revealed new micron-sized grains that were randomly oriented and are highlighted at the joining interface. These small grains were formed via recrystallization at the joining interface during SPS operation at 975 °C. Such small grains have also been observed on lapped surfaces, but their number and size were relatively small.

Consequently, the preliminary results confirmed the possibility of joining PM 2000 small disks using SPS. From previous tests, successful joining was only achieved with samples that had their surfaces lapped or ground. In contrast, SPS joining of samples with a machined surface did not lead to satisfactory joining. Too many defects (non-welded areas and cavities) remained, despite an increase in the SPS temperature or holding time. Distortion along the interface combined with the application of a high SPS temperature resulted in the recrystallization of new micrometric grains. In conclusion, these preliminary tests, despite the observed defects, have demonstrated the potential to join PM2000 samples using SPS. This work shows that the SPS temperature and the holding time have a slight effect on the microstructure of the bonding interface. Bonding coarse grain to coarse grain is known to be difficult [18], however, the micrographic observations revealed some very clean joining areas. SEM and EBSD imaging of the well-bonded interface revealed recrystallized small micron-sized grains that were randomly oriented.

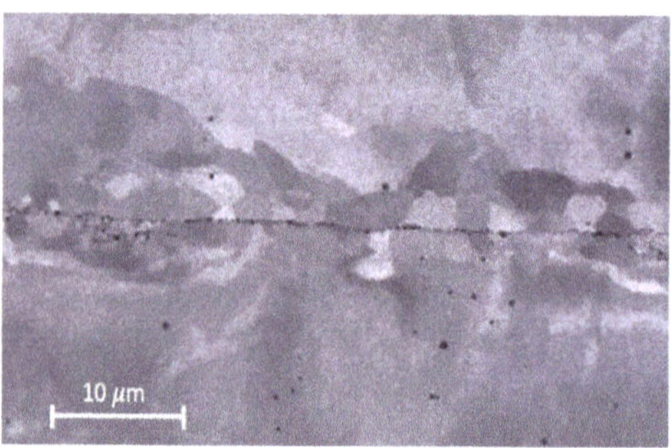

Figure 7. Scanning electron microscopy (SEM) image in backscattered electron (BSE) imaging of the interface between two joined disks with ground surfaces.

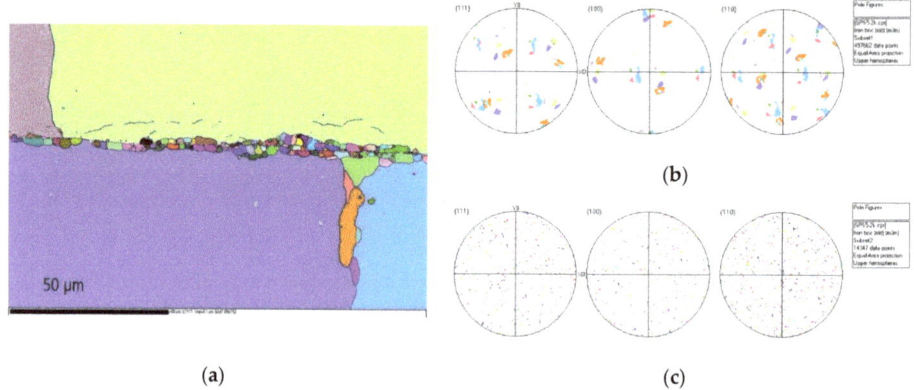

(a)   (b)   (c)

Figure 8. Interface between two joined disks where (**a**) shows the associated electron backscatter diffraction (EBSD) image, (**b**) reveals the pole figures of the large millimetric grains, and (**c**) represents the pole figures of the small micron-size grains at the interface. The same colour code is used in (**a**), (**b**), and (**c**).

3.2. Large Disks: Evaluation of the Mechanical Properties

Five large-size samples were joined with SPS using the welding conditions reported in Table 2. An example of a joined block obtained using SPS is shown in Figure 9.

Table 2. Summary of the SPS-joining conditions.

Samples	Surface Preparation	SPS Conditions
1	Ground	975 °C/20 min
2	Ground	1050 °C/20 min
3	Ground	1100 °C/10 min
4	Lapped	1100 °C/3 min
5	Lapped	1100 °C/10 min

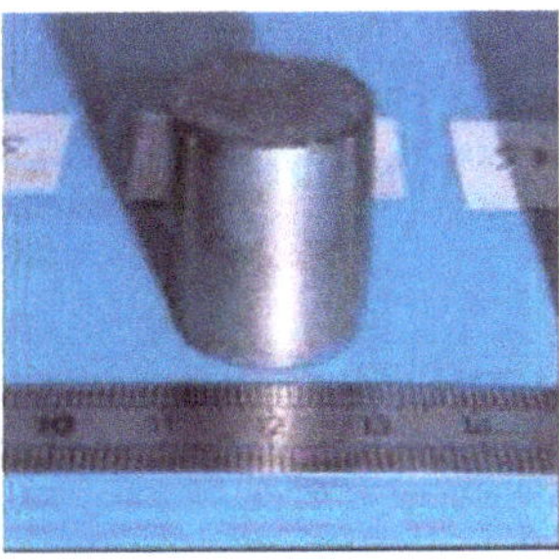

Figure 9. Example of a block joined using SPS.

The sampling of tensile specimens in each block was made using electrical discharge machining (EDM), according to the orientation defined in Figure 10. Two small-sized cylindrical tensile samples with a gauge length of 10.2 mm and a gauge diameter of 2.6 mm, along with two additional flat tensile specimens (2 × 2 mm^2 section and 10 mm gauge length) were machined in each block. The welding bond failed during the machining of numerous tensile test specimens and thus a limited number of tensile specimens were available. Tensile tests were performed at room temperature and at 750 °C in air using an initial strain rate of 5×10^{-4} s^{-1}, according to the NF-EN6892-1 and NF-EN6892-2 standards, respectively. At 750 °C, the tensile specimens were heated with a radiant furnace for 30 min prior to testing.

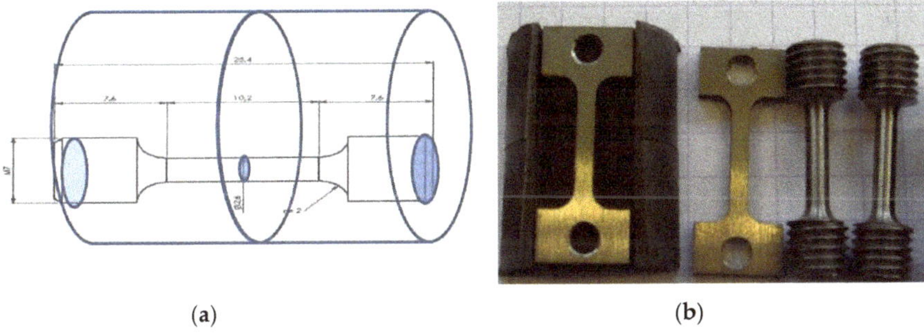

(**a**) (**b**)

Figure 10. (**a**) Tensile testing samples were fabricated using electrical discharge machining (EDM) from the joined blocks. (**b**) Examples of flat and cylindrical tensile testing samples.

Only one tensile specimen from sample 1 could be tested at room temperature. The tensile specimen broke at the welded interface at a stress value of 282 MPa, a value much lower than the elastic yield strength of the parent material (Table 3). At room temperature, due to the anisotropic tensile behaviour of large textured grain, the strain mismatch at the bonding interface in such a small-sized tensile specimen induced high stress concentration at the welding interface.

At 750 °C, the PM2000 tensile behaviour was much more ductile and both tensile specimens extracted from Sample 1 failed at the interface and resulted in an ultimate tensile strength (UTS) of 110 MPa and 114 MPa associated with a total elongation of 6% and 8%, respectively. Sample 2 also demonstrated promising mechanical strength (~112 MPa), however, too many cracks were generated, leading to failure at the joined interface. The tensile rupture properties of all SPS-joined samples are listed in Table 3.

Table 3. Results of the tensile tests performed at 20°C and 750°C for the PM2000 SPS-joined disks

Temperature	At 20 °C			At 750 °C		
Samples	Ultimate Tensile Strength (MPa)	Total Elongation (%)	Failure	Ultimate tensile strength (MPa)	Total Elongation (%)	Failure
1a	282	0	interface	110	6.6	Interface
1b				114	8.4	Interface
2	No tensile specimen available			112	0.4	Interface
3	No tensile specimen available			No tensile specimen available		
4a	200	0		101	53.0	PM 2000
4b	534	2.9		116	44.0	PM 2000
5	No tensile specimen available			No tensile specimen available		

From these results, a higher temperature of 1100 °C and lapped surface preparation was chosen to perform SPS-joining of Sample 4 and Sample 5 (Table 2). Four tensile specimens could be machined from Sample 4, while the entire specimen from Sample 5 broke during final machining. Improved mechanical results compared to the previous were obtained at 20 °C and 750 °C (Table 3). The bonding strength at room temperature of one tensile specimen reached a value of 534 Ma, associated with an elongation of 2.9% of the base material and rupture was located at the welding interface. The second specimen broke at a lower value (200 MPa) and without plastic deformation. After a tensile test at 750 °C, specimens achieved an ultimate tensile strength of 101 and 116 MPa and a total elongation of 53% and 44%, respectively, and they fractured in the bulk material rather than at the joint (Figure 11a,b). In addition, Figure 11b presents the stress–strain curves for the specimens taken from Sample 4 and the as-received PM2000 alloy.

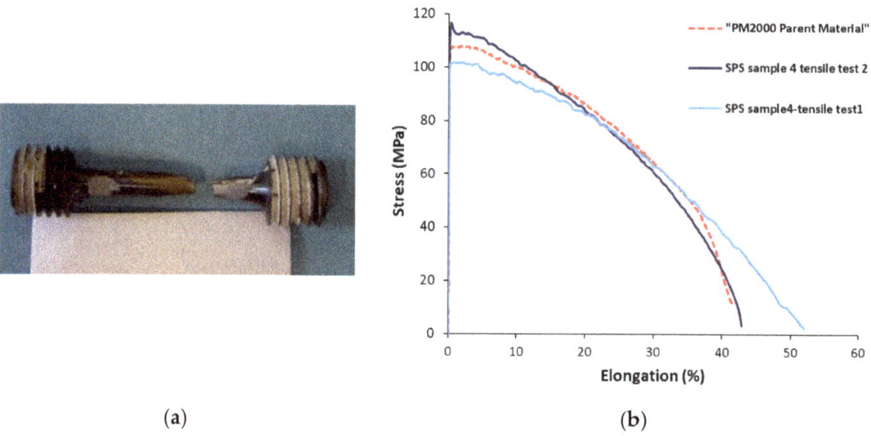

(a)   (b)

Figure 11. (**a**) Feature photographs of cylindrical tensile specimen (Sample 4-tensile test 2) following mechanical testing alongside (**b**) stress–strain curves at 750 °C for Sample 4 (cylindrical and flat samples) and the base PM2000 alloy in the as-received state.

Consequently, Sample 4, joined at 1100 °C with a very short dwell time, exhibited an interface that appeared clean, straight, and undamaged, even after the tensile test at 750 °C, as shown in the micrograph (Figure 12a) of the joint of the flat specimen seen in Figure 12a,b.

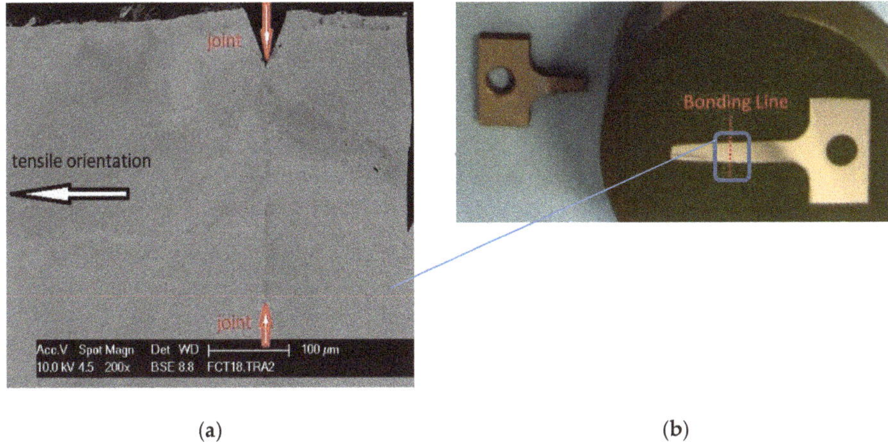

(a)   (b)

Figure 12. (**a**) Micrograph obtained by SEM of the bonding interface of a flat specimen from Sample 4 tested at 750 °C, (**b**) Location of the assembly area on the tensile specimen which is observed by SEM.

Nevertheless, the tensile tests conducted at 20 °C and 750 °C indicate that this joint achieved an interesting value of maximum strength. This study clearly demonstrates the potential to join ODS PM2000 steel to itself using a SPS technique.

4. Conclusions

Twenty wt.% chromium ODS steel PM2000 exhibiting a microstructure of coarse grain was joined to itself via SPS. Different surface preparation techniques, joining temperature, and dwell times were tested. Microstructural observations of the bonding interface were carried out and tensile tests were performed on various samples at room temperature and at 750 °C. The main results are as follows:

- Successful joining was achieved with samples that had lapped joining surfaces.
- Micrographic observations revealed some very clean joining areas. SEM and EBSD imaging of the well-bonded interface revealed recrystallized small micron-sized grains that were randomly oriented.
- The sample prepared with a lapped surface and joined by SPS at 11,000 °C during 3 min exhibited nearly the same tensile behaviour as the base material at elevated temperatures and did not fracture at the joint, but rather failure occurred in the bulk material.

This shows the importance of surface roughness and surface preparation. Indeed, the joining of ODS steel involving the raw machining of its surfaces did not result in a junction free of defects and unassembled areas. The ODS steel joined using SPS with ground surfaces did not exhibit sufficient mechanical strength. The tensile strength at 750 °C reached the nominal values (i.e., 114 MPa) of the base material, but its ductility did not exceed 8% despite the presence of many defects. However, SPS-joining of ODS steel with lapped surfaces demonstrated good performance during tensile testing at 750 °C with failure occurring in the base metal and not at the interface (i.e., an ultimate tensile strength of 101 MPa and a total elongation of 53%). The results of the mechanical tests clearly indicate that these samples joined using SPS gives rise to satisfactory mechanical strength.

Author Contributions: Methodology F.N., J.C.; Modelling, M.A.; Formal analysis, F.N., J.-C.N., J.C. and C.C.; Writing—original draft preparation, F.N., J.-C.N. and C.C.; Writing—review and editing, J.-M.G., M.A. and F.B.; Supervision, F.B. and J.-M.G. All authors have read and agreed to the published version of the manuscript.

Funding: The direction of the Nuclear Energy Department of the CEA is acknowledged for their financial support through the ASTRID project.

Acknowledgments: The authors would like to thank the MATEIS laboratory in INSA Lyon. We would also like to acknowledge the advice provided by G. Bonnefont (INSA Lyon) and G. Fantozzi (INSA Lyon). Specific thanks to N. Al-Mufachi from AEC for his valuable comments.

Conflicts of Interest: The authors declare no conflicts of interest.

References

1. Yvon, P.; Carré, F. Structural materials challenges for advanced reactor systems. *J. Nucl. Mater.* **2009**, *385*, 217–222. [CrossRef]
2. De Carlan, Y.; Béchade, J.-L.; Dubuisson, P.; Seran, J.-L.; Billot, P.; Bougault, A.; Cozzika, T.; Hamon, J.; Doriot, S.; Henry, J.; et al. CEA developments of new ferritic ODS alloys for nuclear applications. *J. Nucl. Mater.* **2009**, *386*, 430–432. [CrossRef]
3. Klueh, R.L.; Maziasz, P.J.; Kim, I.S.; Heatherly, L.; Hoelzer, D.T.; Hashimoto, N.; Kenik, E.A.; Miyahara, K. Tensile and creep properties of an oxide dispersion-strengthened ferritic steel. *J. Nucl. Mater.* **2002**, *307*, 773–777. [CrossRef]
4. Ukai, S.; Okuda, T.; Fujiwara, M.; Kobayashi, T.; Mizuta, S.; Nakashima, H. Characterization of High Temperature Creep Properties in Recrystallized 12Cr-ODS Ferritic Steel Claddings. *J. Nucl. Sci. Technol.* **2002**, *39*, 872–879. [CrossRef]
5. Steckmeyer, A.; Vargas Hideroa, R.; Gentzbittel, J.M.; Rabeau, V.; Fournier, B. Tensile anisotropy and creep properties of a Fe–14CrWTi ODS ferritic steel. *J. Nucl. Mater.* **2012**, *426*, 182–188. [CrossRef]
6. Inoue, M.; Kaito, T.; Ohtsuka, S. Research and Development of Oxide Dispersion Strengthened Ferritic Steels for Sodium Cooled Fast Breeder Reactor Fuels. In *Materials Issues for Generation IV Systems*; Springer: Dordrecht, The Netherlands, 2008; p. 311.
7. Sketchley, P.; Threadgill, P.L.; Wright, I.G. Rotary friction welding of an Fe$_3$Al based ODS alloy. *Mater. Sci. Eng. A* **2002**, *329*, 756–762. [CrossRef]
8. Mathon, M.H.; Klosek, V.; De Carlan, Y.; Forest, L. Study of PM2000 microstructure evolution following FSW process. *J. Nucl. Mater.* **2009**, *386*, 475–478. [CrossRef]
9. Corpace, F.; Monnier, A.; Poulon Quintin, A.; Manaud, J.-P. Resistance upset welding of an ODS steel fuel cladding—Identification of the thermal and mechanical phenomena using a numerical simulation. *Matériaux Tech.* **2012**, *100*, 291–298. [CrossRef]
10. Krishnardula, V.G.; Aluru, R.; Sofrijon, N.I.; Fergus, J.W.; Gale, W.F. Transient liquid phase bonding of ferritic oxide-dispersion-strengthened alloys. *Mater. Trans. A* **2006**, *37A*, 497–504. [CrossRef]
11. Noel, D. *Les Nanomatériaux et Leurs Applications Pour L'énergie Nucléaire*; Lavoisier: Paris, France, 2014; pp. 73–76.
12. Orru, R.; Licheri, R.; Locci, A.M.; Cincotti, A.; Cao, G.C. Consolidation/synthesis of materials by electric current activated/assisted sintering. *Mater. Sci. Eng. R-Rep.* **2009**, *63*, 127–287. [CrossRef]
13. Minier, L.; Le Gallet, S.; Grin Yu Bernard, F. A comparative study of nickel and alumina sintering using Spark Plasma Sintering (SPS). *Mat. Chem. Phys.* **2012**, *134*, 243–253. [CrossRef]
14. Boulnat, X.; Perez, M.; Fabregue, D.; Douillard, T.; Mathon, M.-H.; De Carlan, Y. Microstructure Evolution in Nano-reinforced Ferritic Steel Processed by Mechanical Alloying and Spark Plasma Sintering. *Metall. Mater. Trans. A* **2014**, *45A*, 1485–1497. [CrossRef]
15. Auger, M.A.; De Castro, V.; Leguey, T.; Muñoz, A.; Pareja, R. Microstructure and mechanical behavior of ODS and non-ODS Fe–14Cr model alloys produced by spark plasma sintering. *J. Nucl. Mater.* **2013**, *436*, 68–75. [CrossRef]
16. Naimi, F. Approches Scientifiques et Technologiques du fritta4ge et de L'assemblage de Matériaux Métalliques par Frittage Flash. Ph.D. Thesis, Université de Bourgogne, Dijon, France, 26 November 2013.
17. Tatlock, G.; Dyadko, E.; Dryenpondt, S.; Wright, I. Pulsed Plasma-Assisted Diffusion Bonding of Oxide Dispersion-Strengthened FeCrAl Alloys. *Metall. Mater. Trans. A* **2007**, *38A*, 1663–1665. [CrossRef]
18. Krishnardula, V.G.; Aluru, R.; Sofyan NB, I.; Fergus, J.W.; Gale, W.F. Solid-state diffusion bonding of MA956 and PM2000. In Proceedings of the 7th International Conference, Pine Mountain, GA, USA, 16–20 May 2005; Volume 2005, pp. 885–888.

Article

Processing and Characterization of Bilayer Materials by Solid State Sintering for Orthopedic Applications

Jorge Sergio Téllez-Martínez [1], Luis Olmos [2], Víctor Manuel Solorio-García [1], Héctor Javier Vergara-Hernández [1,*], Jorge Chávez [2] and Dante Arteaga [3]

[1] Tecnológico Nacional de México/ITMorelia, Av. Tecnológico # 1500, Colonia Lomas de Santiaguito, Morelia C.P. 58120, Mexico; D16121682@morelia.tecnm.mx (J.S.T.-M.); v_ms_g@hotmail.com (V.M.S.-G.)
[2] INICIT, Universidad Michoacana de San Nicolás de Hidalgo, Fco. J. Mujica S/N, Morelia C.P. 58060, Mexico; luisra24@gmail.com (L.O.); jomachag@gmail.com (J.C.)
[3] Centro de Geociencias, Universidad Nacional Autónoma de México, Blvd. Juriquilla No. 3001, Querétaro C.P. 76230, Mexico; darteaga@geociencias.unam.mx
* Correspondence: hector.vh@morelia.tecnm.mx; Tel.: +52-443-335-7888

Abstract: A new processing route is proposed to produce graded porous materials by placing particles of Ti6Al4V with different sizes in different configurations to obtain bilayer samples that can be used as bone implants. The sintering behavior is studied by dilatometry and the effect of the layers' configuration is established. To determine pore features, SEM and computed microtomography were used. Permeability is evaluated by numerical simulations in the 3D real microstructures and the mechanical properties are evaluated by compression tests. The results show that a graded porosity is obtained as a function of the size of the particle used. The mechanical anisotropy due to the pore size distribution and the sintering kinetics, can be changed by the particle layer arrangements. The Young modulus and yield stress depend on the relative density of the samples and can be roughly predicted by a power law, considering the layers' configuration on the compression behavior. Permeability is intimately related to the median pore size that leads to anisotropy due to the layers' configuration with smaller and coarser particles. It is concluded that the proposed processing route can produce materials with specific and graded characteristics, with the radial configuration being the most promising for biomedical applications.

Keywords: Functional Graded Materials; sintering; porosity; permeability; mechanical properties

Citation: Téllez-Martínez, J.S.; Olmos, L.; Solorio-García, V.M.; Vergara-Hernández, H.J.; Chávez, J.; Arteaga, D. Processing and Characterization of Bilayer Materials by Solid State Sintering for Orthopedic Applications. *Metals* **2021**, *11*, 207. https://doi.org/10.3390/met11020207

Received: 31 December 2020
Accepted: 19 January 2021
Published: 23 January 2021

Publisher's Note: MDPI stays neutral with regard to jurisdictional claims in published maps and institutional affiliations.

1. Introduction

Nowadays, industrial processes demand materials with specific properties and localized microstructures to improve material performance. To satisfy particular needs, the development of materials with changing mechanical properties and/or microstructures along a preferential direction has been developed. These are called Functional Graded Materials (FGMs). Among these materials, a variation on the porosity along the part is very useful for different industrial applications, such as microfiltration [1], biomedical applications [2–5] and microelectronic devices [6]. Research into materials that can be used as bone implants has received a lot of attention in the past decade due to their capacity to improve human health. Some earlier investigations focused on solving the mismatch stiffness between metallic implants and human bones [6]. Therefore, there is a need to develop porous materials by using different techniques showing that the stiffness of Ti and its alloys can be reduced to close to that of human bones [7–10]. Nevertheless, it is worth noting that bones have anisotropic mechanical properties and that stiffness will change depending on the direction of analysis in the structure [11,12]. Thus, it is necessary for the design of FGMs to satisfy the requirements to produce an improved and specialized bone implant.

199

The powder metallurgy route is better suited to producing materials with graded properties. It allows for control over the sintering parameters and/or the introduction of space-holder particles; with that, it is possible to design the pore characteristics needed in a sample. Processes oriented toward the elaboration of FGMs with a graded porosity of Ti and its alloys have been reported [4,13–16]. M. Dewidar and Kim [13] and Lee et al. [14] produced Ti compacts consisting of a solid core and a porous outer shell and studied the compression behavior of such components. On the other hand, Ahmadi and Sadrnezhaad [15] elaborated on a component with a bonelike configuration, composed of foamy cores of different diameters surrounded by compact shells. The porous core diameter varied from 6 to 14 mm, whereas the outer diameter remained constant to 16 mm, with a linear relation found between Young's modulus and the porous core diameter. Recently, Trueba et al. [16] concocted a device to produce materials with graded porosity in the radial direction that was used to produce FGMs of Ti by varying the quantity and pore size in the radial direction, from the core to the surface or vice versa. These works focused mainly on the mechanical response of the FGMs obtaining materials with stiffness and strength close to that of human bones. Nevertheless, biocompatibility and osteoconductivity are very important to producing a successful bone implant. The first one refers to the acceptance of the material inside the human body, which is related to the release of dangerous ions. For acceptance of materials in the body, after Ta, Ti and its alloys have superior biocompatibility [17]. Osteoconductivity, on the other hand, is the ability of a porous implant to facilitate bone growth through itself. Permeability plays a major role, as it favors the pass of nutrients and minerals to the osteocytes while removing metabolic waste as they are carried out through the porosity of the bones; this is associated with enhanced bone adaptation and regeneration [18]. Also, interconnected pores enhance anchoring and vascularization, obtaining better mechanical adhesion between bone and implant [19–22].

This work aims to develop Ti6Al4V materials with graded porosity by sintering particles with different sizes and distributions located in two main configurations: axial and radial directions. The sintering kinetic of each configuration is evaluated by dilatometry tests. To determine the pore size distribution of the sintered materials, characterization by scanning electronic microscopy (SEM) and computed microtomography is performed. The mechanical properties for graded porous materials are evaluated by compression tests, while permeability is estimated by numerical simulations in the different axes of the sample to determine the anisotropy created by the distribution of the pore sizes.

2. Materials and Methods

2.1. Sample Fabrication

Spherical Ti6Al4V powders furnished by Raymor (Boisbriand, QC, Canada) were used to develop this work. The powder was sieved into two different particle size distributions: 20–45 μm (Figure 1a) and 75–106 μm (Figure 1b). The chemical composition of powders furnished by the fabricant is as follows: 6.35% Al, 4% V, 0.21% Fe, 0.0045% H, 0.02% N, 0.02% O, 0.01% C and the balance of Ti.

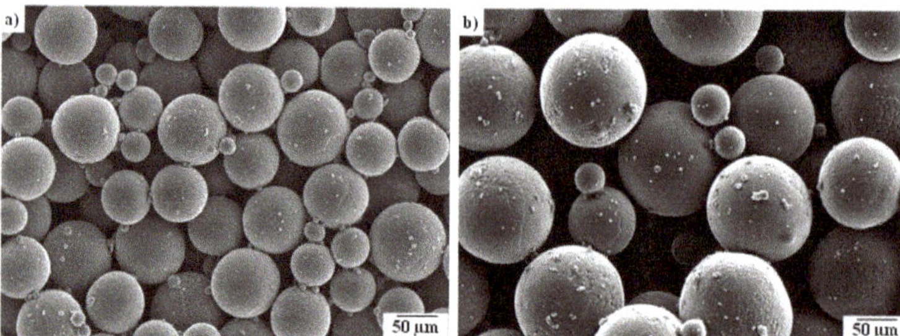

Figure 1. Scanning electronic microscopy (SEM) micrographs of Ti6Al4V powders sieved into two different particle size distributions: (**a**) 20–45 μm and (**b**) 45–75 μm.

To fabricate the samples, the powder's mass was estimated, to obtain samples of 12 mm in height and 10 mm in diameter. The main objective of this study is to obtain porous materials. The powders were poured into a zirconia crucible which was lightly tapped. Four different samples were sintered in a L75 Linseis vertical dilatometer (Linseis Messgeraete GmbH, Selb, Germany) at 1260 °C with 5 min of holding under high purity Ar (99.999%) atmosphere. The cooling rate was the same for all samples 20 °C/min. Sample composition and sintering schedules are detailed in Table 1, as is the nomenclature for each sample.

Table 1. Characteristics and nomenclature of samples used for SEM and compression analysis with dimensions of 10 mm in diameter and 12 mm in height.

Sample Type	Particle Size (μm)	Heating Rate °C/min	Nomenclature	Fabrication Methodology
	20–45	5	MSP-5	
	(Volume Fraction	15	MSP-15	Monolithic samples were elaborated by
	Particles:1)	25	MSP-25	pouring the particles into the crucible; then
	75–106	5	MCP-25	they were flattened by hand pressing with
	(Volume Fraction	15	MCP-25	zirconia punch.
	Particles: 1)	25	MCP-25	
	20–45	5	BA-5	To create a gradient of porosity in the axial
	(Up Volume Fraction	15	BA-15	direction, the sample was prepared by first
	Particles: 0.5)			pouring the smaller particles into a zirconia
	75–106			crucible. Then the surface was flattened by
	(Down Volume	25	BA-25	hand pressing with zirconia punch. The
	Fraction Particles: 0.5)			coarser particles were poured onto the smaller
				ones and the surface was again flattened.
	75–106	5	BR-5	To create a gradient of porosity in the radial
	(Core Volume Fraction	15	BR-15	direction, this sample was fabricated by using
	Particles: 0.25)			a quartz tube of 5 mm inner diameter that
	20–45			was placed in the center of the zirconia
	(Shell Volume Fraction			crucible, where the coarser powders were
	Particles: 0.75)	25	BR-25	poured. Then the smaller powders were
				poured around the quartz tube. Finally, the
				quartz tube was pulled out and the surface
				was flattened.

Two additional samples of 1 and 2 mm in diameter were prepared to achieve the computed tomography analysis. The characteristics and composition of the samples are detailed in Table 2.

Table 2. Characteristics and nomenclature of the samples used for computed microtomography analysis with dimensions of 1 and 2 mm in diameter.

Sample Type	Particle Size (μm)	Heating Rate °C/min	Nomenclature	Fabrication Methodology
	75–106			This sample is fabricated by first pouring the
	(Up Volume Fraction			smaller particles into a quartz crucible of 1 mm in
	Particles: 0.5)	25	TBA-25	diameter. Then the surface is flattened by hand
	20–45			pressing with a glass rod. Next, the coarser
	(Down Volume			particles are poured over the smaller ones and the
	Fraction Particles: 0.5)			surface is again flattened.
	75–106			This sample was prepared by using two quartz
	(Core Volume Fraction			capillaries. The first one, with a 1 mm diameter,
	Particles: 0.25)			was used for the coarser powders. Then it was
	20–45	25	TBR-25	carefully introduced and placed on the center of a
	(Shell Volume Fraction			2-mm-diameter capillary, where the smaller
	Particles: 0.75)			particles were poured around it. Finally, the
				smaller capillary was pulled out.

2.2. Sample Characterization

2.2.1. Scan Electronic Microscopy (SEM) Observation

Samples used for SEM observation were cut in half. Roughing was carried out with different SiC papers and polishing was performed by using alumina powders with a final size of 50 nm. After surface preparation, samples were cleaned in an ultrasonic bath for 30 min to remove alumina particles. The microstructure was observed with a Tescan MIRA 3 LMU field emission scanning electron microscope (FE-SEM, TESCAN ORSAY HOLDING, a.s., Brno-Kohoutovice, Czech Republic).

2.2.2. X-ray Microtomography (microCT)

To obtain detailed information about the porosity and its tridimensional distribution inside the samples, 3D images were acquired with a Zeiss Xradia 510 Versa 3D X-ray microscope (ZEIZZ, Oberkochen, Germany). The beam intensity was set at 120 kV, which was enough to pass through the 1- and 2-mm-diameter Ti6Al4V samples. Nearly 1600 projections were recorded around all 360° of the sample with a CCD camera (ZEIZZ, Oberkochen, Germany) of 1024 × 1024 pixels with a voxel size of 1 and 2 μm, respectively. The voxel size depends on the diameter of the samples, although it is possible to acquire inner volumes without destroying the sample. Those kinds of acquisitions take more time when the exposure time to radiation is longer. Because this work is aimed at analyzing the interparticle pores left between the particles after sintering, a 2 μm voxel size is needed to accurately assess the microstructure features.

2.3. Mechanical Property Evaluation

To obtain the mechanical properties, simple compression tests were performed on the 12-mm-height samples. For that, the bottom and top surfaces of the samples were polished and compression tests were performed following ASTM D695-02 with an Instron 1150 universal mechanical testing machine (INSTRON, Norwood, MA., USA) at a strain rate of 0.5 mm/min. From the stress-strain curves, the Young modulus and the yield strength were estimated in the elastic zone. The stress is measured by correcting the contact surface area of the sample, assuming that the volume was constant during compression. The axial strain is calculated as the ratio of the real axial displacement (after machine stiffness correction) to the initial height of samples.

2.4. Numerical Simulation Permeability

The flow properties of porous samples were estimated by numerical simulations of permeability performed using Avizo® software (V8.0, ThermoFisher SCIENTIFIC, Waltham, MA, USA) on the 3D binary images containing porosity. Before the numerical simulations were run, the minimum representative volume (MRV) was estimated by cropping the image in small volume cubes (20 × 20 × 20 voxels) at the center of the image. Then the pore volume fraction for that volume was calculated. These operations were repeated by increasing the cube volume by 20 voxels per side until constant relative density was reached. The optimal minimal volume to obtain enough accuracy in our results had to be calculated considering computational limitations, to save time for the numerical simulations. It was found that the volume at which the pore volume fraction reached a constant value was near 300^3 voxels^3. Therefore, a volume of 300 × 300 × 300 voxels was used for the numerical simulation in the three main directions of the cube, where x and y represent the horizontal plane and z the vertical axis.

Simulations on Avizo are based on the Darcy law by solving Navier-Stokes equations with a finite volume method. The simulation considered a single-phase incompressible Newtonian fluid with a steady-state laminar flow and a viscosity of 0.0045 Pa s, which represents the viscosity of the blood. The boundary conditions used were the inlet and outlet pressure, with values of 130 and 100 kPa, respectively.

3. Results and Discussion

3.1. Sintering Kinetics

The axial strain for the different samples studied, as a function of the time and temperature during the entire thermal cycle, is depicted in Figure 2a. All samples show the same behavior. First, a positive deformation is due to the thermal dilation of the samples. Then a sharp negative deformation is found, which corresponds to the pore elimination by sintering. Finally, when the temperature reaches the sintering plateau, the axial deformation continues to be negative but with an exponential behavior. It is observed that smaller powders shrank at lower temperatures than coarser ones, which indicates that sintering was activated before. This is explained due to the smaller particles that show a higher surface energy that promotes the atomic diffusion during solid-state sintering [23]. The final deformation is 2.5 times larger for the sample with smaller particles in comparison to that reached by the coarser ones. A similar effect was reported for yttria-stabilized zirconia powder consolidated by flash sintering [24]. The behavior of bilayer samples falls between both monolith samples. However, the onset of sintering is driven by the coarser layer in both bilayers' samples, as the shrinkage generated by the sintering is activated at the same temperature as that of the sample with coarser powders. Despite the fact that sintering is activated at the same time, a clear effect of the position of the layer is noted because the final deformation is larger when the radial configuration is used. This indicates that smaller powders induce stress at the interphase between both layers, which accelerates the shrinkage. On the other hand, each layer sinters more independently, although the axial strain is driven by the layer with the coarser powders.

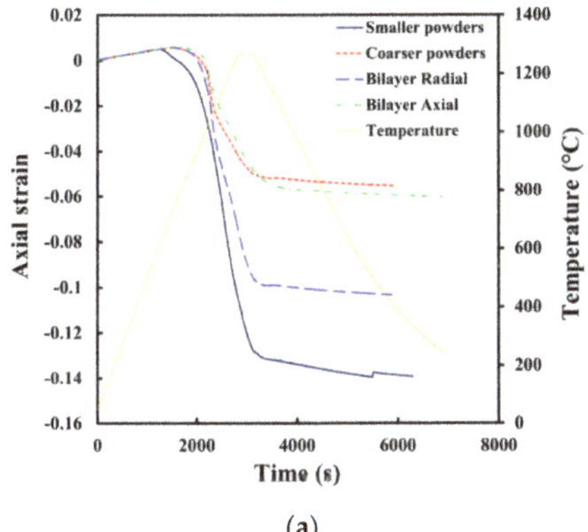

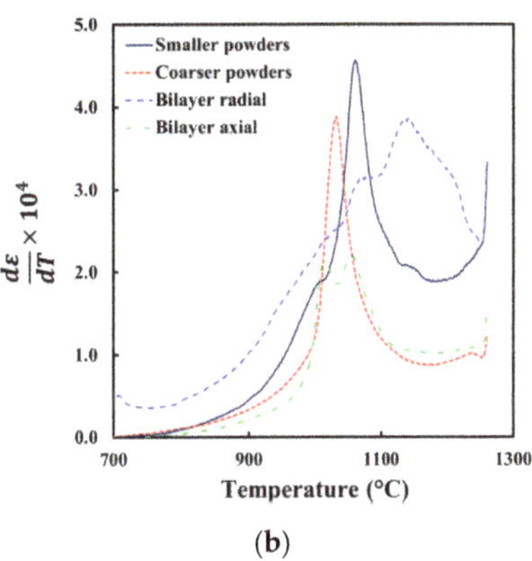

Figure 2. (a) Axial strain as a function of time and temperature during the whole sintering cycle and (b) deformation rate as a function of the temperature during heating with a rate of 25 °C/min. Samples MSP-25, MCP-25, BA-25, BR-25.

The deformation rate as a function of the temperature during the heating stage is presented in Figure 2b for all samples heated at 25 °C/min. It is found that a maximum value of deformation rate is reached for the monolithic samples, which is higher for the sample with smaller particles, confirming the results discussed above. The bilayer samples show maximum values of the deformation rate, which correspond with the maximal ones of monolithic samples. For the axial configuration, the two peaks on the curve correspond approximately with the temperature of the peaks of the monolithic samples, indicating that each layer sinters independently with a low effect of the interphase. On the other hand, for the radial configuration, the maximal values of the deformation rate are reached later than the monolithic samples and the values are higher than those found in the axial configuration. This suggests that sintering is affected by the interphase along the vertical axis, though the maximal values of the deformation rate are retarded and the final deformation reached is larger than that of the bilayer with the axial configuration.

The effect of the heating rate on the axial strain is shown in Figure 3. As was expected, a slower heating rate increases the final strain and as has been discussed, the bilayers in the radial configuration show larger strains. However, the ratio between the maximal values of the strain diminishes from 2.5 to 1.9 times as the heating rate decreases from 25 to 5 °C/min.

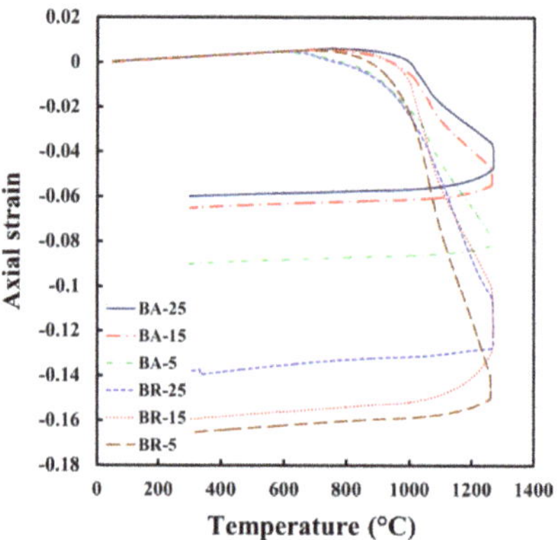

Figure 3. Axial strain as a function of the temperature during the whole sintering cycle of bilayer samples sintered at 1260 °C with different heating rates from 5 to 25 °C/min.

The green and sintered relative densities of all samples studied were estimated by weighing the mass and measuring the volume of cylinders. The obtained values are listed in Table 3. The MSP-25 sample shows a lower green density than the MCP-25 sample, which is expected because smaller particles are more difficult to pack than coarser ones. However, the sintered density is 8% higher for the MSP-25 because sintering advances rapidly as the particle size diminishes. This confirms the findings about the axial strain. The bilayer samples have lower green relative densities than those obtained for monolithic samples, which indicates that packing is affected by the interphase of layers. Nonetheless, the green density of all bilayer samples has the same value, which is not expected because differences in packing between each layer should affect the final packing. The numeric values on the Table 3 show that the samples with the axial configuration (BA-X) have lower densities by sintered process respect at the samples with radial configuration (BR-X); independent of the heating rate. It can be noticed that the lower heating rates produce high densification but BA-5's density is still less than BR-25. The highest value of the sintered relative density is obtained for the BR-5 sample, with a value of 78.59%. It is established that relative densities of bilayer samples BA-25 and BR-25 are close to the monolithic ones MCP-25 and MSP-25, respectively. The results confirm that volumetric densification has a strong relationship with the axial strain measured by dilatometry as is observed in Figure 2a.

Table 3. Green and sintered relative densities of the different samples consolidated at different heating rates.

Sample	Green Density (D_0)	Sintered Density (D_f)
MSP-25	0.6121	0.7413
MCP-25	0.6257	0.6902
BA-25	0.6008	0.6949
BA-15	0.6033	0.7003
BA-5	0.6004	0.7178
BR-25	0.6000	0.7308
BR-15	0.6016	0.7516
BR-5	0.6021	0.7859

3.2. Microstructural Characterization

3.2.1. Scan Electronic Microscopy Observation

Microstructures of the bilayer samples are presented in Figure 4. The sample BA-25, Figure 4a,b, shows a clear division between both layers (interphase). The top layer is composed of coarser particles with a higher pore density. Despite this definition, through these 2D images, it is not possible to clearly establish the difference in the pore characteristics. Figure 4b is an approach of the interphase between both layers and it can be noticed that necks between particles are better developed for the smaller particles and the larger pores in the top layer between coarser particles. Because of surface preparation by cutting and polishing with alumina particles, some pores seems to be closed. Figure 4c,d show the polished surface of the BR-25 sample. The picture of Figure 4c shows a higher density of pores in the core composed of coarser particles; a more detailed difference about this observation will be exposed in the next section by using digital cutting planes (X-ray microtomography analysis). Figure 4d shows a section of the upper outline of the sample between the exterior edge and the core. On this borderline, it is possible to notice that the layer of smaller particles has undergone larger shrinkage in the axial direction compared to the core of the sample. From these images, we can determine that no fractures at the interphases were formed during sintering, despite the large difference in the deformation and deformation rate of both layers. This could be attributed to the lower densification required for the possible biomedical application.

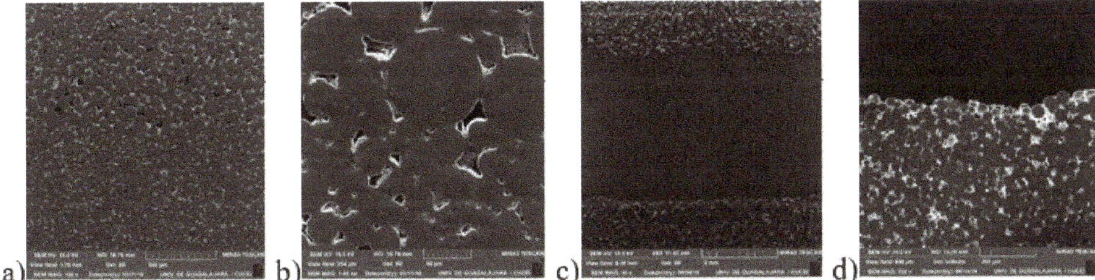

Figure 4. SEM micrographs of bilayer samples sintered at 1260 °C with a heating rate of 25 °C/min; (**a**) and (**b**) BA-25 and (**c**) and (**d**) BR-25.

Thus, the constraint stresses generated at the interphase between layers are weaker than those reported for bilayer ceramic processes by sintering [25].

3.2.2. X-ray Microtomography Analysis

To obtain a deep analysis of the porosity generated by the radial and axial configuration of the bilayers, 3D images of the TBA-25 and TBR-25 samples were acquired using the Median and Unsharp mask filters with the open-source software Image J. Figure 5 shows a 3D and 2D virtual slice of both samples after reconstruction. Figure 5a shows the TBA-25 3D sample; it can be observed that the top layer is composed of coarser particles and the bottom layer of smaller particles. It can be seen that the interphase is not completely flat due to the fragile nature of the quartz capillary. Besides it is possible to better observe the porosity without surface modifications as the ones needed for SEM observation. The TBR-25 2D sample is shown in Figure 5b. We can observe that coarser particles are found in the core of the sample and that interphases are almost vertical. This indicates that the process method is optimal for elaboration of this type of configuration.

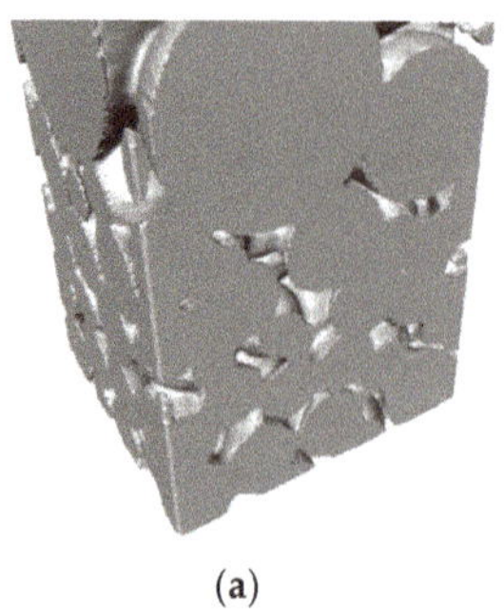
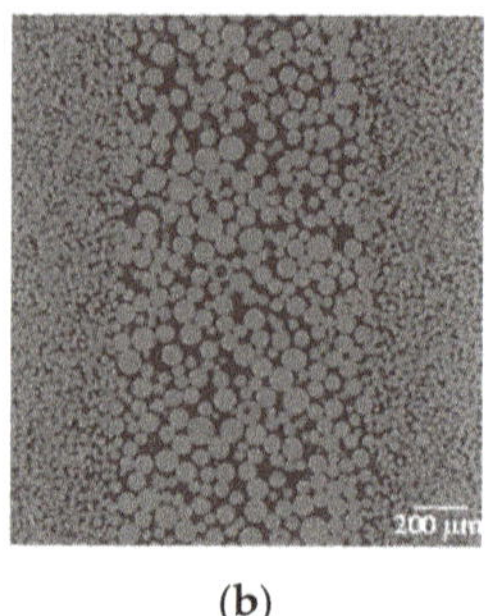

(**a**) (**b**)

Figure 5. 3D and 2D virtual slices of samples of 1 and 2 mm in diameter, respectively, sintered at 1260 °C with a holding time of 5 min and a heating rate of 25 °C/min; (**a**) TBA-25 and (**b**) TBR-25.

To enhance the contrast between the solid and void phases, 3D images were filtered with median and unsharp mask filters, obtaining good quality in the images, as we can observe in Figure 6a. The greyscale images were transformed into binary images to obtain quantitative information. A thresholding method was applied, separating the solid from the void spaces, in which the solid phase gets a value of 1 and void of 0, as can be observed in Figure 6b. This method allowed us to measure the pore volume fraction and size distribution and to run the permeability simulations. Finally, to obtain information like coordination number, neck size and distribution on the volume, a segmentation process was performed by following the watershed method used by Olmos et al. [26,27]. After this process, the particles are considered individual objects in the volume analyzed, as is shown in Figure 6c.

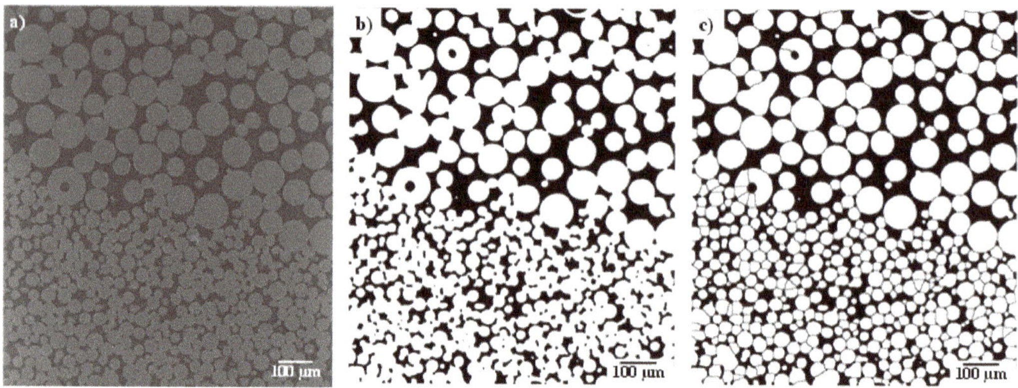

Figure 6. 2D slice of the TBA-25 sample; (**a**) filtered image, (**b**) binary, (**c**) binary segmented.

A 3D rendering of both samples TBA-25 and TBR-25 is shown in Figure 7a,b. Inner volumes were reconstructed due to the size of the sample. However, from these volumes, it can be observed that large particles are located at the bottom of Figure 7a and at the center of Figure 7b. Additional quantitative information will be presented from 3D images. Through this method, a coarse particle located at the interphase of the TBR-25 sample and the smaller particles that are in contact with it can be observed, as can be seen in Figure 7c. This helps to illustrate information at the particle level.

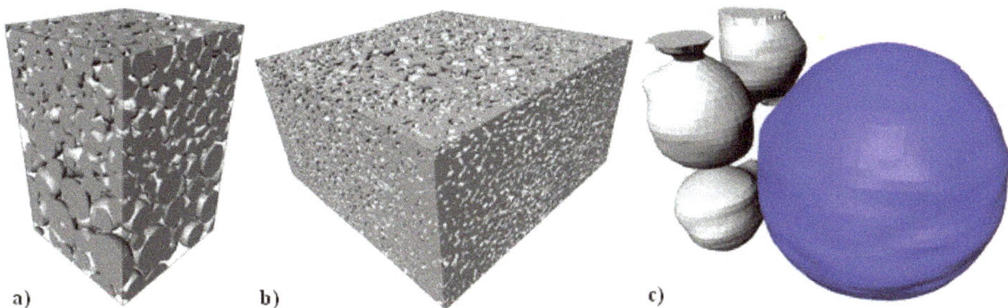

Figure 7. Representation of the tridimensional internal volumes of the samples; (**a**) TBA-25 and (**b**) TBR-25 and (**c**) 3D particles at the interphase of TBR-25 after particle segmentation.

The quantitative data of the packing of particles is separated by layers; taking the whole sample into account, the 3D images allowed us to virtually select and crop the volume of interest inside the sample. The objective of this procedure was to establish the differences obtained after each layer of the bilayer samples was sintered. Table 4 lists the results involving particles and their environment, such as relative density, neck size (a/R), coordination number and particle sphericity. The relative density values calculated from the 3D images for TBA-25 and TBR-25 correspond to those measured from the volume and dimensions of the samples; see Table 3. The local values for each layer in both samples show that the layer with smaller particles is denser than the one with coarser particles. It is observed that the relative density for each layer in the TBA-25 sample is lower than that obtained in the monolithic samples for the same sintering conditions; see Table 3. This suggests that the stresses developed at the interphase reduce the densification in the corresponding layer, as was reported for sintering on solid substrates [28]. On the other hand, the values of the density measured in each layer of the TBR-25 sample correspond with those obtained for the monolithic samples, which indicates that radial configuration reduces the effect of interphase during densification.

Table 4. Quantitative data of the packing of particles of the two bilayer samples.

Nomenclature	Layer	Relative Density	$\frac{a}{R}$	Z	Sphericity
	Smaller	0.7221	0.43	7.52	0.7484
TBA-25	Coarser	0.6564	0.25	5.83	0.8847
	Complete	0.6909	-	6.77	-
	Smaller	0.7494	0.48	7.95	0.7088
TBR-25	Coarser	0.7063	0.39	6.46	0.7927
	Complete	0.7315	-	7.54	–

The average $\left(\frac{a}{R}\right)$ ratio measured from the 3D data for each layer in both TBA-25 and TBR-25 samples increases as the relative density of the layer does, which is expected because the neck size is responsible for the densification. To validate the values obtained from the 3D images, the $\left(\frac{a}{R}\right)$ can be estimated from the geometrical relationship between axial shrinkage and the neck ratio proposed by German [29] as follows:

$$\frac{a}{R} = \left(\frac{\Delta l}{l_0} b\right)^{\frac{1}{2}},$$

(1)

where:

$\left(\frac{a}{R}\right)$, relationship between the neck's radius and the particle's diameter; $\left(\frac{\Delta l}{l_0}\right)$, sample axial shrinkage; (b), a parameter determined by experimental procedures with a value of 3.6.

Likewise, with the assumption that the volume changes are isotropic during sintering $\left(\frac{\Delta V}{V_0}\right)$ and because the mass remains constant during sintering, the relative density value reach (D) depends on the initial green density and the shrinkage as follows [29]:

$$D = \frac{D_0}{\left(1 + \frac{\Delta l}{l_0}\right)^3},$$

(2)

where:

(D_0), is the initial relative density or green value.

Thus, an estimation of the neck size and the particle's radius ratio can be calculated if one knows the relative density values by combining Equations (1) and (2), respectively. To estimate the $\left(\frac{a}{R}\right)$ ratio from Equations (1) and (2), the values of the (D_0) for each layer were the ones measured for the MSP-25 and MCP-25 samples and the (D) values listed in Table 4. The $\left(\frac{a}{R}\right)$ obtained for the BA-25 smaller and coarser layers are 0.44 and 0.24. Meanwhile, the BR-25 smaller and coarser layers are 0.49 and 0.38, respectively. This demonstrates that the geometrical relationship of Equation (1) is a good approximation for lower values of relative densities, which is represented at early sintering stages.

The particle coordination number was calculated for each layer of both TBA-25 and TBR-25 samples after sintering by following the methodology used for Vagnon et al. [30]. The Z values increase as the relative density does for each layer, no matter their configuration. The Z value for the complete samples follows the same tendency.

The particles' sphericity is another indicator of the sintering advance because, the lower the particle sphericity is, the higher the relative density, as is shown in Table 4, where the lower average value of sphericity is found for the layer that shows the highest densification.

The pore size distribution of both TBA-25 and TBR-25 samples is shown in Figure 8. To show the differences in the porosity that remained after sintering in each layer, the pore size distributions are estimated in subvolumes that contain only the characteristics of the smaller and coarser layers for both samples. As was expected, the smaller pore sizes are found in the layers with smaller particles and the pore size distribution of the complete sample falls in between both layers. To get a proper comparison of the porosity after sintering on both samples, the pore volume fraction, the median pore size (d_{50}) and the tortuosity were measured from the 3D volumes, which are listed in Table 5.

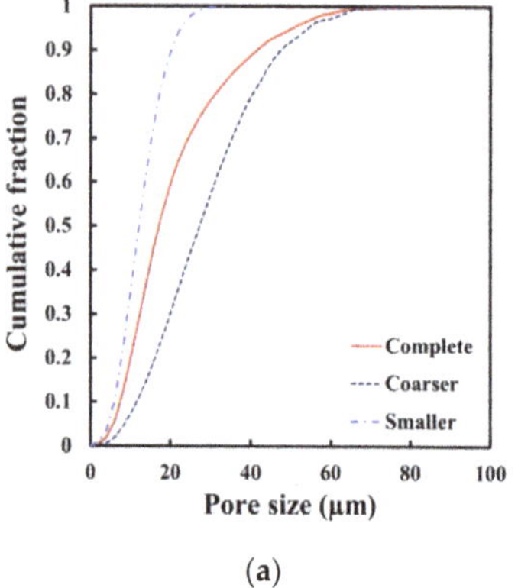
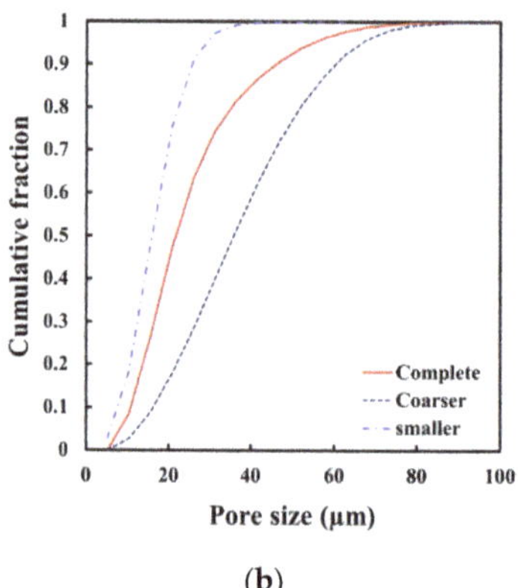

(**a**)   (**b**)

Figure 8. Pore size distribution of bilayer samples, (**a**) TBA-25 and (**b**) TBR-25.

Table 5. Porosity characteristics and properties.

Layer	Pore Volume Fraction (%)	Median Pore Size (μm)	Tortuosity	Axial Permeability (1×10^{-12} m²)	Radial Permeability (1×10^{-12} m²)
			TBA-25		
Smaller	27.78	14.01	1.13	0.4233	0.4380
Coarser	34.36	29.27	1.22	2.2956	2.5903
Complete	30.90	17.24	1.18	0.8110	1.3368
			TBR-25		
Smaller	25.05	16.03	1.27	0.4814	0.3985
Coarser	29.39	35.86	1.08	3.0523	2.9721
Complete	26.84	21.77	1.17	2.0911	0.4854

The highest values for the pore volume fraction are found in the TBA-25 sample. Nevertheless, the larger median pore size was found for the TBR-25 sample, which indicates that the packing of particles in such a configuration is more complicated compared to the axial configuration. Qualitatively, the pore size distribution of the TBR-25 is slightly larger than that of the TBA-25, as can be noticed from Figure 8. The d_{50} obtained for the layer with smaller particles ranges between 14–16 μm and the d_{50} of particles is 36 μm, which means that the d_{50} pore size is close to 38% of the particle size. Similarly, the d_{50} pore size for the coarser particles ranges between 29–35 μm. Meanwhile, the particle size of d_{50} is 76 μm. The same ratio was found for the pore size, which represents 38% of the particle size. This indicates that the median pore size can be predicted from the initial particle size distribution when sintering takes place only in the early stages, as was also mentioned [26]. The values of tortuosity are very close for all layers ranging from 1.08–1.27, indicating that porosity shows a similar path in each layer. This happens because the particles are spherical and the sintering is in the early stage, which means that porosity is completely interconnected, as can be confirmed in the 3D rendering shown in Figure 9a,c.

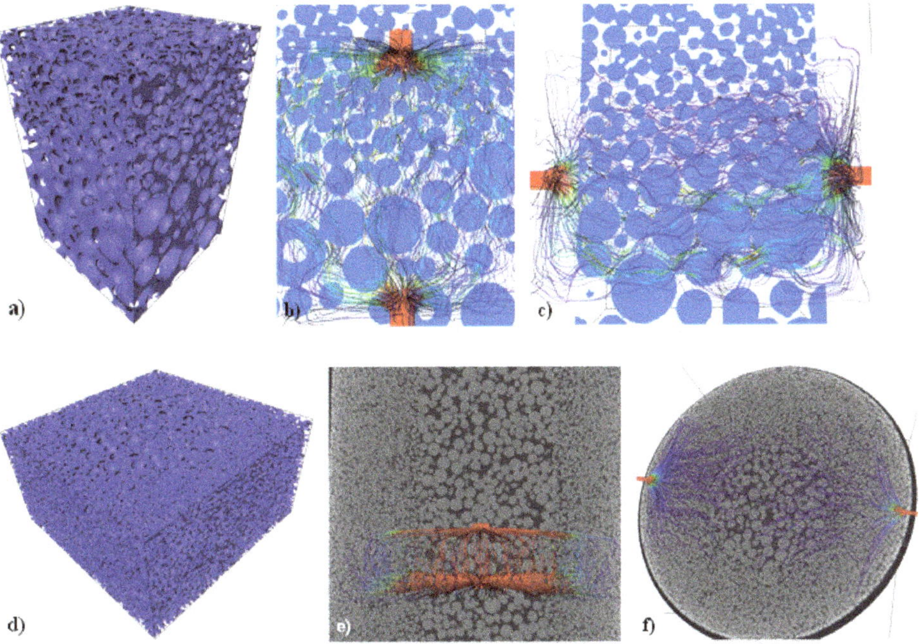

Figure 9. 3D volume rendering of the porosity and streamlines for the fluid flow through pores in the axial and radial directions for TBA-25, (**a–c**) and TBR-25, (**d–f**).

3.3. Permeability Evaluation

Figure 9 is a representation of a 3D rendering of the porosity found in the TBA-25 (Figure 9a) and TBR-25 (Figure 9d) samples, where it is possible to distinguish larger pores in the bottom and center of the samples as a consequence of the configuration of the layers, axial and radial, respectively. A pseudo color code to identify the connectivity of pores was used, obtaining only one color, which means that porosity is fully connected in both samples. Figure 9b,c show the flow path lines throughout the TBA-25 sample obtained from the numerical simulations in the axial and radial directions, respectively. The color code indicates the velocity of the fluid at any point in the volume, with red representing the highest velocity and blue the slowest. It is found that permeability in the radial direction is higher than in the axial one, for the TBA-25 sample, when the simulation is carried out in the whole sample. See Table 5.

To understand the behavior of the permeability, simulations were carried out in subvolumes containing only characteristics of each layer in both directions, axial and radial. The value of permeability in the layer with coarser particles is five times higher than the one with smaller particles. However, the permeability values estimated in the axial and radial directions are barely the same for each layer, which indicates a homogeneous porosity distribution in the sample. Therefore, when the fluid passes through the sample in the axial direction, the permeability is driven by the layer with the lowest permeability, as the fluid is forced to pass through that layer. For the radial direction, most of the fluid passes through the layer with the largest pores and the permeability is higher than in the axial direction. Qualitatively, the above mentioned can be deduced from the flow path lines shown in Figure 9b,c, where it can be noticed that the fluid can achieve a more tortuous and slower velocity when it is passing throughout the layer with smaller pores. In the radial direction, Figure 9c allows us to see more path lines in the bottom layer.

The values of permeability for the TBR-25 sample are slightly higher than those estimated for the TBA-25 sample but are in the same range. As was discussed above, the permeability shows a higher anisotropy when the complete sample is evaluated. For this case, the permeability higher values are obtained when the axial direction is evaluated. This is due to the fluid passing throughout the layer with larger pores that are located in the center of the sample, as can be noticed in Figure 9e. The permeability value in the axial direction is reduced by one-third of the value obtained for the layer with coarser particles. Figure 9f shows the flow path lines in the radial direction. As was mentioned, the paths become more tortuous in the layer with smaller particles. In that case, as the flow must pass twice throughout the layer with smaller particles, the permeability is close to the estimated value for that layer.

The values of permeability measured for the samples fabricated in this work are underestimated by three orders of magnitude compared to those measured by Grimm and Williams [31] for the human trabecular bone (0.4–11×10^{-9} m^2) and those experimentally measured by Nauman et al. [11] in human vertebral and proximal femur bones, which have a large range of permeability values for human bones (2.68×10^{-11}–2×10^{-8} m^2). On the other hand, the permeability values are in good agreement with the range measured for the natural cancellous bone (3×10^{-13}–7.4×10^{-12} m^2) [32], which also has a maximal anisotropy in the permeability (axial/radial) of five times, similar to what was obtained by the radial configuration of layers. This is in agreement with that reported by Nauman et al. [11], who pointed out that permeability in human bones can reach high anisotropy in the permeability values, as much 22 times of difference between axial and radial directions.

3.4. Mechanical Properties

The compression behavior of the bilayer samples is shown in Figure 10. Stress-strain curves of the MSP-25, MCP-25, BA-25 and BR-25 samples are plotted in Figure 10a. All samples show similar behavior, a linear increase in the stress as the strain increases, elastic zone and then a maximum value of the stress is reached until failure. The MSP-25 sample is the strongest among the others and the MCP-25 is the weakest. This is mainly because the relative density is higher for the MSP-25 sample, as well as the a/R ratio, which indicates that necks have a stronger connection between particles. The BA-25 and BR-25 samples show resistance in between the monolithic samples, which was expected. Qualitative analysis indicates that the BA-25 sample behaves as the MCP-25 one and the BR-25 shows similar behavior to the MSP-25 one. For the case of the axial configuration of layers, it is clear that the mechanical resistance is driven by the weakest layer with a small increment in the resistance due to the combination of both layers. Nevertheless, the mechanical values are closer to those of the MCP-25 sample. For the radial configuration of layers, it can be noticed that the resistance of the BR-25 is influenced by both layers and the values are closer to those of the MSP-25 sample. This indicates that the strongest layer can support more load during the compression test, as the weakest layer is in the core of the sample. These results show a similar trend to that reported by Ahmadi and Sadrnezhaad [15] for the compression behavior of foamy core samples, in which they suggest that mechanical resistance is linked to the foamy core diameter.

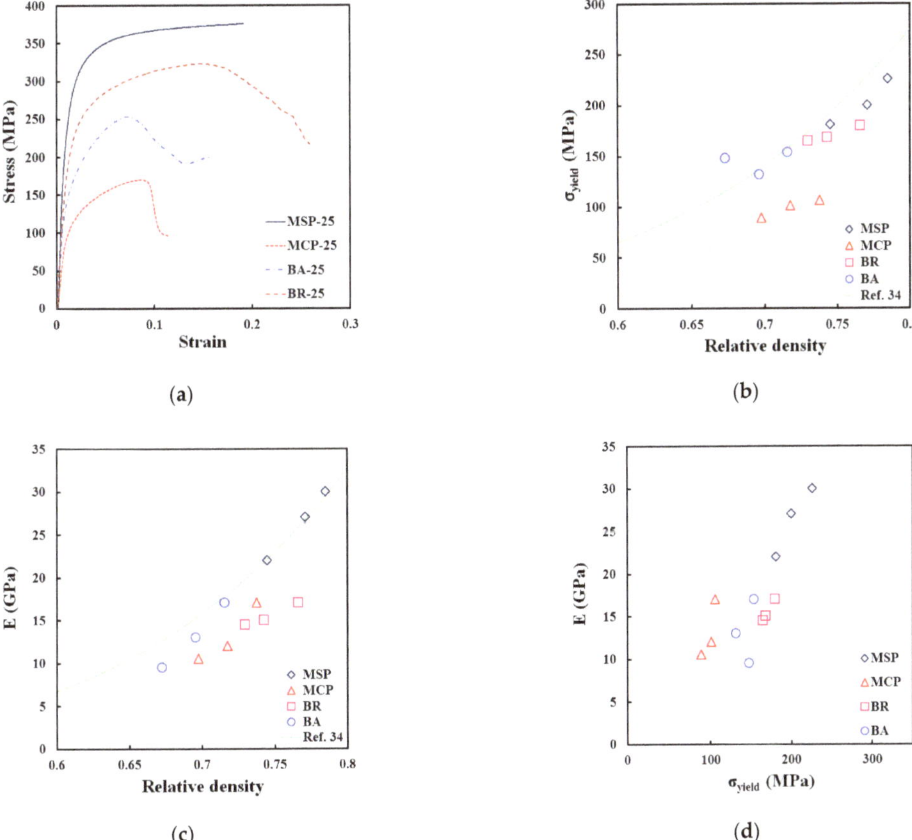

Figure 10. Compression behavior of the different samples tested; (**a**) stress-strain curves, (**b**) Young's modulus as a function of the relative density, (**c**) yield stress as a function of the relative density and (**d**) Young's modulus as a function of the yield stress.

The Young's modulus is estimated from the elastic zone of the stress-strain curves for all samples by fitting a slope in the zone mentioned. The results are depicted as a function of the relative density for the sample, Figure 10b. The higher value of E is 30 GPa and corresponds to the MSP-5 sample, which is the one with the highest relative density. We obtained a reduction of the E values for the monolithic samples when the relative density diminishes, which has been previously reported. However, the E values follow a power law as a function of the relative density that could be associated with that proposed by Gibson and Ashby [33] and modified by Cabezas et al. [34]. The E values calculated for the BA-X samples show a good agreement with the power law, which indicates that the mechanical response of the bilayer samples with axial configuration corresponds to the microstructure of the sample, with a normal effect of porosity. Also, the values of E for BR-X samples are lower than those calculated by the power law, with respect to the relative density of the whole sample. This suggests that stiffness is controlled by the sample with higher porosity. The E values obtained for the bilayer samples ranged between 9–17 GPa, agreeing with those reported for the compact bone [9].

The compressive strength is determined by the yield stress of samples and the values obtained from the stress-strain curves are plotted as a function of the samples' relative density, Figure 10c. The values for the monolithic samples show a similar trend to E, which means the σ_{yield} diminishes as the relative density does. Nevertheless, it can be observed that samples with smaller particles have a better fit with the power law proposed by Cabezas et al. [34]. Meanwhile, the samples with coarser particles are overestimated by the power law. This can be attributed to the advance in the sintering because the a/R ratio is about two times smaller for coarser particles, for an increment of 3% on the relative density. This suggests that necks between particles are weaker and the failure of such samples is due to the necks breaking, which is shown in Figure 10a. Bilayer samples show good agreement with respect to the power law. It can be determined that BR-X samples with around the same relative density as MSP-X samples show a lower σ_{yield}, a result of the combination of layers that has an effect associated with either the a/R ratio or interphase wall effect.

To establish whether this kind of sample could perform well in orthopedic applications, the E values are depicted as a function of the σ_{yield}, Figure 10d. The combination of Young's modulus and mechanical strength called "admissible strain" is defined as the strength-to-modulus ratio and the numerical study of the effect of several metallic elements was performed by Song et al. [35]. In this case, the higher the admissible strain, the more desirable the materials are for implant applications. It is found that stiffness and strength engage in near-linear behavior, in which the E increases as σ_{yield} does. The admissible strain values for all samples range from 6 to 15×10^{-3}. Those values are higher than the ones reported by Zhou and Niinomi [36] for casting Ti-Ta alloys. On the other hand, those values are lower than those reported for porous materials [8], mainly because the addition of large pores reduces the stiffness of materials but the solid phase is more stable, which reduces the effect of porosity in the strength of the porous material. The highest value of the admissible strain was obtained for the BA-25 sample (15×10^{-3}) and the second one with 11×10^{-3} is the BR-25 sample.

4. Conclusions

A novel experimental methodology to fabricate porous gradient materials with controlled porosity by powder metallurgy was proposed. The pore size distribution is determined by the initial particle size of powders and the sintering parameters of monolithic samples. Ti6Al4V bilayer components with two main configurations, axial and radial, were designed by placing particles with different size distributions in a determined place in the sample but any desired configuration can be achieved following this methodology. It was found that no cracks are developed at the interphase of bilayers components because the neck formation between particles is strong enough to support the stresses generated by different densification rates of layers.

It was assessed that mechanical properties and permeability are highly influenced by the configuration of layers, radial or axial one. The radial configuration of the layers is most suitable for bone implants because it is the natural configuration of bone and the Young's modulus, yield stress and permeability values obtained are in good agreement with those reported for human bones. To reduce the E values, additional pores could be added to mimic the real porosity included in bones. This work is in progress by controlling the quantity and pore characteristics.

Author Contributions: J.S.T.-M.: conceptualization, formal analysis, investigation, writing—original draft preparation, writing-review and editing. L.O.: methodology, investigation, supervision, validation, writing-review and editing. V.M.S.-G.: methodology and investigation. H.J.V.-H.: advisor to graduate students, financial administration of the project and supervision as group head of the dilatometry laboratory. J.C.: methodology and investigation. D.A.: methodology, investigation, resources, software, visualization. All authors have read and agreed to the published version of the manuscript.

Funding: This research was funded by Tecnológico Nacional de México (TecNM)/Instituto Tecnológico de Morelia (ITM), through the Doctoral Program in Engineering Sciences. This support allowed the obtaining of J.S. Téllez-Martínez and V.M. Solorio-García. Additional funding was obtained by the CIC of the UMSNH and the Laboratorio Nacional SEDEAM. The Consejo Nacional de Ciencia y Tecnología (CONACyT) supported the posdoctoral program of J. Chávez (postdoctoral fellow, 000614).

Institutional Review Board Statement: Not applicable.

Informed Consent Statement: Not applicable.

Data Availability Statement: The manuscript *"Processing and characterization of bilayer materials by solid state sintering for orthopedic applications"* is an original work and has not been sent elsewhere. All the results are part of the research of the doctoral student (J.S. Téllez-Martínez), and the master's degree student (V.M. Solorio-García). The experiments reported in this work focus on obtaining mechanical properties and permeability data in novel samples with a concentric bilayer structure of sintered particles. These data will be the basis for future works in the authors' field of research.

Acknowledgments: The authors like to thank to the Laboratory "LUMIR" Geosciences of the UNAM, Juriquilla, for the 3D image acquisition and processing with microCT technology, and for the use of the Avizo®software. Finally, thank CONACyT for the support given to graduate students.

Conflicts of Interest: The authors declare no conflict of interest.

References

1. Meulenberg, W.A.; Mertens, J.; Bram, M.; Buchkremer, H.-P.; Stöver, D. Graded porous TiO_2 membranes for microfiltration. *J. Eur. Ceram. Soc.* **2006**, *26*, 449–454. [CrossRef]
2. Reig, L.; Amigó, V.; Busquets, D.J.; Calero, J.A. Development of porous Ti6Al4V samples by microsphere sintering. *J. Mater. Process. Technol.* **2012**, *212*, 3–7. [CrossRef]
3. Bender, S.; Chalivendra, V.; Rahbar, N.; el Wakil, S. Mechanical characterization and modeling of graded porous stainless steel specimens for possible bone implant applications. *Int. J. Eng. Sci.* **2012**, *53*, 67–73. [CrossRef]
4. Torres, Y.; Trueba, P.; Pavón, J.J.; Chicardi, E.; Kamm, P.; García-Moreno, F.; Rodríguez-Ortiz, J.A. Design, processing and characterization of titanium with radial graded porosity for bone implants. *Mater. Des.* **2016**, *110*, 179–187. [CrossRef]
5. Miao, X.; Sun, D. Graded/gradient porous biomaterials. *Materials* **2010**, *3*, 26–47. [CrossRef]
6. Bahraminasab, M.; Sahari, B.B.; Edwards, K.L.; Farahmand, F.; Arumugam, M.; Hong, T.S. Aseptic loosening of femoral components—A review of current and future trends in materials used. *Mater. Des.* **2012**, *42*, 459–470. [CrossRef]
7. Tirta, A.; Baek, E.R.; Chang, S.S.; Kim, J.H. Fabrication of porous material for micro component application by direct X-ray lithography and sintering. *Microelectron Eng.* **2012**, *98*, 297–300. [CrossRef]
8. Cabezas-Villa, J.L.; Olmos, L.; Bouvard, D.; Lemus-Ruiz, J.; Jiménez, O. Processing and properties of highly porous Ti6Al4V mimicking human bones. *J. Mater. Res.* **2018**, *33*, 650–661. [CrossRef]
9. Wang, X.; Xu, S.; Zhou, S.; Xu, W.; Leary, M.; Choong, P.; Xie, Y.M. Topological design and additive manufacturing of porous metals for bone scaffolds and orthopaedic implants: A review. *Biomaterials* **2016**, *83*, 127–141. [CrossRef]
10. Torres, Y.; Rodríguez, J.A.; Arias, S.; Echeverry, M.; Robledo, S.; Amigo, V.; Pavón, J.J. Processing, characterization and biological testing of porous titanium obtained by space-holder technique. *J. Mater. Sci.* **2012**, *47*, 6565–6576. [CrossRef]
11. Nauman, E.; Fong, K.; Keaveny, T. Dependence of intertrabecular permeability on flow direction and anatomic site. *Ann. Biomed. Eng.* **1999**, *27*, 517–524. [CrossRef] [PubMed]

12. Daish, C.; Blanchard, R.; Gulati, K.; Losic, D.; Findlay, D.; Harvie, D.J.E.; Pivonka, P. Estimation of anisotropic permeability in trabecular bone based on microCT imaging and pore-scale fluid dynamics simulations. *Bone Rep.* **2017**, *6*, 129–139. [CrossRef] [PubMed]
13. Dewidar, M.M.; Lim, J.K. Properties of solid core and porous surface Ti–6Al–4V implants manufactured by powder metallurgy. *J. Alloys Compd.* **2008**, *454*, 442–446. [CrossRef]
14. Lee, J.H.; Park, H.J.; Hong, S.H.; Kim, J.T.; Lee, W.H.; Park, J.M.; Kim, K.B. Characterization and deformation behavior of Ti hybrid compacts with solid-to-porous gradient structure. *Mater. Des.* **2014**, *60*, 66–71. [CrossRef]
15. Ahmadi, S.; Sadrnezhaad, S.K. A novel method for production of foamy core@ compact shell Ti6Al4V bone-like composite. *J. Alloys Compd.* **2016**, *656*, 416–422. [CrossRef]
16. Trueba, P.; Chicardi, E.; Rodríguez-Ortiz, J.A.; Torres, Y. Development and implementation of a sequential compaction device to obtain radial graded porosity cylinders. *J. Manuf. Process.* **2020**, *50*, 142–153. [CrossRef]
17. Matsuno, H.; Yokoyama, A.; Watari, F.; Uo, M.; Kawasaki, T. Biocompatibility and osteogenesis of refractory metal implants, titanium, hafnium, niobium, tantalum and rhenium. *Biomaterials* **2001**. [CrossRef]
18. Woodard, J.R.; Hilldore, A.J.; Lan, S.K.; Park, C.J.; Morgan, A.W.; Eurell, J.A.C.; Clark, S.G.; Wheeler, M.B.; Jamison, R.D.; Johnson, A.J.W. The mechanical properties and osteoconductivity of hydroxyapatite bone scaffolds with multi-scale porosity. *Biomaterials* **2007**. [CrossRef]
19. Takemoto, M.; Fujibayashi, S.; Neo, M.; Suzuki, J.; Kokubo, T.; Nakamura, T. Mechanical properties and osteoconductivity of porous bioactive titanium. *Biomaterials* **2005**, *26*, 6014–6023. [CrossRef]
20. Dabrowski, B.; Swieszkowski, W.; Godlinski, D.; Kurzydlowski, K.J. Highly porous titanium scaffolds for orthopaedic applications. *J. Biomed. Mater. Res. B Appl. Biomater.* **2010**, *95*, 53–61. [CrossRef]
21. Le, H.; Wang, C.A.; Huang, Y. *Biomimetics Learning from Nature*; InTechOpen: London, UK, 2010. [CrossRef]
22. Kienapfel, H.; Sprey, C.; Wilke, A.; Griss, P. Implant fixation by bone ingrowth. *J. Arthroplasty* **1999**, *14*, 355–368. [CrossRef]
23. Coble, R.L. Effects of particle-size distribution in initial-stage sintering. *J. Am. Ceram. Soc.* **1973**, *56*, 461–466. [CrossRef]
24. Francis, J.S.; Cologna, M.; Raj, R. Particle size effects in flash sintering. *J. Eur. Ceram. Soc.* **2012**, *32*, 3129–3136. [CrossRef]
25. Morsi, K.; Keshavan, H.; Bal, S. Processing of grain-size functionally gradient bioceramics for implant applications. *J. Mater. Sci. Mater. Med.* **2004**, *15*, 191–197. [CrossRef] [PubMed]
26. Olmos, L.; Takahashi, T.; Bouvard, D.; Martin, C.L.; Salvo, L.; Bellet, D.; di Michiel, M. Analysing the sintering of heterogeneous powder structures by in situ microtomography. *Philos. Mag.* **2009**. [CrossRef]
27. Olmos, L.; Bouvard, D.; Cabezas-Villa, J.L.; Lemus-Ruiz, J.; Jiménez, O.; Arteaga, D. Analysis of compression and permeability behavior of porous Ti6Al4V by computed microtomography. *Met. Mater. Int.* **2019**, *25*, 669–682. [CrossRef]
28. Martin, C.L.; Bordia, R.K. The effect of a substrate on the sintering of constrained films. *Acta Mater.* **2009**, *57*, 549–558. [CrossRef]
29. German, R.M. Sintering theory and practice. *Sol.-Terr. Phys.* **1996**, 568.
30. Vagnon, A.; Rivière, J.P.; Missiaen, J.M.; Bellet, D.; Di Michiel, M.; Josserond, C.; Bouvard, D. 3D statistical analysis of a copper powder sintering observed in situ by synchrotron microtomography. *Acta Mater.* **2008**, *56*, 1084–1093. [CrossRef]
31. Grimm, M.J.; Williams, J.L. Measurements of permeability in human calcaneal trabecular bone. *J. Biomech.* **1997**, *30*, 743–745. [CrossRef]
32. Syahrom, A.; Kadir, M.R.A.; Harun, M.N.; Öchsner, A. Permeability study of cancellous bone and its idealised structures. *Med. Eng. Phys.* **2015**, *37*, 77–86. [CrossRef] [PubMed]
33. Gibson, L.J.; Ashby, M.F. *Cellular Solids: Structure and Properties*; Cambridge University Press: Cambridge, UK, 1999; p. 52.
34. Cabezas-Villa, J.L.; Lemus-Ruiz, J.; Bouvard, D.; Jiménez, O.; Vergara-Hernández, H.J.; Olmos, L. Sintering study of Ti6Al4V powders with different particle sizes and their mechanical properties. *Int. J. Miner. Metall. Mater.* **2018**, *25*, 1389–1401. [CrossRef]
35. Song, Y.; Xu, D.S.; Yang, R.; Li, D.; Wu, W.T.; Guo, Z.X. Theoretical study of the effects of alloying elements on the strength and modulus of β-type bio-titanium alloys. *Mater. Sci. Eng. A* **1999**, *260*, 269–274. [CrossRef]
36. Zhou, Y.L.; Niinomi, M. Ti–25Ta alloy with the best mechanical compatibility in Ti–Ta alloys for biomedical applications. *Mater. Sci. Eng. C* **2009**, *29*, 1061–1065. [CrossRef]

Article

Influence of Successive Chemical and Thermochemical Treatments on Surface Features of Ti6Al4V Samples Manufactured by SLM

Jesús E. González [1,*], Gabriela de Armas [1], Jeidy Negrin [1], Ana M. Beltrán [2], Paloma Trueba [2], Francisco J. Gotor [3], Eduardo Peón [1] and Yadir Torres [2]

[1] Departamento de Biomateriales Cerámicos y Metálicos, Centro de Biomateriales, Universidad de La Habana, La Habana 6323, Cuba; gdearmasortiz@gmail.com (G.d.A.); jeidy930924@gmail.com (J.N.); epeon@biomat.uh.cu (E.P.)

[2] Departamento de Ingeniería y Ciencia de los Materiales y el Transporte, Escuela Politécnica Superior, Universidad de Sevilla, 41011 Sevilla, Spain; abeltran3@us.es (A.M.B.); ptrueba@us.es (P.T.); ytorres@us.es (Y.T.)

[3] Instituto de Ciencia de Materiales de Sevilla (CSIC-US), 41011 Sevilla, Spain; francisco.gotor@icmse.csic.es

* Correspondence: jgonzalezr1961@gmail.com; Tel.: +53-58454530

Citation: González, J.E.; Armas, G.d.; Negrin, J.; Beltrán, A.M.; Trueba, P.; Gotor, F.J.; Peón, E.; Torres, Y. Influence of Successive Chemical and Thermochemical Treatments on Surface Features of Ti6Al4V Samples Manufactured by SLM. *Metals* **2021**, *11*, 313. https://doi.org/10.3390/met11020313

Academic Editor: Eric Hug

Received: 24 January 2021
Accepted: 8 February 2021
Published: 11 February 2021

Abstract: Ti6Al4V samples, obtained by selective laser melting (SLM), were subjected to successive treatments: acid etching, chemical oxidation in hydrogen peroxide solution and thermochemical processing. The effect of temperature and time of acid etching on the surface roughness, morphology, topography and chemical and phase composition after the thermochemical treatment was studied. The surfaces were characterized by scanning electron microscopy, energy dispersive X-ray spectroscopy, X-ray diffraction and contact profilometry. The temperature used in the acid etching had a greater influence on the surface features of the samples than the time. Acid etching provided the original SLM surface with a new topography prior to oxidation and thermochemical treatments. A nanostructure was observed on the surfaces after the full process, both on their protrusions and pores previously formed during the acid etching. After the thermochemical treatment, the samples etched at 40 °C showed macrostructures with additional submicro and nanoscale topographies. When a temperature of 80 °C was used, the presence of micropores and a thicker anatase layer, detectable by X-ray diffraction, were also observed. These surfaces are expected to generate greater levels of bioactivity and high biomechanics fixation of implants as well as better resistance to fatigue.

Keywords: selective laser melting; Ti6Al4V; acid etching; chemical oxidation; thermochemical treatment; surface features

1. Introduction

Titanium and its alloys are widely used in the manufacturing of biomedical devices, especially dental and orthopedic implants, which operate under high biomechanical loads [1–3]. Titanium has a moderate capacity to osseointegrate, excellent mechanical characteristics and great resistance to corrosion in biological fluids [4–6]. However, there are significant differences between the chemical and phase composition presented by these materials and those of bone tissues. Therefore, their insertion into the human skeleton may result in the absence of strong bonds between the bone and implant. Commonly, the Ti6Al4V alloy is one of the most widely used in the medical world, since it significantly increases the strength, ductility and fatigue resistance of the implants, which could prevent their fracture [2,7,8].

In the last two decades, a kind of technological revolution involving advanced biomaterials, structure designs and new manufacturing methods for implantable medical devices has notably improved the clinical success of surgical operations for the treatment of hard tissue affections [9]. Additive manufacturing (AM) is a new concept involving

215

the industrial production of objects through which the material is deposited layer by layer [10,11]. Using this technique, which is also known as three-dimensional (3D) printing, custom geometric shapes can be produced depending on the needs of each patient [11]. This process is suitable for producing low-volume parts with great shape complexity. AM has beneficial features, such as high precision, freedom of design, minimization of waste, production of components directly from digital files, as well as lightweight parts with complex scales. It also reduces the cost of product development and cycle time [12].

One of the advantages of AM, is that it provides extensive customization for medical applications based on individual patient data and requirements, and enables the design and manufacture of patient-specific implants [12–14], which are modeled in 3D sections. For this reason, in recent years there has been notable progress in the implementation of AM in the field of biofabrication. Selective laser melting (SLM) and electron beam melting (EBM) have been selected in most research studies as suitable methods for fabricating scaffolds for bone tissue engineering (BTE), due to their good controllability and high level of precision [11,15]. The SLM technique is used in biomedicine to print complex geometries and lightweight structures and, since printed components can have thin walls, deep cavities, and hidden channels, it has a high potential for being used for the manufacturing of porous objects, such as metal scaffolds. For instance, by SLM, it is possible to manufacture implants with porous 3D structures known as lattice structures [12].

In the last 30 years, a great deal of effort has been devoted to obtaining a biological answer related to the topology and chemistry of the surface of implants [16]. Increasing cellular activity on the implant surface is of great importance to accelerate the growth of bone tissue. The relationship between surface topography and cell viability has drawn increasing attention to a wide variety of surface modification approaches [16,17]. Nanoscale profiles may play an important role in the adsorption of extracellular matrix (ECM) proteins and in cell adhesion properties [2,18]. Micro/nanoscale surface topography has been confirmed to modulate cellular functions and have positive effects on the differentiation, orientation, adhesion of osteoblasts and implant osseointegration [19,20]. In different studies, human osteoblasts were found to prefer surfaces with nanometric topologies [6].

Some works have addressed the manufacturing and characterization of specimens of the commercially pure (c.p.) titanium and Ti6Al4V alloys, made by SLM [21–23]. SLM titanium and SLM Ti6Al4V samples have shown good biocompatibility both in vitro and in vivo [24]. However, their topography and surface chemistry are not the most appropriate for achieving rapid osseointegration [17,25]. In this context, to improve the biocompatibility and osseointegration of SLM Ti6Al4V implants, several surface modification treatments have been proposed in the literature [26–30], obtaining roughness surface at submicro and nanoscale level. However, no studies have been found in which these surface treatments were able to obtain and control roughness features ranging from the macroscale to the nanoscale. Therefore, the main goal of this investigation is to evaluate the role of different consecutive surface treatments: acid etching (influence of temperature and time), chemical oxidation and thermochemical treatment on the surface features of SLM Ti6Al4V samples.

2. Materials and Methods

2.1. Fabrication of Ti6Al4V Samples

The samples were designed with Inventor Professional three dimensional computer-aided design (3D CAD) Inventor software (Autodesk Inc, Mill Valle, CA, USA). Plates with a size of around 10 mm × 5 mm × 5 mm were made on the SLM250 selective laser melting machine (SLM 250^{HL}, SLM solutions GmbH, Lübeck, Germany). The samples were printed in the z-direction (parallel to the longitudinal direction of the plates) in an argon atmosphere. A Ti6Al4V alloy powder (grade 5 supplied by SLM Solutions GmbH) with spherical morphology and an average diameter of 31 ± 12 µm, with dimensions between 10 and 65 µm, was used. Details of the raw powder have already been reported in a previous work [31]. An ytterbium fiber laser, with a power of 200 W, line scanning strategy using antiparallel stripes, hatch distance of 120 μm and thickness of the powder layer of 50 μm, were used.

2.2. Surface Modification Treatments

The surface modification process of the Ti6Al4V samples used during the development of this research can be summarized as described in Figure 1. First, the samples were subjected to an acid etching (AE) treatment. A mixture of HCl/H_2SO_4 at 67% with a *v/v* 1:1 ratio was prepared according to the procedure described by Zhang et al. [32] and the samples were immersed in this mixture using one of the four treatment regimens shown in Table 1. Then, they were washed using distilled water in an ultrasonic bath for 10 min and dried in an oven at 90 °C for 1 h. Later, the chemical oxidation (C) treatment of samples in a mixture of H_2O_2 and HCl was carried out according to the procedure reported by Wang et al. [33,34]. Test tubes, with 5 mL of a mixture of H_2O_2 with a concentration of 8.8 mol/L and 0.1 mol/L of HCl with a *v/v* 1:1, were placed in a thermostatic water bath at 80 °C and the samples were submerged in the oxidizing mixture for 30 min. The specimens were then washed again with distilled water and dried in an oven at 90 °C for 1 h. Finally, a thermochemical (T) treatment was carried out in a furnace at 400 °C for 1 h, using a heating regime of 10 °C/min. The successive chemical oxidation and thermochemical treatments were applied to the samples obtained with the four AE regimens shown in Figure 1. Table 1 summarizes the label of the samples and the successive superficial treatment of each of them (AE+C+T, AECT). For instance, AECT-1 means this substrate was subjected to a sequential acid etching (AE) a 40 °C during 8 min, followed by a chemical oxidation (C) and, then, thermochemical treatment (T).

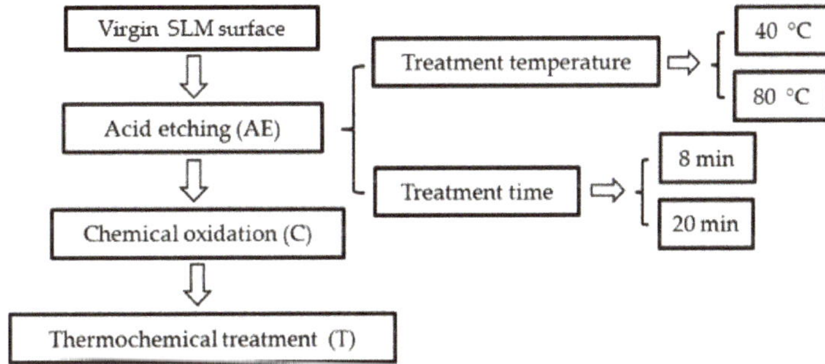

Figure 1. Schematic representation of the surface treatments used. SLM: selective laser melting.

Table 1. Acid etching treatments. All samples were superficially modified with successive chemical oxidation treatment (at 80 °C for 30 min) and thermochemical treatment (at 400 °C for 1 h).

Sample Denomination	Acid Etching (AE)	
	T (°C)	t (min)
AECT-1	40	8
AECT-2		20
AECT-3	80	8
AECT-4		20

2.3. Surface Characterization

Surface morphology of different samples was studied by field emission scanning electron microscopy (FE-SEM) (S-4800, Hitachi, Japan). The semi-quantitative elemental composition was measured using energy dispersive X-ray spectroscopy coupled with scanning electron microscopy (EDX-SEM) (Quantax EDS, Bruker Corporation, Mannheim, Germany). Surface topographical features (i.e., protuberances, size of the pores) were measured using the software ImageJ version 1.44. The X-ray diffraction (XRD) patterns of the surfaces were acquired with a PANalytical X'Pert Pro diffractometer, using Cu Kα radiation (λ = 0.1542 nm) with 40 kV and 40 mA, a step size of 0.05°, a counting time of 80 s/step and a diffraction angle 2Theta between 20° and 80°. The surface roughness (R_a) of the samples before and after the thermochemical treatment was measured using a contact profilometer (Surftest SJ-210, Mitutoyo, Japan) at 0.5 mm/s, using the standard UNE-EN-ISO 4287:1999 [35]. Two replicas of each surface were used, and data were acquired five times on each sample.

2.4. Statistical Analysis

All experimental measurements are presented as the mean value ± standard deviation (SD). At least one replica of each experimental run was used, with more than 50 measurements of the micropore, sub micropore and protrusion sizes. In addition, more than 5 measurements of the R_a of each sample were performed. The data were analyzed using StatGraphics Centurion XV software (Statpoint Technologies, Warrenton, VA, USA). Multiple-sample comparison tests (multiple range tests, Tukey HSD) were used to determine significant differences among groups. A value of $p < 0.05$ was taken as a statistically significant difference.

3. Results

3.1. Acid Etching of SLM Surface

Different AE regimens (Table 1) were performed to determine the influence of the temperature and time on the topographical surface features of Ti6Al4V samples after the full surface modification process (Figure 1). As an example, the SEM images of the Ti6Al4V sample before and after the acid etching at 80 °C for 20 min are displayed in Figure 2. The as-prepared SLM samples showed surface morphological features according to those reported in previous studies (Figure 2a) [25,36]. Residual partially melted powder particles (protrusions) were found on the surface of the native SLM samples that showed a rough wavy surface without nano-topographic characteristics. On the surfaces subjected to the AE process, a significant variation of the topography and morphology (Figure 2a,b) was observed in comparison with the SLM surfaces. In samples etched at 80 °C, micropores were revealed on the surfaces, which were not observed when etched at 40 °C. Higher micropores content was noticed on the surfaces of samples etched at 80 °C for 20 min with an average size of 6.1 ± 3.2 µm. Furthermore, some grooves with a width of around 9 µm and parallel orientation between them were observed. On the other hand, the mean size of the protrusions on the AE surface decreased from 31 ± 12 µm to 28 ± 10 µm according to the dimensional losses observed in Figure 2.

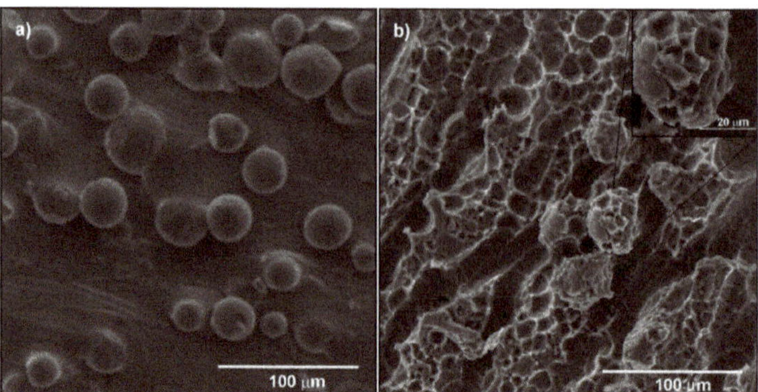

Figure 2. SEM micrographs of the surface of the Ti6Al4V alloy samples obtained by SLM before and after the acid etching (AE). (**a**) SLM surface, (**b**) AE surface (etched at 80 °C for 20 min). Inset: higher magnification image.

The superficial elemental composition of the SLM and AE surfaces was determined by EDX-SEM (Figure 3). In general, the spectra acquired after etching are similar to those observed on the surface of the SLM samples without treatment. Peaks associated with Ti, Al and V were identified. Only a slightly higher Al content was observed on the AE surfaces. Moreover, a new peak was detected on these surfaces, which was assigned to S. The presence of this element must be related to the existence of H_2SO_4 in the acid mixture.

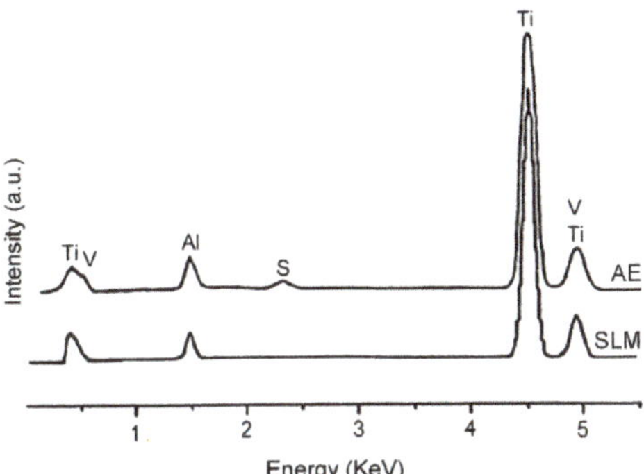

Figure 3. EDX-SEM spectra of the surface of SLM and AE samples.

3.2. Successive Chemical Oxidation and Thermochemical Treatments of the Acid Etched Ti6Al4V Surfaces

Figure 4 shows the SEM images of the surface of the Ti6Al4V samples subjected to successive acid etching, chemical oxidation (in H_2O_2/HCl mixture) and thermochemical (400 °C for 1 h) treatments (AECT surfaces). AECT surfaces differed by the AE regimen used (Table 1) and two surface topographies were observed. The AECT-1 and AECT-2 samples presented a similar topography to the SLM samples, while the surface of the AECT-3 and AECT-4 samples additionally showed micropores. In Figure 4, it is possible to appreciate the size and the size distribution of the spheroidal protrusions present on the surfaces of these samples.

The AECT surfaces showed protrusions with diameters between 6 and 60 μm and an average diameter of about 30 μm, as could be observed in the histograms (insets Figure 4). In general, no statistically significant differences were found when comparing the diameter of the protrusions of the AECT surfaces with the AE surfaces. However, a slight increase in this parameter was observed in the AECT surfaces. Neither statistically significant differences were found between the diameter of the protrusions on the four AECT evaluated surfaces. On the other hand, on AECT-3 and AECT-4 surfaces, the presence of micropores can be observed with an average diameter of 4.8 ± 2.8 μm and 5.2 ± 2.0 μm, respectively. This parameter decreased in about 1 μm compared to that presented by the AE surfaces and, furthermore, statistically significant differences were found when the micropore diameter on AECT-3 and AECT-4 surfaces was compared with AE surfaces. The micropores appeared not only on the protrusions, but also on the rest of the surface and had a concave configuration with the tendency to adopt a spheroidal shape. Note that the pores cover a greater surface area in the sample subjected to the treatment at 80 °C–20 min (AECT-4), while this was not observed in the samples treated at 40 °C (AECT-1 and AECT-2). Porous surfaces on implants provide high frictional resistance between the host bones and high primary stability. After implantation, the bone tissues can grow into the pores, and biological fixation is achieved [26].

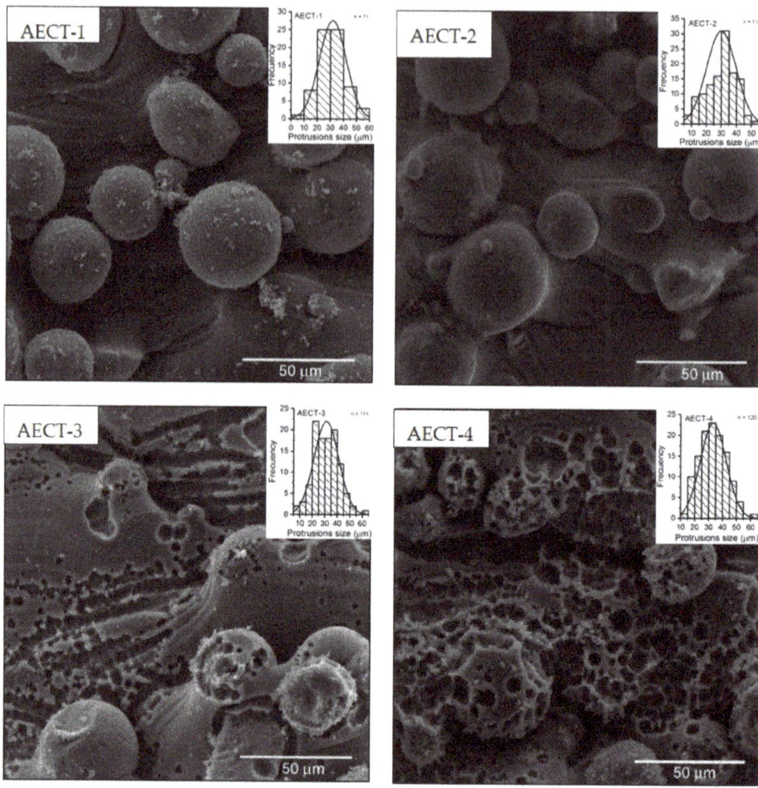

Figure 4. SEM micrographs AECT surfaces at low magnification and protrusions size distribution on AECT surfaces.

The SEM images at higher magnification of the AECT samples show the presence of structures at the submicrometric and nanometric scale (Figure 5), as a result of the chemical reaction between the titanium and the H_2O_2/HCl mixture [34]. Specifically, on all surfaces, a three-dimensional interconnected network with an open porous structure, formed by nanosheets that surround nano-submicropores was observed. The aforementioned network structure could be seen both on the surface of the protrusions and inside the pores previously formed in the AE process. The nano-submicropores (dark areas, while lighter areas were the sheets; see Figure 5) exhibited an irregular shape with an average size of around 130 nm. No statistically significant differences were found when comparing the average size of these pores for all studied surfaces. Some microcracks were also observed, which probably originated during the heating and cooling steps of the thermochemical treatment of the samples. In addition, structures with spheroidal shape and size of about one micrometer were detected.

In general, two multiscale topographies were observed after the successive chemical and thermochemical oxidation treatments. These topographies are related to the time and temperature used during acid etching. In the samples with acid etching at 40 °C (AECT-1 and AECT-2), macro, submicro and nanoscale structures were observed, while in the samples treated at 80 °C (AECT-3 and AECT-4), micropores were additionally seen. In recent years, the combination of micro- and nano-features has attracted the attention of researchers [14,18,19,37]. Xu et al. have reported that the presence of micro-nano topography in SLM titanium allowed significantly higher osteoblast proliferation, total protein contents, bone–implant contact (BIC) and bone-bonding force than in as-built SLM [25]. In this sense, the nanotopographical features increase the surface area of biomaterials and may contribute to increased protein adsorption [18], the adhesion of osteoblasts and the osseointegration of the implant surface [17]. Therefore, it is expected that the surfaces could improve the cell response and the osseointegration of the implants.

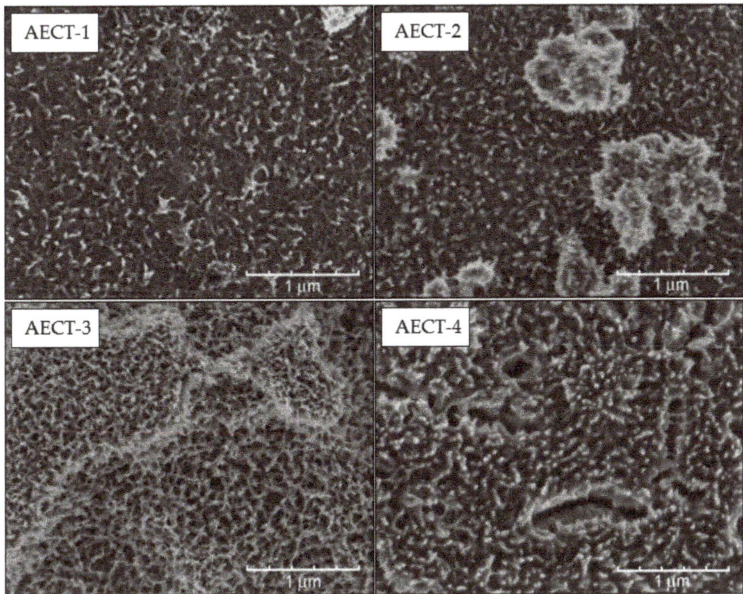

Figure 5. SEM micrographs of the surface of AECT samples at high magnification.

EDX-SEM spectra of the surfaces subjected to the chemical oxidation and thermo-chemical treatments are shown in Figure 6. Peaks of Ti, Al and V corresponding to the starting composition and additional O were observed. The presence of high oxygen content (around 40%) must be related to the formation of an oxide scale layer. The spectra did not show the peak corresponding to S detected on the AE surfaces that was probably eliminated during the H_2O_2/HCl chemical oxidation. In general, the Al, V and Ti contents on these surfaces remained lower than on the SLM and AE surfaces, but the Ti/Al/V ratio was similar to that of the AE surfaces.

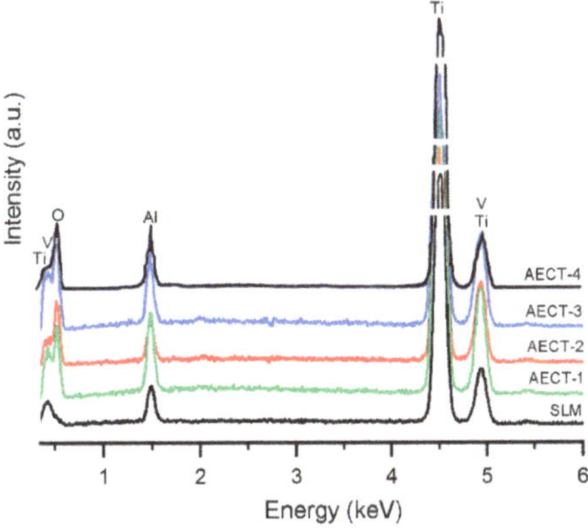

Figure 6. EDX-SEM spectra of SLM and AECT surfaces.

Figure 7 shows the XRD patterns of the surface of the SLM samples before and after the successive surface modification treatments evaluated in this work. The presence of the α phase, with hcp crystal structure, was clearly observed. The existence of a small amount of the cubic β phase cannot be excluded. The low intensity XRD peaks also observed in Figure 7 were assigned to anatase TiO_2 (2 Theta = 25.28 and 75.03). The low intensity of these peaks was related to the small thickness of the oxide layer (according with the results of Wang et al., the anatase layer thickness was estimated to be about 30 nm [33]). The titania gel obtained during the treatment of Ti in H_2O_2 solution was transformed into an anatase crystal structure after heating between 400 and 500 °C [34]. Su et al. reported that protein adsorption and subsequent cellular responses could be also affected by the surface functional groups [38]. Specifically, anatase has excellent bioactivity and significant differences have been found in the percentage of BIC between the anatase layer and control implants during the early stages of bone regeneration [39]. On the other hand, the formation of a titanium oxide layer increases the corrosion resistance of the titanium alloy and prevents the release of ions into the body fluid [40]. In this sense, accumulation of aluminum has been observed around Ti6Al4V implants, which could be harmful; therefore, a proper passivating layer would reduce the risk of aluminum release [41].

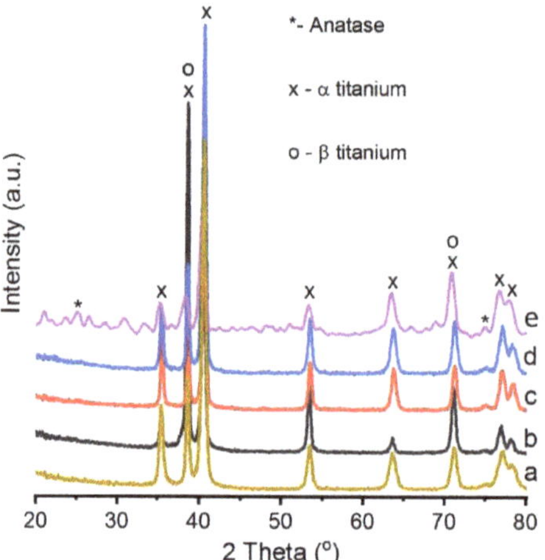

Figure 7. XRD patterns of SLM and AECT surfaces. (**a**) SLM, (**b**) AECT-1, (**c**) AECT-2, (**d**) AECT-3 and (**e**) AECT-4.

Figure 8 shows R_a values of the AE and AECT surfaces. In general, rough surfaces were observed in all the samples (R_a average between 6.8 and 7.5 μm). These R_a values were slightly lower than those obtained for the SLM surfaces (R_a = 9.35 ± 0.47 μm) and were like those reported by Benedetti et al. [42]. On the other hand, different studies have reported R_a values in the range of 6 and 40 μm on as built SLM Ti6Al4V parts [42–45]. The surface roughness of samples manufactured by SLM depends on several factors: material, processing parameters, laser inclination angle to the build platform, build direction, particle size distribution and parts spacing [46–50]. Both the transition boundaries between layers and partially melted powder particles contribute to the overall roughness of the top surfaces. In the top surface, the roughness differs strongly from the roughness of side surfaces [43]. The influence of the partially melted particles on the surface roughness increases as the inclination angle increases, and it is the primary cause of surface roughness when the inclination angle is close to 90° [50]. In this sense, it has been reported that specimens built in the 45° direction show higher levels of roughness than the vertically built specimens.

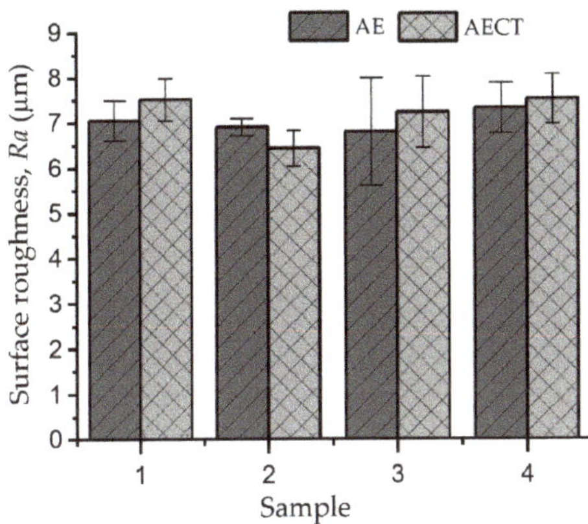

Figure 8. Surface roughness in AE and AECT samples. (**1**) AECT-1, (**2**) AECT-2, (**3**) AECT-3, (**4**) AECT-4.

In general, larger R_a values were observed in AECT surfaces compared to the AE surfaces. Nevertheless, no statistically significant differences were found between the Ra values of the AE and AECT surfaces ($p > 0.005$). On the other hand, the AECT surfaces only showed slight differences between their R_a values. Several studies have demonstrated the influence of the surface roughness of titanium implants on their osseointegration rate and biomechanical fixation [3]. The rough surface could increase the anchorage possibility of the bone cells and, according to Bose et al., the intrinsic roughness of the AM surfaces can increase tissue integration, implant fixation and, also, the mechanical adherence of coatings [16]. Benedetti et al. investigated the effect of shot peening and electropolishing in SLM Ti6Al4V samples on cell growth at different times [51]. The surface roughness decreased because of these treatments, but they found no influence of these surfaces on cell growth at 90 h of incubation. Additionally, Tsukanaka et al. stated that a rough surface was beneficial for early mechanical stability, but for osteoblast differentiation and bone formation, the surface must undergo a bioactive treatment [26].

Endosseous implants under load bearing must maintain high mechanical properties, biocompatibility and osseointegration over a time scale exceeding at least two decades [51]. The multiscale topography, chemical and phase composition obtained in the AECT-3 and AECT-4 surfaces must generate adequate biocompatibility and fast osseointegration of the implants. Although the mechanical properties of SLM Ti6Al4V samples are superior, compared to conventionally manufactured parts, this is not the case in the high-cycle fatigue regime [52,53]. The fatigue performance of as built SLM Ti6Al4V components is over 75% lower than wrought materials, due to their surface finish, porosity and residual stresses [54]. In this sense, it was reported that as built Ti6Al4V parts manufactured by SLM after a stress relief treatment have a fatigue resistance of 240 MPa at 5×10^6 cycles [51]. The fatigue crack initiation life depends on different factors, such as residual stresses, surface roughness, internal defects, microstructure and microstructural inhomogeneities. For built SLM specimens, surface roughness has been found to be the most influential factor in reducing fatigue life [55,56]. The mean fatigue life of SLM Ti6Al4V parts decreases with increasing surface roughness due to the stress concentration at the surface [52,57]. Different post-melting treatments, such as heat treatment, machining, acid etching, polishing, shot peening, hot isostatic pressing (HIP) and electropolishing, were used to increase the resistance of SLM Ti6Al4V parts to fatigue [44,51,52,58]. The best results were obtained when stress–relief treatments were used, followed by at least one of the following processes: machining, shot peening or HIP [44,51,59]. The machining processes reduce surface roughness and subsurface defects, but its use in complex geometries is difficult. On the other hand, shot peening and HIP reduce surface defects and create a surface compressive residual stress layer [51].

Porous coatings have also been associated with decreases in the fatigue life of medical devices [60]. Smith considered that the decrease in endurance limited, to sintered porous coatings, was related to pores and cracks in the layer [61]. Apachitei et al. found a significant increase in fatigue resistance by decreasing the thickness of porous coatings by plasma electrolytic oxidation (PEO) on titanium alloys [60]. They also observed that, in the coatings in which anatase prevails over rutile, the fatigue strength values were increased. According to Khan et al., the anatase coatings induced compressive stresses [62], which must improve the fatigue performance of the treated implant.

As previously stated, it is expected that the surfaces acid etched at 80 °C (AECT-3 and AECT-4), which present macro-micro-submicro and nanoscale structures and, in addition, an anatase layer, would generate greater levels of bioactivity and high biomechanic fixation of implants to the bone. However, these surfaces had greater porosity, which could also affect their resistance to fatigue. Thus, future research should determine the influence of the two topographies obtained on the biological behavior of the endosseus implants, using in vitro and in vivo tests. In addition, the influence of the surface features obtained in the AECT samples on the fatigue resistance of SLM biomedical devices should be evaluated.

4. Conclusions

In this work, the effect of successive processes of acid etching, chemical oxidation and thermochemical treatment of Ti6Al4V samples (AECT surfaces) manufactured by selective laser melting was evaluated. It was found that AECT surfaces showed significant differences in their topography and elemental composition in comparison with the AE surfaces. The temperature used in the AE process had a greater influence on the surface features of the samples. Two topographies were obtained on the AECT surfaces as a function of the temperature used during acid etching. After the thermochemical treatment at 400 °C for 1 h, the samples subjected to acid etching at 40 °C (AECT-1 and AECT-2 samples) showed macro structures combined with submicro and nano scale topographies, characterized by the absence of micropores. The samples with acid etching at a temperature of 80 °C (AECT-3 and AECT-4 samples) also showed a multiscale topography in which additionally micropores were observed. A network shape structure was observed on all surfaces, both on their protrusions and inside their pores previously formed in the acid etching. In addition, thermochemical treatment caused an increase in oxygen content on the Ti6Al4V surface, the formation of an anatase thin layer and a micropore size decrease.

Author Contributions: Conceptualization, project administration, supervision, methodology, J.E.G., E.P., F.J.G. and Y.T., investigation, formal analysis, validation, G.d.A., J.N., A.M.B. and P.T., discussion and writing—original draft preparation, all the authors. All authors have read and agreed to the published version of the manuscript.

Funding: This work was supported by the Ministry of Science and Innovation of Spain under the grant PID2019-109371GB-I00, by the Junta de Andalucía–FEDER (Spain) through the Project Ref. US-1259771.

Institutional Review Board Statement: Not applicable.

Informed Consent Statement: Not applicable.

Data Availability Statement: Not applicable.

Acknowledgments: The authors are sincerely grateful to the to the "Luces" project of IMRE, University of Havana for its collaboration in conducting tests.

Conflicts of Interest: The authors declare no conflict of interest.

References

1. Gagg, G.; Ghassemieh, E.; Wiria, F.E. Effects of sintering temperature on morphology and mechanical characteristics of 3D printed porous titanium used as dental implant. *Mater. Sci. Eng. C* **2013**, *33*, 3858–3864. [CrossRef]
2. Tsai, P.-I.; Lam, T.-N.; Wu, M.-H.; Tseng, K.-Y.; Chang, Y.-W.; Sun, J.-S.; Li, Y.-Y.; Lee, M.-H.; Chen, S.-Y.; Chang, C.-K. Multi-scale mapping for collagen-regulated mineralization in bone remodeling of additive manufacturing porous implants. *Mater. Chem. Phys.* **2019**, *230*, 83–92. [CrossRef]

3. Shibata, Y.; Tanimoto, Y. A review of improved fixation methods for dental implants. Part I: Surface optimization for rapid osseointegration. *J. Prosthodont. Res.* **2015**, *59*, 20–33. [CrossRef] [PubMed]
4. Bai, L.; Gong, C.; Chen, X.; Sun, Y.; Zhang, J.; Cai, L.; Zhu, S.; Xie, S.Q. Additive manufacturing of customized metallic orthopedic implants: Materials, structures, and surface modifications. *Metals* **2019**, *9*, 1004. [CrossRef]
5. Yadroitsava, I.; du Plessis, A.; Yadroitsev, I. Bone regeneration on implants of titanium alloys produced by laser powder bed fusion: A review. In *Titanium for Consumer Applications*, 1st ed.; Elsevier: Amsterdam, The Netherlands, 2019; pp. 197–233.
6. Miranda, G.; Sousa, F.; Costa, M.; Bartolomeu, F.; Silva, F.; Carvalho, O. Surface design using laser technology for Ti6Al4V-hydroxyapatite implants. *Opt. Laser Technol.* **2019**, *109*, 488–495. [CrossRef]
7. Chang, J.; Tsai, P.-I.; Kuo, M.; Sun, J.-S.; Chen, S.-Y.; Shen, H.-H. Augmentation of DMLS Biomimetic Dental Implants with Weight-Bearing Strut to Balance of Biologic and Mechanical Demands: From Bench to Animal. *Materials* **2019**, *12*, 164. [CrossRef]
8. Sui, Q.; Li, P.; Wang, K.; Yin, X.; Liu, L.; Zhang, Y.; Zhang, Q.; Wang, S.; Wang, L. Effect of Build Orientation on the Corrosion Behavior and Mechanical Properties of Selective Laser Melted Ti-6Al-4V. *Metals* **2019**, *9*, 976. [CrossRef]
9. Yuan, L.; Ding, S.; Wen, C. Additive manufacturing technology for porous metal implant applications and triple minimal surface structures: A review. *Bioact. Mater.* **2019**, *4*, 56–70. [CrossRef]
10. Bartolomeu, F.; Dourado, N.; Pereira, F.; Alves, N.; Miranda, G.; Silva, F. Additive manufactured porous biomaterials targeting orthopedic implants: A suitable combination of mechanical, physical and topological properties. *Mater. Sci. Eng. C* **2020**, *107*, 110342. [CrossRef]
11. Zhang, X.-Y.; Fang, G.; Zhou, J. Additively manufactured scaffolds for bone tissue engineering and the prediction of their mechanical behavior: A review. *Materials* **2017**, *10*, 50. [CrossRef]
12. Dhiman, S.; Sidhu, S.S.; Bains, P.S.; Bahraminasab, M. Mechanobiological assessment of Ti-6Al-4V fabricated via selective laser melting technique: A review. *Rapid Prototyp. J.* **2019**, *25*, 1266–1284. [CrossRef]
13. Javaid, M.; Haleem, A. Additive manufacturing applications in medical cases: A literature based review. *Alex. J. Med.* **2018**, *54*, 411–422. [CrossRef]
14. Bouet, G.; Cabanettes, F.; Bidron, G.; Guignandon, A.; Peyroche, S.; Bertrand, P.; Vico, L.; Dumas, V. Laser-Based Hybrid Manufacturing of Endosseous Implants: Optimized Titanium Surfaces for Enhancing Osteogenic Differentiation of Human Mesenchymal Stem Cells. *ACS Biomater. Sci. Eng.* **2019**, *5*, 4376–4385. [CrossRef]
15. Oliveira, T.T.; Reis, A.C. Fabrication of dental implants by the additive manufacturing method: A systematic review. *J. Prosthet. Dent.* **2019**, *122*, 270–274. [CrossRef]
16. Bose, S.; Robertson, S.F.; Bandyopadhyay, A. Surface modification of biomaterials and biomedical devices using additive manufacturing. *Acta Biomater.* **2018**, *66*, 6–22. [CrossRef] [PubMed]
17. Xiong, Y.; Gao, R.; Zhang, H.; Li, X. Design and fabrication of a novel porous titanium dental implant with micro/nano surface. *Int. J. Appl. Electromagn. Mech.* **2019**, *59*, 1097–1102. [CrossRef]
18. Wang, G.; Moya, S.; Lu, Z.; Gregurec, D.; Zreiqat, H. Enhancing orthopedic implant bioactivity: Refining the nanotopography. *Nanomedicine* **2015**, *10*, 1327–1341. [CrossRef] [PubMed]
19. Ferraris, S.; Bobbiob, A.; Miola, M.; Sprianoa, S. Micro- and nano-textured, hydrophilic and bioactive titanium dental implants. *Surf. Coat. Technol.* **2015**, *276*, 374–383. [CrossRef]
20. Cohen, D.J.; Cheng, A.; Sahingur, K.; Clohessy, R.M.; Hopkins, L.B.; Boyan, B.D.; Schwartz, Z. Performance of laser sintered Ti–6Al–4V implants with bone-inspired porosity and micro/nanoscale surface roughness in the rabbit femur. *Biomed. Mater.* **2017**, *12*, 025021. [CrossRef] [PubMed]
21. Bartolomeu, F.; Costa, M.; Gomes, J.; Alves, N.; Abreu, C.; Silva, F.; Miranda, G. Implant surface design for improved implant stability–A study on Ti6Al4V dense and cellular structures produced by Selective Laser Melting. *Tribol. Int.* **2019**, *129*, 272–282. [CrossRef]
22. Wysocki, B.; Idaszek, J.; Zdunek, J.; Rożniatowski, K.; Pisarek, M.; Yamamoto, A.; Święszkowski, W. The influence of selective laser melting (SLM) process parameters on in-vitro cell response. *Int. J. Mol. Sci.* **2018**, *19*, 1619. [CrossRef] [PubMed]
23. Wally, Z.J.; Haque, A.M.; Feteira, A.; Claeyssens, F.; Goodall, R.; Reilly, G.C. Selective laser melting processed Ti6Al4V lattices with graded porosities for dental applications. *J. Mech. Behav. Biomed. Mater.* **2019**, *90*, 20–29. [CrossRef] [PubMed]
24. Wang, H.; Zhao, B.; Liu, C.; Wang, C.; Tan, X.; Hu, M. A comparison of biocompatibility of a titanium alloy fabricated by electron beam melting and selective laser melting. *PLoS ONE* **2016**, *11*, e0158513. [CrossRef] [PubMed]
25. Xu, J.-Y.; Chen, X.-S.; Zhang, C.-Y.; Liu, Y.; Wang, J.; Deng, F.-L. Improved bioactivity of selective laser melting titanium: Surface modification with micro-/nano-textured hierarchical topography and bone regeneration performance evaluation. *Mater. Sci. Eng. C* **2016**, *68*, 229–240. [CrossRef] [PubMed]
26. Tsukanaka, M.; Fujibayashi, S.; Takemoto, M.; Matsushita, T.; Kokubo, T.; Nakamura, T.; Sasaki, K.; Matsuda, S. Bioactive treatment promotes osteoblast differentiation on titanium materials fabricated by selective laser melting technology. *Dent. Mater. J.* **2016**, *35*, 118–125. [CrossRef] [PubMed]
27. Chen, Z.; Yan, X.; Chang, Y.; Xie, S.; Ma, W.; Zhao, G.; Liao, H.; Fang, H.; Liu, M.; Cai, D. Effect of polarization voltage on the surface componentization and biocompatibility of micro-arc oxidation modified selective laser melted Ti6Al4V. *Mater. Res. Express* **2019**, *6*, 086425. [CrossRef]
28. Zhao, D.-P.; Tang, J.-C.; Nie, H.-M.; Zhang, Y.; Chen, Y.-K.; Zhang, X.; Li, H.-X.; Yan, M. Macro-micron-nano-featured surface topography of Ti-6Al-4V alloy for biomedical applications. *Rare Met.* **2018**, *37*, 1055–1063. [CrossRef]

29. Luo, Y.; Jiang, Y.; Zhu, J.; Tu, J.; Jiao, S. Surface treatment functionalization of sodium hydroxide onto 3D printed porous Ti6Al4V for improved biological activities and osteogenic potencies. *J. Mater. Res. Technol.* **2020**, *9*, 13661–13670. [CrossRef]

30. Torres, Y.; Sarria, P.; Gotor, F.J.; Gutiérrez, E.; Peon, E.; Beltrán, A.M.; González, J.E. Surface modification of Ti-6Al-4V alloys manufactured by selective laser melting: Microstructural and tribo-mechanical characterization. *Surf. Coat. Technol.* **2018**, *348*, 31–40. [CrossRef]

31. Sarria, P.; Torres, Y.; Gotor, F.J.; Gutiérrez, E.; Rodríguez, M.; González, R.; Hernández, L.; Peón, E.; Guerra, H.; González, J.E. Processing and characterization of Ti-6Al-4V samples manufactured by selective laser melting. *Key Eng. Mater.* **2016**, *704*, 260–268. [CrossRef]

32. Zhang, E.; Wang, Y.; Gao, F.; Wei, S.; Zheng, Y. Enhanced bioactivity of sandblasted and acid-etched titanium surfaces. *Adv. Mater. Res.* **2009**, *79–82*, 393–396. [CrossRef]

33. Wang, X.X.; Hayakawa, S.; Tsuru, K.; Osaka, A. A comparative study of in vitro apatite deposition on heat-, H_2O_2-, and NaOH-treated titanium surfaces. *J. Biomed. Mater. Res.* **2001**, *54*, 172–178. [CrossRef]

34. Wang, X.-X.; Hayakawa, S.; Tsuru, K.; Osaka, A. Bioactive titania gel layers formed by chemical treatment of Ti substrate with a H2O2/HCl solution. *Biomaterials* **2002**, *23*, 1353–1357. [CrossRef]

35. ISO. *ISO 4287*; ISO: Geneva, Switzerland.

36. Wu, F.; Xu, R.; Yu, X.; Yang, J.; Liu, Y.; Ouyang, J.; Zhang, C.; Deng, F. Enhanced Biocompatibility and Antibacterial Activity of Selective Laser Melting Titanium with Zinc-Doped Micro-Nano Topography. *J. Nanomater.* **2019**, *2019*, 1–14. [CrossRef]

37. Xu, R.; Hu, X.; Yu, X.; Wan, S.; Wu, F.; Ouyang, J.; Deng, F. Micro-/nano-topography of selective laser melting titanium enhances adhesion and proliferation and regulates adhesion-related gene expressions of human gingival fibroblasts and human gingival epithelial cells. *Int. J. Nanomed.* **2018**, *13*, 5045. [CrossRef]

38. Su, Y.; Luo, C.; Zhang, Z.; Hermawan, H.; Zhu, D.; Huang, J.; Liang, Y.; Li, G.; Ren, L. Bioinspired surface functionalization of metallic biomaterials. *J. Mech. Behav. Biomed. Mater.* **2018**, *77*, 90–105. [CrossRef] [PubMed]

39. He, F.M.; Yang, G.L.; Li, Y.N.; Wang, X.X.; Zhao, S.F. Early bone response to sandblasted, dual acid-etched and H_2O_2/HCl treated titanium implants: An experimental study in. *Int. J. Oral Maxillofac. Surg.* **2009**, *38*, 677–681. [CrossRef]

40. Cao, H.; Liu, X. Activating titanium oxide coatings for orthopedic implants. *Surf. Coat. Technol.* **2013**, *233*, 57–64. [CrossRef]

41. Zaffe, D.; Bertoldi, C.; Consolo, U. Accumulation of aluminium in lamellar bone after implantation of titanium plates, Ti–6Al–4V screws, hydroxyapatite granules. *Biomaterials* **2004**, *25*, 3837–3844. [CrossRef] [PubMed]

42. Benedetti, M.; Fontanari, V.; Bandini, M.; Zanini, F.; Carmignato, S. Low-and high-cycle fatigue resistance of Ti-6Al-4V ELI additively manufactured via selective laser melting: Mean stress and defect sensitivity. *Int. J. Fatigue* **2018**, *107*, 96–109. [CrossRef]

43. Bagehorn, S.; Wehr, J.; Maier, H. Application of mechanical surface finishing processes for roughness reduction and fatigue improvement of additively manufactured Ti-6Al-4V parts. *Int. J. Fatigue* **2017**, *102*, 135–142. [CrossRef]

44. Cao, F.; Zhang, T.; Ryder, M.A.; Lados, D.A. A review of the fatigue properties of additively manufactured Ti-6Al-4V. *JOM* **2018**, *70*, 349–357. [CrossRef]

45. Vayssette, B.; Saintier, N.; Brugger, C.; El May, M. Surface roughness effect of SLM and EBM Ti-6Al-4V on multiaxial high cycle fatigue. *Theor. Appl. Fract. Mech.* **2020**, *108*, 102581. [CrossRef]

46. Bourell, D.; Stucker, B.; Spierings, A.; Herres, N.; Levy, G. Influence of the particle size distribution on surface quality and mechanical properties in AM steel parts. *Rapid Prototyp. J.* **2011**, *17*, 195–202.

47. Günther, J.; Leuders, S.; Koppa, P.; Tröster, T.; Henkel, S.; Biermann, H.; Niendorf, T. On the effect of internal channels and surface roughness on the high-cycle fatigue performance of Ti-6Al-4V processed by SLM. *Mater. Des.* **2018**, *143*, 1–11. [CrossRef]

48. Vandenbroucke, B.; Kruth, J.P. Selective laser melting of biocompatible metals for rapid manufacturing of medical parts. *Rapid Prototyp. J.* **2007**, *13*, 196–203. [CrossRef]

49. Krol, M.; Tański, T. Surface quality research for selective laser melting of Ti-6Al-4V alloy. *Arch. Metall. Mater.* **2016**, *61*, 1291–1296. [CrossRef]

50. Strano, G.; Hao, L.; Everson, R.M.; Evans, K.E. Surface roughness analysis, modelling and prediction in selective laser melting. *J. Mater. Process. Technol.* **2013**, *213*, 589–597. [CrossRef]

51. Benedetti, M.; Torresani, E.; Leoni, M.; Fontanari, V.; Bandini, M.; Pederzolli, C.; Potrich, C. The effect of post-sintering treatments on the fatigue and biological behavior of Ti-6Al-4V ELI parts made by selective laser melting. *J. Mech. Behav. Biomed. Mater.* **2017**, *71*, 295–306. [CrossRef]

52. Fotovvati, B.; Namdari, N.; Dehghanghadikolaei, A. Fatigue performance of selective laser melted Ti6Al4V components: State of the art. *Mater. Res. Express* **2018**, *6*, 012002. [CrossRef]

53. Leuders, S.; Meiners, S.; Wu, L.; Taube, A.; Tröster, T.; Niendorf, T. Structural components manufactured by selective laser melting and investment casting—impact of the process route on the damage mechanism under cyclic loading. *J. Mater. Process. Technol.* **2017**, *248*, 130–142. [CrossRef]

54. Edwards, P.; Ramulu, M. Fatigue performance evaluation of selective laser melted Ti–6Al–4V. *Mater. Sci. Eng. A* **2014**, *598*, 327–337. [CrossRef]

55. Kahlin, M.; Ansell, H.; Moverare, J. Fatigue behaviour of notched additive manufactured Ti6Al4V with as-built surfaces. *Int. J. Fatigue* **2017**, *101*, 51–60. [CrossRef]

56. Zhang, J.; Fatemi, A. Surface roughness effect on multiaxial fatigue behavior of additive manufactured metals and its modeling. *Theor. Appl. Fract. Mech.* **2019**, *103*, 102260. [CrossRef]

57. Chan, K.S.; Koike, M.; Mason, R.L.; Okabe, T. Fatigue life of titanium alloys fabricated by additive layer manufacturing techniques for dental implants. *Metall. Mater. Trans. A* **2013**, *44*, 1010–1022. [CrossRef]
58. Vayssette, B.; Saintier, N.; Brugger, C.; El May, M.; Pessard, E. Numerical modelling of surface roughness effect on the fatigue behavior of Ti-6Al-4V obtained by additive manufacturing. *Int. J. Fatigue* **2019**, *123*, 180–195. [CrossRef]
59. Witkin, D.B.; Patel, D.N.; Helvajian, H.; Steffeney, L.; Diaz, A. Surface treatment of powder-bed fusion additive manufactured metals for improved fatigue life. *J. Mater. Eng. Perform.* **2019**, *28*, 681–692. [CrossRef]
60. Apachitei, I.; Lonyuk, B.; Fratila-Apachitei, L.; Zhou, J.; Duszczyk, J. Fatigue response of porous coated titanium biomedical alloys. *Scr. Mater.* **2009**, *61*, 113–116. [CrossRef]
61. Smith, T. The effect of plasma-sprayed coatings on the fatigue of titanium alloy implants. *JOM* **1994**, *46*, 54–56. [CrossRef]
62. Khan, R.H.; Yerokhin, A.; Matthews, A. Structural characteristics and residual stresses in oxide films produced on Ti by pulsed unipolar plasma electrolytic oxidation. *Philos. Mag.* **2008**, *88*, 795–807. [CrossRef]

MDPI

St. Alban-Anlage 66

4052 Basel

Switzerland

Tel. +41 61 683 77 34

Fax +41 61 302 89 18

www.mdpi.com

Metals Editorial Office

E-mail: metals@mdpi.com

www.mdpi.com/journal/metals

Manufacturing Processes

Manufacturing Processes

Santosh Bhatnagar
BSc, BSc (Engg) Mech AMA eSI PGDQM
Head of Department
Aeronautical Engineering
Feroze Gandhi Institute of Engineering and Technology
Raebareli.

BS Publications
An unit of **BSP Books Pvt., Ltd.**

4-4-309, Giriraj Lane, Sultan Bazar,
Hyderabad - 500 095
Phone : 040 - 23445605, 23445688

Published by :

 **BS Publications**

An unit of **BSP Books Pvt., Ltd.**

4-4-309, Giriraj Lane, Sultan Bazar,
Hyderabad - 500 095
Phone : 040 - 23445605, 23445688
e-mail : info@bspbooks.net

ISBN : 978-93-52300-71-6 (HB)

Dedicated to
My Beloved Parents

Preface

This book is written specially for the Engineering students of U.P. Technical University, Lucknow. This subject "Manufacturing Processes" is a compulsory subject, which is common to all branches of engineering in First year Engineering exam of UPTU.

The book is written as per UPTU syllabus. Care has been taken to explain the basic manufacturing process in a simple language with diagrams wherever necessary with special stress on materials for aeronautical use.

Questions have been compiled at the end of each chapter. This will enable the students to review their syllabus at the end of the course.

Suggestions from readers are most welcome and shall be acknowledged.

- Author

Acknowledgement

I would like to start by thanking my honorable Director, Shri R.P. Sharma for his encouragement and blessings which he has bestowed upon me for writing this book.

While writing this book my beloved mother 'Shanti' had been a guiding light to me all these days. Her demise had left me all heart broken. None other than my loving wife 'Rashmi' stood by me and gave me moral support to complete this book. My son 'Saumya', his wife 'Bhawana', my granddaughter 'Arushi' and my sister-in-law Divyashri, also encouraged me from time to time during the preparation of this book.

I would also like to thank my publishers B.S. Publications and their staff especially Shri Naresh Davargave and Mrs Kalpana who helped me in publishing this book.

I also want to thank my colleagues and Mr. Ajay Trivedi of F.G.I.E.T, Raebareli for helping me to complete this book.

Looking forward for more support in near future.

- Author

Contents

Preface .... (vii)

Acknowledgement .... (ix)

Unit I - Basic Metals and Alloys: Properties and Applications

Chapter 1

Properties of Materials

1.1 Introduction 1

1.2 Classification of Properties 1

    1.2.1 Mechanical Properties 1

        1.2.1.1 Brittleness 1

        1.2.1.2 Creep 2

        1.2.1.3 Ductility 3

        1.2.1.4 Elasticity 3

        1.2.1.5 Fatigue 4

        1.2.1.6 Hardness 4

        1.2.1.7 Strength 5

        1.2.1.8 Toughness 6

        1.2.1.9 Fracture 6

        1.2.1.10 Malleability 6

    1.2.2 Physical Properties 6

    1.2.3 Chemical Properties 6

Questions 6

Chapter 2

Ferrous Materials

2.1 Introduction 8

2.2 Carbon Steels 8

2.3 Alloy Steels 9

2.3.1   Effect of Individual Elements on Alloy Steels 9
        2.3.1.1   Manganese 9
        2.3.1.2   Chromium 10
        2.3.1.3   Nickel 10
        2.3.1.4   Molybdenum 10
        2.3.1.5   Tungsten 11
        2.3.1.6   Vanadium 11
        2.3.1.7   Titanium 11
        2.3.1.8   Silicon 11
2.3.2   Special Alloy Steels 11
        2.3.2.1   Stainless Steel 11
        2.3.2.2   High speed steels 12
        2.3.2.3   Tool Steels 13
2.4   Cast Iron 13
    2.4.1   Types of Cast Iron 14
2.5   Heat Treatment of Carbon Steels 15
    2.5.1   Iron-Carbon Diagram 15
    2.5.2   Different Forms of Iron and Steel 17
2.6   Heat Treatment Processes of Steels 18
Questions 20

Chapter 3

Non-ferrous Metals and Alloys

3.1   Introduction 22
3.2   Aluminum Alloys 22
    3.2.1   Duralumin 23
    3.2.2   Y-alloy 23
    3.2.3   Aluminium casting alloys 23
3.3   Nickel alloys 23
    3.3.1   Inconel 23
    3.3.2   Monel 24
    3.3.3   K-monel 24
3.4   Copper Alloys 25
    3.4.1   Brass 25

3.4.1.1   Muntz metal 25

3.4.1.2   Manganese brass 25

3.4.1.3   Naval brass 25

3.4.1.4   Admirality brass 25

3.4.1.5   Delta Brass 25

3.4.2   Bronze 26

3.4.2.1   Gun metal 26

3.4.2.2   Phosphor bronze 26

3.4.2.3   Phosphor bronze casting alloy 26

3.4.2.4   Aluminium bronze........................... 26

3.4.2.5   Aluminium bronze casting alloy................... 26

3.5   Magnesium Alloys 26

Questions 27

Unit II - Introduction to Metal Forming, Casting Processes and its Applications

Chapter 4
Metal Forming

4.1   Introduction... 29

4.1.1   Recrystallisation Temperature 29

4.1.2   Hot Working.. 29

4.1.2.1   Advantages of Hot Working............ 29

4.1.2.2   Disadvantages of Hot Working 30

4.1.2.3   Main Hot-working Processes.................. 30

4.1.2.3.1   Forging............................ 30

4.1.2.3.2   Hot Rolling 32

4.1.2.3.3   Extrusion.......................... 37

4.1.2.3.4   Wire Drawing 40

4.1.2.3.5   Sheet Metal Working............. 40

Questions 41

Chapter 5
Casting

5.1   Introduction... 42

5.1.1   Advantages of Casting Processes 42

5.1.2   Terms Used in Castings.. 43

5.2   Pattern Making..44

    5.2.1   Pattern Materials..44

        5.2.1.1   Wood 44

        5.2.1.2   Metals 45

            5.2.1.2.1   Cast Iron......................................45

            5.2.1.2.2   Brass 45

            5.2.1.2.3   Aluminium....................................45

            5.2.1.2.4   White Metal 45

        5.2.1.3   Plastic 46

        5.2.1.4   Plaster of Paris or Gypsom Cement............46

        5.2.1.5   Wax..46

        5.2.1.6   Polystyrene or Expanded Thermocole..........46

    5.2.2   Pattern Making Tools 46

    5.2.3   Pattern Allowances...46

        5.2.3.1   Shrinkage Allowance (Contraction) 47

        5.2.3.2   Draft Allowance 47

        5.2.3.3   Machining Allowance................................47

        5.2.3.4   Distortion Allowance (Camber) 47

        5.2.3.5   Rapping Allowance (Shaking)....................48

5.3   Types of Patterns...48

    5.3.1   Solid or single piece pattern 48

    5.3.2   Spilt Pattern 48

    5.3.3   Sweep Pattern 49

    5.3.4   Gated Pattern 49

    5.3.5   Match Plate Pattern 49

    5.3.6   Loose Piece Pattern 49

5.4   Moulding Sand...50

    5.4.1   Properties of Moulding Sand..50

        5.4.1.1   Porosity or Permeability 50

        5.4.1.2   Plasticity 50

        5.4.1.3   Adhesiveness 50

        5.4.1.4   Refractoriness 50

        5.4.1.5   Cohesivness 50

        5.4.1.6   Collapsibility 51

5.4.2   Types of Moulding Sand 51

5.5   Cores 52

    5.5.1   52

        5.5.1.1   Horizontal Core 52

        5.5.1.2   Vertical Core 52

        5.5.1.3   Balanced core 52

        5.5.1.4   Stop off Core or Drop Core 53

        5.5.1.5   Hanging Core or Cover Core 53

    5.5.2   Core Making 53

    5.5.3   Core Setting 53

5.6   Moulding Processes 54

    5.6.1   Types of Moulds 54

        5.6.1.1   Green Sand Mould 54

        5.6.1.2   Dry Sand Mould 54

        5.6.1.3   Skin Dried Mould 54

        5.6.1.4   Loam Mould 55

        5.6.1.5   Metal Mould 55

        5.6.1.6   Shell Mould 55

    5.6.2   Methods Used for Preparation of Mould 55

5.7   Moulding Procedure 55

5.8   Gating System 56

    5.8.1   Pouring Basin / Cup 56

    5.8.2   Sprue 56

    5.8.3   Runners 56

    5.8.4   Gates 56

    5.8.5   Risers 57

5.9   Casting Defects and Remedies 57

    5.9.1   Blow Holes 57

    5.9.2   Hot tears (Pulls) 57

    5.9.3   Cold Shuts (Misrun) 58

    5.9.4   Wrapage 58

    5.9.5   Hot Spots 58

    5.9.6   Fins 58

5.9.7   Run Out 58

5.9.8   Swell... 58

5.9.9   Buckles Or Rattail 59

5.9.10   Scabs (Metal Penetration).. 59

5.9.11   Pin Holes (Porosity) 59

5.9.12   Inclusions... 59

5.9.13   Poured Short 59

5.9.14   Shrinkage Or Hot Cracking... 59

5.9.15   Flow Marks... 59

5.10   Cupola Furnace 60

5.11   Die Casting.. 61

Questions 63

Unit III - Introduction to Machining Welding and Its Applications

Chapter 6

Machining

6.0   Basic Principles of Lathe Machine and Operations Performed on it 65

6.1   Lathe 65

6.1.1   Main parts of a Centre Lathe Machine 65

6.1.2   Lathe Operations 67

6.1.2.1   Plain Turning 68

6.1.2.2   Step Turning 68

6.1.2.3   Facing 68

6.1.2.4   Taper Turning 69

6.1.2.5   Eccentric Turning 69

6.1.2.6   Drilling 69

6.1.2.7   Boring 70

6.1.2.8   Reaming... 70

6.1.2.9   Knurling... 70

6.1.2.10   Chamfering 70

6.1.2.11   Forming 71

6.1.2.12   Under Cutting or Grooving........................... 71

6.1.2.13   Threading or Thread Cutting or
                  Screw Cutting 72

	6.1.3		Types of Lathes	72
		6.1.3.1	Speed Lathe	72
		6.1.3.2	Engine or Centre Lathe	72
		6.1.3.3	Tool Room Lathe	73
		6.1.3.4	Copying Lathe	73
		6.1.3.5	Capstan and Turret Lathes	73
		6.1.3.6	Automatic Lathes	73
		6.1.3.7	Special Lathes	73
	6.1.4		Difference between Capstan and Turret Lathes	74
6.2	Shaper			74
	6.2.1		Principal Parts of A Shaper	75
	6.2.2		Classification of Shaper	75
	6.2.3		Operations Performed on Shaper Machine	76
	6.2.4		Advantages and Disadvantages of Shaper	76
6.3	Planar			77
	6.3.1		Types of Planning Machines	77
	6.3.2		Main Parts of a Planar	77
	6.3.3		Working Principle of a Planar	78
	6.3.4		Difference between Planar and Shaper	78
6.4	Drilling Machine			78
	6.4.1		Types of Drilling Machines	78
	6.4.2		Parts of a Drilling Machine	79
	6.4.3		Geometry of Twist Drills	80
	6.4.4		Operations on Drilling Machines	80
6.5	Milling Machine			81
	6.5.1		Types of Milling Machine	81
	6.5.2		Main Parts of a Vertical Milling Machine	81
	6.5.3		Milling Cutters	82
	6.5.4		Difference Between up Milling and Down Milling	83
	6.5.5		Advantages and Disadvantages of Milling	83
6.6	Grinding Machines			84
	6.6.1		Types of Grinding Machines	84
	6.6.2		Common Grinding Operations	84
		6.6.2.1	Cylindrical Grinding	84

6.6.2.2	Surface Grinding....................................85

6.6.2.3	Internal Grinding 85

6.6.2.4	Centreless Grinding 86

6.6.3	Grinding Wheels 86

6.6.4	Grinding Wheel Specification 88

Questions 89

Chapter 7

Welding

7.1	Importance and Basic Concepts of Welding...........................91

7.2	Classification of Welding Processes 91

7.3	Gas Welding..92

7.3.1	Oxyacetylene Welding 92

7.3.2	Oxyacetylene Welding Equipment.......................92

7.3.3	Types of Flames 94

7.3.4	Advantages and Disadvantages of Gas Welding.................95

7.4	Electric Arc Welding 96

7.4.1	Metallic Arc Welding...96

7.4.2	Carbon Arc Welding 97

7.4.3	Atomic Hydrogen Welding 97

7.4.4	Inert Arc Welding..98

7.4.5	Multi Arc Welding 99

7.4.6	Metallic Inert Gas Welding (MIG).........................99

7.5	Electric Resistance Welding 100

7.5.1	Advantage and Disadvantages of Resistance Welding......101

7.5.2	Types of Electric Resistance Welding............................101

7.5.2.1	Spot Welding.......................................101

7.5.2.2	Seam Welding 102

7.5.2.3	Butt Welding 102

7.6   Soldering (or Soft Soldering) 103

7.7   Brazing 104

    7.7.1   Copper Brazing 104

    7.7.2   Silver Brazing 105

    7.7.3   Aluminium Brazing 105

7.8   Difference Between Soldering and Brazing 105

Questions 105

Unit IV - Manufacturing

Chapter 8

Miscellaneous Topics

8.1   Importance of Materials and Manufacturing towards Technological and Socio economic Developments 107

8.2   Plant Location 108

8.3   Plant Layout 109

    8.3.1   Objectives of Good Plant Layout 109

    8.3.2   Advantages of a Good Plant Layout 109

    8.3.3   Types of Plant Layout 110

        8.3.3.1   Product or Line Layout 110

        8.3.3.2   Process or Functional Layout 110

        8.3.3.3   Combination or Group Layout 111

        8.3.3.4   Fixed Position Layout 111

8.4   Types of Production 113

8.5   Production Verses Productivity 113

Questions 114

Chapter 9

Non-Metallic Materials

9.1   Common Types and Uses of Wood   115

    9.1.1   Types of Wood 115

    9.1.2   Uses of Wood 116

9.2   Cement Concrete 116

9.3   Ceramics 116

9.4   Rubber.. 117

    9.4.1   Types of synthetic Rubbers 119

        9.4.1.1   Buna S 119

        9.4.1.2   Buna N... 119

        9.4.1.3   Neoprene.. 119

        9.4.1.4   Butyl 119

        9.4.1.5   Thiokol 120

9.5   Composite Materials 120

    9.5.1   Types of composite materials... 120

        9.5.1.1   Laminated Materials............................. 120

        9.5.1.2   Agglomerated Materials 120

Questions.. 121

Chapter 10

Miscellaneous Processes

10.1   Powder Metallurgy Process and its Applications................................. 122

    10.1.1   Main Characteristics of Powders..................................... 122

    10.1.2   Steps in Powder Metallurgy 123

    10.1.3   Production of Metal Powders 124

    10.1.4   Advantages and Disadvantages of Powder Metallurgy..... 125

    10.1.5   Applications of Powder Metallurgy 126

10.2   Plastic Products Manufacturing   126

    10.2.1   Introduction 126

    10.2.2   Properties of Plastics 127

    10.2.3   Types of Plastics.. 127

        10.2.3.1   Thermoplastics 127

        10.2.3.2   Thermosetting Plastics................................ 129

10.2.4	Plastic Processing Methods	129
10.2.4.1	Molding	129
10.2.4.2	Calendering	132
10.2.4.3	Casting	133
10.2.4.4	Laminating	133
10.3	Electroplating	133
10.4	Galvanizing	134
Questions		135

Unit I

Basic Metals and Alloys: Properties and Applications

CHAPTER **1**

Properties of Materials

1.1 Introduction

Manufacturing processes depend on materials, which in turn depend on their properties. While selecting a material, a manufacturer has to decide on the type of material based on the end use of the product. Designer's recommendations in the engineering drawing are to be adhered to or mandatory.

In the case of aeronautical industry, special attention is to be paid on the weight. Light materials such as aluminium, magnesium and its alloys are preferred above than other materials. Use of composite materials is also recommended.

For any production process the knowledge of various materials and their properties is essential.

In this chapter an effort is made in this direction.

1.2 Classification of Properties

The properties of materials may be broadly classified into three main categories -

Mechanical, Physical and Chemical

1.2.1 Mechanical Properties

These include brittleness, creep, ductility, elasticity, elastic limit, Hookes's law, fatigue, fatigue stress, fatigue limit, fatigue strength, toughness and fracture.

1.2.1.1 Brittleness

It is the property of resisting a change in the relative position of molecules or the tendency to fracture without appreciable deformation and under low stress.

Examples of brittle materials are cast iron, glass etc.

1.2.1.2   Creep

It is a time dependant strain occurring under stress at high temperatures below elastic limit and continuous deformation at steady load. Some of the terms related to creep such as creep limit, creep strength and creep tests should also be understood.

Creep limit

The maximum tensile stress to which a material can be subjected at a given temperature without resulting in measurable creep.

Creep Strength

The ability of a material to resist deformation under constant stress. It is measured as the amount of creep induced by a constant stress acting for a given time and temperature.

Creep tests

These are the methods of measuring the resistance of metals to creep. Time extension curves under constant loads are determined. It is very similar to tensile test carried out at higher temperatures.

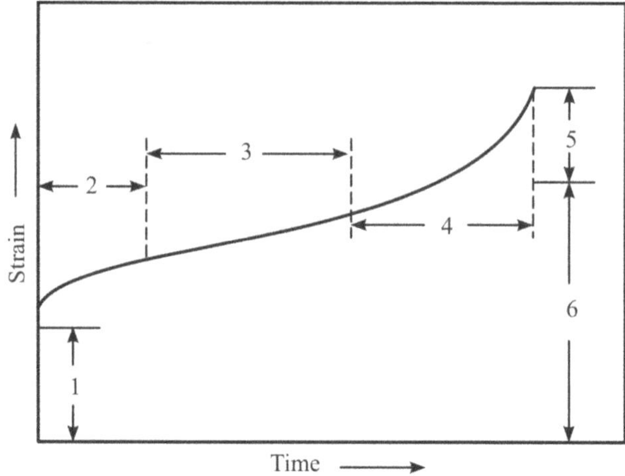

Fig. 1.1 Creep curves

A creep curve is shown in Fig. 1.1. When a load is applied following are the details:
1. Elastic extension occurs.
2. Specimen gradually stretches at decreasing rate i.e. first stage creep.
3. The rate becomes constant for a certain period of time i.e. second stage creep.

4. The rate of creep increases till the specimen fails i.e. third stage creep.

5. This is elastic contractions.

6. This is permanent change of shape.

1.2.1.3   Ductility

Ductility is the property of materials which allows them to be drawn out into wires through a die without breaking on application of load

or

Ability of metals and alloys to retain strength and freedom from cracks when shape is altered

or

The property of material that can be deformed without fracture when a tension load is applied.

Ductility is similar to malleability.

A ductile material should have both strength and plasticity.

Some examples of ductile material are gold, silver and copper. Lead is difficult to draw because its strength is low.

1.2.1.4   Elasticity

It is the tendency of a body to return to its original size or shape after removal of load after having been stretched, compressed or deformed under load

or

The extent to which a structure may be elongated by deforming stress without causing any permanent change.

Elastic limit

It is the limit or value of the deforming force beyond which a body does not return to its original shape or dimensions when the force is removed. It is the highest stress that can be applied to a metal without producing a measurable amount of plasticity i.e. permanent deformation as in the case of limit of proportionality.

Hooke's law

It states that stress is proportional to strain upto the elastic limit or stress is proportional to strain.

$$\text{(Modulus of elasticity) E} \quad \frac{\text{Stress}}{\text{Strain}} \quad \text{(upto elastic limit)}$$

$$\text{Stress} \quad \frac{\text{Force}}{\text{Area}} \quad \frac{F}{A}$$

$$\text{Strain} \quad \frac{\text{Change in length}}{\text{Original length}} \quad \frac{l}{L}$$

$$\therefore \quad E \quad \frac{F/A}{l/L}$$

$$E = \frac{F.L}{A.l}$$

Thus if we know the above parameters we can find out the modulus of elasticity.

1.2.1.5   Fatigue

It is the failure in metals due to application of a cycle of stresses

or

A material that has been subjected to severe or moderate straining due to fluctuating stresses acting on it for a long time and thereby weakens in strength and finally breaks under loads that were safe.

Fatigue Strength

The measured resistance of a body to failure caused by repeated applications of stress.

Fatigue Stress

It is the stress which the material will endure without failure no matter how many times the stress be repeated.

Fatigue Limit

It is the upper limit of the range of stress that a metal can withstand indefinitely. If this limit is exceeded failure will occur.

1.2.1.6   Hardness

It is the property of resistance to plastic deformation by indentation, penetration, scratching or bending.

Hardness Measurement

Hardness can be measured by indentation tests. These are performed on mainly three types of machines namely:

1. Brinell hardness machine.   2. Rockwell hardness machine.

3. Vicker hardness machine.

The principle of these type of machines is that a hardened steel ball is sunk under a definite load into the material being tested. The impression made by the ball is then measured and recorded. Smaller the impression harder the material. For each material there is a definite relationship between the depth of penetration (represented by hardness number) and the ultimate strength of the material.

Various symbols for hardness number is given depending on the type of machine.

Machine	Symbol
Brinnel hardness machine	BHN
Rockwell hardness machine	C
Vicker hardness machine	VHN

Hardness tables have been drawn up based on experiments on different materials showing relation between BHN, C and VHN.

The advantage of hardness testing machines is that hardness can be obtained without destroying the part.

1.2.1.7 Strength

It is the ability of a material to sustain loads without undue distortion or failure. The material should have adequate strength when subjected to tension, compression, shear, bending or torsion to resist the strain.

Tensile Strength

It is the strength necessary to enable a bar or structure to resist a tensile strain or the pulling stress required to break a given specimen.

Compressive Strength

It is the resistance provided against forces that tend to squeeze from the end or compress, shorten or crush.

Shear Strength

It is the value of load applied tangentially to shear it off across the resisting section.

Bending Strength

It is the ability of a beam or other structural member to resist a bending force.

Torsional Strength

It is the strength of a material in such a direction that opposes a twisting force.

1.2.1.8   Toughness

It is the measure of the ability of a material to absorb energy before fracture takes place. It is a condition between brittleness and softness. It is indicated in tensile tests by high ultimate tensile stress and low to moderate elongation and reduction in area. It is also associated with high values in impact strength.

1.2.1.9   Fracture

It is the failure of a material under load to break apart or to separate the continous parts of an object by sudden shock or by excessive strain.

Fracture tests are conducted by deliberately breaking metal part to allow the examination of the fracture face in order to determine the grain size, case depth etc.

1.2.1.10   Malleability

It is the property of permanently extending in all directions without fracture by pressing, hammering, rolling etc.

1.2.2 Physical Properties

Those properties of a body which may be determined by methods other than chemical, including shape, size, weight and behaviour in certain conditions.

Physical tables listing the physical properties and characteristics of materials are used for studies and selection of material. There are many physical properties; some of them are density, conductivity, resistivity, specific heat, reflection and refraction in light, permeability and magnetism etc.

1.2.3 Chemical Properties

The chemical properties of material depend on composition and structure. Corrosion, resistance is one of the major property of a material to decide its use.

Questions

1. What are main categories in which materials can be classified?
2. What are the properties of materials? Which property is most essential for aircraft materials.
3. What is Hooke's law? Define modulus of elasticity?
4. What is a Creep? Explain the details with the help of creep curve.
5. Differentiate between
   (a) Malleability and Ductility   (b) Brittleness and Fracture
6. Explain the term "Fatigue". What is the difference between "Fatigue stress" and "Fatigue limit"

7. Define Hardness. What are the various hardness machines in use for measuring hardness.

8. What are main mechanical properties? Explain any five of them.

9. Define Stress and Strain. How are they related

10. Define Strength, compressive strength, shear strength, bending strength, Torsional strength.

CHAPTER 2

Ferrous Materials

2.1 Introduction

Ferrous metal is a term applied to a group of metals having iron as their principal constituent. Ferrous is a term used in a metal alloy in which iron is the major ingredient or base metal especially.

2.2 Carbon Steels

When a small percentage of one to two percent of carbon is added to iron the produce becomes far superior to iron and is known as carbon steels. It becomes the base metal for many of the alloys. Besides iron and carbon, carbon steels contain manganese (up to 1.6%), silicon (up to 0.6%), sulphur and phosphorous (upto 1%) but no chromium, molybdenum, nickel etc.

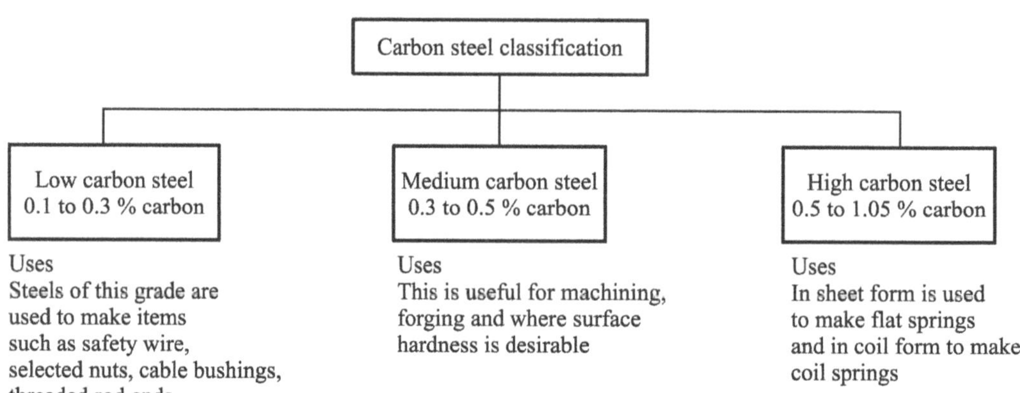

2.3   Alloy Steels

An alloy steel is a carbon steel to which other metallic elements such as nickel, chromium, molybdenum, vanadium, tungsten, titanium, silicon and manganese are added.

The new alloy steel has different properties from the original carbon steel.

Alloy steels are usually named after the principal alloying element e.g. nickel steels, chromium steels, molybdenum steels and manganese steels.

When two alloying elements are present they are named as:

Nickel-chromium steel

Chromium molybdenum steel

Chromium vanadium steels

Silicon-Manganese steels

One alloy steel which is commonly used for propeller hubs contains chromium-nickel molybdenum.

The advantages of alloy steels are as follows:
- It increases corrosion resistance
- It increases hardenability
- It improves wear resistance
- It increases toughness
- It increases strength
- It improves ductility and machining properties
- It also increases magnetic and electrical properties

Some of the disadvantages of alloy steels are:
- They are costly as compared to corrosion steels
- Sulphur and phosphorous are undesirable impurities which cannot be eliminated completely but are kept as low as possible.

2.3.1   Effect of Individual Elements on Alloy Steels

2.3.1.1   Manganese

It is the most important constituent in steel. The main purpose of manganese is to deoxidise and desulfurise the steel. It deoxides by eliminating ferrous oxide which is a harmful impurity. It combines with sulphur to form manganese sulphide which is harmless in small amounts. Manganese possesses the property known as "Penetration hardness" which means that in heat treatment of large sections, the hardness is not merely on surface but penetrates to the core also. The presence of manganese greatly improves the forging qualities of steel.

Upto 1% manganese in steel produces hardening and toughening. More than 1% manganese will increase brittleness. However steel containing about 13% manganese is hard and ductile.

2.3.1.2 Chromium

Chromium when added to steel imparts hardness, wear resistance, magnetic qualities and corrosion resistance to steel. Chromium possesses excellent "Penetration hardness". Corrosion resistant steels contain large amounts of chromium. Most common of these is 18-8 steel i.e. 18% chromium and 8% nickel. Chromium alloys are used where wear resistance is required like ball bearings. Tungsten-chromium alloys are used for high speed cutting tools. It is also the main alloying element of stainless steel. The main use of chromium in alloys is when combined with nickel, molybdenum and vanadium. About 1% of chromium is present in these alloys which are strong, hard and have fair ductility.

It is almost non-magnetic. However some of chromium steels are used for magnets.

2.3.1.3 Nickel

Steel added with nickel increases the strength, yield point and hardness without affecting ductility and also increases corrosion resistance. it is used for structural parts of valves. It is non-magnetic and in the form of wire, was much used in computers.

When combined with chromium it produces nickel chromium steel (Ni 1.5 to 4% and Cr 0.5 to 2%) to produce an alloy of high tensile strength, hardness and toughness. Used for highly stressed automobile and aero engine parts.

The commonly used nickel steels contain from 3 to 5% nickel. In heat treatment the presence of nickel in steel slows down the critical rate of hardening, which in turn increases the depth of hardening and produces a fine grain structure. There is also less warpage and scaling of heat treated nickel steel parts. Nickel increases corrosion resistance of steel. It is also one of the principal element of stainless steel.

Various nickel alloys are Inconel, Monel and K-Monel which will be dealt in non-ferrous metals.

2.3.1.4 Molybdenum

It improves homogeneity of metal, elastic limit, impact strength and wear resistance fatigue strength and reduces grain size. Its air hardening property is very useful in aircraft materials as it gives better welding joint. Molybdenum steel is readily heat treated. It is highly resistant to heat and conducts electricity easily.

2.3.1.5 Tungsten

Tungsten is used as alloying element in steel in the industries for making the cutting tools due to its hardness. Tungsten steels have no direct application in aircraft construction. It has the property of "Red hardness" and so used as high speed steel which will retain its hardness even when it is red hot and as such used as cutting tool in lathe work. Tool steels contain from 14-18% tungsten and 2-4% chromium.

2.3.1.6 Vanadium

Vanadium is an intensive deoxidising agent and improves the grain structure, impact strength, fatigue strength, resistance to vibration and stress reversal. Vanadium alloys generally contain 1% of chromium and are called chrome-vanadium steel and have good ductility and high strength.

2.3.1.7 Titanium

It is highly resistant to corrosion and is used widely in making alloys because it unites with nearly every metal except copper and aluminium. It is characterised by strength, lightness and corrosion resistance. It is used as a deoxidiser in aircraft construction and repair. Titanium is used for fuselage skins, engine shrouds, fire walls, frames, fittings, air ducts and fasteners. Titanium is often added in small quantities to 18-8 corrosion resisting steel to reduce embrittlement at the operating temperatures of exhaust stacks and collectors.

2.3.1.8 Silicon

It is used in steel as deoxidising and hardening agent. A small amount 0.3% or less if present in steel improves ductility. An alloy which has 0.5-4.5% silicon is used in electrical transformers in the form of lamination to make up the core.

Silicon and manganese in large amounts are used as alloying elements in the form of silico-manganese steels. These steels have good impact resistance.

2.3.2 Special Alloy Steels

These alloy steels are used for specific purpose only. Some of these are:

- Stainless steels or corrosion resisting steel.
- High speed steels (H.S.S)
- Tool steels

2.3.2.1 Stainless Steel

These steels are classified into three groups as follows:

Group 1 (Chromium nickel steel): These steels contain 0.2% carbon or less, 17-25% chromium and 7-13% nickel. 18-8 steel belongs to this group. This steel with some changes are used in aircraft construction. Strength of these steels cannot be increased by heat treatment but only by cold working.

Group 2 (Hardenable chromium steels): These steels contain from 12% to 18% chromium with varying amounts of carbon up to as high as 1%. This type is used for household cutlery. It is also used for manufacture of aircraft bolts and fittings requiring good corrosion resistance.

Group 3 (Non hardenable chromium steels): These steels contain from 15% to 30% chromium and up to 35% carbon. They are not hardenable by heat treatment.

2.3.2.2 High speed steels

It is a hard steel used for metal cutting tools. It retains its hardness at a low red heat and hence the tools can be used in lathes etc. operated at high speeds. It usually contains 12-18% tungsten, up to 5% chromium 0.4-0.7% carbon and small amounts of other elements such as vanadium, molybdenum etc. Some of the high speed steels are as follows:

18-4-1 HSS

Constituents	Uses
18% Tungsten 4% Chromium 1% Vanadium 0.75% Carbon Rest Iron	It is widely used for making cutting tools for lathes, shapers, milling cutters, planers, slotters, broaches etc.

Cobalt High Speed Steel (HSS)

Constituents	Uses
12% Cobalt 20% Tungsten 4% Chromium 2% Vanadium 0.8% carbon Rest iron	It is principally used for heavy cutting operation only, since it is costly

Vanadium High Speed Steel (HSS)

Constituents	Uses
Carbon 0.4-0.5% Vanadium 0.15-2% Chromium 1.1-1.5% Rest Iron	This is superior to 18-4-1 HSS. It is preferred for machining for difficult to machine materials

Molybdenum High Speed Steel (HSS)

Constituents	Uses
6% Tungsten 6% Molybdenum 4% chromium 2% vanadium	Used for drilling and tapping operations it possesses excellent toughness and cutting ability

2.3.2.3 Tool Steels

It is steel suitable for use in tools usually for cutting and shaping of wood and metals. Main qualities required are hardness, toughness, ability to retain a cutting edge, it contains 0.6-1.6% carbon. Many cutting tool steels contain high percentages of alloying metals such as tungsten, chromium, molybdenum etc. It is usually quenched to obtain the required properties such as HSS. These are of two types.

Low alloy tool steels: These contain silicon, tungsten, chromium and manganese as alloying elements and retain hardness upto 200 °C approximately.

High alloy tool steels: These contain vanadium, chromium, tungsten. Primarily they retain hardness upto 600 °C approximately. These steels have good abrasion resistance and wear resistance.

2.4 Cast Iron

It is an alloy of iron and carbon in which carbon content exceeds the solubility of carbon in austenite at eutectic temperature. The carbon content is above 2%, it also contains silicon, manganese, phosphorous and sulphur also. It is available in different forms such as grey cast iron, white cast iron, malleable cast iron, nodular cast iron etc.

It has low melting point about 1200 °C, can be easily melted, hence can be cast into various shapes and easily machined.

Cast iron is produced by remelting the pig iron in a cupola furnace together with a definite amount of limestone (flux), steel scrap and spoiled castings. Remelting of pig iron as above enables it to be cast into moulds to give suitable castings (It is then known as cast iron).

Cast iron contains 2.2% to 4.5% carbon. It is very brittle has high wear and abrasion resistance, can be easily machined has lower melting temperature than steel, have high yield strength, high ductility and toughness. It is low cost material and can be easily cast. Lappers are made of cast iron which are used for superfinishing method (lapping) for machined parts.

It has wide applications for producing castings for mechanical / aeronautical parts such as pump housing, steam turbine housing, engine frames, beds and slides of Lathe, milling, planing machines etc.

2.4.1 Types of Cast Iron

The various types of cast iron which are used widely are:

- Grey cast iron
- White cast iron
- Malleable cast iron
- Nodular or ductile cast iron
- Alloy cast iron

The following table shows a comparison between all the five types of cast iron.

Types of cast iron	Percentage of carbon	Properties	Applications
Grey cast iron	2 to 4% carbon 1% silicone rest iron	Grey colour carbon present as free graphite low tensile strength high compressive strength, no ductility, easily machined. Free graphite acts as lubricant	Base for erection of machinery, cylinder blocks and heads of IC engines, piston rings, gas / water pipes
White cast iron	2 to 3.5% carbon 0.8 to 1.5% silicone up to 0.2% sulphur up to 0.5% manga-nese rest iron	White surface on fracture because of iron carbide i.e. cementite. It is wear resistant, very hard and brittle, has high compressive strength. It is not machinable. It has high abrasion resistance	It is used for making malleable cast iron and wrought iron. It is used for making pump liners and electrical line hardware
Malleable cast iron	2.2 to 3.6% carbon 1 to 1.8% silicone rest iron	Obtained from white cast iron. It bends without breaking i.e. it is ductile. It can be hammered or rolled into different shapes. It has good machinability and shock resistance, excellent wear resistance and vibration damping capacity	It is used for making automobile parts such as crankshafts, chains etc. It is also used for making hinges, doors, keys, spanners, mountings, agricultural parts etc.
Ductile or nodular cast iron	3 to 4% carbon 2 to 3% silicone 0.3 to 1.2% magnesium rest iron	It is produced by adding magnesium to molten cast iron. Magnesium converts the graphite of cast iron from flake form to spheroidal or nodular form. It possesses good	It is widely used for making earthmoving equipments, agricultural implements, hydraulic parts such as cylinders, valves etc.

Contd...

		machinability, excellent wear resistance, good castability, ductility and impact strength is as good as mild steel.	
Alloy cast iron	Produced by adding nickel, chromium, molybdenum, copper, silicone and manganese to plain cast iron	It has increased strength, high wear resistance, corrosion resistance heat resistance, shock and impact resistance	It is widely used for automobile parts such as cylinders, piston, piston rings, crank case crank shafts etc.

2.5   Heat Treatment of Carbon Steels

Any heating operation performed on a solid metal e.g. heating for hot working or annealing after cold working. Heat treatment consists of a series of operations which improve the physical properties of a material. In the case of carbon steels or say steels these are annealing, normalisiing, hardening and tempering. The actual heat treatment process involves heating, soaking and quenching operations. To understand the heat treatment process it is mandatory to know the internal structure based on relationship of steel and iron and the iron-carbon diagram. (This shows the critical points of steel).

2.5.1   Iron-Carbon Diagram

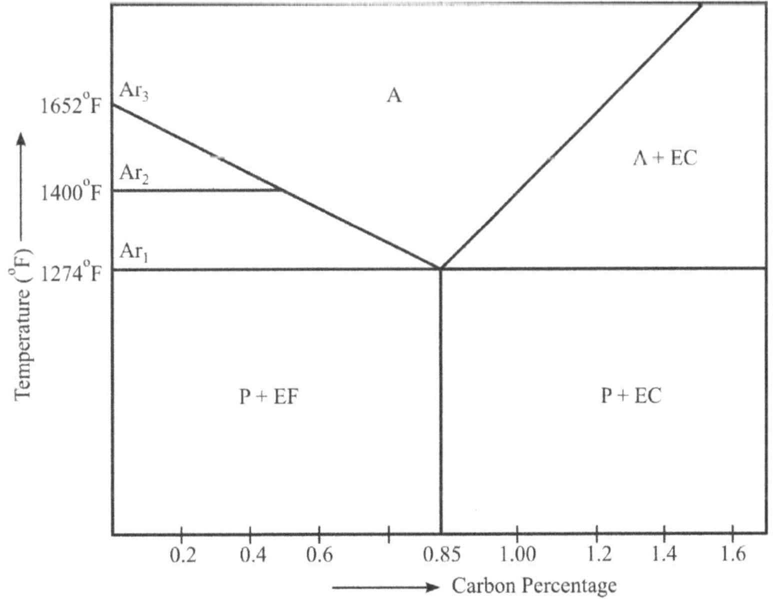

Code:

A = Austenite

P = Pearlite

F = Ferrite

C = Cementite

EC = Excess Cementite

EF = Excess ferrite

Pure iron exists in three forms – alpha, beta and gamma. Each of these are stable only between definite temperature limits

- Alpha iron up to 1400 °F

- Beta iron between 1400 °F to 1652 °F

- Gamma iron above 1652°F

Carbon steels have definite critical points and a critical range. The exact temperature at which these points occur and the number of points depend upon the carbon content of steel. Low carbon steels have three critical points.

Ar_3 (Upper critical point) 1652°F

When molten iron solidifies and is permitted to cool at uniform rate, the cooling stops for a moment at 1652 °F.

$$\text{Gamma iron} \xrightarrow{\text{transformed}} \text{Beta iron}$$

The structure has changed slightly due to evolution of heat. This accounts for retardation of cooling, this point is Ar_3.

Ar_2 (Second critical point) 1400 °F

As cooling continues second retardation occurs at 1400 °F. This point is Ar_2.

Ar_1 (Lowest critical point) 1274 °F

When small amount of carbon is added to the iron Ar_1 occurs.

In the heating of pure iron similar points occur in which heat is absorbed without a rise in the metal temperature. These heat absorption points are 20 °F higher than the above three points and are designated as Ac_1, Ac_2 and Ac_3 respectively.

2.5.2 Different Forms of Iron and Steel

Ferrite: It is pure alpha iron in steels. It is very ductile and has low tensile strength. In alloy steels containing nickel, molybdenum or vanadium these alloying elements are in solid solution in ferrite. It does not have any hardening properties because it has maximum of 0.025% carbon in dissolved state.

Austenite: A solid solution of carbon in gamma iron now includes all solid solutions based on gamma iron when carbon steel is heated to a temperature of about 1341 °F (727 °C). A solid solution of iron and carbon forms that consists of 0.76% carbon this is austenite. If the steel contains less than 0.76% carbon then a mixture of ferrite and austenite is formed. These steels are of low tensile strength, soft, ductile and non-magnetic.

Cementite: The iron carbide constituent of steel and cast iron such as white cast iron. It is very hard and brittle. In steels containing 0.85% carbon the cementite forms a perfect mixture with the pure iron (ferrite) present. It has high compressive strength, low tensile strength and is used in cutting tools.

Pearlite: Is is a mixture of ferrite and cementite in the ratio of 6 : 1. When austenite is cooled below 1341 °F (727 °C) a compound of iron carbide (Fe_3C) called cementite precipitates from the solution at the crystal boundaries. As cooling continues the austenite continues to decompose into alternate platelets of ferrite and cementite in a form called pearlite. Pearlite is a layered structure of platelet of iron carbide in a ferrite matrix. The amount of pearlite formed and the nature of its structure will depend upon the rate of cooling and the carbon content of steel.

Steels with less than 0.85% carbon are composed of pearlite and excess ferrite are called hypo eutectoid. Steels with more than 0.85%   carbon are composed of pearlite and excess cementite are called hyper eutectoid Steels with 0.85% carbon are called eutectoids.

Pearlite is strong, hard and ductile.

Martensite: It is formed in steel when it is cooled rapidly to suppress the change from austenite to pearlite. It is the main constituent of hardened steel. It is an intermediate form of cementite in alpha iron during the transition from austenite to pearlite. It is the hardest structure obtained in steel. It results from the decomposition of austenite at low temperatures. In consists of solid solution of carbon in iron and is responsible for the hardness of quenched steel. It has a needle like structure.

Troosite: It is another intermediate form, similar to martensite which is often present in hardened steels. It is also present in drawn or tempered steel whereas martensite is not. It is softer than martensite.

Sorbite: It is an intermediate form between austenite and pearlite. It is the main constituent of drawn steel and gives maximum strength and ductility.

2.6 Heat Treatment Processes of Steels

Process	Method	Objectives
Annealing	Heating the metal to just above the upper critical temperature for a definite time and cooling very slowly in the furnace itself. Time of soaking is 1 hour per inch thickness of material	It makes the metal fine grained, soft, ductile and without internal stresses or strains, so it can be easily machined. It prepares the material for further heat treatment. The ductility of annealing steel is utilised in tube and wire drawings and in rolling sheet
Process annealing	It consists in heating ferrous alloy below the lower critical temperature i.e. between 1020 °F to 1200 °F (549 °C to 649 °C)	It is used in the sheet and wire industry to restore ductility and to remove the effects of cold working i.e. internal stresses
Spherodise annealing (Spherodising)	It is performed by heating the steel slightly above critical point 730 °C to 770 °C with subsequent holding at this temperature, followed by slow cooling at the rate of 25 °C to 30 °C per hour at a temperature of 550 °C to 600 °C.	It is applied particularly to high carbon steels to improve machinability. It provides a structure of steel which is composed of grained (or globular) pearlite in alloy steels and ball bearing steels. This is known as spherodising.
Diffusion annealing (homogenising)	It is carried out at temps from 1000 to 1200 °C (optimum temperature is 1150 °C) at which diffusion proceeds quite easily and to some extent equalises the composition of steels. After the required heating temperature is achieved the metal is held for a very short period and is followed by cooling within the furnace for 6 to 8 hrs to 800 °C to 850 °C and the further cooling in air.	It is applied to steel ingots (both carbon and alloy steels) and heavy complex casting for eliminating chemical inhomogeneity within the separate crystals by diffusion.

Contd...

Normalising	It consists in heating the steel above the critical range followed by cooling in still air	It produces harder and stronger steel. It gives material of uniform physical characteristics. It removes internal stresses in forgings. It relieve stresses, refines grain structure and makes the steel more uniform. All welded parts are normalised after fabrication. It is done to improve the machining qualities in low carbon steels and reduces its distortion in further heat treating operations. It is recommended for aircraft steels. It increases the yield point, ultimate tensile strength and impact values.
Hardening	It consists of heating the steel above the critical temperature (AC_3) and then rapidly cooling it by quenching in oil, water or brine.	It produces a fine grained structure, greater hardness, maximum tensile strength, minimum ductility and internal strains. It is a carefully controlled process in industry. Rate of cooling, rate of heating and the time of soaking are all regulated to achieve desired results. More rapidly steel is cooled from critical temperature, greater will be the hardness.
Tempering	It consists of heating hardened steel well below AC_1, soaking at that temperature and then quenching in oil or air or water or brine. The steel is then reheated to lower temperature; held at this temperature for sometime and then cooled at room temperature. This reheating and cooling is tempering.	It relieves the strains in hardened steel, decreases brittleness and restores ductility and reduces hardness. It makes the steel tough

Contd...

Quenching	It means cooling of the steel piece after heating to required temperature. Hardness obtained during heat treatment depends upon quenching.	It may be done by dipping the steel piece in oil, water bath or exposing to air
Case hardening	In this process a hard wear resistant and shock resistant surface is produced on steel having a tough core inside. It consists of carburising a piece of steel, quenching, either mildly or rapidly reheating to refine the core quenching rapidly, reheating again to refine and harden the case, quenching rapidly tempering at a low temperature and cooling slowly. It is carried out in two stages. • Carburising i.e. addition of carbon to outer skin to make it rich in carbon content so that it can be hardened • Hardening this skin and refining the structure of the core to make it tough. There are three common methods of carburising • Pack carburising • Liquid carburising • Gas carburising	Metal hardness is imparted only to relatively small depth below the outer skin and soft and tough core is obtained because of low carbon content the outer surface gives as good service against wear shock and abrasion as a hardened piece of high carbon steel.

Questions

1. Define ferrous metals? How are plain carbon steels classified?
2. What are alloy steels? What are the advantages of alloy steels?
3. What is the effect of following alloying elements on steel.
   (a) Manganese   (b) Chromium   (c) Nickel
4. What are the different alloying elements in steel? Explain the effect of any two.
5. What is stainless steel? How are they classified?

6. What are high speed steels? Briefly describe about 18-4-1, steel and give its uses.

7. What is the difference between

    (a) Cobalt high speed steel and vanadium high speed steal

    (b) 18-4-1 steel and molybdenum high speed steel.

8. What are alloy tool steels. How they are classified?

9. What are alloy cast irons? Name the various types of cast iron, Explain any one of them in detail.

10. What is the difference between

    (a)   Grey cast iron and white cast iron

    (b)   Malleable cast iron and nodular cast iron.

11. What is iron-carbon diagram? Explain the different forms of iron.

12. What is the difference between (a) Ferrite and Austenite (b) Cemntite and pearlite

13. Name the various heat treatment processes for steels. What is the purpose for normalising.

14. Write short notes on (a) Hardening, (b) Tempering (c) Annealing

15. What is case hardening ? How is it different from simple hardening

CHAPTER 3

Non-ferrous Metals and Alloys

3.1 Introduction

All those metals except iron are non-ferrous metals. Alloys formed from the metals other than iron are non-ferrous alloys.

Some of the most useful nonferrous metals are aluminum, copper, nickel, zinc, tin and lead etc.

Some of the most common non-ferrous alloys are Duralumin, Y-alloys, Brass, Bronze, Monel metal, Inconel etc.

Non-ferrous metals are costlier than ferrous metals but are widely used because of its various mechanical, physical and chemical properties. Some of these are as follows:

- High corrosion resistance
- Can be fabricated easily
- High electrical and thermal conductivity
- Can be easily cast
- Strength and hardness at elevated temperature
- Attractive colour

3.2 Aluminium Alloys

At present aluminium alloys are used almost exclusively in the construction of aircraft.

This metal has light weight, high strength, ease of fabrication, and is available in various forms. It is one third as heavy as steel.

It is obtained from its ore Bauxite (Largely Al_2O_3 mixed with impurities). The impurities are removed by chemical process and pure alumina (aluminum oxide) is obtained. It is cast into pig form. This is later remelted to form commercial ingots used in rolling, forging, extruding etc. The common shapes used in aircraft construction are sheet, tubing, wire, bar, angles, channels, z-section, u-section, etc.

		Nomenclature	Composition in %	Properties and uses
3.2.1		Duralumin	Cu 3.5-4% Mn 0.5% Mg 0.5% Rest A*l*	It has high tensile strength, is light as aluminum, and has low corrosion resistance. A Further improvement is carried on duralumin sheets by giving a thin coating of A*l* or another A*l* alloy. This alloy is known as A*l* clad. A*l* clad 2014, 2017 are used in manufacture of aircraft.
3.2.2		Y-alloy	Cu 4% Ni 2% Mg 1% Rest A*l*	It is a casting alloy. It maintains strength at elevated temperature. It is used in strip and sheet form. It is useful for casting purpose for pistons, cylinder heads and other components of Aero Engines.
3.2.3		Aluminum casting alloys	**Alloy 212** Cu 8% Iron 1% Si 1.2% Zn 2.5 max Rest Al	It has good strength, hardness and machinability, good casting properties and is used as a general purpose alloy when high strength is not important. It may be sand, gravity or pressure die casting. It is used for aircraft castings.
			Alloy 13 Si 12% Rest A*l*	It has good castability, low shrinkage. It has good corrosion resistance. It is used for die castings for aircrafts.
			Alloy 43 Si 5% Rest A*l*	It remains fluid down almost to solidification point and for this reason can be used for complicated and thin walled aircraft castings. It also makes a dense, leak proof casting with good corrosion resistance. It is used for making carburetors, fuel line fittings, fuel and oil flanges for aircrafts. When cast in a permanent mold it provides better finish and closer dimensional tolerances than when sand cast. The cost of machining can thus be saved.
3.3		**Nickel alloys**		
3.3.1		Inconel	Ni 80% Cr 14% Iron 6% Mn 0.25% Si 0.25%	It is a corrosion and heat resisting metal. It is used in aircraft work for exhaust collectors, jet pipes etc. It has good workability. It is similar to stainless steel. It cannot be hardened by heat treatment.

Contd...

		Cu 0.2% Rest Carbon	It is classified as non-ferrous because of low content iron, has the property of retaining high strength at elevated temperatures. Inconel bends easily machining is difficult and must be done at low speeds with carefully treated and sharpened tools. It welds easily and gives a strong sound ductile weld which resists corrosion. It can be used for making springs. It is available in various forms such as • Sheets, strip, rods (hot rolled or cold drawn) • Tube (cold drawn welded) • Wire (cold drawn) • Castings
3.3.2	Monel	Ni 67% Cu 30% Iron 1.4% Mn 1.0% Si 0.1% Rest Carbon	It has high strength and excellent corrosion resistance. It cannot be hardened by heat treatment, except by cold working. It is not used generally in aircraft construction but used for industrial and chemical applications. Machining can be done without difficulty. It is similar to mild steel in its cold working properties such as drawing, bending and forming. It can be readily welded. It is used in the manufacture of oil coolers, pump impellers, strainers and rivets.
3.3.3	K-monel	Ni 66% Cu 29% Al 2.75% Iron 0.9% Mn 0.85% Si 0.5% Rest Carbon	It is corrosion resistant, and can be hardened by heat treatment. It is used for instrument parts and for structural parts in the vicinity of compasses because of its non-magnetic quality. In aircrafts it is used for making machined parts that are subject to corrosion, gears and chains for landing gear. It can be welded and brazed easily. It is available in Strips, wires, rods and forgings.

Contd…

3.4   Copper Alloys

3.4.1 Brass

It is a copper alloy consisting of a solid solution of zinc in copper. In addition to zinc and copper they also contain small amounts of aluminum, iron, lead, manganese, phosphorous, tin, magnesium and nickel. Brasses with zinc content of about 30-35% is very ductile and with 45% zinc content has relatively high strength.

Brasses with zinc content upto 37% are called alpha brass which is good for cold working while brasses with 33-46% zinc are good for hot working and are known as beta brass.

Depending on the percentage of zinc, brasses are known with different names. Some of these are:

3.4.1.1	Muntz metal	Cu 60% Zn 40%	It has excellent corrosion resisting properties in contact with salt water. Strength can be increased by heat treatment. It is used in the manufacture of bolts and nuts pump parts, valves, etc.
3.4.1.2	Manganese brass	Cu 57-60% Zn 38-42% Tin upto 1.5% Iron upto 2% Mn Max 0.5% Al Max 0.25% Pb Max 0.2%	It can be forged, extruded, drawn or rolled to any desired shape. It is used in rod form for machined parts in aircraft construction. It is strong tough and corrosion resistant. It is also called manganese bronze
3.4.1.3	Naval brass	Cu 59-62% Zn 39% Tin upto 1.5%	It is similar in construction to muntz metal and manganese brass but has greater strength, toughness and corrosion resistance than commercial brass. It is used for turnbuckle barrels, bolts, studs, nuts and parts in contact with salt water.
3.4.1.4	Admirality brass	Cu 70% Zn 29% Si 1%	It is resistant to sea water action. It is widely used for marine castings, marine fittings. It is used for nut and bolts etc.
3.4.1.5	Delta Brass	Cu 60% Zn 37% Iron 3%	It is easily hot worked, forged cast and rolled. It has good tensile strength and corrosion resistance. It can replace steel castings.

3.4.2 Bronze

Bronzes are copper alloys containing tin. Lead, zinc and phosphorous are also present in some bronzes. Bronzes have excellent bearing qualities due to the fact that the tin is in a hard delta solution in copper. This hard delta solution distributed through the alpha metal gives ideal bearing properties. Delta solution is present in bronzes with over 9% tin content.

3.4.2.1	Gun metal	Cu 86-89% Tin 9-11% Zn 1-3%	It is a hard bronze casting material. It has fair machinability. It is used for gears, bearings, bushes, glands etc.
3.4.2.2	Phosphor bronze	Cu 94% Min Tin 3.5% Min Phosphorous upto 0.5%	It can be obtained as rod, bar, sheet, strip and spring wire. It increases strength, ductility. It is used for bolts, valve discs, electrical contacts and springs.
3.4.2.3	Phosphor bronze casting alloy	Cu 86-89% Sn 7.5-11% Zn 1.5-4.5% Lead upto 0.3%	It is sometimes called a leaded phosphor bronze or leaded gun metal. It machines more easily than gun metal. It is used for bearings and gears, bushings. It is used where good strength is required. It is resistant to salt water corrosion
3.4.2.4	Aluminum bronze	Al 6.5-11% Iron 4.0% Max Mn 2.0% Max Tin 0.6% Max Ni 5.5% Max Si 2.25% Max Rest Cu	It possesses greater resistance to corrosion than manganese bronze, good strength high corrosion resistance and good heat resistance. It is available in rods and bars, plates sheets and strips. It has good bearing qualities also. It is readily forged. It is used for making gears, bearings, valve seats, guides, cams, rollers etc.
3.4.2.5	Aluminum bronze casting alloy	Cu (min) 78% Al 10.5-12% Iron 2.5 % Mn (Max) 5% Ni Max 5% Tin Max 0.2 %	This alloy is as hard as manganese bronze and has great resistance to shock and fatigue. It is used for worm gears, valve seats, bearings and propeller hub cones.

3.5   Magnesium Alloys

Magnesium is the lightest of structural metals available for aircraft construction. Magnesium is commonly alloyed with aluminum, zinc and manganese to create usable structural materials. The light weight and relatively high strength of magnesium alloys results in a strength/weight ratio which is good for aircraft design.

It is used for making ducts, doors, brackets buckheads, fairings, etc. where strength is secondary and only thickness of material is necessary. This results in weight saving.

Magnesium alloys are non-magnetic. Machining can be done easily. They are easily welded and are available as sand, permanent mold and die castings, press and hammer forgings, extruded bar, rod, various shapes, tubing, rolled sheet, plate and strip.

Magnesium alloy castings are used extensively in aircraft construction for making wheels, brake pedals, control columns, bell cranes instrument housings, engine housings gear box housings, brackets etc. Magnesium alloys can be machined, sheared, blanked, punched, formed by bending, drawing, spinning, pressing or stretching.

Riveting, gas welding, arc welding and spot welding are the joining methods for magnesium alloys.

Questions

1. What are non-ferrous metals and Alloys ? Name five of them. What are their properties.

2. Why Aluminum alloys are used in aircraft industry ? Name some of the aluminum alloys. Give properties and uses of two of them.

3. Write short notes on
    (a)   Duralumin   (b)   Y-alloy

4. What are the properties of Aluminum casting alloys? Why alloy 212 is used for aircraft castings.

5. What are the properties of Nickel alloys ? Name two of them. Explain any one in detail.

6. Write short notes on any two
    (a)   Inconel   (b)   Monel   (c)   K-monel

7. What is brass ? Describe the composition, properties and uses of any two brasses.

8. Write down the composition and applications of the following.
    (a)   Muntz metal   (b)   Manganese brass
    (c)   Naval brass   (d)   Delta brass

9. What are Bronzes ? Give composition and properties of the following:
    (a)   Gun metal   (b)   Phsphor bronze   (c)   Aluminum bronze

10. What are Magnesium alloys? Give properties and uses. Why they are used in Aeronatical Industry for making Aircraft parts.

Unit II

Introduction to Metal Forming, Casting Processes and its Applications

CHAPTER 4

Metal Forming

4.1 Introduction

Metal forming processes are the processes subjected on a material to make it into the desired shape i.e., shaping process is called forming. By this the desired size and shape are obtained through the deformation of metals under the action of externally applied force by means of mechanical working such as hot working or cold working. Recrystallistation temperature plays a major role in the metal forming processes.

4.1.1 Recrystallisation Temperature

During the plastic deformation process plastic flow of metal takes place and the shapes of the grains are changed. The process of formation of new grains is known as recrystallisation and is complete when old grains structure is destroyed.

The temperature at which the formation of new grains is complete is called recrystallisation temperature.

Hot working is done above recrystallisation temperature. While cold working is done below recrystallisation temperature.

4.1.2 Hot Working

It is the process of deformation which is done above recrystallisation temperature of the metal or alloy. It is the process of shaping the metal by rolling, extrusion, forging or hot drawing at elevated temperatures. The hot working range varies from metal to metal, but it is in general, a range in which recrystrllisation proceeds concurrently with the working so the no strain hardening occurs. For Aluminum hot working is performed at 343 to 482 °C depending on the alloy. This is above the temperature required for recrystallisation but well below the melting point or other heat treatment temperatures.

4.1.2.1 Advantages of Hot Working

    (a)   Grain structure is refined,

    (b)   No residual stresses and strain hardening,

(c)   Porosity of metal is minimized,

(d)   Impurities are evenly distributed,

(e)   Physical properties are improved such as ductility, impact resistance, strength, shock resistance etc.

(f)   High rate of production.

4.1.2.2 Disadvantages of Hot Working

(a)   Poor surface finish due to scaling,

(b)   Close tolerances cannot be obtained,

(c)   Cost of tooling is high,

(d)   Fatigue failure may occur at later stage due to fatigue crack.

4.1.2.3 Main Hot Working Processes

Main hot working processes are forging, rolling, extrusion and drawing.

4.1.2.3.1 Forging

The operation of shaping (hot) malleable metals by means of hammers or presses. It includes hand hammer, steam hammer and press and drop forging. Forging of steel is a mechanical working above the critical range to shape the metal as desired. Due to the pressure exerted, the grain of the metal is refined and the metal is made more dense and homogeneous. Forging should be completed by just above the critical point to prevent grain growth or distortion. Forging is done either by pressing or hammering the heated steel until the desired shape is obtained. Pressing is adopted when parts are large and heavy and it transforms the metal uniformly and gives the best possible structure throughout. The parts are heated in a closed furnace before the forging. The portion of a work in which forging is done is called forge. The main forging operations are:

Drawing down: In this the length of the metal gets elongated and the diameter of a bar gets reduced.

Upsetting: In this a hot piece of metal is increased in thickness and decreased in length by hammering on the end i.e., increasing the cross sectional area. It is the reverse of drawing.

Squeezing: It increases the length but does not increase the cross sectional area.

The forging is classified as open die, closed die and impression die.

Open die forging: It is a hot forging process in which metal is pressed between two flat surfaces or simple contoured dies. It is also called smith forging. In this process the compressive force is applied slowly on the metal stock. It is used where the components to be forged are not large. The operations performed on open die processes are:

(a)   Drawing out,

(b)   Upsetting,

(c)   Combination or both, by hand forging and power forging. In hand forging the hot work piece is given the desired shape by holding it with tongs and then beating it over anvil by using portable tools and hammer.

Portable Tools are as follows:

Fullers, punches and dies for cutting. Fullering, flattening and swaging can be done by open die forging method. This method can produce step shafts, solid shafts, bars, slabs, billets, hollow cylindrical shapes.

In power forging a large number of identical forgings, such as machines which work on forgings by blow are called hammers and those which work by pressure are called presses. There are two main advantages of open die forging

  (a) Less costly,

  (b) Tooling is simple.

Limitations of open die forging are:

  (a) Only simple shapes can be produced,

  (b) Close tolerances could not be obtained,

  (c) Further machining is required,

  (d) Skilled labour is necessary.

Closed Die Forging

In this impressions or cavities are formed in die block. The compressive force is applied to the surface uniformly. In this the metal flows into the die cavity and so there is no loss of material. Presses used are either mechanical or hydraulic. It is used basically for automobile parts such as fittings, elbows etc.

Limitations of closed die forgings are as follows:

  (a) Dies are of complicated shapes,

  (b) Dies are costly,

  (c) Tooling is costly,

  (d) Used for parts where batch size is large.

Advantages of close die forgings are mainly as follows:

  (a) Require less time than open die forgings,

  (b) Production is high,

  (c) Tolerances achieved are better,

  (d) Complicated shapes can be produced.

Impression Die Forging

When two or more dies containing impression of the part shape are brought together, the work piece undergoes plastic deformation. These forgings are better than open die forgings. A better product can be obtained at reduced cost. The finished part closely resembles the die impression. These are of three types-press forging, machine forging and drop forging.

Press Forging: It uses closed impression dies. In this method a slow continuous pressure is applied to the area to be forged. It is used when the parts to be forged are large and heavy. The force is transmitted uniformly to the centre of the section and thus the interior grain structure is affected as well as the exterior giving the best possible structure throughout.

Drop Forging: In this method two dies are used, one of which is attached to the hammer and other to the anvil. When the upper die block falls on the lower die the block metal is squeezed into the die cavity due to impact force. At the intersection of the two dies there is a relieved section all around the edge to take care of surplus metal squeezed out when the dies are brought together. The operation is completed by means of mechanically or hydraulically operated presses. During the forging operation dies are kept clean with a high pressure steam or air to prevent scale formation on the forging. It is used for production of invidual parts in large quantities and for complicated shapes. Aircraft forgings are produced by this method. Chrome-molybdenum and chrome-nickel 'Molybdenum' steels are commonly used for aircraft forgings.

Machine Forging (upset forging): This is done by upsetting operation. It consists of applying pressure longitudinally on a hot bar, gripped firmly between grooved dies to upset a desired portion of its length. The metal is increased in thickness and decreased in length by hammering on the end (upsetting). Heads or bolts are formed by this. An upset head is stronger than machined head. It is done in steps; many steps may be required to get the final shape.

Comparison of Drop Forging, Press Forging and Machine Forging (upset forging)

Drop forging	Press forging	Upset forging
Close tolerances cannot be produced	Close tolerances are produced	Better dimensional accuracy
No need of maintenance	Maintenance is simple	Maintenance of machines is cheaper. Saving in material
Draft is necessary	Less draft	Dies carry no draft
Not suitable for large production	High production	Higher productivity

4.1.2.3.2 Hot Rolling

It is the process of plastic deformation that reduces the thickness of metal by subjecting it to compressive forces by passing it between driven cylindrical rollers by means of various arrangements. Many ingots are first rolled into plates or bars to provide easy handling of materials for future processing and provide finished materials such as plate, sheet, bars and some other shapes.

Principle of Rolling

AB and A'B' are the contact arcs on the rolls. The coarse grain structure is converted into elongated grain structure as the material passes through the rolls. The upper and lower rolls are moving in opposite direction. The velocity at exit is greater than the velocity at entry. The thickness of the metal gets reduced.

In rolling, the quantity of metal going into a roll and out is same but area and velocity are changed.

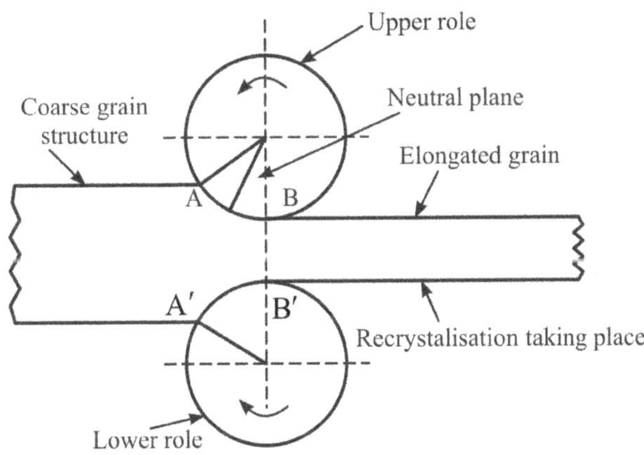

Types of Rolling Mills

A rolling mill consists of one or more rolls arranged differently in rolling stands, such as:

- (a)   Two high rolling mill
- (b)   Three high rolling mill
- (c)   Four high rolling mill
- (d)   Cluster rolling mill
- (e)   Continuous rolling mill
- (f)   Planetary rolling mill.

Two High Rolling Mill

It is of 2 types

1. Reversible

2. Non-reversible

1.   Reversible

Two high reversible mill consists of two rollers rotating in opposite direction. The direction of these rollers can be reversed. The metal is passed through the rolls. When all the metal reaches to one side the direction of roll is reversed. It is more expensive than nonreversible type. To reduce a big ingot into a bloom several passes are required.

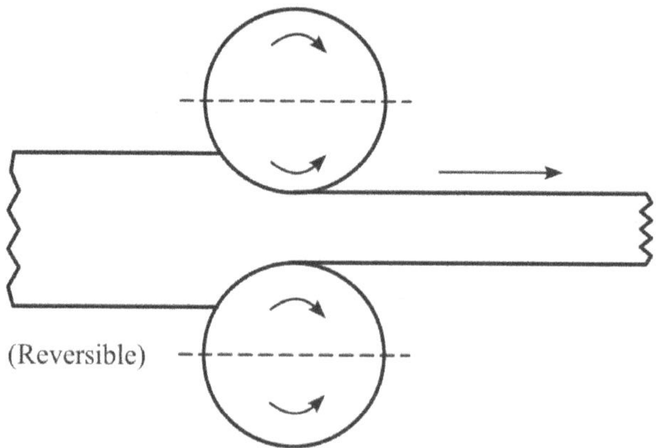

2. Non-reversible

In this, the rolls move in one direction only. Upper roller may be raised or lowered to change the distance between rollers.

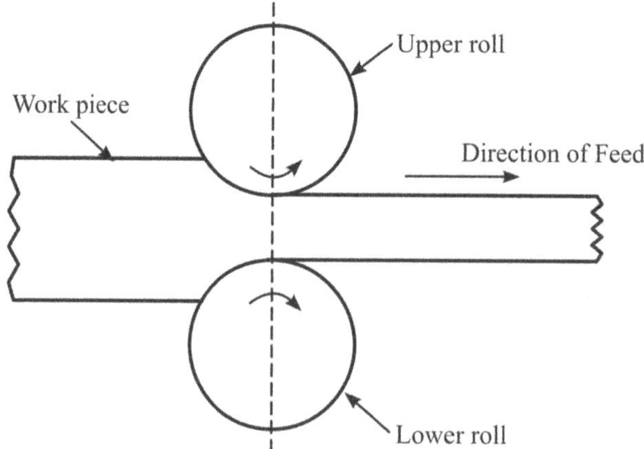

Three High Rolling Mill

It consists of three rolls. The rotational direction of top and bottom rollers is same. The middle roller moves in opposite direction. The work piece is fed between top and middle roller and the reverse pass is between middle and bottom roller in opposite direction. It is less expensive and has high efficiency than two high rolling mill. It is used for rolling structural steel shapes e.g., I beam, angles, channels.

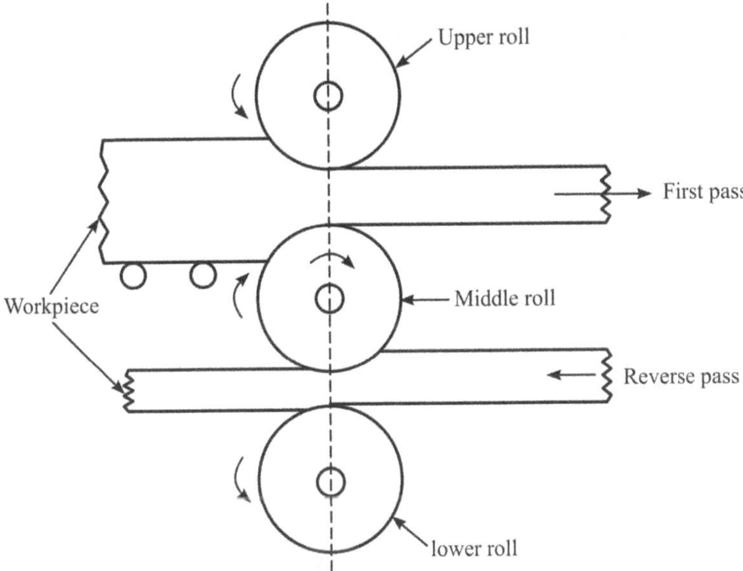

Four High Rolling Mill

It consists of four rolls, two each of different sizes arranged one above the other. The bigger rolls are called back up rolls and they reduce the roll deflection. Thin plates and sheets are rolled by hot and cold rolling.

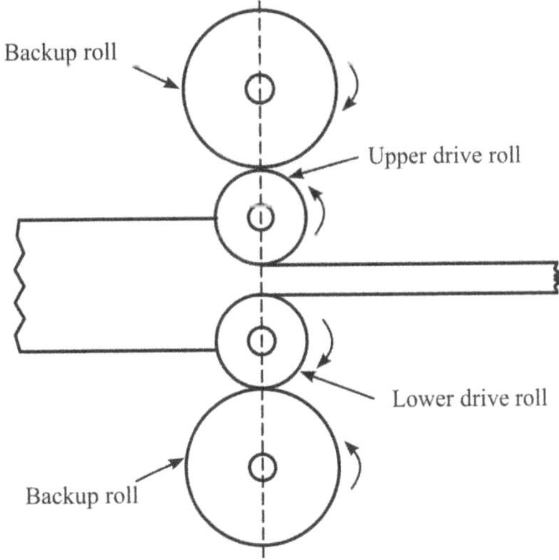

Cluster Rolling Mill

It consists of two working rolls and four or more back up rolls of large size. It is used for cold rolling.

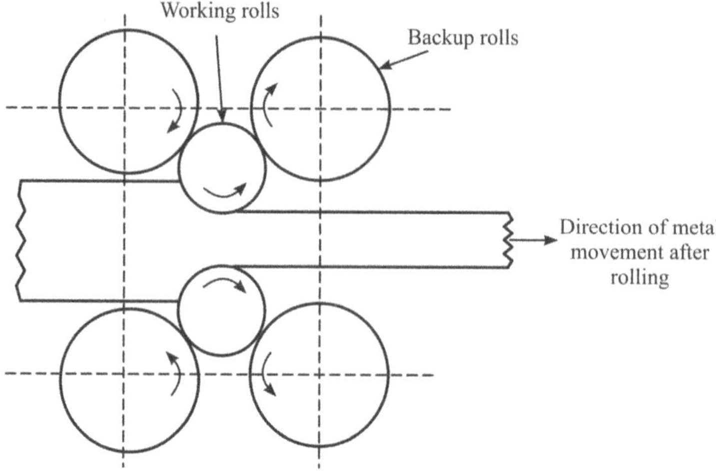

Continuous Rolling Mill

It consists of a number of non-reversing 2 – high rolling mills arranged one after the other. The material is passed through these combination of rollers. This is used for mass production as the process is fast.

Planetary Rolling Mill

It is used when large reduction is required. It consists of 2 backing rolls surrounded by small diameter. Rolls are mounted in cages. In this thickness reduction of 25:1 is possible.

Rolling Defects

The various rolling defects found in rolled plates, sheets and structure may be as follows:

Edge Cracks: These occur due to deformation or limited ductility of metal.

Structural Defects: These affect the rolled product thereby distorting.

Wavy Edges: These are caused by bending of rolls thereby causing buckling. The edges are thinner, while centre is thicker.

Surface Defects: These are caused by impurities in the material mainly due to rust, dirt etc.

Zipper Cracks: These are caused by low ductility.

Alligatoring: It is caused by deformation of the metal due to non-homogenousity of metal thereby splitting the metal on rolling.

Applications of Rolling

Rolling is used to produce parts having constant cross-section throughout its length. The rolled products are plates and sheets, rings, balls, wheels, tubes, round bars, square bars, hexagonal bars, channels, H and I beams, rail sections etc.

4.1.2.3.3 Extrusion

It is the process that forces metal to flow though a die under a compressive force resulting in plastic deformation of the metal. The shape of the die will be the cross-section of an angle, channel, tube or some intricate shape.

The equipment consists of a cylinder into which a dummy block and heated metal billet are placed into position. The die plate with necessary opening is fixed on one end of cylinder. The shape of the opening is the same as the desired shape of the job. The hot billet is pressed against the container walls.

Combined Extrusion

It is a combination of both forward and backward extrusion. The metal is kept inside between lower and upper punches. Metal flows in between i.e., both up and down, The extruded part is lifted from the die on upward stroke of the slide by a lift out on the bed of the press.

Hydrostatic Extrusion

In this method the chamber is filled with a liquid. There is no need of lubricant and force transmitted is uniform on all sides. It is cold extrusion process and produces good surface finish and dimensional accuracy. High strength super alloys, molybdenum, grey cast iron can be extruded. (Reactor fuel rods also)

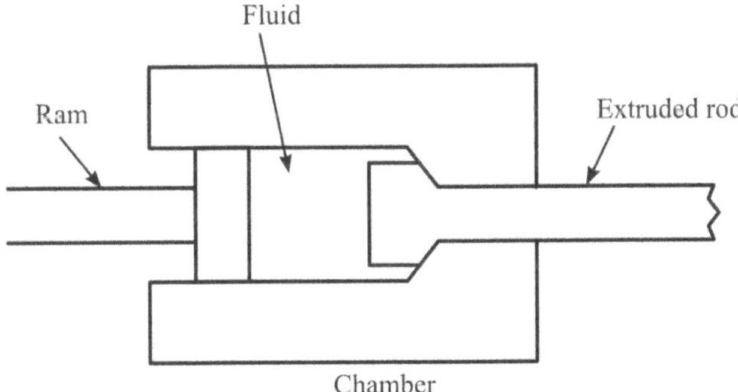

Tube Extrusion

It is a form of direct extrusion. Hollow objects such as tubes and other shapes can be produced by this method.

In this the heated metal is put into the cylinder. The die containing the mandrel is pushed through the heated metal. After the billet is placed inside the cylinder the plunger along with mandrel advances and extrudes the metal though the die. The operation is performed rapidly. Low carbon steels can be extruded cold. Its principle is used for production of seamless tubes. By drawing a large roughly formed tubular piece of material through dies of progressively decreasing size.

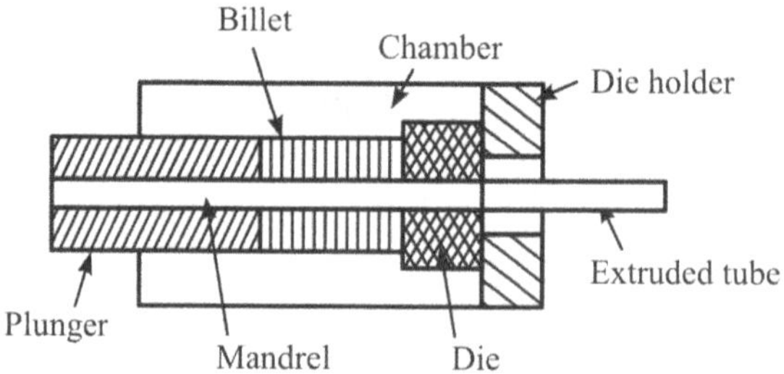

Impact Extrusion

It is a method of forcing cold metal through a die by hard blows by a punch driven at high velocity. It is useful for producing hollow pipes. It is applicable to soft and ductile materials such as aluminum and its alloys. It is performed on a high speed mechanical press and die plate forging it to flow through the die.

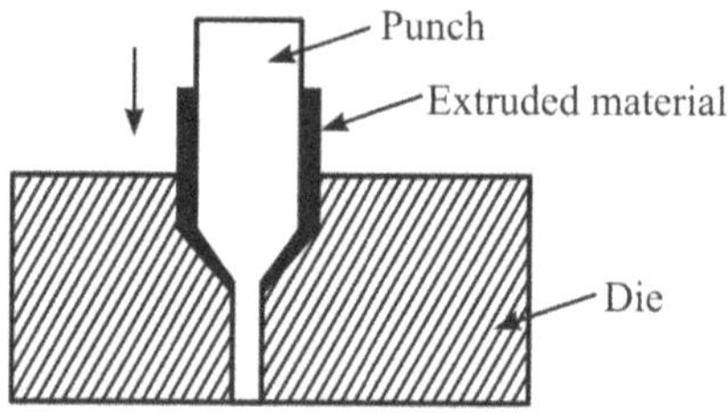

Some of the advantages of the extrusion process are as follows:

   (a) Dies can be easily manufactured and can be replaced or removed without any difficulty,

   (b) Complicated shapes can be extruded with accuracy and finish,

   (c) It is better than rolling,

   (d) Hardness and yield strength are increased,

   (e) Tooling cost is low.

Some of the Limitations of Extrusion are:

   (a) It is limited to ductile materials,

   (b) Surface cracking is noticed,

   (c) Internal cracking also occurs.

Types of Extrusion

Extrusion may be classified as hot extrusion and cold extrusion, which is further divided into direct (forward) and indirect (backward) extrusion. In addition to this, there is combined extrusion, hydrostatic extrusion, impact extrusion and tube extrusion.

Hot Extrusion

Direct (forward) Extrusion

In direct extrusion a ram forces the preheated billet through a die. The flow of metal is in the direction of movement of ram. The billet slides against the cylindrical walls of the chamber. The ram forces the hot billet though the die. This process is used to produce straight tubular shapes, hollow bars, thin walled tubular parts, bolts, screws or stepped shafts.

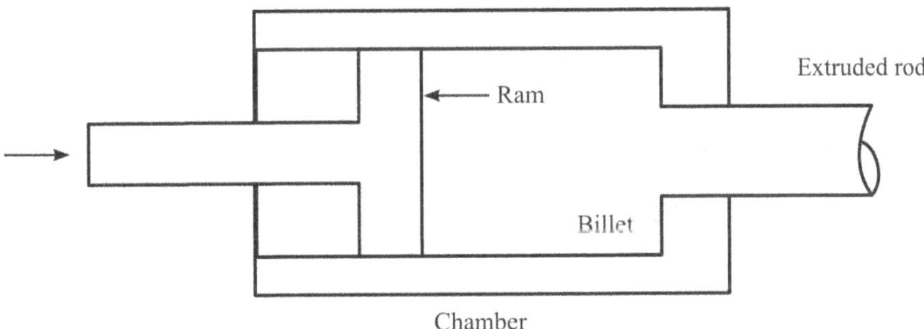

Indirect (backward) Extrusion

In indirect extrusion the metal inside the cavity is forced to flow in a direction opposite to the ram travel. The frictional force between the billet and container are less and so power required for extrusion is less than direct extrusion. The hollow ram carries a die, while the other end of the container is closed with a plate.

Many cylindrical shapes, nuts, sleeves and tubular rivets are produced by backward extrusion.

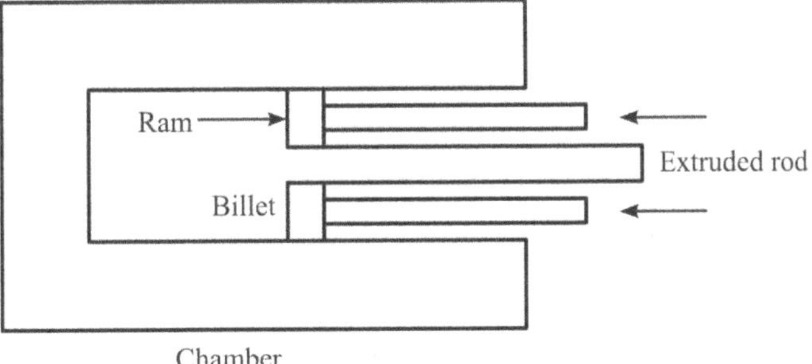

This set up consists of a die and a punch. Molten metal is kept in the die. The outside die of tube is same as diameter of hole in die. The punch descends at a high speed and strikes the slag causing the metal to flow between punch and die. The metal is extruded in opposite direction to the movement of punch because of force of impact. The process is done at room temperature. It is used for production of components for automobiles, cars, domestic appliances of aluminum, copper, and alloy steels.

Defects in Extrusion

(a) Surface cracking caused by high temperature,

(b) Extrusion defect caused by flow of metal when surface oxides and impurities are drawn towards centre of billet,

(c) Internal cracking caused due to hydrostatic tensile stress at the center line of the deformation zone of die, also known as centreburst.

Applications of Extrusion

The extrusion process is used for tubing, railings, fire extinguishers, window frames, and aircraft structural parts.

I – beams, channels, angles, pipes and hollow bars can be produced.

4.1.2.3.4 Wire Drawing

To manufacture wires and tubes, the material is drawn through a die.

It is an operation to produce wires of various sizes within the required tolerances. The process involves reducing the diameter of rods or wires in steps by passing them through a series of wire drawing dies with each successive die having smaller bore diameter than the one preceding the other. The maximum reduction in area is less than 50%.

A wire drawing die is a tool that consists of highly polished hole through which wire is drawn to reduce its diameter.

All the wires are produced by cold drawing through dies. Dies are made of chilled cast iron, hardened alloy steel, cemented tungsten carbide and diamonds.

4.1.2.3.5 Sheet Metal Working

It is the process of working with metal sheets to fabricate parts for various uses. The metals are used in the form of sheets, and plates. This is utilized for automobile, aircraft, house hold parts, decoration pieces etc.

Various stresses are caused in sheet metal operations such as shearing, tension, compression of bending. Metals alloys used for sheet metal works are as follows:

(a) Aluminium / aluminum alloys,

(b) Magnesium / magnesium alloys

(c) Tin

(d) Lead

(e) Copper

(f) G.I. Sheets

(g) Stainless steel.

Various sheet metal operations are:

(a) Punching	(b) Blanking	(c) Notching	(d) Bending
(e) Forming	(f) Coining	(g) Trimming	(h) Shaving
(i) Embossing, and	(j) Nibbling.		

Questions

1. What do you understand by metal forming? What is the role of recrystallisation temperature in mechanical working of materials?

2. What is hot working? What are its advantages and disadvantages?

3. What are the main hot working processes? Explain any one of them in detail.

4. What is forging? What is difference between open die forging and closed die forging?

5. Write short notes on

   (a) Press forging   (b) Drop forging

   (c) Impression die forging   (d) Machine forging

6. What is the difference between drop forging, press forging and upset forging give in tabular form.

7. What is the principle of rolling? Show by means of sketches arrangements for two high rolling mills for reversible and irreversible rolling mills.

8. Describe in brief the various types of rolling mills used in hot rolling.

9. What is the difference between

   (a) Two high rolling mill and three high rolling mills

   (b) Cluster rolling mill and continuous rolling mills

10. Briefly describe the various rolling defects.

11. What is extrusion? What is the difference between direct and indirect extrusion.

12. Write short notes on

    (a) Tube extrusion   (b) Impact extrusion

    (c) Combined extrusion   (d) Hydrostatic extrusion

13. How many types of extrusions are there? What are the applications of extrusion.

14. Write short notes on

    (a) Wire drawing   (b) Sheet Metal Working

CHAPTER 5

Casting

5.1 Introduction

The casting process consists of making solid metal objects by pouring molten metals into sand or metal moulds in which they solidify. It is then taken out from the mould by breaking the mould or taking the moulds apart. This solidified object is called *CASTING*. It is also known as *FOUNDRY PROCESS*.

The mould is made of heat resisting materials such as

- **Metal/Alloys** (Grey cast iron, Steel, Aluminium Alloys, Brass, Bronze, Magnesium, Zinc Alloys, Nickle Alloys etc)
- **Non–Metals** (Moulding Sands, Ceramics, Wax, Plaster of Paris, Silicon Carbide etc.)

The factors that affect the selection of the material, depends on following factors

- Size / shape of casting
- Cost of material
- Quantity required
- Quality of casting
- Dimensional accuracy

The most common material used for making moulds is sand and cast Iron.

5.1.1 Advantages of Casting Processes

- Most complicated shapes are produced
- Any shape can be produced
- Tools used are very simple

- Machining can be done easily and in less time.
- Identical casting can be produced in large numbers.

Limitations in Casting Processes

- It is uneconomical for small batches
- Accuracy can not be obtained by sand castings
- Defects like shrinkage and porosity cannot be removed permanently.

5.1.2 Terms Used in Castings

Moulding Flask :

It is Box like frame without top and bottom base into which sand is rammed it is called "Flask". It holds the sand mould. It is made of wood or metal according to use. It has Drag, Cope and Cheek.

Parting Line: It is the line that separate Cope and Drag of mould.

Bottom Board : It is the Board (made of wood) used at starting of moulding sand.

Pattern: It is the duplicate of final casting to be made with some modifications.

Core: It is placed in moulding cavity to make hole through the castings. It is made up of core sand.

Core Box: They are used to give desired shape to core sand.

Core Prints: They are provided to locate and support the core.

Chaplets: They are the metal supports kept inside the mould cavity to support the core.

Chills: These are metallic objects of high heat capacity and are placed in the mould to provide uniform rate of cooling.

Gating System: These are those elements which are related to flow of molten metal from ladle to the mould cavity such as pouring cup or pouring basin, sprue, runner, gates and riser.

Basic Steps in Casting Process

- Pattern making
- Core making
- Mould making (Moulding)
- Melting and pouring
- Cleaning and finishing

5.2   Pattern Making

Pattern making is the process of *making patterns*. The pattern is wood, metal or plaster copy in one piece or in sections, of an object to be made by casting. It is made slightly larger than the finished casting to allow for contraction of casting while cooling and suitably to facilitate withdrawal from the mould.

Patterns are required to make moulds. The mould is made by packing suitable moulding material such as moulding sand around the pattern. When the pattern is withdrawn its impression provides mould cavity, which is finally filled with molten metal to become casting.

Core Making

Cores are made of sand. It is a sand shape that is kept inside the moulded cavity to produce hollow cavities in casting.

Mould Making

It consists of all operations necessary to prepare mould for receiving molten metal.

Melting and Pouring

Melting means preparation of molten metal. The methods are pit furnace, open hearth furnace, cupola furnace etc.

Pouring means transferring the molten metal from furnace to mould.

Solidification and Mould Breaking

The casting is made to solidify after pouring the molten metal in the mould.

The mould breaking is done after complete cooling.

Cleaning and Finishing

These operations are required to remove excess material.

5.2.1 Pattern Materials

Most commonly used materials used for making pattern are wood, metals (Cast Iron, Brass, Aluminium, White Metal), Plaster of Paris, Plastics and Wax. The various advantages, disadvantages and usage/applications are given as a comparative table

	Material	**Advantages**	**Disadvantages**	**Applications / Usages**
5.2.1.1	**Wood** (Teak, Sal Shisham, Pine, Deodhar, Mahogany etc)	Cheap, Light, Readily available, can be machined, can be cut and fabricated into	Due to low resistance to sand abrasion it wears out easily so cannot be used again. It is easily affected	Most Common Material

Contd...

		various shapes such as glueing, bending etc. Good surface finish is obtained for longer time by varnishing / polishing	by moisture it has low strength and used only when few casting required.	
5.2.1.2	**Metals**			
5.2.1.2.1	**Cast Iron**	Cheap, can be cast into any shape. Good machinability' high resistance to abrasive action of sand, good strength.	Castings are heavy and brittle. Easily prone to rust, hence require dry storage area. Edges of thin sections, cool quickly, hence may be difficult to machine	The patterns for heavy parts are made of cast iron
5.2.1.2.2	**Brass**	Strong, tough, rust proof, better surface finish than cast iron, it is wear resistant, can be cast into any shape.	Casting are heavier and costlier than cast iron.	It is used for small patterns in bench and machine moulding
5.2.1.2.3	**Aluminium**	Cheap and light weight, soft and easy to work, resistant to corrosion. Its melting point is low and be cast into any shape.	Handling is difficult because they are soft	Used for production moulding
5.2.1.2.4	**White Metal**	Has low melting point (it is an alloy of Pb, Cu and Sb) of 260°C. It can be cast into narrow cavities between master pattern and frame of stripping plate.	Pattern are soft and easily worn out by moulding sand	Used for lining stripping plates, used for die casting

Contd....

5.2.1.3	Plastic	Light in weight, high strength, high wear resistance, high corrosion resistence, low solid shrinking and smooth surface finish.	Patterns are weak, not suitable for machine moulding	Thermo settings resins usually phenolic resins are being used for patterns used in large production
5.2.1.4	**Plaster of Paris or Gypsom Cement**	High compressive strength, easily worked with wood tools It expands during solidification hence it can neutrilise the shrinkage of casting during solidification	Big castings can not be made as they are difficult to handle	Used for small making small patterns and core boxes of irregular or intricate shape
5.2.1.5	**Wax**	High degree of surface finish, high accuracy are obtained	They are used as split mould	Used for investment castings
5.2.1.6	**Polystyrene or Expanded Thermocole**	Disposable type, soft. It changes to gaseous state on heating	Used for limited used	Used for single castings

5.2.2 Pattern Making Tools

The tools used for pattern making are almost the same as used in carpentry shop. However some extra skill and accuracy is required in the use of following hand tools

- Marking and Measuring Tools (Bevel Square, Try Square, Marking Gauge etc.)
- Cutting Tools (Tenon Saw, Rip Saw, Dove Tail Saw etc.)
- Striking Tools (Mallet, Hammer)
- Planning Tools (Smoothing Plane, Trying Plane etc.)
- Holding and supporting Tools (Carpenter's Vice)
- Boring Tools (Ratchet Brace, Hand Drills etc.)

5.2.3 Pattern Allowances

The pattern is almost same as the casting, but pattern allowances are kept for metallurgical and mechanical reasons as follows:

5.2.3.1 Shrinkage Allowance (Contraction)

When the molten metal is poured in the mould it cools and solidifies and contracts. To compensate for shrinkage the pattern is made slightly larger in every dimension. The rate of contraction is different for different materials.

Hence there are different shrinkages for different metals as per table for Ferrous and Non-ferrous material

Sl.	Non Ferrous	Shrinkage Allowance (mm / meter)	Ferrous	Shrinkage Allowances (mm / meter)
1.	Copper, Brass, Aluminium	16	Grey C.I.	10.5
2.	Magnesium	18	White C.I.	21
3.	Lead	26	Malleable C.I.	15
4.	Zinc	24	Steel	20

5.2.3.2 Draft Allowance

It is a taper which is given to vertical walls of the pattern for easy and clean withdrawal of the pattern from sand. It is usually expressed in degrees. This is known as draft. This depends on the method of moulding and drawing of the pattern, the material the pattern is made of, the degree of precision, the surface smoothness of the pattern.

It ranges from 1.5° for external surfaces and upto 3° for inner surface as per table.

Sl.	Pattern	Outer Draft Angle (Degree)	Inner Draft Angle (Degree)
1.	Plastic	0.25 to 1.00	0.35 to 2.25
2.	Wood	0.25 to 3.00	0.5 to 3.00
3.	Metal	0.35 to 1.5	0.5 to 3.00

5.2.3.3 Machining Allowance

The finish on sand castings are poor. Thus in order to obtain better finish machining is carried on the castings. For this machining allowance is kept. This is in addition to shrinkage allowance. This amount varies from 1.5 to 12 mm depending on the type of metal, size and shape of casting, type of machining and nature of finish required.

5.2.3.4 Distortion Allowance (Camber)

This allowance is used for castings in which contraction is not uniform throughout, such as **I** beams, **V** sections or **U** shaped castings. Extra material is provided to reduce the distortion of such objects.

5.2.3.5 Rapping Allowance (Shaking)

This allowance is provided because when the pattern is withdrawn from the mould it is first rapped or shaken by striking it from side to side, so that the surfaces may be free from adjoining walls of the mould. As a result of this the cavity in the mould is slightly increased. To compensate for this a negative allowance is to be provided for, by making the pattern slightly smaller in large castings. For small and medium size castings, this allowance can be neglected.

5.3   Types of Patterns

The type of patterns selected for a particular casting depends on

- No. of castings to be produced
- Method of moulding process adopted
- Shape and size of casting
- The cost factor

The following types of castings are commonly used. They are shown in the table below

Sl.	Type of Casting	Figure	Brief Description / Advantages
5.3.1	Solid or single piece pattern		It is made in one piece without any joints. They are made of wood and are cheap. Stuffing box and gland of steam engine are made by using this type.
5.3.2	Spilt Pattern		It made up of two parts or more. One part produces lower part of mould while other the upper half. The parts are joined by means of dowel pins. The line of separation is called parting line. They are used for casting of spindles Taps, bearings, cylinders, wheels, pulleys etc. It can also be made in three pieces. It is known as multi piece pattern.

Contd...

5.3.3	Sweep Pattern	Curved sweep	The moulds of large sizes and symmetrical in shape particularly of circular section can be prepared by using a sweep pattern instead of full pattern. Sweep is a wooden template which has the same size and shape of the casting. It is used for making bells or very large castings.
5.3.4	Gated Pattern	Gates Runner Pattern	These are used where many castings are required. Patterns may be made of wood or metal. Such a casting system ensures full supply of molten metal into every part of mould, thereby producing good castings. The number of castings can be produced by this method.
5.3.5	Match Plate Pattern		They are widely used in machine moulding for mass production of small and accurate castings. The advantage of these patterns are high rate of production and accuracy and minimum machining.
5.3.6	Loose Piece Pattern	Main pattern Pin Loose piece	Sometimes it is difficult to remove the pattern from the mould due to projecting parts. Such projections are made in loose pieces and are fastened loosely to the main pattern by means of wooden or wire dowel pins. These pins are taken out during moulding operations.

5.4   Moulding Sand

The main material used in moulding is sand. Such sand is found in many natural deposits and is suitable for moulding. It comprises of 50 to 95% of total material used in moulding sand. The composition of moulding sand is as follows:-

Silica Sand (SiO_2)   80.8%

Alumina   (Al_2O_3) 14.9%

Iron Oxide   (Fe_2O_3) 1.3%

Combined Water   2.5%

Other Inert materials 1.5%

5.4.1 Properties of Moulding Sand

Moulding sand should possess the following properties:

5.4.1.1 Porosity or Permeability

It is that property of sand which permits the steam and other gases to pass through the sand mould. When hot metal is poured into the sand mould it evolves a great amount of steam and other gases while coming in contact with the moist sand. If these gases do not escape completely through the mould it will cause defects such as gas holes and pores. The sand should be porous and permeable.

5.4.1.2 Plasticity

It is the property of sand due to which it flows to all portions of the mould box or flask during ramming and acquires a predetermined shape under ramming pressure and retains this shape when pressure is removed. The sand must have sufficient plasticity to produce a good mould. The plasticity is increased by adding water and clay to sand.

5.4.1.3 Adhesiveness

It is the property of sand by which it adheres or clings or sticks together to the sides of the moulding box. Good sand must have sufficient adhesiveness so that heavy sand masses are successfully held in the moulding box with out any danger of its falling out when the box is removed.

5.4.1.4 Refractoriness

It is the property of moulding sand to withstand high temperature of the molten metal so that fusion does not occur.

5.4.1.5 Cohesivness

It is the property of sand due to which the particles of sand stick or cling together during ramming operation. Due to this pattern withdrawal from the mould is easy. It is also known as strength of the moulding sand.

It is of three types

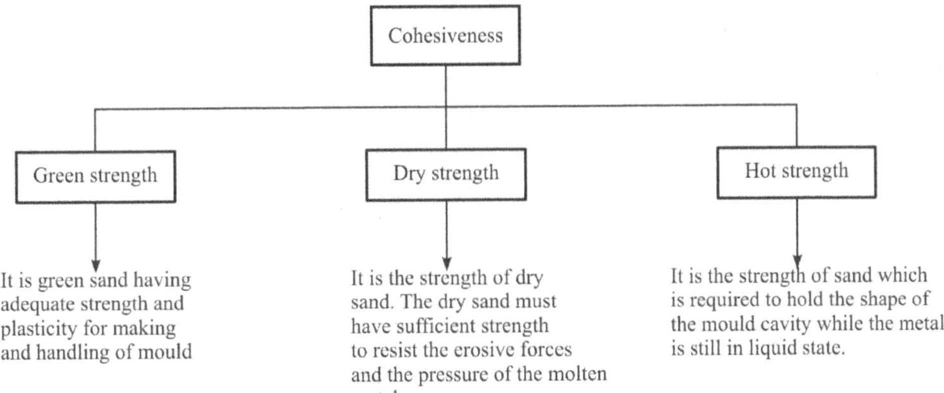

5.4.1.6 Collapsibility

It is the property of sand due to which the sand mould collapse after solidification of the casting in order to allow free contractions of the metal.

5.4.2 Types of Moulding Sand

Moulding Sand may be classified into three different types as follows –

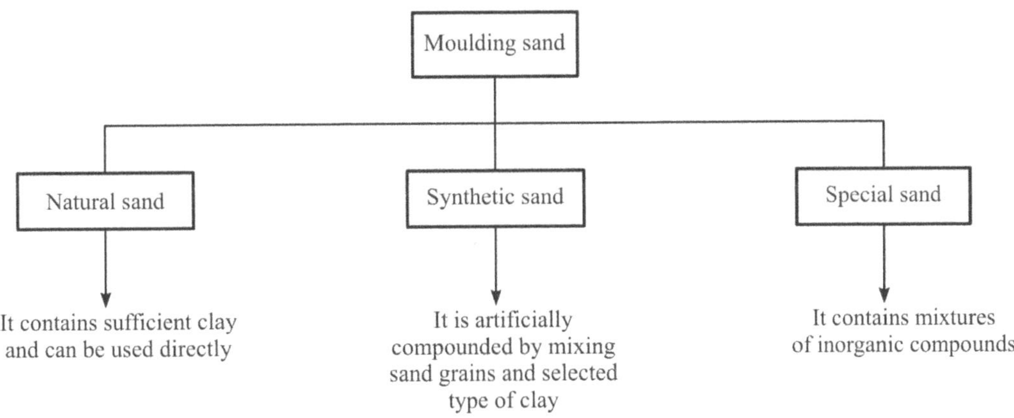

Moulding sands can be further categorized according to use as follows:

Green Sand : Sand containing moisture is known as green sand. It is a mixture of silica sand with 20-30% clay and water content of 6-8%. No baking is required with this sand. It is used foe ferrous and non-ferrous metals.

Core Sand : It is used for preparing cores, it is also called oil sand. It is silica sand mixed with linseed oil or any other oil as binder.

Dry Sand :   The green sand moulds when baked or dried before pouring the molten metal into moulds are called dry sand moulds. They are used for large and heavy casting.

Facing Sand : Sand used for facing of the mould is called facing sand. It is rammed around the pattern surface. Thickness of facing sand is 20 to 30 mm.

Parting Sand : A sand used on the faces of the pattern before moulding is called parting sand. It consists of dried silica sand, sea sand or burnt sand. It is sprinkled on the pattern and the parting surfaces of the mould.

Backing or Floor Sand : It is the sand which is left on the floor after the casting have been removed from the mould.

5.5   Cores

These are defined as sand bodies used to form the hollow portions or cavities of desired size and shape in a casting. The various types of cores are detailed as below –

5.5.1	Type of Core	Figure	Description / Use
5.5.1.1	Horizontal Core		This type is used on a split or two piece pattern. It is placed horizontally on the parting line so that one half remains in cope and the other half in the drag.
5.5.1.2	Vertical Core		This type is placed vertically in the mould. Major portion of the core remains in the drag.
5.5.1.3	Balanced Core		It is also a horizontal core which is supported at one end only and the other end remains free. Balanced cores are used to produce blind holes along a horizontal axis in a casting.

5.5.1.4	**Stop off Core** **or** **Drop Core**	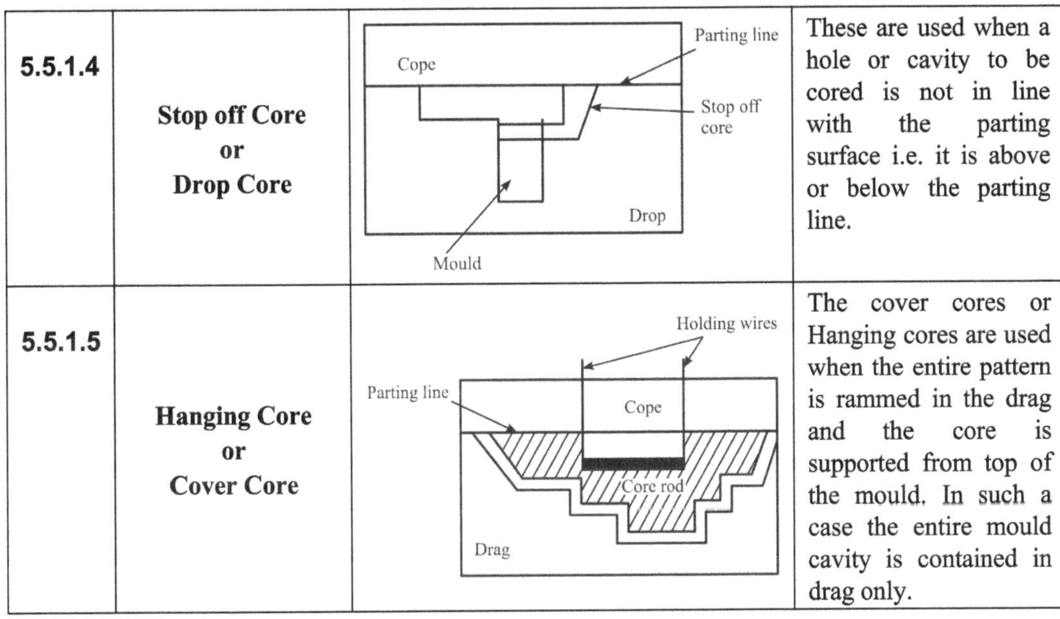	These are used when a hole or cavity to be cored is not in line with the parting surface i.e. it is above or below the parting line.
5.5.1.5	**Hanging Core** **or** **Cover Core**		The cover cores or Hanging cores are used when the entire pattern is rammed in the drag and the core is supported from top of the mould. In such a case the entire mould cavity is contained in drag only.

5.5.2 Core Making

The core is formed by being rammed into a core box or by the use of sweeps. The cores are mostly made of a core sand mixture consisting of sand grains and organic binders which provide green strength cured strength and collapsibility. Fragile and medium sized cores are reinforced with wires for added strength to withstand deflection and the hydrostatic action of the metal. In large cores perforated pipes or arbors are used.

The core making procedure consists of following steps –

(a) Mixing of core sand

(b) Ramming of core sand

(c) Venting of core

(d) Reinforcing of core

(e) Baking of core

(f) Cleaning of cores

(g) Sizing of cores

(h) Joining of cores

5.5.3 Core Setting

Core Setting is the operation of placing cores in moulds. The cores must be of correct size and shape and positioned properly in the mould cavity. Small cores are placed in the mould by hand whereas a large core requires hoist or crane service. The cores must be properly held when the metal is poured.

5.6   Moulding Processes

Moulding is the process of making a mould with the desired cavity to pour the molten mtal. Moulding processes are usually classified as

(a) Hand Moulding   (b) Machine Moulding

The former is used where production is less i.e. lot size is less and the latter is used for high production.

5.6.1 Types of Moulds

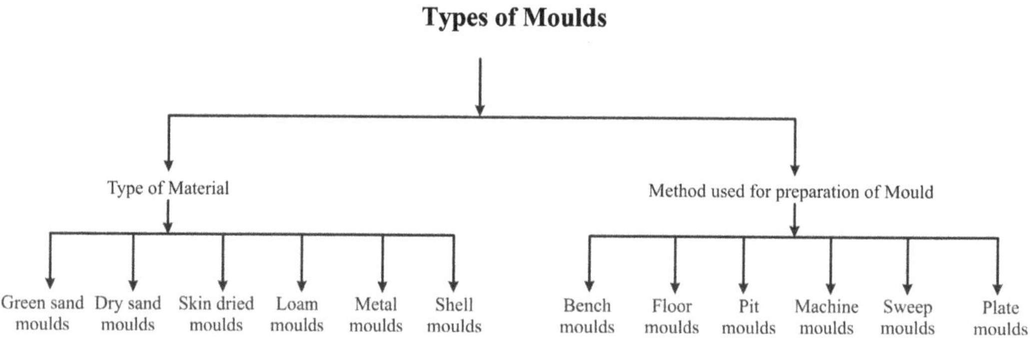

Types of moulds according to the type of material

5.6.1.1 Green Sand Mould

The sand mould prepared from natural moulding sand in its green state is called green sand mould. These are mixture of sand, clay and water. The mould contains moisture at the time of pouring. It is used for small and medium casting of ferrous and non ferrous alloys. Defects like blow holes may occur in castings due to moisture.

5.6.1.2 Dry Sand Mould

Moulding sand mixed with a suitable building material is used. These are oven baked hence it is free from gas troubles due to moisture. These are used in small and medium sized castings. For large castings dry sand moulds may be made in sections.

5.6.1.3 Skin Dried Mould

The sand mould with a dry sand facing and a green sand backing is called skin dried mould. It is used for ferrous and non ferrous castings. These moulds are not as strong as dry sand mould.

5.6.1.4 Loam Mould

The mould made with loam sand is called loam mould. Loam sand is a mixture of sand and clay. The mould is first made up with bricks or large iron parts, these parts are then plastered over with a thick loam mortar. The shape of the cavity is obtained with sweeps or skeleton patterns. The mould is then allowed to dry completely. It requires long time to make and so is not used extensively.

5.6.1.5 Metal Mould

These are mainly used in die casting of low temperature alloys. It is made of metal. It has a smooth finish therefore much of the machining after casting is reduced.

5.6.1.6 Shell Mould

The moulds made of two or more thin shell like parts consisting of thermosetting resin bonded sand is called shell mould. It is used in shell moulding

5.6.2 Methods Used for Preparation of Mould

1 **Bench Moulding**

It is done on the bench. It is used for small castings which are light in weight.

2. **Floor Moulding**

It is done on the floor. It is used for medium and large size castings.

3. **Pit Moulding**

It is done in a pit. It is used for big castings.

4. **Machine Moulding**

It is done by a machine. The machine performs the operations of ramming of sand, rolling the mould over, forming the gate and drawing out of the pattern.

5. **Sweep Moulding**

Moulding done by a sweep pattern is called sweep moulding. It is used for making moulds of large size and symmetrical in shape particularly of circular sections.

6. **Plate Moulding**

In this method the pattern is divided into two halves and mounted on a plate. By this method patterns can be handled easily and rapidly.

5.7   Moulding Procedure

Sand moulds are prepared in specially made boxes of wood or metal called flask. These boxes are open at top and bottom. The lower box is called the drag while the upper box is called the cope. In making the mould the flask and pattern are placed over a smooth board called the mounting board. Properly conditioned sand is pressed around and over the pattern with fingers and then rammed with hand rammers. After ramming is finished excess sand is levelled off with strike bar. Small vent holes are made for escape of gases. Then another board called bottom board is placed on top and the mould is turned over. The moulding board is removed. Surface is first smoothened with a trowel and then coated with dry parting sand. The cope is then placed on the drag and held properly by dowel pins. A passage for pouring molten metal is made by using a tapered pin called sprue pin. The mould is filled with sand, rammed and completed with proper vent holes. The sprue pin is withdrawn before withdrawing the sand around the edges is moistened with a swab, Pattern is shaken slightly in all directions such that it becomes loose and then lifted with a draw spike. A small passage known as gate is cut at the bottom of the sprue hole and the mould is closed.

5.8   Gating System

Gating system refers to all those elements which are connected with the flow of molten metal from ladle to mould cavity or in other words it is the passage way for the molten metal. The various elements are

- Pouring basin / cup
- Sprue
- Runner
- Gate
- Riser

5.8.1 Pouring Basin / Cup

The molten metal is poured into the pouring basin. It is a small funnel shaped cavity at the top of the mould which receives the molten metal from the pouring vessel called ladle. It acts as a reservoir from which molten metal flows smoothly into sprue.

5.8.2 Sprue

Sprue helps in feeding metal to the runner, which in turn reaches the cavity through gates. Sprue can be straight or tapered in shape. It is usually tapered as straight shape may cause blow holes and inclusions.

5.8.3 Runners

Runner is a horizontal channel, which connects the sprue to the gates and allows the molten metal to enter cavity. Runners are trapezoidal in cross-section.

5.8.4 Gates

These are channels which connects runner with mould cavity and these are actual entry points through which molten metal enters mould cavity.

Gates can be of various types depending upon their applications such as

- Parting Gate
- Bottom Gate
- Top Gate
- Step Gate

5.8.5 Risers

Risers are added reservoirs provided in mould to feed molten metals into the mould cavity to compensate for shrinkage. The risers are of various types such as

- Top Risers
- Side Risers
- Blind Risers
- Open Risers

5.9   Casting Defects and Remedies

Various casting defects are specified below:

Casting Defect	Probable Causes	Remedies
5.9.1 Blow Holes		
These are smooth, round cavities produced in a casting due to trapping of gas bubbles in the metal. These are visible on machining of the castings.	• Excessive moisture in moulds or cores • Improper venting of moulds and cores • Low sand permeability • Excessive amounts of additives containing hydrocarbons • Insufficient baking of cores • Core binders which liberate large amount of gas	• Increase permeability of mould and cores • Make enough provision for extraction of air and gas from mould.
5.9.2 Hot tears (Pulls)		
These are internal or external cracks having ragged edge	• High dry and hot strength of mould • High sulphur content • Too low pouring temperature. • Too much shrinkage of metal during solidification.	• Design of casting to be improved

Contd...

5.9.3 Cold Shuts (Misrun)

When molten metal floats to fill the entire cavity of the mould, this defect occurs.	• Improper Gating system • Faulty design • Lack of fluidity in molten metal	• Gating system to be improved. • Design to be studied.

5.9.4 Warpage

It is undesirable deformation in a casting produced during solidification of metal.	• Large sections are prone to warpage. • No directional solidification of casting	• Design of casting to be reviewed

5.9.5 Hot Spots

Hardening of metal generally developed in iron castings rich in silicon content.	• Faulty metal composition • Faulty design	• Change in metal composition by adding Nickel content

5.9.6 Fins

It is a thin projection of metal in the casting. It occurs at the parting of the mould or core section.	• Improper clamping of flasks • Mould and cores not assembled properly	• Modification in casting design. • Proper clamping to be done. • More care during assembly of mould and cores.

5.9.7 Run Out

This occurs when molten metal leak out of the mould during pouring.	• Faulty mould making. • Excessive pouring pressure • Faulty moulding flask	• Mould making to be improved. • Pouring processes to be reviewed. • Moulding flask design to be checked.

5.9.8 Swell

It is enlargement of mould cavity after the metal has been poured on the casting surface.	• Improper ramming of mould • Inadequate supporting of mould and core. • Low strength of mould and core.	• Ramming to be improved. • Supports on mould and core to be further strengthened.

5.9.9 Buckles Or Rattail

It is a long fairly shallow, broad V shaped depression occurring on the surface of a flat casting of a high temperature metal.	• Excessive mould hardness • Excessive temperature of molten metal	• Strength of mould to be increased. • Mould hardness to be reduced. • Molten metal temperature to be reduced.

5.9.10 Scabs (Metal Penetration)

These are rough, irregular projection on the casting surfaces with embedded sand.	• Too fine sand • Slow or intermittent running of molten metal.	• Sand composition to be checked • Pouring of metal to be improved

5.9.11 Pin Holes (Porosity)

These are small holes less than 2 mm, visible on the surface of the casting.	• Sand with high moisture content • Faulty metal • Slow rate of solidification	• Moisture content of sand is reduced. • Impurities in metal to be removed. • Solidification process to be improved

5.9.12 Inclusions

Any foreign material present in cast metal	• Oxides, slag, dirt which enters during pouring • Breaking of sand from mould	• More care during pouring • Ramming to be improved

5.9.13 Poured Short

The upper portion of casting is missing	• Insufficient quantity of liquid metal in ladle • Incorrect pouring	• Keep sufficient metal in the ladle. • Care to be taken during pouring.

5.9.14 Shrinkage Or Hot Cracking

A crack which is generally not visible. Discoloration of casting results.	• Damage of casting due to rough handling • Non-uniform solidification of casting.	• Care to be taken during handling of castings. • Cooling to be improved.

5.9.15 Flow Marks

This defect occurs as lines which trace the flow of streams of liquid metal	• Oxide films which lodge at the surface partially marking the path of metal flow through the mould	• Increase mould temperature. • Lower the pouring temperature

5.10   Cupola Furnace

It is a furnace which is used for melting cast iron such as grey cast iron, malleable cast iron, modular cast iron.

Constructional Details

It is a vertical furnace, consisting of a cylindrical steel shell (About 7 to 8 feet diameter) lined with refractory material. It consists of a drop door at the bottom. The openings for introducing the air to coke bed are called tuyeres.

Tuyeres are surrounded by wind box for air supply. It consists of the following main parts such as spark arrestor, shell, refractory lining, charging door, charging floor, tuyeres, tap hole, slag spout , wind box, blast pipe etc. These are shown in the figure.

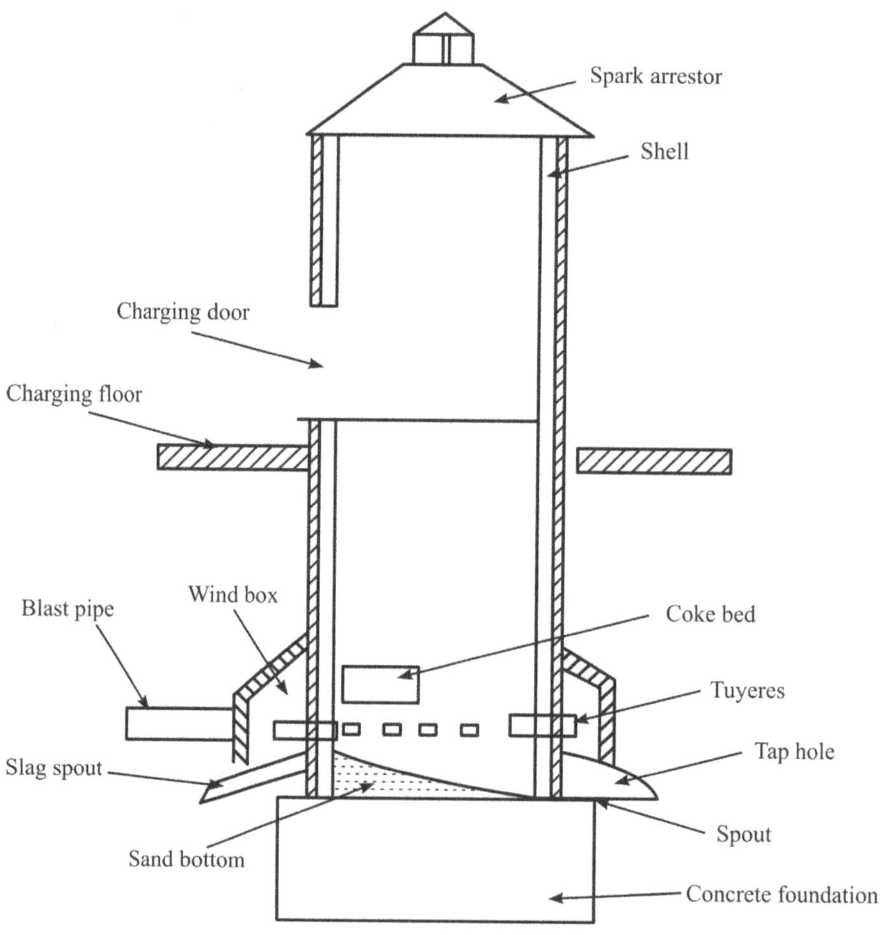

Cupola Operation

The cupola operation is as follows:

A sand bottom is prepared by ramming the moulding sand which slopes gently towards the tap hole. The drop door is closed. A coke bed is prepared above the sand bottom. The coke bed is ignited through the tap hole. When coke bed is properly ignited the charge consisting of metal, limestone, pig iron, cast iron scrap is fed through the charging door. Limestone is added to react with impurities in the metal as it melts. The limestone / impurities combination is called slag. Air blast is turned on and it enters through tueyeres. This causes combustion. The coke is consumed and molten metal is collected near its tap hole. As the material is consumed additional charges can be added to the furnace. A continuous flow of iron emerges from the bottom of the furnace. The slag is drawn off through slag spout which is slightly higher than the tap hole. The exchange gases emerge from the top of the cupola.

Advantages

- It is very simple and economical to operate
- It refines the metal charge
- They are more efficient and less harmful to the environment.

Disadvantages

- Carbon monoxide fumes
- Less oxygen

5.11   Die Casting

It is a precision casting method. In this the metal is forced under pressure. Molten metal flows into the mould or die. It is suitable for non-ferrous metals such as lead, zinc, aluminum, magnesium etc. which have low melting points. Iron, steel, ferrous metals can also be cast by this method but the die life is reduced. There are two types of die casting machines

  (i) Hot Chamber   (ii) Cold Chamber

The methods are shown as below. The difference can be seen by sketches

Hot Chamber	Cold Chamber
• It is known as submerged plunger Die casting process • It has a suitable furnace for melting and holding metal	• In this moulding metal is not integral with die casting machine. • The metal is melted in a separate furnace and then poured into casting method with the help of ladle.

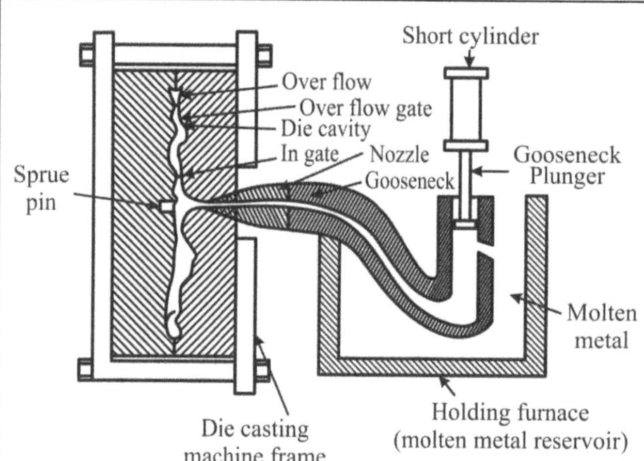

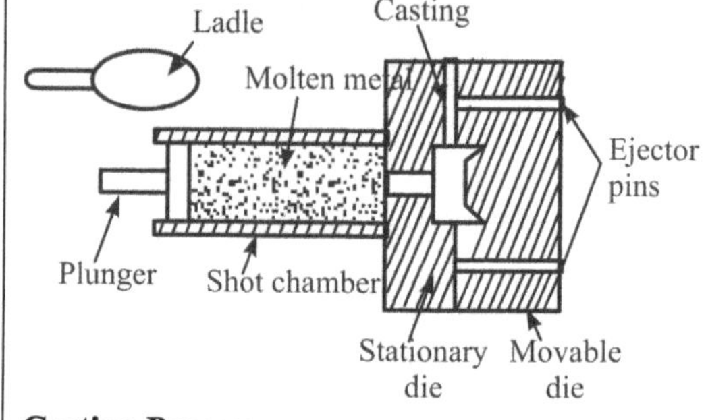

Casting Process

- When the plunger is raised the opening of cylinder is uncovered, the molten metal enters the cylinder from the molten metal reservoir.

- When the plunger is forced downward it closes the cylinder opening and forces the molten metal through a nozzle in a die.

- As soon as Casting is solidified the pressure is relieved the Dies are opened and casting is ejected by means of knock out pins. The Plunger again moves up uncovering the holes on cylinder and allows the liquid metal from furnace to cylinder.

Casting Process

- Molten metal is poured into the shot chamber with the help of ladles.

- Die is in 2 halves of a) stationary Die and b) movable Die and are also closed and clamped at the time of pouring.

- Plunger forces the metal into Die after solidification the Dies are opened and casting is ejected.

- The Plunger again returns to its original positioin.

Difference Between the Two Processes

• Molten metal is injected into the same chamber in which it is melted.	• Already molten metal is poured with the help of ladle into the shot chamber.
• No handling or transfer of metal.	• Handling time is extra.
• Less operating cycle.	• Longer operating cycle.

Advantages of die Casting

- Good accuracy, surface finish can be produced.
- High production is possible.
- Cleaning, machining and finishing costs are low.
- Complex castings can be produced.
- Wastage is less.

Limitations of Die Casting

- Costs for equipments and Dies are high.
- Not suitable for high melting alloys such as Iron, Steel and other ferrous metals.
- Casting size is limited to say 20 Kg.

Questions

1. What are the various factors that affect the selection of material for castings? Why are castings prepared over other manufacturing processes?

2. Briefly explain the fallowing terms used in castings

   (a) pattern   (b) core

   (c) gating system   (d) core box

   (e) chaplets   (f) core prints

   (g) chills   (h) partins line

3. What are the basic steps in casting process? Briefly describe the process of pattern making.

4. Describe the advantages, disadvantages and applications of the following pattern materials

   (a) Brass   (b) aluminum

   (c) plastics   (d) plaster of paris

5. What are the various pattern making tools. How are they different from tools used in a carpentry shop?

6. Write short notes on

   (a) shrinkage allowance   (b) draft allowance

   (c) machining allowance   (d) distortion allowance

   (e) rapping allowance

7. Explain the following patterns with the help of neat sketches

   (a) split pattern   (b) sweep pattern

   (c) gated pattern   (d) match plate pattern

   (e) loose piece pattern

8. Write short notes on

   (a) porosity or permeability   (b) plasticity

   (c) adhesiveness   (d) cohesiveness

   (e) refractoriness

9. Name and describe the various types of moulding sand.

10. Briefly describe the following cores with sketches

    (a) Horizontal core
    (b) vertical core

    (c) balanced core
    (d) hanging core

11. What is core making? Briefly describe its process.

12. Write short notes on

    (a) Green sand mould
    (b) Dry sand mould

    (c) loom mould
    (d) shell mould

13. What are the various casting defects and their remedies?

14. What is die casting? What are its advantages and limitations?

15. Describe with diagram the couple furnace.

Unit III

Introduction to Machining, Welding and its Applications

CHAPTER 6

Machining

6.0 Basic Principles of Lathe Machine and Operations Performed on it

6.1 Lathe

It is a machine tool that is used in manufacturing. Various operations can be performed on it. During the operation on the Lathe the work piece or job or component to be manufactured is held in a chuck or face plate or held between two rigid supports called centres. The cutting tool is held on a tool post which is mounted on Lathe bed in cross-slides. The job rotates while an angle point cutting tool advances in a direction parallel to the axis of rotation of the spindle. The spindle is driven by an electric motor through a system of belt drives or gear trains.

6.1.1 Main parts of a Centre Lathe Machine

The main parts are as follows:-

- Bed

- Head Stock

- Tail Stock

- Carriage

- Feed Mechanism

- Legs

The Sketch shows a simple Lathe Machine

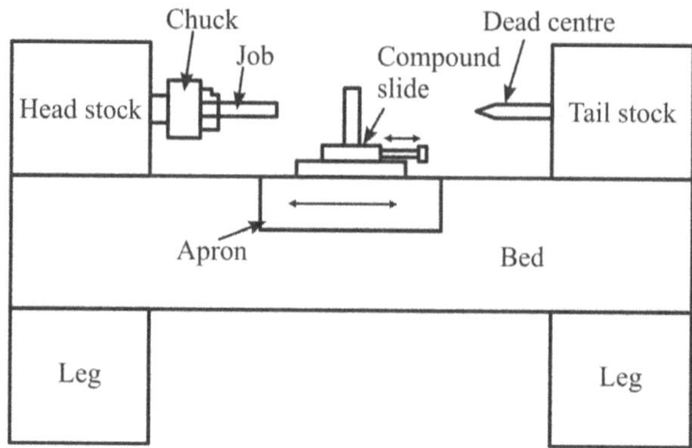

1. Bed

It is the base on which different parts such as Head Stock, Tail Stock, Carriage assembly etc. are mounted. It is made of grey or modular cast iron.

2. Head Stock

It is on the left side of the Lathe. It acts as a housing for cone pulleys or a set of transmission gear or combination of both. Different spindle speeds are required to form various operation. The head stock spindle protrudes from gear box and contains means for fastening holding devices such as chuck, face plate etc. for holding the job. The spindle is mounted on bearing and are hollow. The power for the spindle is supplied from an electric motor through chain drives.

3. Tail Stock

It is on the right side of the Lathe. It is mounted on the bed and slides along it on ways. It can accommodate jobs of different lengths and can be clamped in any position. It is a tool holder mounted directly on the spindle axis. It is made of cast iron.

4. Carriage

It is mounted on the bed of the Lathe. It's main function is to support guide and feed the cutting tool during the operations. It provides necessary motion (i.e. parallel to the axis of rotation of spindle) to the cutting tool. It has the following four parts i.e. cross slide, compound rest, tool post and apron.

5. Feed Mechanism

It is the mechanism which provides feed. Feed is the distance the tool advances into the job or work piece through one revolution of head stock speed. It can be longitudinal cross feed or angular.

Longitudnal Feed: In this the tool travels along the work piece or job parallel to the direction of Lathe bed.

Cross Feed: In this the tool travels perpendicular to the direction of Lathe Bed.

Angular Feed: It is used in taper turning operation and is done with the help of compound rest. It is hand operated.

Lead screw is long threaded which allows fixed transverse motion of carriage for single rotation of the spindle and produces accurate threads.

6. **Legs**

They are hollow cast iron supports and bear the entire weight of the Lathe such as bed, carriage, head stock, tail stock etc. Both the legs are firmly secured to floor by means of foundation bolts.

6.1.2. Lathe Operations

The different operations performed on Lathe machine are

(a) Standard Operations such as

- Plain Turning
- Step Turning
- Facing
- Taper Turning
- Eccentric Turning
- Drilling
- Boring
- Reaming
- Knurling
- Chamfering
- Forming
- Undercutting or Grooving
- Threading

Special Operations

Such as milling, grinding, spherical and elliptical training, copying etc. These operations are performed with the help of special attachments.

Brief description of the standard operations along with sketches is shown as follows

Operation	Sketch
6.1.2.1 Plain Turning	
It is the operation of removing excess material from the surface of the cylindrical job or component. The tool is held between the centre on a chuck. As the workpiece rotates the tool is made to move in the direction parallel to the Lathe axis. This movement of tool is called feed.	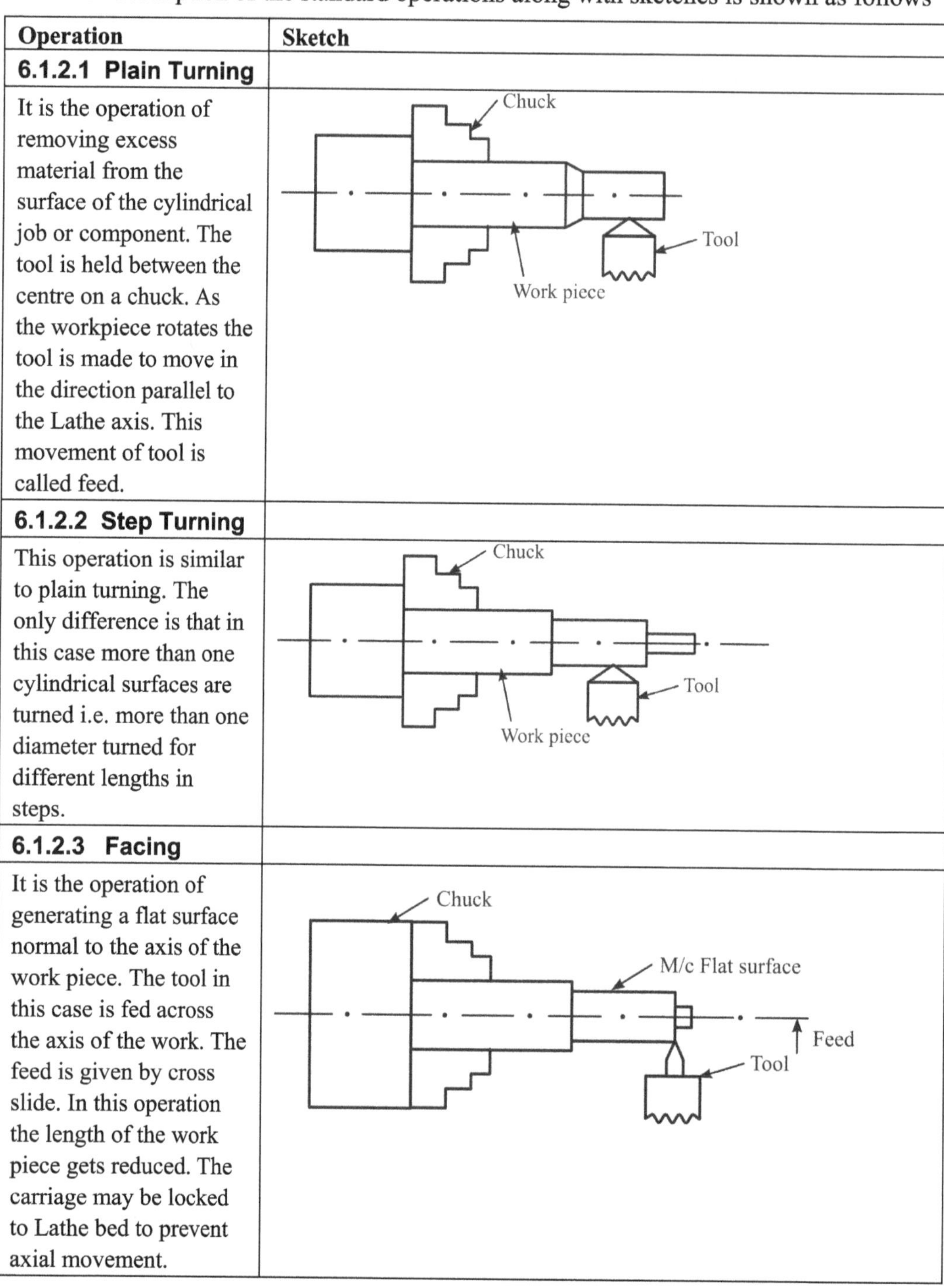
6.1.2.2 Step Turning	
This operation is similar to plain turning. The only difference is that in this case more than one cylindrical surfaces are turned i.e. more than one diameter turned for different lengths in steps.	
6.1.2.3 Facing	
It is the operation of generating a flat surface normal to the axis of the work piece. The tool in this case is fed across the axis of the work. The feed is given by cross slide. In this operation the length of the work piece gets reduced. The carriage may be locked to Lathe bed to prevent axial movement.	

6.1.2.4 Taper Turning

It is the operation of gradual reduction in the diameter of a workpiece either along full length or part. There are four methods

- Offsetting the tailstock
- Using taper turning attachment
- Using form tool
- Using compound rests

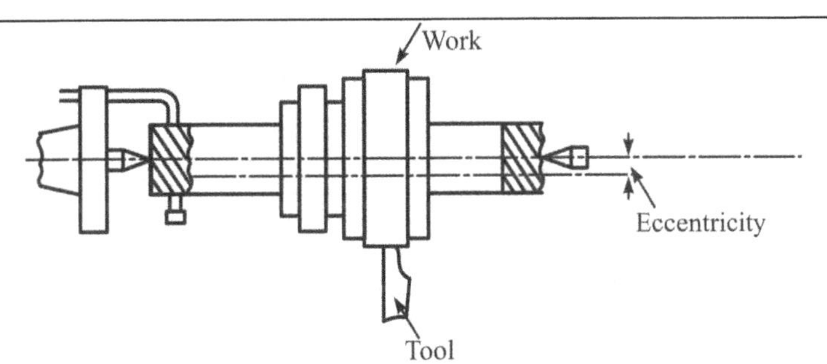

6.1.2.5 Eccentric Turning

When two cylindrical surfaces, which are eccentric with each other are turned on Lathe is called eccentric turning. In this two different pairs of center are used for turning different surfaces.

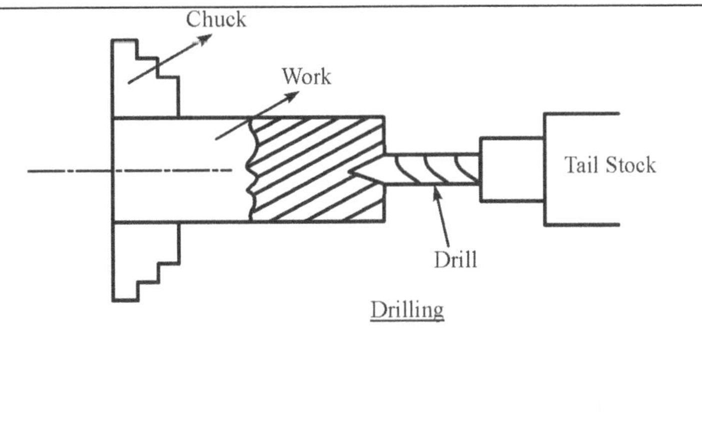

6.1.2.6 Drilling

It is the operation of making a hole in a work piece with the help of a drill. In this operation the work piece is held in the chuck and the drill is held in the tail stcok. First a centre drill is used to drill a small hole at the end face. The drill is fed manually into the rotating work piece by rotating the tail stock hand wheel.

### 6.1.2.7 Boring	
It is the operation of enlarging a hole which is already made in the work piece. In this a boring tool is used. It can be fed by hand or power as in turning.	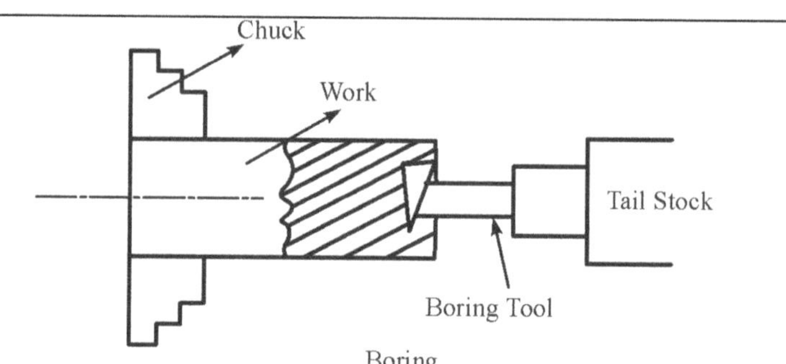
### 6.1.2.8 Reaming	
It is the finishing operation performed to provide a superior surface finish to a drilled or bored hole.	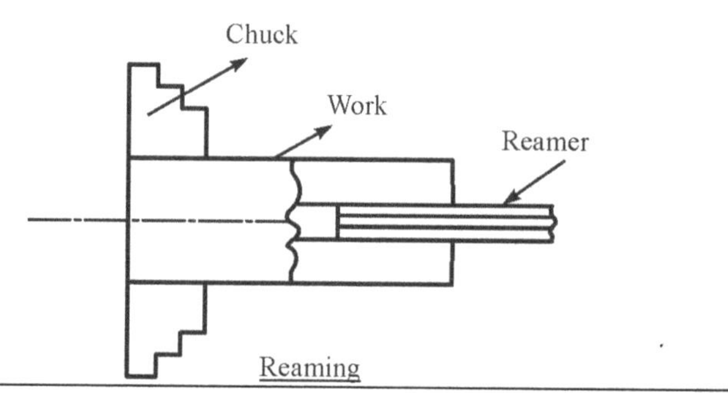
### 6.1.2.9 Knurling	
It is an operation of providing knurled surface on the workpiece. A knurling tool is used. It is moved longitudinally to a revolving work piece surface. The projections on the knurled tool reproduce depression on the work piece. This operation is required on handles, screws, nuts and instruments for better gripping.	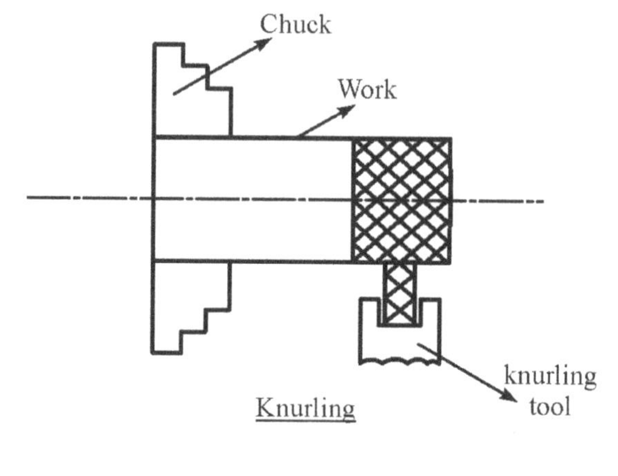

6.1.2.10 Chamfering	
Turning a very small taper at the end of a work piece is known as chamfering. It is done to remove sharp edge or corners with the help of a chamfering tool.	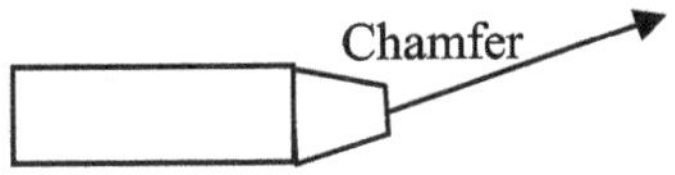
6.1.2.11 Forming It is also known as form turning. In this a form tool is used in which the cutting edge is of the desired shape. It is the process of machining radial and irregular shapes.	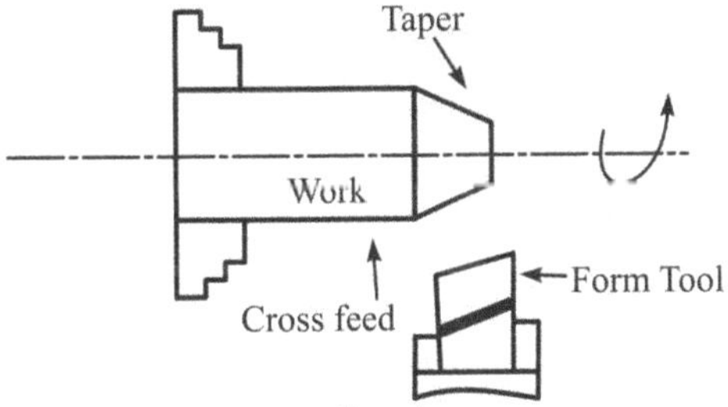
6.1.2.12 Under Cutting or Grooving	
Undercuts are reductions in diameter machined onto the centre portion of the work piece. It is usually formed to hold a seal. Under cutting is done by feeding the tool bit into the work piece while moving the carriage back and forth slightly. It is also called grooving (or necking). Common	

<table>
<tr>
<td>
grooves are square, round or V-shape. It is formed on a round bar, cylinder or work piece.
</td>
<td>
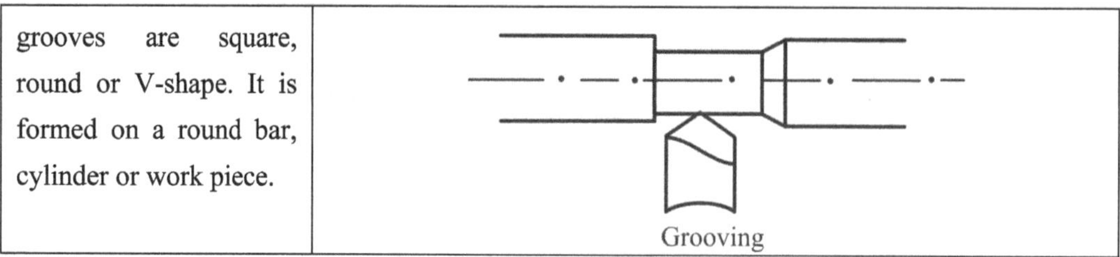

Grooving
</td>
</tr>
</table>

6.1.2.13   Threading or Thread Cutting or Screw Cutting

Threading can be

a)   External : on outer surface

b)   Internal: on inner surface

It is uniform helical groove cut on inside of a cylindrical surface or on outside of a tube or shaft.

Principle of Thread Cutting

It is cut by establishing a definite ratio of speeds between the work piece and lead screw. Thread cutting can be as follows

a)   Metric thread cutting : by using Lathe designed for same

b)   Tapered Screw threads : by use of attachment taper

c)   Cutting Internal Threads: same as external threads

d)   Cutting External Acme threads: with the use of compound rest.

6.1.3 Types of Lathes

Commonly used Lathes are as follows

6.1.3.1 Speed Lathe

It is one of the simplest type of Lathes. It is driven by power and consists of a bed, a head stock, tail stock and an adjustable slide for supporting the tool. Since the tool is fed into the work by hand and cuts are very small these lathes are driven at high speeds between 1000 to 3000 rpm. They are used mainly for wood working, centering, metal spinning, polishing etc.

6.1.3.2 Engine or Centre Lathe

It is most common Lathe used in workshops. It is called centre Lathe because the work piece is clamped between the centers. It is called engine Lathe because in older days it was run by steam engines. Now it is run by electric motors. It has additional mechanism for controlling the spindle speed and for supporting and controlling the feed of the fixed cutting tool. The cutting tool may be fed both in cross and longitudinal direction with reference to the Lathe axis with the help of a carriage.

6.1.3.3 Tool Room Lathe

It is used basically for making tools, dies, gauges, jigs, fixtures and other small parts. It is similar to engine lathe but is equipped with accessories needed for accurate work. It is expensive and requires a skilled operator. It also has a higher range of speeds and accuracy.

6.1.3.4 Copying Lathe

It is accurate Lathe with copying attachment mounted on it. It is needed to machine complex contours. In this Lathe a template or master piece is used which is similar to the shape of work piece to be machined. The size of the master piece way vary but shape remains same. The movement of the cutting tool is driven by a stylus which traces the master piece for actual contour to be produced.

6.1.3.5 Capstan and Turret Lathes

These are the modification of engine lathes. They are used for mass production of identical items in less time. These lathes are fitted with capstan and turret heads which hold the tools. Several operations can be performed in one setting without disturbing the tools. Difference between capstan and turret Lathes are described Later.

6.1.3.6 Automatic Lathes

These lathes are so designed that the tools are fed automatically into the work piece and withdrawn after completion of all the work. These are used for mass production of identical items.

6.1.3.7 Special Lathes

These lathes are used for special applications which cannot be handled by other lathes.

6.1.4 Difference Between Capstan and Turret Lathes

Capstan Lathe	Turret Lathe
The Turret is mounted on an auxiliary slide (ram) which slides on the saddle as shown in fig **Capstan lathe**	The Turret is mounted on the saddle which directly slides on the bed guide ways Movement of turret is on the entire length of bed **Turret latthe**
Movement of turret is short	Movement of turret is in entire length of bed
Light in weight. Used for medium work.	Heavier than capstan lathe Used for heavier work.
Less fatigue to operates due to light ram	Feeding requires heavy force hence more tiresome
Power chucks not required	Power chuck required
Ram is fed into the work and overhangs from saddle	The Turret is mounted on the saddle

6.2 Shaper

Shaper is also called a shaping machine. It is superseded by milling and grinding machines. It is used for producing flat surface. In this the single point cutting tool moves in a straight line motion in relation to work piece to generate flat surfaces. It rarely used for production operations. It is a low cost machine which is basically used for initial roughing operation.

Working Principle

In this a single point tool reciprocates over the stationary work piece. The tool is held in the tool post of the reciprocating ram and performs the cutting operation during its forward stroke. In the backward stroke of the ram, the tool does not remove the material from the work piece. Both these strokes form the working cycle of the cylinder.

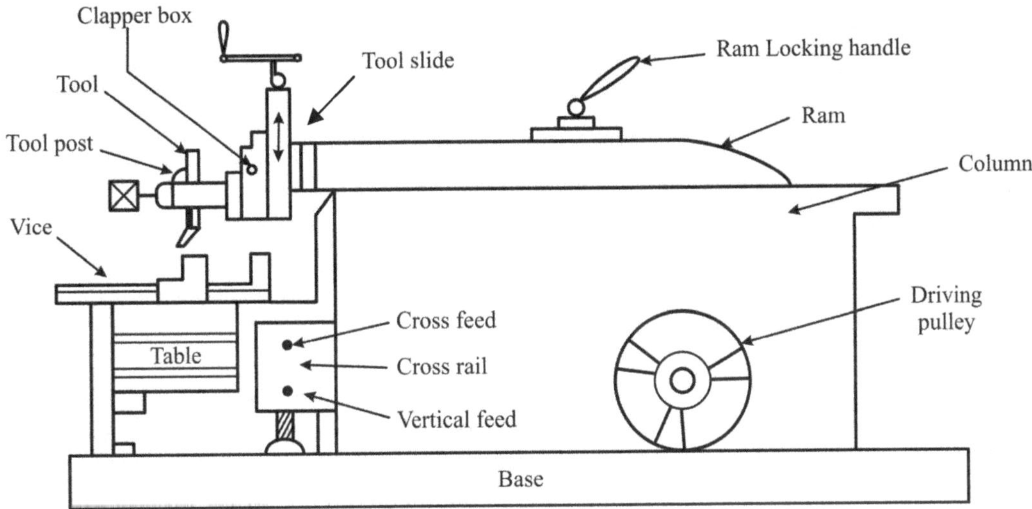

Principal parts of a shaper

6.2.1 Principal Parts of a Shaper

The principal parts of a shaper are shown in the sketch of shaper as above

Base: It is a heavy structure of cast iron which supports other parts of a shaper.

Column: The column supports the ram and the rails for saddle. The mechanism for moving the ram and table is housed inside the column.

Ram: The ram reciprocates on the guide ways provided in the column. It provides a tool slide and a mechanism for adjusting stroke length.

Tool Head: The tool head is fastened to the ram on a circular plate so it can be rotated for making angular cuts. It is fastened with the help of nut and bolt. The tool head can also be moved up or down by its hand crank for precise depth adjustments.

Cross Rail: It is attached to the front vertical portion of the column. It is used in elevating the table over the column in the upward direction. The table can also be moved in a direction perpendicular to the axis of the ram over the cross-rail.

Table: It can be moved left and right by hand or automatically. The table is moved side ways to feed the work under the cutter at the end or beginning of each stroke. It holds the work piece.

6.2.2 Classification of Shaper

Shapers can be classified accordingly as follows:

(a) **Ram Driving Mechine**

Crank Shaper: A Crank and slotted lever quick return mechanism is used

Geared Shaper: Ram carries a rack, driven by spur gear

Hydraulic Shaper: Hydraulic system is used to drive ram.

(b) **Position and Travel of Ram**

Horizontal: Ram reciprocates in a horizontal direction.

Vertical: Ram reciprocates vertically

(c) **Direction of Cutting Stroke**

Push Cut Shaper: Ram pushes the tool across the work during cutting operation.

Draw Cut Shaper: Ram draws or pulls the tool across the work during cutting operation.

(d) **Design of the Table**

Standard or Plain Shaper: The table has only two movements i.e. horizontal and vertical.

Universal Shaper: In this the table can be swivelled about a horizontal axis parallel to the ram and the upper table can be tilted about the other horizontal axis perpendicular to the first axis.

6.2.3 Operations Performed on Shaper Machine

The following operations can be performed on shaper machine :-

- Shaping of a rectangular job, regularly angled part and irregularly curved surfaces.
- Machining of angular surfaces or thin jobs
- Cutting of an angle of large job, dovetail bearing or keyway.

6.2.4 Advantages and Disadvantages of Shaper

Advantages	Disadvantages
It provides economical and simple tooling. Set up is easy	Due to single point tooling it is not suited for production work and metal removal takes time close tolerance jobs could not be taken up for machining.

6.3   Planar

The planer is used for producing horizontal, vertical or inclined flat surfaces on jobs which are too big to be handled on a shaper. The planar has also similar work as that of a shaper.

6.3.1 Types of Planing Machines

The various types of planars that are mostly used are as follows
- Standard or Double housing planer
- Open side planar
- Edge or flat planar

6.3.2 Main Parts of a Planer

The main parts of a planar are shown in the figure

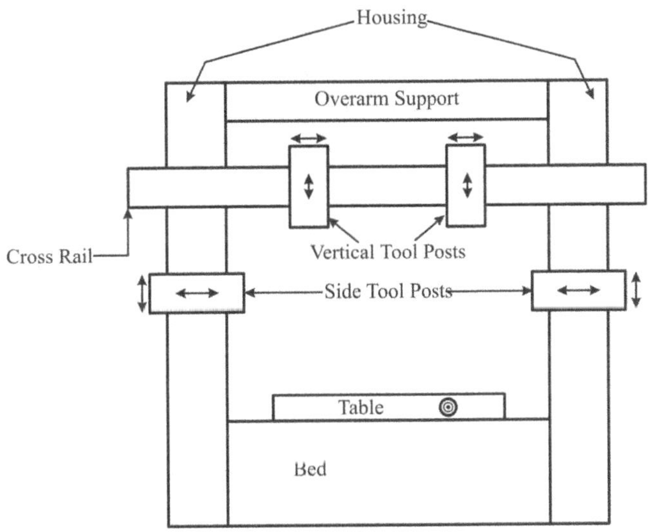

Main Parts of a Planar

They are as follows:

Cross Rail: The cross rail is a heavy box. It slides up and down on a vertical or flat ways. After being positioned they are clamped.

Rail Head : The rail heades can be moved left or right across the cross rail. The rail heads can be rotated and vertically adjusted for depth of cut.

Table: The table is a heavy casting. It carries the work part the cutting heads. It runs on vertical or flat ways.

Bed: The bed must be twice as long as the table.

Tool Posts: The tool posts hold the tool. The tools are high speed steel or carbide tipped cutting tools.

Housing: The housing or frame is basically two heavy columns fastened together at the top and bottom.

6.3.3 Working Principle of a Planar

The job is held rigidly on the work table of the machine. The tool is held vertically in the tool head mounted on the cross rail. The work table together with the job is made to reciprocate past the vertically held tool.

6.3.4 Difference between Planar and Shaper

Planar	Shaper
• It is a heavy machine tool.	• It is a light machine tool.
• It requires greater floor area.	• It is requires less space.
• In this the tool is fixed and the job moves.	• In this job is fixed and tool moves.
• More than one cutting tool can be used at a time.	• Only one cutting tool can be used at a time.
• These are designed for holding big jobs.	• These are designed for holding small jobs.
• Production rate is more.	• Production rate is less.
• Cost of machine is high.	• Cost of machine is less.
• Setting of Job takes time.	• Setting of job requires less time.
• It can take deeper cut.	• It can not take deeper cut.
• Feed movement is provided after every cutting stroke.	• Feed movement is provided at right angles to the motion of cutting tool after every cutting stroke.
• Used for mass production of small components.	• Used for batch or for production
• Planar tools are heavy.	• Shaper tools are simple.

6.4   Drilling Machine

The drilling machine is a single purpose machine for the production of cylindrical holes in the jobs. This is the most common and widely used machine used in workshops or machine shops in any Industry. The holes are cut with the help of a tool called a drill. The drill is fixed in a rotating spindle and can be fed towards the work piece which may be fixed to the table or base of the machine. The speed of the spindle and the feed can be adjusted according to the type of job.

6.4.1 Types of Drilling Machines

- Portable Drilling Machine

- Upright Drilling Machine
- Radial Drilling Machine
- Bench Drilling Machine
- Multiple Spindle Drilling Machine
- Deep hole Drilling Machine
- Gang Drilling Machine
- Horizontal Drilling Machine
- Automatic Drilling Machine

6.4.2 Parts of a Drilling Machine

The various parts of a drilling machine are as follows :

- Head
- Spindle
- Drill Chuck
- Table
- Base
- Column

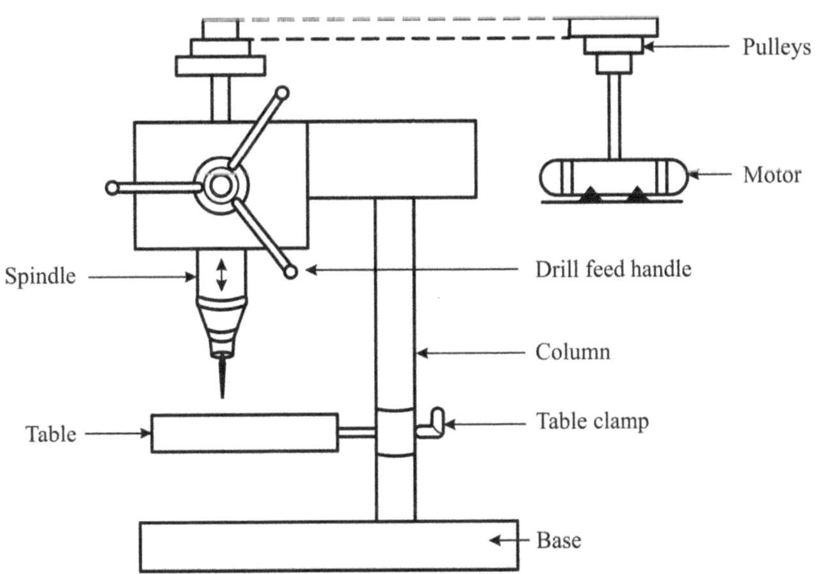

6.4.3 Geometry of Twist Drills

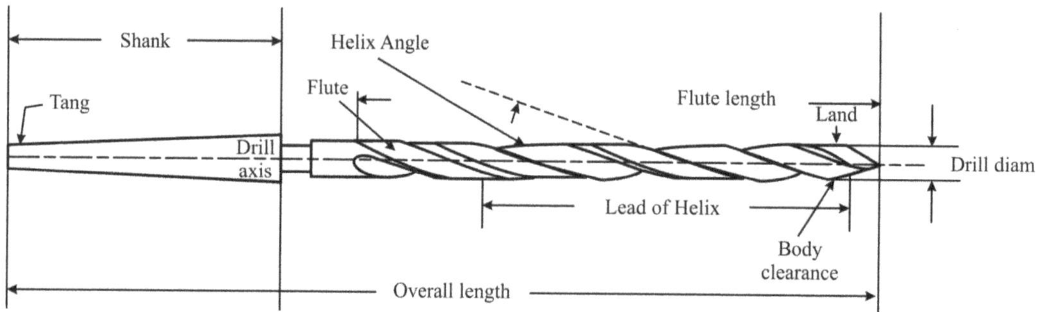

The figure shows a twist drill and its various parts:-

Axis: It is the imaginary straight line which forms the longitudinal centre of the drill.

Body: The portion of the drill extending from the shank or the neck to the outer corners of the cutting tips.

Drill Diameter: The diameter over the margins of the drill measured at the point.

Helix Angle: The angle made by the leading edge of the land with a plane containing the axis of the drill.

Flute Length: The length from the outer corners of the cutting tips to the extreme back end of the flutes.

Overall Length: The length from the extreme end of the shank to the outer corners of the cutting tips.

Shank: The part of the drill by which it is held and driven.

Neck: The reduced diameter between the body and the shank of a drill.

Tang: The flattened end of a taper shank.

6.4.4 Operations on Drilling Machines

The operations performed on drilling machines are as follows:

Operations	*Details*
Drilling	It is the operation of producing a cylindrical hole in a solid body by means of a tool, which is called a drill.
(a) Core Drilling	When sand castings are made, cores are used to displace the metal where holes are desired. During casting the molten metal flows around the core. After the metal solidifies, the casting is removed from the mould and cores disintegrate leaving the desired holes. These holes are cleaned by drill.

Contd.....

(b) Step Drilling	More than one diameter can be ground on drill body, which saves an operation.
Boring	It is the process of enlarging a already drilled hole by a single point tool.
Reaming	It is the process of obtaining a smooth finish on a already drilled hole. Reaming is called a finishing operation.
Tapping	This is the operation of producing internal threads in a hole by means of a tool called 'tap'.
Counter Boring	It is the operation of enlarging the mouth of a drilled hole to set bolt heads and nuts below the surface so that they may not project from the surface. It is done with the help of a counter boring tool.
Counter Sinking	It is the operation of enlarging a drilled hole to give a conical shape in order to accommodate flat head screws. It is done with counter sinking tool.
Spot facing	It is the operation to finish off a small portion of rough surface around a drilled hole to provide smooth seat for bolt head.

6.5   Milling Machine

Milling is used in machine shop for cutting, shaping and finishing a piece of metal. It is used to create dovetails, threads, bevels, slots and ridges. The machine for holding the workpiece, rotating the cutter and feeding, it is known as the milling machine. It is the process of cutting away material by few and a workpiece past a rotating multiple tooth cutter.

6.5.1 Types of Milling Machine

Milling machines can be classified as follows:

- Column and Knee type Milling Machine or vertical Milling Machine
- Bed Type Milling Machine
- Special Type Milling Machine

6.5.2 Main Parts of A Vertical Milling Machine

Base: It takes the complete load of the machine.

Head: It holds the cutter.

Column: It is the vertical member of the machine.

Knee: Support the Saddle.

Saddle: Supports the table and moves transversely.

Table: On which work piece is clamped.

T-Slots: Used to clamp work piece.

Cutter: Tool used for milling.

Work Piece: It is kept on table.

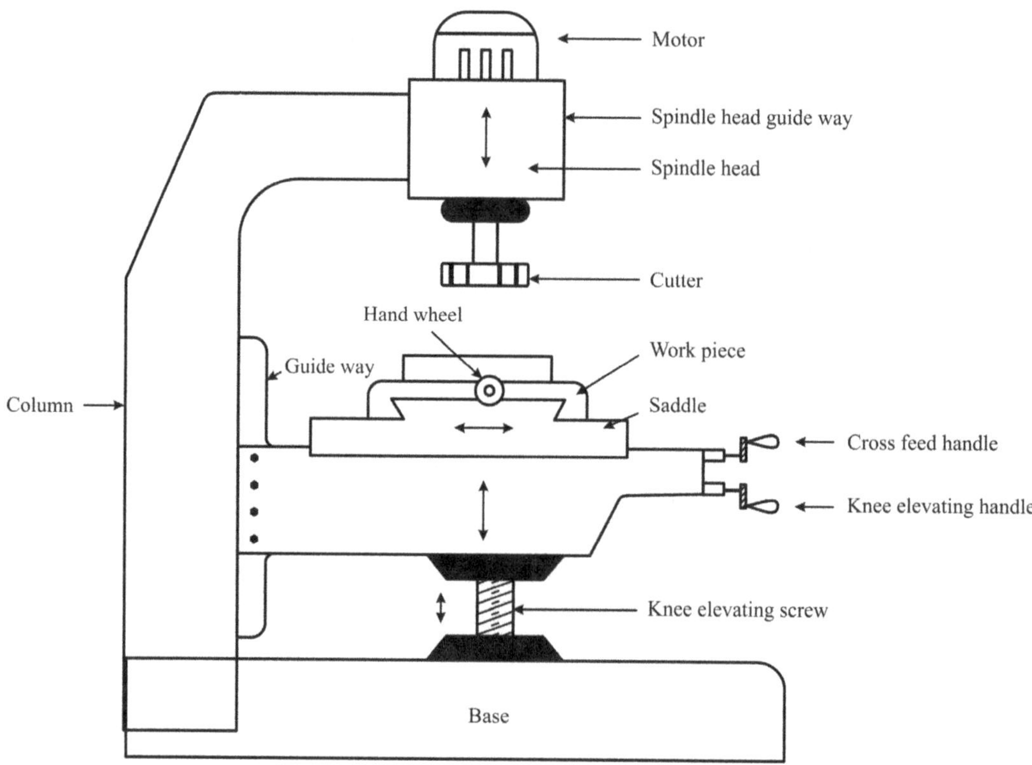

Vertical Milling Machine

6.5.3 Milling Cutters

It is the tool used for milling operation. Milling cutters are used in various sizes, diameter widths and shapes. In this each tooth removes material in the form of chips. There are 3 types of cutting operation (a) peripheral milling (up and down milling), (b) face milling, and (c) end milling.

6.5.4 Difference Between Up Milling and Down Milling

Up Milling	Down Milling
Principle The direction of the cutter rotation opposes the feed motion. In this the table moves towards the cutter opposing the cutter direction. **Advantages** • Cutting process is smooth • Contamination or scale does not affect tool life. • Tooth engagement is not a function of work piece surface characteristics. **Disadvantages** • Chattering is noticed on tool • Proper clamping is required	**Principle** The direction of cutter rotation is same as the feed motion. **Advantages** • Better surface finish is obtained • Down ward component of cutting force holds the work piece in place. • Less radial pressure on arbor **Disadvantages** • It is not suitable for forging / castings • Scale is hard and thus results in less tool life.

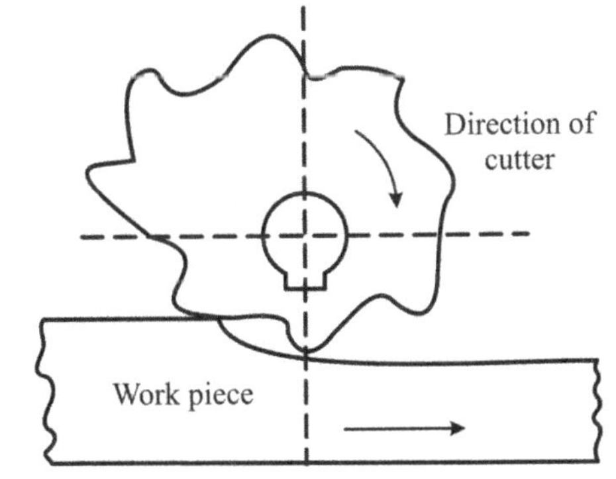

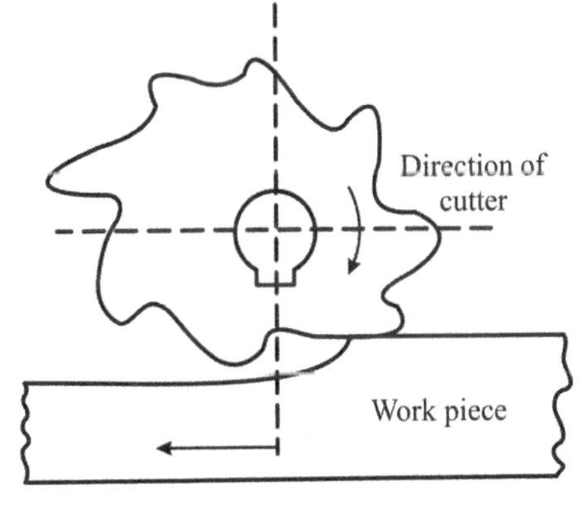

6.5.5 Advantages and Disadvantages of Milling

Advantages	Disadvantages
• Ideal for Small batches	• It is a costly process
• It can produce complex shapes	• It uses special purpose machines to obtain accurate jobs.

6.6   Grinding Machines

It is a machine tool used for producing very fine finish or making light cuts on a job with the help of a tool known as "Abrasive Wheel" or 'Grinding Wheel". It consists of a sharp crystal called abrasives held together by a binding material or "Bond. A very small amount of material is removed during grinding.

6.6.1   Types of Grinding Machines

Different types of grinding machines are as follows :

- Roughing or Non precision grinders.
  - (a) Bench Pedestal or floor grinders.
  - (b) Swing frame grinder
  - (c) Portable and flexible shaft grinder
- Precision grinders.
  - (a) Cylindrical grinders
  - (b) Surface grinders
  - (c) Internal grinders

6.6.2   Common Grinding Operations

Some of the common grinding operations are as follows:-

6.6.2.1 Cylindrical Grinding

It means grinding of outside cylindrical surfaces as in the case of Lathe Machine. The tool used is a grinding wheel. A cylindrical grinder may have many grinding wheels. The job is rotated and fed past the wheels to form a cylinder. It is used for precision work.

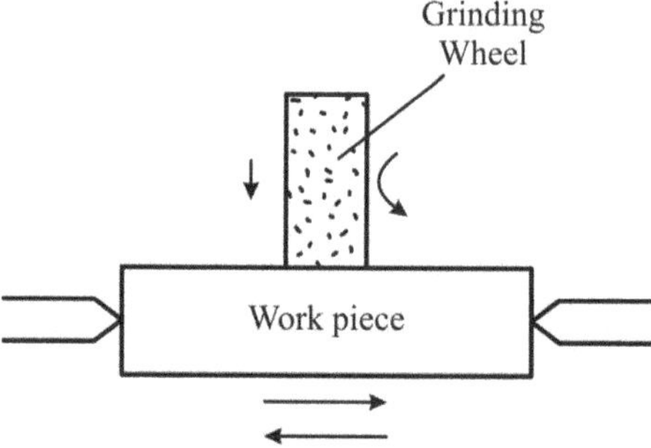

Principle

It involves holding the job rigidly on centres in a chuck or in a suitable holding fixture, rotating about its axis and a fast rotating grinding wheel is fed past the job.

Cylindrical grinding wheels are of three types:
- Plain cylindrical grinder
- Centreless grinder
- Universal cylindrical grinder

6.6.2.2 Surface Grinding

It is a method of grinding horizontal flat surface. This is similar to horizontal milling but removes small amount (thickness) of material. It has a head that is lowered and the job is moved back and forth past the grinding wheel on a table that has a permanent magnet for use with magnetic stock. These can be manually operated or numerically controlled.

Surface grinders can be of the following types:
- Horizontal spindle with reciprocating table
- Horizontal spindle with Rotary table
- Vertical spindle with reciprocating table
- Vertical spindle with rotary table.

6.6.2.3 Internal Grinding

These grinders are used for grinding the internal surface of different type of holes i.e. tapered, cylindrical and formed. The hole to be ground may be through and through or blind.

Various type of internal grinders are used such as
- Plain internal grinders
- Universal grinders
- Chucking grinders
- Planetary grinders
- Centreless grinders

Principle

The main principle of internal grinding is shown in fig.

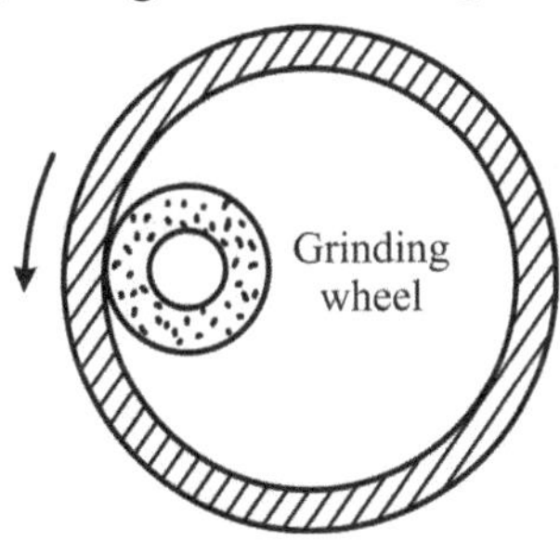

6.6.2.4 Centreless Grinding

This type of grinding is most frequently used in machine shop. It is a method of grinding external cylindrical surfaces. In this method there is a grinding wheel and a regulating wheel. The job is held between the two and the grinding work is done by the grinding wheel. The regulating wheel acts to slow down the rotation of the job so that it does not spin at the same speed as the grinding wheel and reduces the surface speed of the grinding operation.

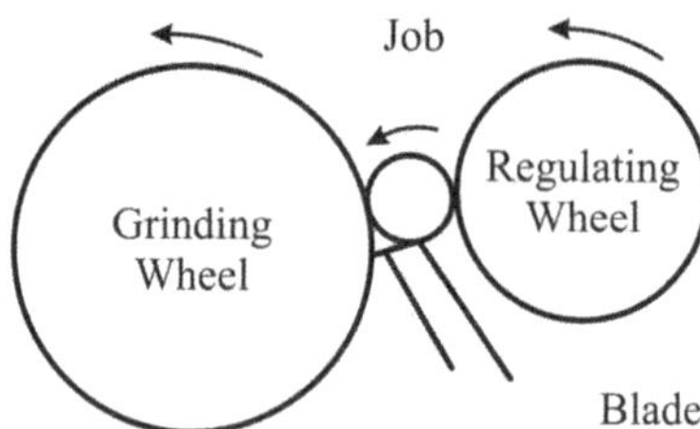

It is similar to centre grinding except that it does not have a spindle. There are 3 types of centreless grinding such as :

Through Feed Grinding

In this type of grinding the job rotates between the grinding wheel and the regulating wheel. In through feed grinding center wheel may be used. One or both the wheels of the centreless grinding may be canted out of the horizontal plane.

In Feed Grinding

The part is not fed axially so that ground surface need not be a cylinder. The grinding wheel can be dressed to accommodate the job. In this the width of grinding wheel should be more than the length of the job. The wheel is fed radially.

End Feed Grinding

In end feed grinding the part of job moves in axially between the grinding wheels, stops for grinding and then moves out again. The wheel can be dressed to form more complex shapes, but the part can only get smaller in diameter.

6.6.3 Grinding Wheels

Grinding wheels is a tool of grinding. It carries an abrasive compound on its periphery.

The wheels are used on grinding machines. The wheel is made from coarse particles pressed and bonded together to form various shapes and sizes according to the type of grinding.

Function

- To remove material from the job.

- It acts as a cutting tool.

- It is a self sharpening tool.

The Grinding Wheel consists of ABRASIVES

The Type of Abrasives are

Artificial Abrasive: Aluminum Oxide, Silicon Carbide, Tungsten Carbide, Artificial Diamond, Zirconia, Cubic Boron Nitride

Natural Abrasive: Sand Stone, Emery Corundum, Quartz, Diamond

The selection of abrasives depends on the properties of the material to be ground such as hardness, toughness etc. Two types of abrasives as shown below are generally used.

Silicon Carbide: Cast iron, Aluminum, Copper, Rubber, Leather, Plastics, Soft Bronze and Bronze, Glass and Stone.

Aluminum Oxide: Malleable Iron, Heat Treated Carbon and Alloys, Hard Brass and Bronze, Mild Steel and Wrought Iron.

Bond: The material used as glue to hold the abrasive grains together. It does not have any cutting action.

The principal bonding materials are as follows :

Silicate Bond (S): Its base material is silicate of soda. Wheels are grey in colour. Used where less wear is needed.

Oxychloride Bond (O): It is a mixture of oxide and chloride of magnesium. Grinding is done dry.

Vitrified Bond (V): It is a clay bond and reddish brown, in colour. The base metal is 'Felspar' which is fusible clay.

Resinoid Bond (B): It is a synthetic organic compound which is strong and flexible. Used for cutting bar stocks, precision grinding and com grinding.

Shellac Bond (E): They are used on hardened tool steel and thin sections. Good surface finish can be obtained by these wheels.

Rubber Bond (R): They are composed of hard vulcanized rubber. They are used for obtaining good surface finish and accuracy.

The wheels are represented by different alphabets as mentioned.

Grain or Grit	(Gives an idea of coarseness or fineness of the grinding wheel (it is denoted by a No.)	
Coarse	10-24	
Medium	30-60	Used in grinding work
Fine	80-180	
Very Fine	220-600	Used in honing work
Grade	(It indicates the strength of bond in a wheel) They are denoted by English alphabets. A being softest and Z being hardest.	
Soft	A – H	
Medium	I – P	
Hard	Q – Z	
Structure	(It is the spacing between abrasive grains or the density of wheel). It depends on hardness of work material, type of grinding operation and surface finish needed.	
Open	12 or more	
Close	6	

6.6.4 Grinding Wheel Specification

Grinding wheel specifications is a must for ordering the type of grinding wheel from the market. It depends on the type of job accuracy and the material to be ground. BIS (Bureau of Indian Standards) has devised a standard system to be followed by all manufacturers in IS-551-1954.

The details are specified in a particular order.

Abrasive, Grain size, Grade, Structure, Bond

300 x 25 x 30 MC 50 Q 6 E 15

e.g. a grinding wheel carrying the markings

Details are as wheel Diameter = 300 mm.

Thickness of wheel   = 25 mm

Base Diameter   = 30 mm

M – Manufacturing prefix to abrasive

C - Abrasive Silicon carbide

50 – Grain Size Medium

Q – Hard Grade

6 – Closed Structure

E – Shellac Bond

15 – Suffix denoting bond type of manufacturer

Questions

1. Make a Line Diagram of a simple Lathe Machine. Outline the main parts. Describe briefly.

2. What are the different operations performed on Lathe machine? What is the difference between plain turning and step turning?

3. Explain the following Lathe operations in brief, with sketches

   (a) Facing   (b) Taper Turning

   (c) Drilling   (d) Knurling

4. Write short notes on

   (a) Boring   (b) Reaming

   (c) Chamfering   (d) Forming

5. What is the principle of thread cutting. How is thread cutting performed on Lathe?

6. What is the difference between

   (a) Counter Boring and Counter Sinking

   (b) Knurling and threading

   (c) Reaming and Boring

7. How many types of Lathes are there? What is the difference between Capstan lathe, Turret Lathe and Simple Lathe?

8. Outline the differance between Capstan and Turret Lathe?

9. What is the working principle of a shaper machine? How are shapers classified?

10. Draw a sketch of a planar machine. Indicate various parts.

11. What is the difference between planar and shaper machine.

12. How many types of drilling machines are there? How drilling operation is performed by Drilling Machine?

13. Draw a sketch of a Twist Drill. Indicate and describe the various parts.

14.   What is milling? How it is performed? What is the difference between UP milling and Down milling?

15.   What is grinding operation? Name various types of grinding machines.

16.   What is the difference between the following

    (a) Cylindrical grinding and surface grinding.

    (b) Internal grinding and Centreless grinding.

17.   Describe the various methods of centreless grinding?

18.   What is a grinding wheel? What are the various elements?

19.   Write short notes on

    (a) Grade   (b) Structure

    (c) Bond   (d) Artificial Abrasives

20.   How is grinding wheel specified? Give example.

CHAPTER 7

Welding

7.1 Importance and Basic Concepts of Welding

Welding is a metal joining process. Metal may be joined together by mechanical means (bolting, rivetting etc.) or by welding, brazing, soldering or adhesive bonding.

Welding is a process of joining two similar or dissimilar metals by fusion with or without the application of pressure and with or without the use of filler material. All the methods of welding are used widely in aircraft construction also.

Mechanical joining processes involves temporary or semi permanent joining process such as joining nuts, bolts, screws etc. and by revitting etc.

Atomic Bonding joining process are permanent in nature such as soldering, brazing, welding and adhesive joining.

7.2 Classification of Welding Processes

Welding processes are of two types

Solid State Welding

Forge welding, Friction welding, Diffusion welding, Explosive welding, Ultrasonic welding.

Fusion Welding

Metallic arc welding, Carbon arc welding, Tungsten arc welding, Resistance spot welding, Resistance Seam welding, Resistance butt welding, Gas welding, Electron Beam welding. Atomic bonding process can be further classified as

Auto Geneous (No filler material is used)

Solid State Welding, Resistance welding

Homogeneous (Filler material is same as parent material)

Arc Welding, Gas Welding, Thermite welding

Heterogeneous (Filler Material is different from parent metal)   Soldering, Brazing

7.3 Gas Welding

This type of welding is a fusion or non pressure welding method. It is also called oxyfuel welding. It obtains heat for welding by combustion of fuel gases. There are two types of gas welding in common use i.e. oxyacetylene and oxyhydrogen. In gas welding the edges or surfaces to be joined are melted by the heat of a gas flame. The molten metal is allowed to flow together and a solid continuous joint is obtained as the molten metal solidified. The gas fusion welding suitable for joining metal sheets and plates having thickness from 2 to 50 mm. In order to weld materials having thickness more than 15 mm filler metal from the welding rod is melted into the gap between the two parts to be joined. The material of the welding rod is the same as that of the parts to be joined.

Aircraft fittings fabricated from chrome-molybdenum or mild carbon steel are gas welded. Engine mounts, landing gears and entire fuselages constructed for steel tubing are also gas welded. Aluminum alloy parts made from strain hardened alloys and from some heat treatable alloys are also gas welded.

7.3.1 Oxyacetylene Welding

The oxyacetylene welding can be used for welding almost all metals and alloys used in engineering industry. When mixed with oxygen, acetylene burns to produce a temperature as high as 3200°C. The oxyacetylene flame is produced by the combustion of acetylene gas with oxygen.

Acetylene (C_2H_2) which is produced by chemical reaction between water and calcium carbide (CaC_2) mixes with pure oxygen to produce welding heat. The combustion of acetylene with pure oxygen takes place in two stages. Reaction is as follows

Ist Stage $C_2H_2 + O_2 \rightarrow 2CO + H_2$

IInd Stage $4CO + 2H_2 + 3O_2 \rightarrow 4CO_2 . 2H_2O$

There are 2 systems of oxyacetylene welding.

High Pressure System

Oxygen and Acetylene are derived from High Pressure Cylinders.

Low Pressure System

Oxygen is taken from High Press Cylinder but acetylene is generated by the action of water on carbide (Calcium Carbide) in low pressure Acetylene generator.

7.3.2 Oxyacetylene Welding Equipment

The oxyacetylene gas welding equipment is shown in the Fig. 7.1. The main parts of the equipment are as follows: -

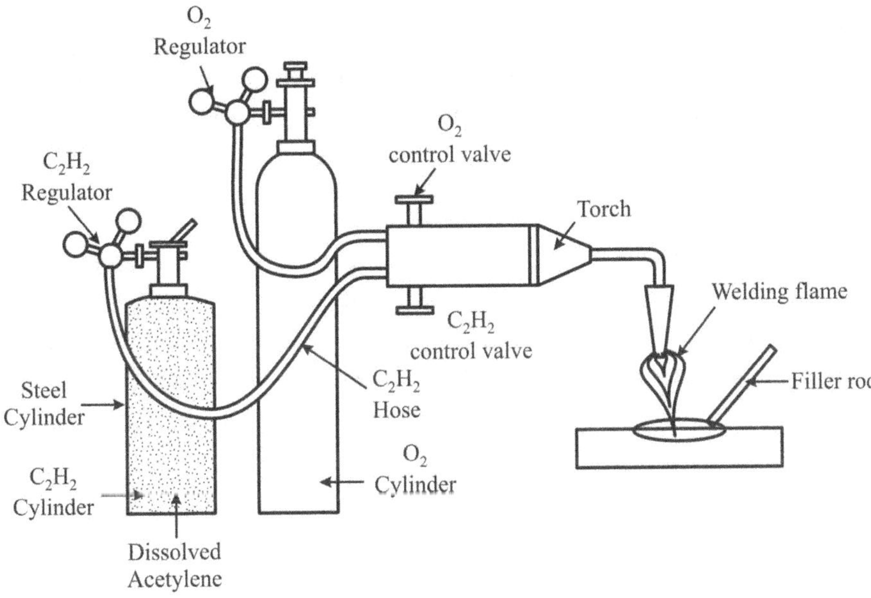

Fig. 7.1

Cylinder (2Nos)

O_2 Cylinder

These are made of steel and are usually painted green. They contain compressed O_2. They are fitted with valves and a steel cap.

Acetylene Cylinder

These are also made of steel. They also contain valves. They are painted black. Acetylene is dissolved in acetone. The cylinders are filled with porous filler such as calcium silicate. Acetylene should never be used at a pressure that exceeds 15 psi.

Regulator

A pressure reducing regulator is used for controlling the pressures in acetylene cylinder and oxygen cylinder. The regulators have two gauges which show the inlet and working pressure.

Full acetylene cylinder pressure is 250psi.

Full oxygen cylinder pressure is 2200psi.

Welding Hoses

The cylinder regulators and torch are connected together by double line rubber hoses. The oxygen line is green and fuel line is red.

Check Valve

For combustion to occur, fuel and O_2 have to mix. This happenes in the torch mixture or torch tip. Check valves when installed between the hoses and torch prevent the back flow as they close if reverse flow starts.

Torches or Blow Pipes

The torch assembly consists of the handle, oxygen and fuel gas valves and mixing chamber. Welding tips or a cutting attachment can be used with the handle allowing it to be used for welding, heating and cutting operations. They are made in different designs and sizes according to the work. The pipes are of two types namely High Pressure and Low Pressure.

In high pressure pipes there are different passages for both O_2 and C_2H_2. Both these gases are mixed in the chamber and then driven out through nozzle or fire with the desired velocity. These nozzles are known as tips and are interchangeable.

The low pressure blow pipe works on the principle of injector. In this oxygen at high pressure is made to pass through injectors type nozzle which in doing so draws the acetylene along with it to the mixing chamber, then to the outlet of tip.

7.3.3 Types of Flames

There are three types of flames possible in gas welding such as-

Neutral Flame

Welding is carried out by using equal amounts of O_2 and C_2H_2. All the acetylene is completely burned and thus all the available heat in the acetylene us released. The neutral flame has a well defined white cone at the tip. The chemical equation is as follows:

$$4CO + 2H_2 + 3O_2 \rightarrow 4CO_2 + 2H_2O + heat$$

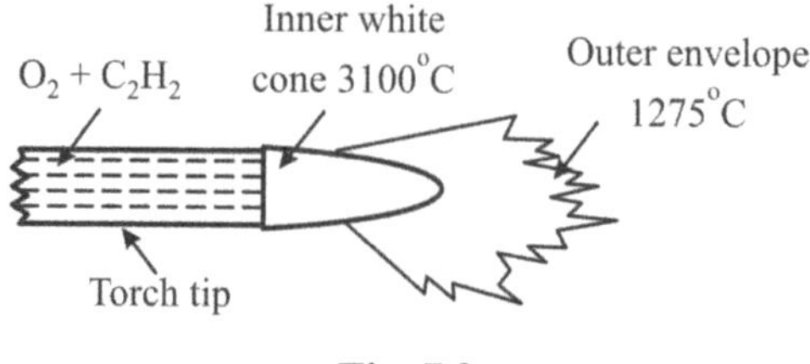

Fig. 7.2

This flame is used for welding and gives a thoroughly fused weld free from burned metal or hard spots. It is used for welding cast iron, aluminium, copper and stainless steel.

Reducing Flame *(Carbonising Flame)*

This type of flame is produced when as excess of acetylene is burned and can be identified by a feathery edge on the white core. The length of the feather is an indication of excess acetylene present. This flame introduces carbon into the weld. It is used for oxygen free copper alloys, high carbon steels, cast irons and high speed steels.

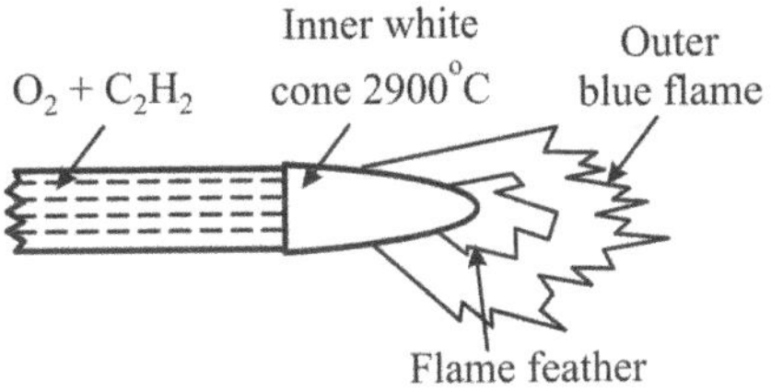

Fig. 7.3

Oxidising Flame

It is produced by an excess of oxygen. It is identified by its small pointed white cone and relatively short envelope of flame. An oxidizing flame will oxidize or burn the metal and result in a porous weld. It is used only in welding brass and bronze. However, some alloys of iron such as cast iron and manganese steel can also be welded.

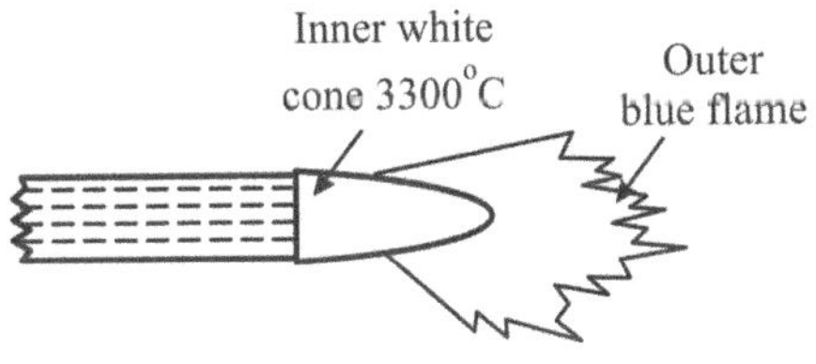

Fig. 7.4

7.3.4 Advantages and Disadvantages of Gas Welding

Advantages	Disadvantages
The equipment is of low cost	Skill is required in welding
It is portable also	Flux is required except low carbon steel
It can be used for different types of jobs	Cannot be used for refractory materials

7.4 Electric Arc Welding

Electric Arc Welding is a fusion welding process. This process is based on the heat generated in an electric arc.

Different Types of Electric Arc welding are as follows:

- Metal Arc Welding
- Carbon Arc Welding
- Atomic Hydrogen Welding
- Inert Arc Welding
- Multi Arc Welding
- MIG Welding

7.4.1 Metal Arc Welding

In this process a metal electrode is used which furnishes the filler material for the weld as it melts. To maintain the arc between the electrode and the work, the metal electrode must be fed at a uniform rate or maintained at a constant distance from the work as it melts. The maintenance of a constant arc length requires an experienced welding operator.

The metallic arc develops a temperature of around 5500°C. This heat is concentrated and causes less buckling and warping of the work than gas welding. Metallic arc welding is applicable to carbon steel, to corrosion and heat resisting steel and to Aluminium.

The basic circuit for metal arc welding is as below:

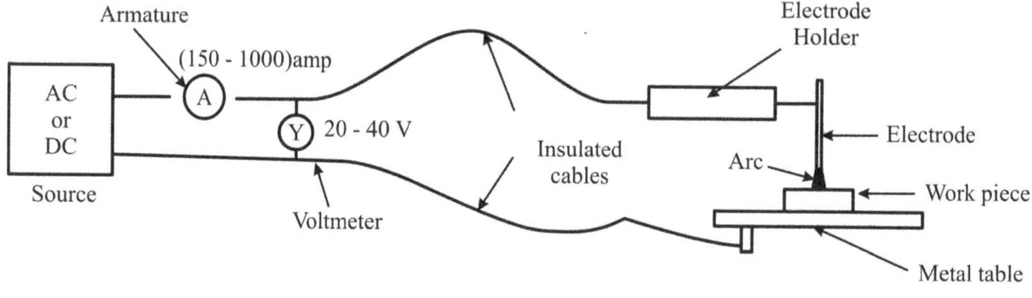

Fig. 7.5

The source of current may be alternating current (AC) or direct current (DC). The work piece or job is kept on the metallic table. One cable from power supply is connected to the electrode holder into which the electrode is gripped. The other lead is connected to the metallic table on which work piece is kept. When the electrode is brought near the workpiece. i.e., in contrast with workpiece then arc generates high temperature and welding takes place.

7.4.2 Carbon Arc Welding

In this process a carbon electrode is used and a filler rod is held in the arc and fused into the joint. In carbon arc welding a current of 20 to 800 amperes is used which depends upon the thickness of the metal to be welded and the diameter of the electrode. Only DC is used in carbon arc welding. Automatic welding plant is used for welding. Welding can be with single carbon electrode or double carbon electrode.

Carbon arc welding is used in welding of tanks, pipes, barrels and ships. It is also applicable to carbon steels, corrosions resisting steel, aluminum and copper alloys. It is not much used in aircraft work. It is also used in welding cast iron, bronze and galvanized iron.

Either carbon or graphite electrodes are used. The graphite and carbon electrodes are consumed slowly during welding because of slow oxidation of carbon. Straight polarity may be used in this method.

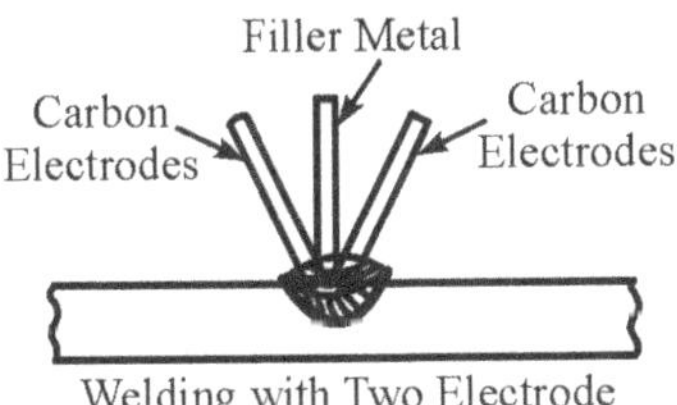

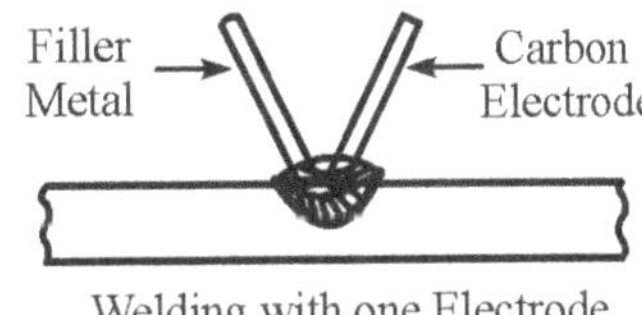

Fig. 7.6(a)   **Fig. 7.6(b)**

7.4.3 Atomic Hydrogen Welding

In this process a stream of hydrogen is directed into an arc drawn between two tungsten electrodes. As the hydrogen enters the arc, the high temperature breaks up the molecules of hydrogen into single atoms i.e. molecular hydrogen into atomic hydrogen. The atomic hydrogen is unstable and recombines to form molecular hydrogen. At the same time giving off heat energy that was used in dissociating it in the tungsten arc.

A temperature of about 3000°C is generated at the work surface. Since the arc is formed between two electrodes in the welding torch it is possible to control the heated zone on the parts being welded by moving the torch towards or away from the weld.

Atomic H_2 welding is frequently used in welding aluminum and its alloys. The joint must be coated with flux and flux coated filler rod is applied manually as in gas welding.

Main advantage is to provide high heat concentration. In addition to this H_2 acts as a shielding gas to protect the electrodes and molten metal from oxidation. This process has been used to weld corrosion and heat resisting steels. Smooth welds are obtained in light gauge sheets.

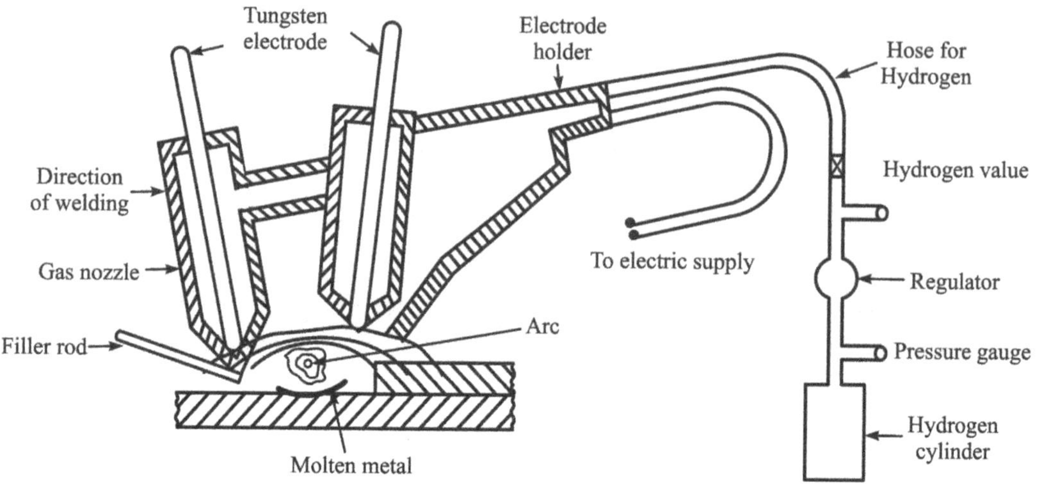

Fig. 7.7

7.4.4 Inert Arc Welding

In this process a tungsten or carbon electrode surrounded by helium or argon gas is used. In tungsten inert gas welding it is known as TIG welding. The helium or argon gas are inert and exclude the O_2 and H_2 present in the air from the area being welded. The heat is produced from an arc between the non consumable tungsten electrode and the workpiece. The welding zone is shielded by an atmosphere of inert gas supplied from a suitable source. In this process DC or AC may be used depending upon the type of metal to be welded. DC with straight polarity is used for Cu alloys and stainless steel and whereas reverse polarity is used for magnesium. This method is best suitable for magnesium. It is also used for welding Aluminium and if argon is used for shielding gas no flux is required. Relatively thin pieces of metal can be welded by this method. This may be done by manual process or automatic welding machine.

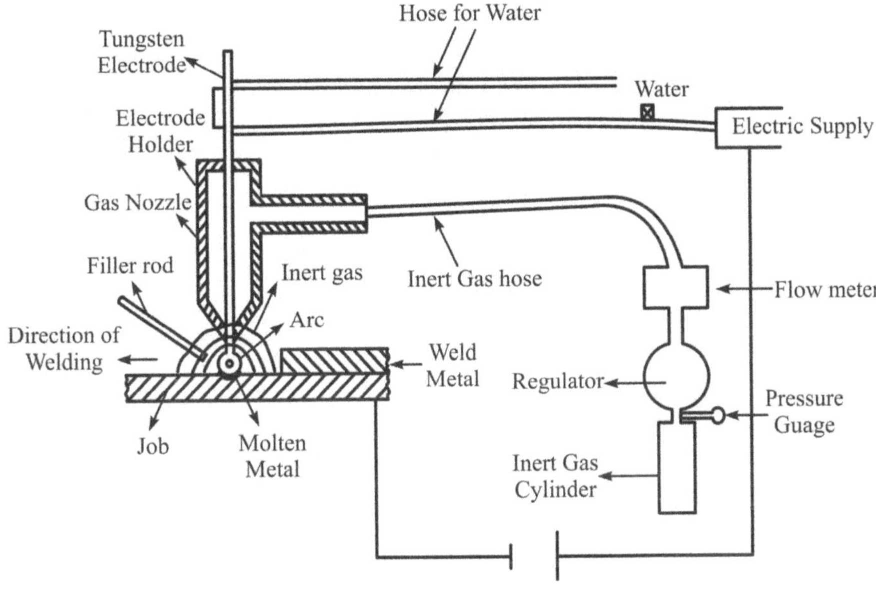

Fig. 7.8

7.4.5 Multi Arc Welding

It consists of a combination of AC & DC together with both metallic and carbon electrodes. A metal electrode is supplied with DC and carbon electrodes are supplied with AC. This arrangement results in supplying of five arcs i.e. three between electrodes, one between metal electrode and the work and one between carbon electrodes and the work. This arrangement gives an accurate control and concentration of heat.

By this method Butt Welding of Aluminium sheets is done. A*l* clad, Magnesium alloys, Brass, Copper and many types of Corrosion resistant steels are also welded.

7.4.6 Metal Inert Gas Welding (MIG)

It is similar to tungsten inert gas welding except that the electrode is of consumable type. The electrode is in the form of a continuous wire in the form of a reel which is fed into the arc by an adjustable speed electric motor at the same speed at which it is melted and deposited in the weld. Use of electrode holder is done which in addition to a passage for wire electrode also incorporates passages for supply of inert gas for shielding the electrode molten weld metal, and the adjacent hot area of base metal from atmospheric contamination and for supply of cooling water. In MIG welding gases like argon, helium, carbon-dioxide or a mixture of these gases are used for providing the inert gas shield. Now CO_2 is most widely used both as a single gas and also in mixture with other inert gases.

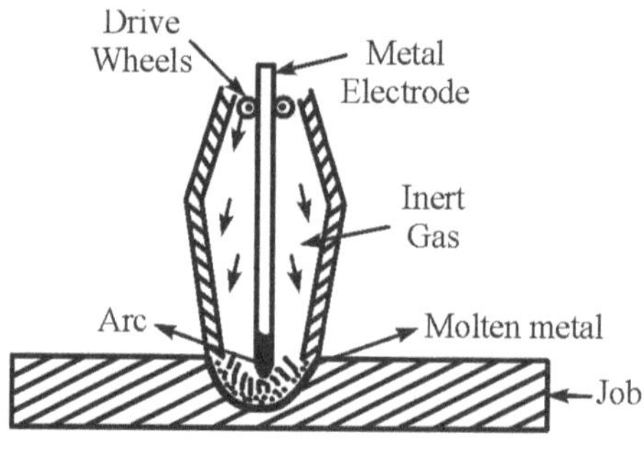

Fig. 7.9

7.5   Electric Resistance Welding

It is a type of pressure welding. It is based on the principle that heat is generated by the resistance offered by a conductor. The heat increases with an increase of resistance. Current is admitted into the work through low resistance copper electrodes. The low voltage high amperage current meets much greater resistance when it enters the work to be welded, intense heat is generated.

The amount of heat generated by the resistance of the metal pieces at the points of contacts is given by

$$H = I^2Rt$$

where

    H = Total heat generated in the work in 'joules'

    I = Electric current in 'Ampere'

    R = Resistance of metal being welded in 'Ohms'

    t = Time in which electric current flows in 'Secs.'

In this method the time is for very short duration, say of the order of 0.001 Secs. Hence time may be controlled by electronic timer.

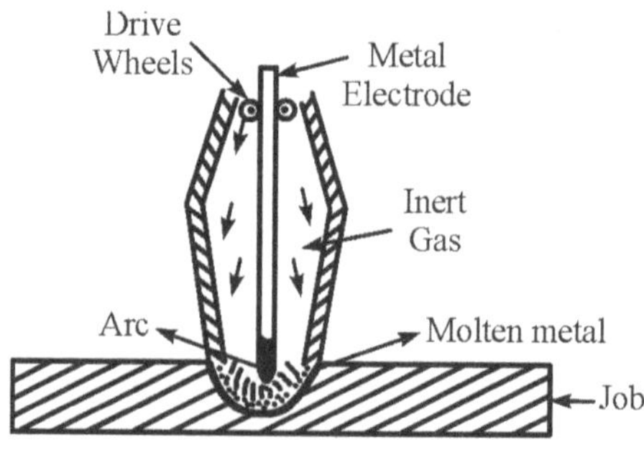

Fig. 7.10

7.5.1 Advantage and Disadvantages of Resistance Welding

Advantages	Disadvantages
- Good mechanical properties. - Welds are quickly made. - High Speed welding. - Suitable for mass production. - Economical in operation. - Possible to weld dissimilar joints. - Skilled operators are not required.	- High equipment and initial costs such as tooling etc. - Irregular shaped welds are produced. - Lower tensile strength and fatigue strengths.

Resistance welding has wide applications in welding of sheet metal material (except Cu, Ag), aircraft and automobile industries for welding of various parts and accessories and production of pipe and tubes.

7.5.2 Types of Electric Resistance Welding

7.5.2.1   Spot Welding

Spot welding is used for welding lap joints, joining components from plate material. This is possible with the help of a resistance spot welding machine. It is used for welding of ferrous/non-ferrous alloys particularly Al alloys. It has wide applications for manufacturing automobile parts, refrigerators, and metal stampings, aircraft construction.

Principle

The plates to be joined together are placed between the two electrode tips of Copper or Copper alloys which are water cooled. A low voltage and high ampere current is passed between electrodes. The interface melts over a spot and form a weld.

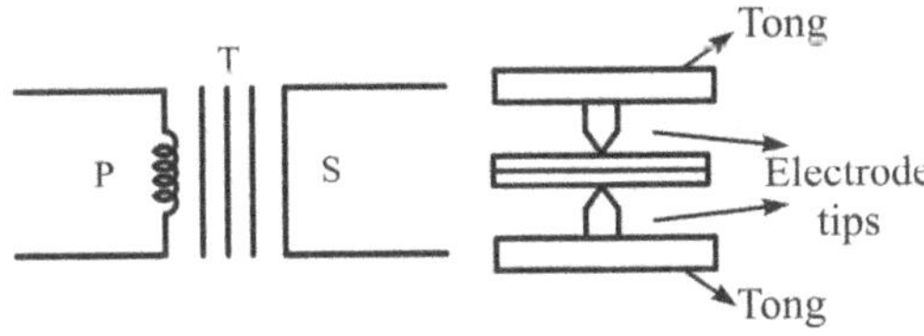

Fig. 7.11

Advantages	Disadvantages
- High production rates - Easy automation. - No filler material required. - Efficient use.	- Strength of weld is lower than other methods and so it is used for specific purpose where low strength is required.

7.5.2.2   Seam Welding

It is a continuous spot welding process where in the spot welds overlaps each other to desired extent to form a leak proof seams (joint). This is called seam welding. Application of seam welding are for Al alloys, radiator and drums etc.

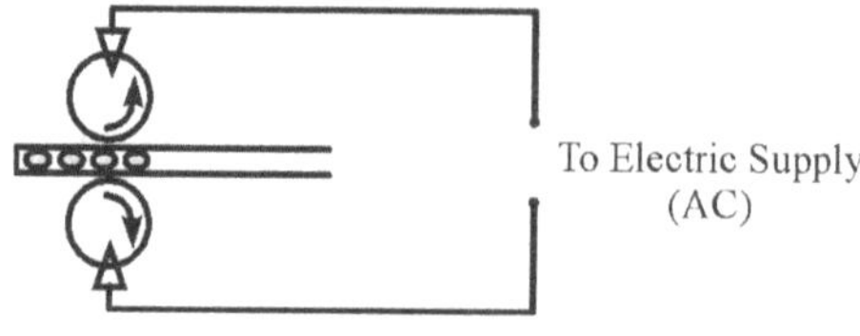

Fig. 7.12

Principle

The workpieces to be welded are passed between the rotating disc shaped electrodes (Cu Alloy). These electrodes maintain a continuous pressure on work. The current is applied electrodes in a series electrodes of pulses at proper interval and timing are so adjusted that the pulses overlap each other and form a continuous joint.

7.5.2.3   Butt Welding

It is a widely used to weld together long sheets, bars, tubes, rods or wires. The welding machine can do only one type of job. It is applicable to almost all metals including steel and *Al*. There are two types of butt welding such as upset butt welding and flash butt welding. Upset Butt welding is carried for rods, pipes etc. and flash butt welding is used for steel containers and welding of high speed shanks to high speed steel drills and reamers.

Principle

The work to be welded is clamped in large copper jaws (electrodes). The electrodes are water cooled.

Upset butt welding: The heat is generated in the contact surface by the electrical resistance when a suitable electric current flows across the joint when the metal at the joint is heated sufficiently the pressure is applied by the upper jaws to force the pieces together.

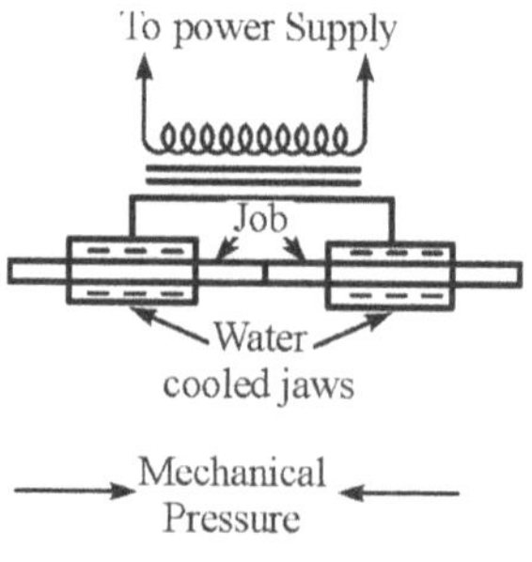

Fig. 7.13(a)

Flash butt welding: In this heat operation is derived from the arcing that occures in the space between the parts to be joined. The ends to be welded are pushed together and immediately after contact made is a separated by a short distance. After the ends are fused, the current is shut off and pieces are forced together.

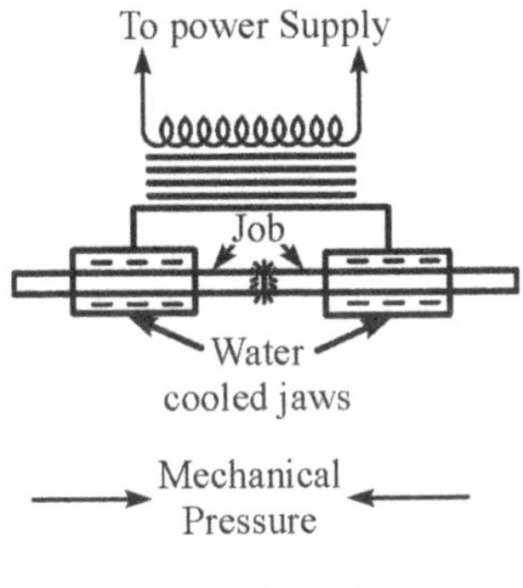

Fig. 7.13(b)

7.6 Soldering (or Soft Soldering)

Soldering is the process of joining similar or dissimilar metals by means of heat and filled material whose melting points is less than 450°C (Pb-Sn Alloy) than the melting point of metals to be joined. This is soldering or 'Soft Soldering'. The filler metal is known as solder. In Soldering the gap between the metals to be soldered is filled by filler material through capillary action. A soldering iron is used for the purpose of heating.

Soldering process involves:

- **Proper Cleaning to remove oil, grease and dirt**

  Cleaning is done to provide chemically clean surface to obtain proper bond and may be done as follows:

For iron and Steel	-	10% H_2SO_4
For Brass	-	10% H_2SO_4
For Copper	-	Dil. H_2SO_4
For stainless steel	-	Mixture of 3 parts H_2SO_4, 1 part HNO_3, 10 parts H_2O

- **Fluxing**

  Fluxing is done to remove the oxides from joint surface and to prevent the filler metal from oxidizing. Fluxes are used in the form of powder, paste or liquid.

- **Heating**

  The bit of soldering iron is heated, cleaned, dipped in flux and thin rubbed on the solder to pick up a small quantity of solder with it and deposited along the joint.

Uses

It is used extensively for making electrical connections, hermetic sealing of metal containers, joining head pipes, copper tubing and for general repair work. The metals which can be joined by Soft solder are iron, steel, copper , brass nickel and monel metal.

Soft Soldering is never used in aircraft work for joints requiring strength. It is used for making electrical connections or to solder the wrapped on sliced ends of aircraft control cables.

Flux used is composed of 75% mineral grease (petroleum), wax and resins combined with 25% $ZnCl_2$. Soldering of Al can not be done by this as it will corrode seriously. Rosin is used as a flux in electrical and electronic soldering. It is non corrosive and non hygroscopic..

Common soft soldering alloys are composed of tin and Lead. Good grade is composed of 50-50 i.e. 50% Sn and 50% Pb.

A fine solder containing 2 parts of Sn and one part of Pb is best suited for soldering steel, iron, cu, and Brass.

When silver is alloyed with Sn it is used as solder to join parts, it is known as hard soldering. However they may be alloy of copper and Zinc also with some silver.

7.7   Brazing

It is the method of joining similar or dissimilar metals with the help of molten filler material whose melting point is more than 450°C but less than the melting point of the parent metals. In brazing there is no fusion (as in welding) of the metals being joined. Brazing is most widely used in aircraft industry.

Three types of brazing are as follows :
- Copper Brazing or simply brazing
- Silver brazing
- Aluminium Brazing

The type of brazing is named after the metal whose alloy is used as the filler metal.

7.7.1   Copper Brazing

It is the process of using metal parts by means of molten brass filler. When the filer metal is applied to the joint it flows throughout the joint by capillary action. Commonly used flux is borax or two parts of borax and one part of boric acid.

Before brazing the parts should be cleaned of oil, grease or paint by means of benzol or hot caustic soda solution. Heavy scale should be removed by pickling, followed by caustic dip to neutralize the acid.

Parts to be brazed must be securely fastened together to prevent any relative movement. Strong brazed joint is one is which molten filler is drawn in by capillary action.

The main brazing (Cu brazing) methods are dip brazing and flame brazing.

7.7.2 Silver Brazing

Silver brazing is based on the fact that practically any metal will surface alloy with another metal that has a higher melting temperature. This latter metal must have a clean surface and be heated to the melting temperature of solder or filler. There are number of hard solders. In aircraft work it is known as Silver Brazing. They can be used to solder metals such as Cu, Brass, Bronze, iron, carbon resistant steel, Inconel, Monel, Nickel, and Silver. This solder will make a strong joint. Solder brazing can be done by gas, dip, furnace, electric resistance or induction hardening.

7.7.3 Aluminium Brazing

This process is applicable to Aluminium alloys for brazing of Aircraft parts. Torch, furnace and dip brazing are used.

Aluminum brazing is used widely as Aircraft construction. It has been used in the manufacture of intercooler ducts, fuel tanks and other applications.

7.8 Difference Between Soldering and Brazing

Soldering	Brazing
- Filler metals arc alloys of Pb and Sn	- Filler metals are alloys of Cu (e.g. Cu-Zn, Cu-Ag)
- The joints are weak	- are stronger than soldered joints
- Soldering is widely used in joining small assemblies, electric and electronic parts	- It is used for joining parts, electrical parts, Aircraft parts, Automobile parts. Carbide tips to tools, heat exchangers, radiators.
- It uses a soldering iron to carry out soldering	- It can be done by torch, furnace, dip brazing
- The melting point of filler material is less than 450°C and also less than melting point of base metal	- The melting point of filler material is above 450°C but less than melting point of base material.

Questions

1.   What is Welding? How it is different from soldering and Brazing?

2.   What is Gas Welding? Give the uses of Oxy Acetylene Welding.

3. Draw a sketch of Oxy Acetylene welding equipment. Name the various parts. Explain briefly.

4. Describe the various types of flames for Gas welding? Draw Sketches.

5. What is electric arc welding? Name the various types in use.

6. Write short notes on

    (a) Metallic Arc Welding   (b) Carbon Arc Welding.

7. Describe the working of

    (a) Atomic hydrogen welding   (b) Inert Arc welding

8. What is the difference between

    Multi Arc welding and Metal Inert Gas Welding (MIG)

9. What is the principle of electric resistance welding? What are its advantages and disadvantages?

10. What is the principle of spot welding? Explain its procedures. What are its advantages.

11. Write short notes on:

    (a) Butt Welding   (b) Seam Welding

12. What is soldering? What are its various uses.

13. What is Brazing? Why it is preferred over other metal joining processes for aeronautical work.

14. Write Short notes on

    (a) Aluminium Brazing   (b) Silver Brazing   (c) Copper Brazing

Unit IV

Manufacturing

Miscellaneous Topics

8.1 Importance of Materials and Manufacturing towards Technological and Socio-economic Developments

Manufacturing has a wide scope in everyday life. It this fast developing world the present day engineers and scientists are busy in innovation and development of new materials. They are searching for materials of greater strength, lightness safety reliability, cheapness, corrosion resistant, heat resistant, etc. The most developing areas are in the field of Aeronautics, Automobiles, space technology and Bio-technology. This new study of processes are being studied under Material Sciences.

The weight, strength and reliability of materials is of utmost importance in aircraft construction. All materials must have good strength / weight ratio and must be free from failures under various atmospheric conditions. The experts who have wide knowledge of materials are employed to carry out research and operations in aircrafts industries. Use of composite materials is widely accepted in modern aircraft.

Thus selection of materials is very important for the development of the product.

Once the material is selected the manufacturing of the product from raw material adds to the cost of the product. Sometimes in the selection of the material a material which is imported from other country may be cheaper than producing in our own country. Therefore various considerations such as socio-economic developments are also to be studied. The two main considerations for production are:

- Economic Aspects

- Engineering or technological aspects

Economic	Technological
Availability Material selected should be available in sufficient quantities to satisfy normal and emergency requirements. Material should be purchased from reputed manufactures	*Strength* The material may be capable of developing the required strength.

Contd...

Cost Cost per price should be calculated before taking up production.	*Weight* Weight is usually considered in conjunction with strength. The strength / weight ratio of a material is of utmost important from structural point of view in aircraft construction.
Shop Equipment Required The initial and maintenance cost of shop equipment required for working of the material selected must be evaluated.	*Corrosion* Corrosion aspect of materials is studied in detail while selecting a material for both aeronautical / navigational purpose. In automobile parts coating also plays an important role.
Standardization of Materials The materials are to be standardised and the stocks to be maintained	*Surface finish* Surface finish plays a major role in most mating parts for automobile / aeronautical purposes
Reliability The reliability of a material to be used should be explored before use so that rejection costs are reduced at the maximum level.	*Close tolerances* Very close tolerances are also required in various engineering, operations such as turning, grinding, milling etc.
Supplementary operations required The supplementary operations such as condition of material, plating, heat treatment, cleaning etc should be studied properly before use and total cost estimated	*Working properties* Various properties such as forming, bending, machining welding, brazing, soldering, forging, casting are also looked into.

8.2 Plant Location

Plant location is an important decision-making for an organisation. The main consideration for plant location is necessary because the customers should be able to utilise the serviceability of the product in minimum time. Thus the organisation must reach to the customer in the shortest possible time. A satisfied customer is a positive sign for the growth of the organization. Thus plant location will depend on many factors. Some of these are as follows:

- Suitability of land and climate
- Transport facilities
- Supply of raw material
- Availability of water
- Nearness to markets
- Availability of man power
- Presence of other supporting industries
- Availability of power

8.3 Plant Layout

Plant layout refers to the arrangement of 3 Ms (men, material, machines), equipment, buildings, work places and other facilities needed for the production of a product inside the factory.

Thus plant layout is a systematic and effective arrangement of different departments, men, machine, equipments and services in an organisation / factory to facilitate the production of the product in the most efficient and economical way in the minimum possible time.

8.3.1 Objectives of Good Plant Layout

The following are the objectives of good plant layout:

- Utilisation of men and machines is optimum
- Material handling and transportation is minimum
- Entire space is fully utilised
- Movement of worker should be minimised
- Supervision is easy, simple and effective
- Provision for future expansion of the plant is made
- The flow of material is smooth and rapid
- Adequate safety measures are taken
- workers should feel joy in working
- Cordial relations between management and workers is maintained
- Adequate light, ventilation should be provided in shop floor
- Maintenance of machines should be done at regular intervals
- Productivity to be increased
- Quality of the product to be maintained.

8.3.2. Advantages of a Good Plant Layout

- Minimised handling time
- Increased production
- Quality product delivered
- Safety of workers
- Efficiency of workers
- Job satisfaction
- Reduces number of accidents
- Better health of employees

- Reduces time delays
- Less supervision
- Reduces cost.

8.3.3 Types of Plant Layout

Layout is an arrangement of various facilities so as to carry out production efficiently and economically. This will depend upon the capacity, size, nature of the product etc.

The various plant layouts are as follows:

8.3.3.1 Product or Line Layout

In this type the machines are arranged according to the sequence of operations of the product.

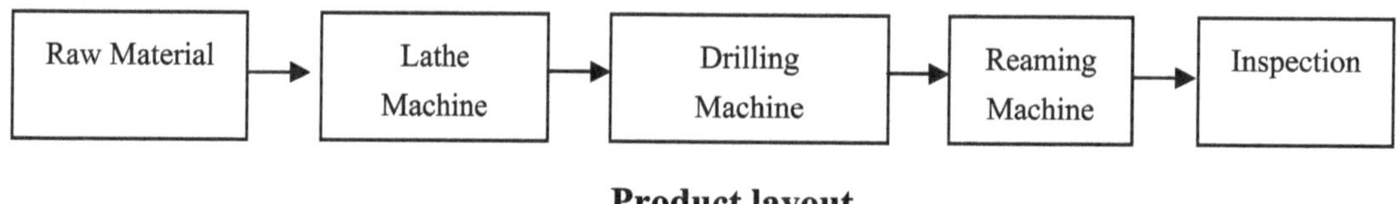

Product layout

This is preferred for mass production.

This type is used for assembly shops, manufacturing shops of automobiles, mass production of cycle industries, rolling mills etc.

Advantages of Product Layout

- Machines are arranged according to sequence of operation
- Reduced material handling due to straight flow of material
- Less floor area is required for same volume of production
- Less amount of work in progress
- Less skilled worker
- Smooth and continuous work flow
- Less time for production.

Disadvantages of Product Layout

- Capital investment is high
- Special purpose machines are required
- It is less flexible to changes in product design
- Breakdown of a single machine may effect production.

8.3.3.2 Process or Functional Layout

In the type machines performing similar type of operations are grouped at one location. This is used for low volume of production.

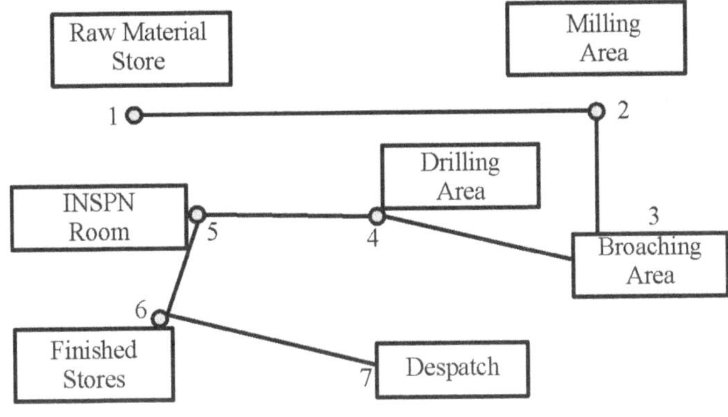

Process Layout

Advantages of Process Layout

- Better utilisation of machines
- Variety of components makes the work interesting to the workers
- Lower initial investment
- Breakdowns of machines can be easily handled
- Since the workers handle same type of machine, better quality can be achieved
- Greater flexibility in utilisation of equipment.

Disadvantages of Process Layout

- Material handling costs are high
- Requires large storage space because of slow movement of materials
- Completion of product takes time
- It needs more stringent inspection.

8.3.3.3 Combination or Group Layout

Nowadays the industries are using combination or group layout which are neither product nor process layout. These combine the good points of both. In this a set of machines or equipment are grouped together in the section so that each set or group of machines or equipment is used to perform similar operations to produce like components. Hence it is called group layout.

It is used for producing files, Hacksaws, wooden circular saws, refrigerators etc.

8.3.3.4 Fixed Position Layout

In this type of layout the material or major components remain in a fixed location and tools, machinery, men and other materials are brought to the location of the material.

This type of layout is suitable where heavy products are manufactured and when assembly consists of heavy parts. In this case it is economical to bring the tools near the job or product. It is used in manufacturing of hydraulic and steam turbines, ship building, aircraft manufacturing, boilers etc.

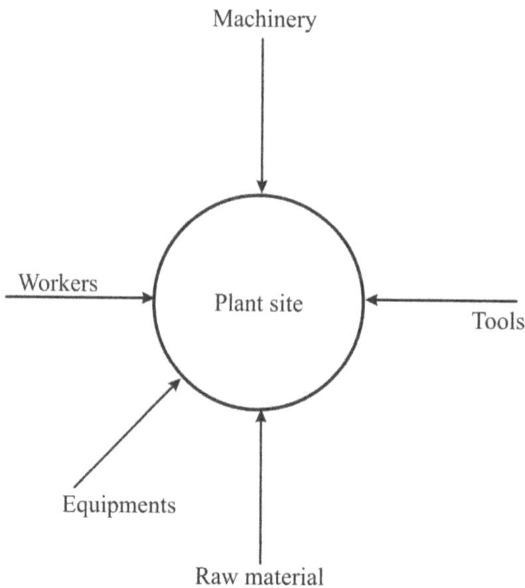

Fixed position layout

Advantages of Fixed Position Layout

- Number of projects can be handled at one time
- Flexibility of working
- Capital investment for layout in lower
- Workers feel a pride in doing the job as their movement is less
- Material handling is reduced.

Disadvantages of Fixed Position Layout

- Movement of machine and equipment to the plant site takes time and money initially,
- Skilled workers are required at site because machine is fixed and every worker cannot handle
- In case of absenteeism machines utilisation is low.

8.4 Types of Production

There are three types of productions as shown below:

Job production	Batch production	Mass production
This is as per special demand of the customers	This is the manufacture of products to meet specific requirements of special orders	This is used for automobile, electrical, electronics, consumer goods
Quantity is small	It is common for aeronautical industries space programmes	The batch size is large
It is for special type of requirements	It is the manufacture of identical items	Investment is more
It is concerned with special projects	It is common for various machining processes such as forging casting grinding, milling drilling etc.	Demand is more
It is concerned with special machinery	Batch size is small for welding brazing etc for plating operations.	Parts produced are of repetitive nature.

8.5   Production Verses Productivity

Production	Productivity
Production is defined as a planned group activity in which a product is produced from raw material to final stage. The product goes through various intermediate stages with the help of men, machines and material. A large infrastructure is also necessary for production. Thus production engineering may be defined as the art of making components. A simple line diagram is detailed below	Productivity is the quantative relation between what we produce (output) and what we use to produce (input) $$\text{Productivity} = \frac{\text{output}}{\text{input}}$$ In manufacturing or production for an industry we do not use the term efficiency but we use productivity. Thus productivity represents the efficiency of the production system. $$\text{Productivity index} = \frac{\text{Value or goods or services produced}}{\text{Value or Qty of given input resources}}$$

Production may be of various types such as	Productivity is the outcome of production
Job production	
Batch production	
Mass production	
This has already been explained earlier	
Production is an absolute concept	Productivity is a relative concept
It reflects the volume of production of goods and services	It reflects operating efficiency
Production can be increased by increasing the input resources	Productivity may remain the same or decline.

Questions

1. What is the importance of materials and manufacturing towards technological and socio-economic development?

2. The two main considerations for production are economic and technological. Please explain.

3. What are the objectives of good plant layout? What are its advantages?

4. Write short notes on

   (a) Product layout   (b) Process layout

   (c) Group layout   (d) Fixed position layout

5. How many types of production are there? Which production is preferred for aeronautical industry?

6. Differentiate between

   (a) Production and productivity

   (b) Batch production and mass production

   (c) Plant location and plant layout.

CHAPTER 9

Non-Metallic Materials

9.1 Common Types and Uses of Wood

Wood is non-metallic material which is widely used in various forms in manufacturing of various items.

Wood is available from trees. The wood that is obtained from fully grown trees is cut into various commercial sizes. This wood is known as timber. It is widely used for engineering or building purposes.

9.1.1 Types of Wood

Wood can be classified as

- Soft wood
- Hard wood
- Plywood

Softwood	Hardwood	Plywood
It is light in weight	It is heavier	It is a material made by gluing a number of plies of thin wood together
It is relatively weaker and less durable	It is stronger and more durable	It is used for furniture
It is light in colour	It is dark in colour	It is used in the construction of box spars for wings, wing and fuselage coverings for aircraft construction. Water-proof plywood is a hot pressed resin plywood assembled with a synthetic resin adhesive.
Annual rings are quite distinct	Annual rings are not distinct	Smooth finish
It is a resinous wood	It is non-resinous wood	It uses glue
It has straight fibers and fine texture	Its fibers are quite close and compact	It is pressed to form thin sheets

9.1.2 Uses of Wood

- Used for flooring, paneling and construction of partition walls
- Used for furniture, beams, lintel, doors, windows
- Used for frame work for concrete, transmission poles and fencing.

9.2   Cement Concrete

Concrete is most widely used material in building construction. It is a mixture of cement, sand, brick ballast or stone ballast and water in different proportions. On hardening the mixture forms a hard stone like material known as cement concrete. It is used widely for construction of houses, roads, bridges, dams etc.

The following main properties of concrete are as follows:

- Strength of cement depends upon water cement ratio,
- It produces excellent resistance to compressive stress, shear stress and abrasion,
- The limitation of cement concrete such as low tensile strength shrinkage and expansion are removed by combining cement concrete with steel to form reinforced cement concrete.

9.3   Ceramics

These are compounds of metallic and non-metallic elements, with strong ionic or covalent bonds between atoms.

They have some of the properties better than those of the metals. Ceramics are good insulators and so are used in electrical industries. Other applications are for cutting tools, abrasives, refractories, structural shapes, building materials etc.

Some of the other properties of ceramics are as follows:

- They are hard and brittle
- They possess low thermal conductivity
- They have high hardness
- They have good creep resistance
- They have high modules of elasticity
- They have low specific gravity
- They have low toughness and low density.
- Ceramics are of two types – traditional ceramics and modern ceramics.

Traditional ceramics	Modern ceramics
<ul><li>They are made up of 3 mutually basic components clay, silical and feldspar.</li><li>The structure is plate like.</li><li>It is hydraded compound of alumina silicate mineral.</li><li>The clay are converted into bricks, tiles, porcelain sanitary wares etc.</li><li>Silica is converted into crystalline quartz.</li><li>Feldspar (Sodium Potassium Aluminum Silicates) is a common mineral. A simple flow diagram represents the process for traditional ceramics.</li></ul>	These are pure compounds such as Magnesium oxide, Aluminium oxide, Barium titanate, silicon carbide, silicon nitride. Modern ceramics are synthesized by chemical reactions.

```
┌──────────────────────────┐
│   Mined raw materials     │
└──────────────────────────┘
         │ By Mlling/Grinding
         ▼
┌──────────────────────────┐
│    Small size particles   │
└──────────────────────────┘
         │ Screening or sizing
         ▼
┌──────────────────────────┐
│ Desired particle size powder │
└──────────────────────────┘
         │ Water + Additives
         ▼
┌──────────────────────────┐
│         Mixture           │
└──────────────────────────┘
         │ Heated
         ▼
┌──────────────────────────┐
│       Melted slag         │
└──────────────────────────┘
         │
         ▼
   Further processing to
   obtain desired shape
```

9.4   Rubber

Rubbers are of two types

- Natural rubber
- Synthetic rubber

Both these rubbers are polymers or copolymers and are chemically similar to plastics. A polymers is a complex material formed by a polymerisation reaction. In this reaction a simple chemical is converted to an extremely complex material with entirely different properties due to reaction with itself. Catalysts are required to aid this chemical reaction. In copolymerization two simple chemicals react to form a single complex product with new properties.

It elongates on stretching and regains its original shape after removal of the stress. The shortage of natural rubber during the war years resulted in large scale developments for the manufacture of synthetic rubbers.

Natural Rubbers	Synthetic Rubbers
It is an elastic material present in the latex (sip or fluid) of certain plants Its structure is as follows 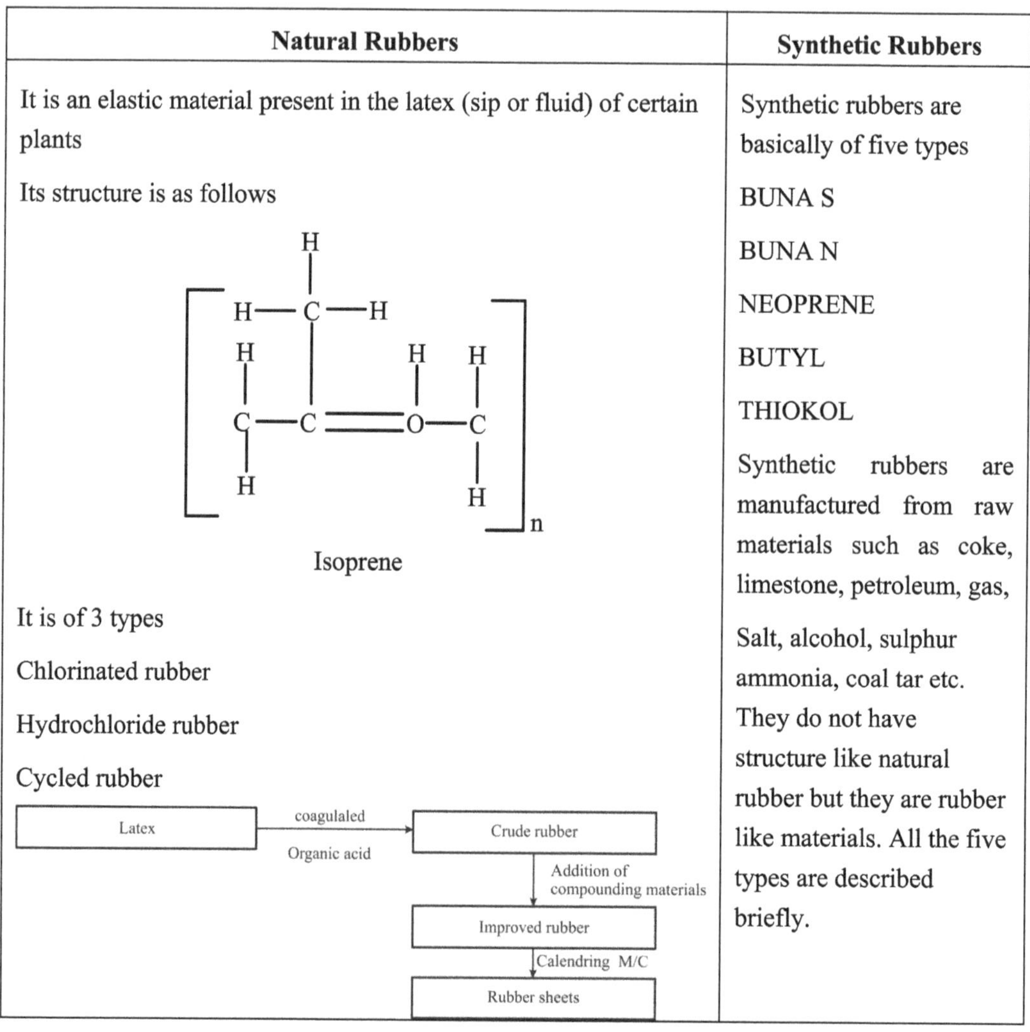 Isoprene It is of 3 types Chlorinated rubber Hydrochloride rubber Cycled rubber	Synthetic rubbers are basically of five types BUNA S BUNA N NEOPRENE BUTYL THIOKOL Synthetic rubbers are manufactured from raw materials such as coke, limestone, petroleum, gas, Salt, alcohol, sulphur ammonia, coal tar etc. They do not have structure like natural rubber but they are rubber like materials. All the five types are described briefly.

9.4.1 Types of Synthetic Rubbers

A brief description of some of the important synthetic rubbers are as follows

9.4.1.1 Buna S

These are also called Butadiene rubbers. It is a copolymer of butadiene and styrene. 'Bu' is the first syllable of butadiene and 'Na' is the symbol for sodium, which was used as catalyst for polymerisation of butadiene, 'S' stands for styrene.

It is most nearly like natural rubber. It can be vulcanised with sulphor and cured to a hardness equal to hard rubber. It must be combined with black pigment such as carbon black to bring out its best physical properties 'Buna S' which is used commercially is in black colour. It is widely used for tyres, belts, tubes, shoe soles, flooring, electric wire insulation etc.

9.4.1.2 Buna N

These are also called nitrile rubbers. It is a copolymer of Butadiene and Acrylonitrle, 'N' is the first letter of nitrile. Some commercial names of 'Buna N' are perbunan, Hycar, Chemigum, Thiokal RD and Butaprene.

'Buna N' is similar to rubber in that it can be vulcanized with sulphor and can be cured to hard rubber. It has excellent resistance to oil and grease. It also possesses good resistance to abrasion and aging.

The properties are improved by addition of carbon black. 'Buna N' can be bonded by vulcanisation or cement to metal surfaces. Phenol resin cements are used for the purpose. It is used for oil and gasoline hose, tank lining; hydraulic accumulator bags, gaskets and seals.

9.4.1.3 Neoprene

It is a polymer of chloroprene. They are also called poly chloroprenes. They are produced by polymerising chloroprene molecules.

It is a good general purpose rubber that has good resistance to oil and excellent resistance to heat, air, light and flame. It has better light resistance than any other rubber. It can be vulcanised without sulpher but cannot be cured to as hard a condition as hard rubber.

It is widely used for parts such as gaskets, oil seals, tank linings, truck tyres, shoe soles, etc.

9.4.1.4. Butyl

It is copolymer of isobutylene and small amounts of unsaturated hydro carbons such as butadiene or isoprene. It is produced cheaply from petroleum by products, one of which is butylenes. It is also known as Flexon.

Butyl can be vulcanised with sulphur but cannot be hardened to the condition of hard rubber. They have great impermeability to gases, high resistance to aging and oxidising agents etc. They are used for gas masks, life jackets, plywood molding bags, tyres, inner tubes, adhesives, tank linings wire and cable insulation etc.

9.4.1.5 Thiokol

It is a polysulphide polymer. It is also called poly sulphide synthetic rubber.

It has highest resistance to deterioration but the lowest physical properties. It is noted for its resistance to aromatics hydrocarbons and aromatic blended gasolines. It can be vulcanized with zinc oxide but not to a hardness comparable with hard rubber.

It is used for oil hose, tank linings for aromatic aviation gasolines, paint spray hose, gaskets, seals etc.

9.5   Composite Materials

Composite materials are obtained when two or more materials are combined together to form a new material which possesses superior properties than the parent materials. Two or more different materials may be bonded mechanically or metallurgically.

9.5.1 Types of composite materials

9.5.1.1 Laminated Materials

These are the materials which are produced by bonding two or more layers of different materials completely to each other. These may be metallic or non metallic.

Plywood, tufnol, sunmica, linoleum etc are common examples of laminates. They are used for furniture, flooring, tiles etc.

9.5.1.2. Agglomerated Materials

These are the materials in which the particles are condensed to form an integral mass. Cement concrete is one of its examples. Other materials are used in grinding wheels.

Fibre Reinforced Composites: Common examples of reinforced composites are (RCC) i.e. Reinforced Cement Concrete, glass fibre, reinforced plastic etc. Nowadays composite materials are being used in the manufacture of LCA (Light Combat Aircrafts) in M/S Hindustan Aeronantics Ltd. Banglore Division. They are also used for ship building industries and chemical process industries.

Questions

1. How woods can be classified? What is the difference between Hard wood and Soft wood?

2. Write short notes on

   (a) Cement concrete   (b) Ceramics

3. How many types of rubbers are in use? What is the difference between Natural rubbers and Synthetic rubber?

4. "Synthetic rubbers are used widely nowadays" Explain.

5. How many types of synthetic rubbers are used. What advantage do they have over Natural Rubbers?

6. Write short notes on following rubbers

   (a) BUTA S   (b) BUNA N

   (c) NEOPRENE

7. Describe in detail

   (a) Thiokol rubbers   (b) Butyl rubbers

8. Briefly describe composite materials and give their uses.

CHAPTER 10

Miscellaneous Processes

10.1 Powder Metallurgy Process and its Applications

It is a process by which various articles are produced from powdered metals both ferrous and non ferrous by placing these powders in moulds and applying pressure.

10.1.1 Main Characteristics of Powders

Morphology

The morphology of powders can be critical in determining both flow characteristics as well as playing an important role in the heat and momentum transfer to the powders during spraying. The morphology of powders can be described as irregular, blocky or spherical.

Powder Size, Range and Distribution

Powder size distribution plays an important role in powder metallurgy in the following ways:

    (a) It has direct impact on finished product quality,

    (b) it provides ease and efficiency of filling a die,

    (c) porosity fine particles leave smaller pores which are easily closed during sintering process.

Powder Flowability: It is the property which enables a powder to flow easily and fill the mould cavity. Production depends on flowability.

Apparent Density: It is the specific gravity of the powder. Reduction will need more powder for filling hence it should be maintained at a constant level.

Compressibility: It is governed by shape and particle size distribution of powder.

Crushing

This method is used for brittle materials. Various ferrous and non ferrous metal alloys can be crushed to powder form. The powder obtained are of angular shape which are finally reduced to powder form by milling, Moving hammers, and jaw crushers are used for crushing.

Electrolysis

Metal plates are placed in a tank containing an acid solution known as electrolyte. These metal plates act as anodes. The other metal plates act as cathodes are also placed in the tank. The powders are deposited on the cathodes when high current is passed. The cathodes are taken out from tank, washed and dried. The deposit is scraped off, milled or ground to produce fine powder. It is used for tin, silver, copper, zinc, and iron mainly.

Agglomeration

In this method the constituents are physically mixed together with an organic binder. The solvent is then driven off and resultant material sized. The binder should be burnt off during spraying.

The use of spray drying has become a common method of agglomeration.

Chemical Composition

Chemical composition varies from lot to lot. Mixing or blending of lots is done to improve the chemical composition.

Sprayability

It shows that the powder provides a consistent product. Normally a spray test is done on the powder prior to use.

10.1.2 Steps in Powder Metallurgy

(a) Powder production,

(b) Blending / mixing of alloy powders,

(c) Compacting in a die,

(d) Sintering,

(e) Secondary or finishing operations,

(f) Finished product

The process is shown in the following flow diagram.

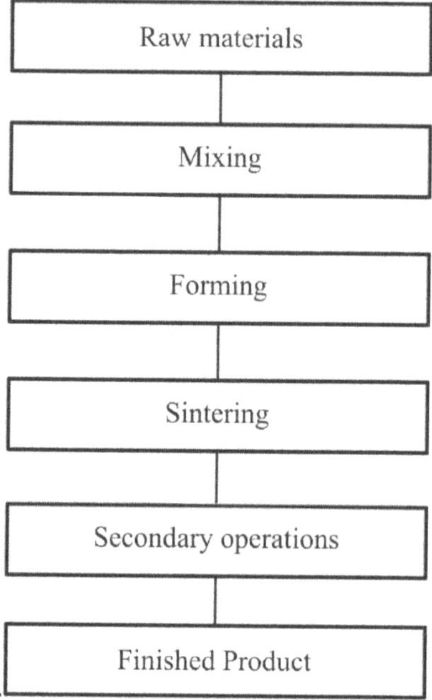

10.1.3 Production of Metal Powders

It involves metal powders of almost all metals and alloys. Powder metallurgy depends upon the production of metal powders of required degree of fineness and purity is being produced by following processes:

(a) Atomisation,

(b) Mechanical comminution,

(c) Chemical reduction,

(d) Crushing,

(e) Electrolysis,

(f) Agglomeration.

Atomization

It is used commercially to produce large quantities of metal powder. The molten metal is forced through a nozzle into a stream of compressed air or water. The metal on cooling solidifies into tiny particles of various sizes and irregular shapes. The size of particle depends on the size of the nozzle, rate of flow of metal, pressure and temperature of water and air stream. A water atomization process is shown in Fig. 10.1

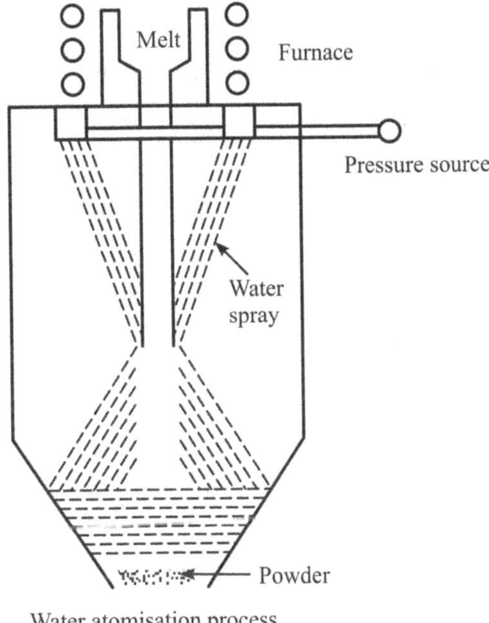

Water atomisation process

Fig. 10.1

Mechanical Comminution

This comprises of various methods such as milling, lathe turning, grinding, chipping scratching, etc. Such methods are used to produce hard and brittle powders.

Chemical Reduction

In this method, the metal powders are obtained by reducing refined ores or oxides in a current of hydrogen gas. Iron, copper, nickel, cobalt, molybdenum and tungsten are produced by this method.

10.1.4 Advantages and Disadvantages of Powder Metallurgy

Advantages

    (a)   It involves saving in material,

    (b)   Part of great accuracy can be produced,

    (c)   Porosity of parts can be controlled,

    (d)   Skilled workers are not required,

    (e)   High production can be achieved,

    (f)   Hard materials such as diamonds, ceramics can be processed,

    (g)   Ensures uniformity of composition,

(h)   Various properties such as purity, density, porosity particle size can be controlled,

(i)   Self lubricated bearings can be produced by this method only

(j)   Layers of different metal powders can be moulded together to obtain multi-metallic products.

Disadvantages

(a)   Metal powders and equipment are costly,

(b)   Not economical for small scale production,

(c)   Storage of powders is a problem,

(d)   Complete dense structure cannot be produced,

(e)   Due to low flowablity of metal powders production is restricted,

(f)   Small holes, knurling, Threads have to be produced by machining.

10.1.5 Applications of Powder Metallurgy

Powder metallurgy has some of the application as follows:

(a)   Magnets,

(b)   Self-lubricated bearings,

(c)   Cemented carbide tools and dies,

(d)   Clutch facings and brake linings,

(e)   Filament wire for electric bulbs,

(f)   Motor brushes,

(g)   Pump roles and gears,

(h)   Porous fillers,

(i)   Metal coating.

10.2 Plastic Products Manufacturing

10.2.1 Introduction

The plastics are synthetic organic compounds which become plastic when heated and can be formed into different shapes when subjected to heat and pressure.

These organic compounds are called polymer. Some natural polymers are starch, resins, shellac, cellulose, proteins etc. The plastics are used in almost all applications. They are used in the manufacture of aeronautical and automobile parts, safety glasses, laminated gears, pulleys, self lubricated bearings etc.

The raw materials for plastics are various agricultural products and numerous minerals and organic materials including petroleum, coal, gas, limestone, silica and sulphur. The natural resins are shellac, bitumen resin, rubber etc.

10.2.2 Properties of Plastics

The properties of plastics are as follows:
   (a)   Light weight,
   (b)   Low thermal conductivity,
   (c)   High corrosion resistance,
   (d)   Good insulators,
   (e)   Good strength and rigidity,
   (f)   Easy formability,
   (g)   High resistance to abrasion,
   (h)   Low fabrication cost,
   (i)   Can be made transparent or coloured,
   (j)   Impermeable to water,
   (k)   Minimum surface finishing operation.

10.2.3 Types of Plastics

An infinite number of plastics can be developed. They can be classified into four different types such as synthetic resin plastics, natural resins, cellulose and protein.

However depending on their reaction to heat, they can be classified into two categories:

Thermo plastics	Thermosetting
These are known as cold setting matcrials. They are plastics which will repeatedly soften when heated and harden when cooled. These materials can be heated until soft, molded into desired shape and when cooled will retain this shape. The same material can be reheated any number of times and reshaped. They are better than thermosetting plastics as they can be used again and again. No chemical change occurs on application of heat and pressure. Used for making toys and other purposes.	These plastics are chemically changed by first application of heat and are therefore infusible. They will not soften on further application of heat and cannot be reshaped after once being fully cured by application of heat. These are also known as heat setting materials. These are soluble in alcohol and other organic solvents when they are in thermoplastic range. They are hard, strong and durable and are manufactured in variety of colours.

10.2.3.1 Thermoplastics

The following types of thermoplastics are shown in tabular form.

Type	Composition	Form	Typical use
Cellulose	Cellulose nitrate	Sheet, rod, tubing ribbon, film	Spectacle frames and novelties
	Cellulose acetate	Sheet, rod, tubing, foil film, molding compounds	Hardware, movie film, cabin windows
	Cellulose acetate butyrate	Molding compounds	Hardware, flash light, cases, pencils
Vinyl	Vinylidene chloride	Tubing, tubing fittings, molding compounds extrusions	Tubing for water chemicals, water resistant fabrics.
	Polyvinyl butyral	Sheet resin, molding compounds	Safety glass, water proof coatings.
	Polyvinyl chloride (PVC)	Molding compounds	Insulation for cables and wires, rain coats, table, cloths, curtains, garden hoses
	Polyvinyl acetate (PVA)	Sheet, film, molding compounds	Making paints, lacquers, plastic emulsions coatings, wrapping paper.
Acrylic	Methyl methacrylate	Sheet, rod, tubing, molding powder	Aircraft enclosures, windows, windshields, lenses.
Polyethylene	Polyethylene	Sheet, rod, tubing, molding powder	Insulators battery boxes, bottle covers, films, bags sacks, for packaging toys, bottle milk crates.
Polystyrene	Polystyrene	Molding compound, extrusions	Toys, combs, buttons, refrigerator parts, radio insulators, lenses, battery boxes bottles.
Polyamides	Polyamide derivatives	Molding compounds, filaments	Gears, bushes, bearings rope.
P.T.F.E.	Polytetrafluora ethylene	Trade name Teflon, can be machined, punched, drilled.	Insulators, gaskets, packings, pump parts, bearings, asbestos fibers

10.2.3.2 Thermosetting Plastics

Type	Composition	Form	Typical use
Phenolic	Phenol formaldehyde	Cast laminated fabric	Knobs, buttons, headless pulleys, wheels, switches, telephone, hand sets (Trade name bakelite).
	Phenol furfural	Fabric asbestos	Brake linings, electrical parts, instrument panels.
Amino	Melamine formaldehydes	Molded cellulose fabric	Electrical applications, telephone sets. Switch panels. Trade name mélange plastron, catalon.
	Urea formaldehyde	Molded cellulose fillers.	Containers, kitchen ware, thermos caps, table ware, dials for instruments, plugs, electrical switches.
		Laminated cotton base	Moulded shapes
Allyl	Allyl	Cast	Aircraft enclosures, lenses
		Laminated glass fabric	Structural applications.

10.2.4 Plastic Processing Methods

The process of making various products of plastics is known as plastic processing or fabrication of plastics. Some of these are:

   (a) Molding,
   (b) Calendering,
   (c) Casting,
   (d) Laminating,

10.2.4.1 Molding

It is one of the most common methods for fabrication of plastics.

Compression Molding

It is used for thermosetting plastics and some thermoplastic plastics. The process consists of placing a correct amount of plastic compound in a heated mould. The mould is closed and heated in the range of 120-250 °C depending upon the material being moulded. Pressure is applied hydraulically. The molten plastic flows uniformly into every corner of the mould. The mould is kept closed for sufficient time to allow polymerization to take place, so that the product is sufficiently hardened. The mould is now opened by releasing the pressure and the product taken out.

It is used for making gears, knobs, electrical switches, fuse boxes, gas and electricity meter housings etc.

When thermo plastic materials are made by compression moulding, the mould should be sufficiently cooled before the product is taken out of it. In the figure the mould is shown in open and closed position. A compression moulding press is used for the purpose.

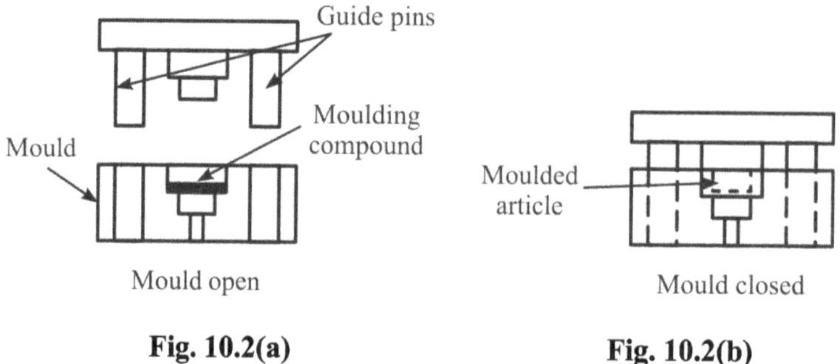

Fig. 10.2(a)   **Fig. 10.2(b)**

Transfer Moulding

It is used for thermo setting plastics and is similar to compression moulding. In this the heat and pressure are applied to the compound separately outside the mould. It is transferred into the mould cavity through one or more sprue. The mould is held under pressure till the product is completely cured. After curing, the mould is opened and product removed. The curing time for transfer moulding is slightly less than that for compression moulding. It is used for production of intricate parts.

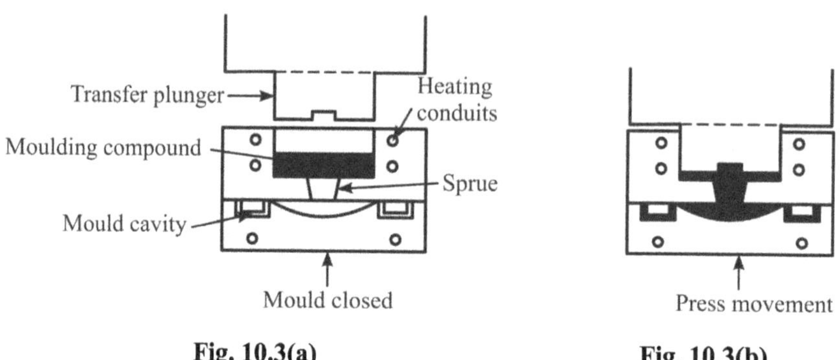

Fig. 10.3(a)   **Fig. 10.3(b)**

Injection Moulding

This method is used for thermoplastic materials which plasticize when heated and solidify when cooled. The powdered plastic compound is first heated to drive off moisture and then fed into the hopper through a metering device which supplies a constant amount of material for each cycle. The hydraulic plunger forces the plastic material into the mold through the heating cylinder, the nozzle, sprue and runners. Temperature ranges between

120-250 °C. After cooling, the mold is opened and the product is removed by ejector system. The torpedo or spreader improves the rate of transfer into the plastic.

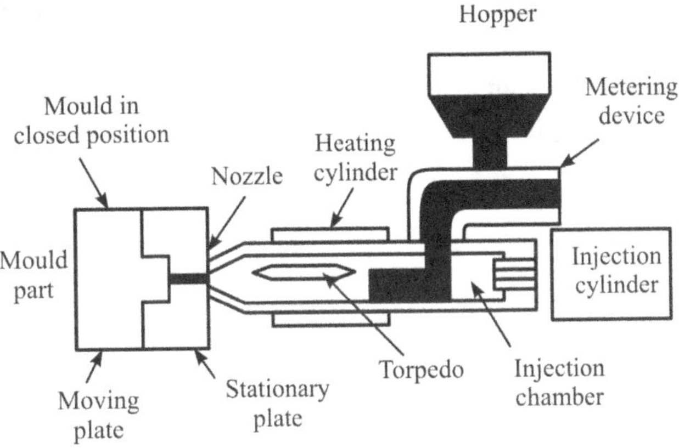

Fig. 10.4

Extrusion Molding

All thermoplastic materials can be extruded into various shapes like tubes, rods, sheets, films, pipes, ropes etc. Extrusion is like squeezing toothpaste out of its tube. It is a continuous process. It is done with the help of an extrusion machine.

Plastic material is first loaded into the hopper and then fed into long heated chamber through which it is moved by the action of a continuously revolving screw. At the end of the heated chamber the molten plastic is forced out through a small opening called a die that is cast in the shape of the finished product. As the plastic extrusion comes out from the die it is fed to a conveyer belt where it is cooled by blowers or by immersion in water.

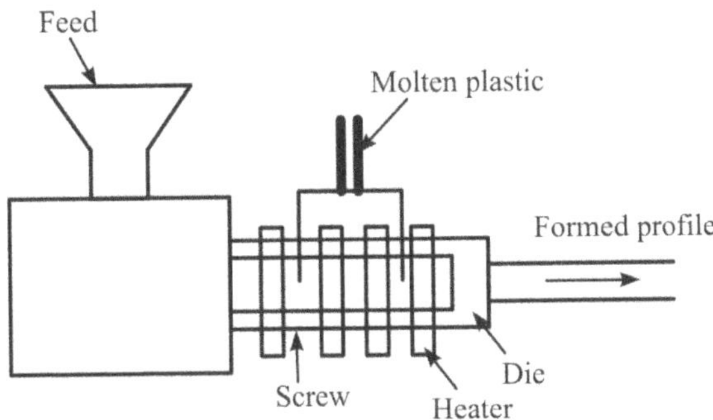

Fig. 10.5

Blow Moulding

It is used for producing narrow neck bottles and plastic containers. In this process a heated closed end thermoplastic tube is placed in the mould and air pressure is applied to inflate it to acquire the shape of the mould cavity. After the plastic is cooled, the mould is opened and the product taken out. This is done with the help of a blow molding machine. The figure shows making a plastic bottle.

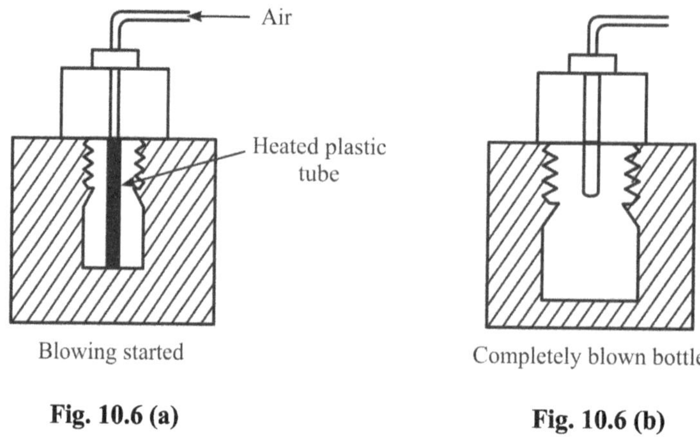

Fig. 10.6 (a)   **Fig. 10.6 (b)**

Slush Molding

It resembles casting in that no pressure is applied. In this method the viscous solution of the thermoplastic material is prepared and then poured into a preheated mould. On account of heat of the mould a uniform layer of resin sets all along the surface of the mould. Excess material, if any is removed. Additional heat is provided for curing the resin set in the mould, followed by hardening of the product by chilling and removing the same from the mould. This process is used for making flexible toys, artificial flowers etc.

10.2.4.2 Calendering

It is the process for making film or thin sheets by squeezing a thermoplastic material between the revolving cylinders as shown in figure (10.7). The material consisting of thermoplastic resin, plasticiser, filler and colour pigments are compounded and heated and being fed into the calendar. The thickness of the sheet depends upon the spacing between the cylinders that stretch the plastic. The last roll is water cooled and is called chilling roll. Vinyl, polyethylene, cellulose acetate films and sheets are produces by calendering.

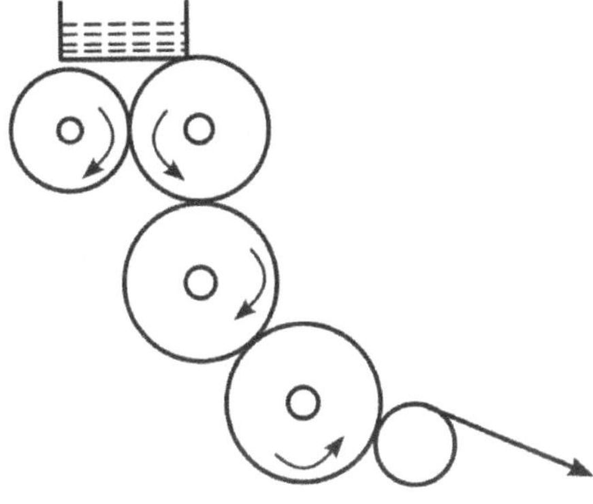

Process of calendering

Fig. 10.7

10.2.4.3 Casting

In this process of producing plastics, only heating is involved, No pressure is applied. Liquid resins, such as phenolics, polyesters, silicones, acrylics, epoxies and some cellulose derivatives are used for casting.

In this method, the plastics are heated to a suitable temperature and poured into the moulds. Moulds are made of lead, rubber, glass, plaster wood or metal. A catalyst is added to help polymerization. After due curing and hardening the moulds are opened and castings removed from moulds.

Examples of cast products are sheets, films, jwellery, dies, punches and jigs.

10.2.4.4 Laminating

It is the process of bonding together a variety of materials by heat and pressure to form a single piece (known as laminates). Sheets of wood, asbestos, fabric can be laminated. These materials are first impregnated or coated with phenolics, silicones epoxies and melamines and then combined together under heat and pressure to acquire the finished product.

10.3 Electroplating

It is the process of depositing a very thin layer of metal coating by means of electrolysis. The article to be coated may be metallic or non-metallic. In this method a thin layer of metal coating is deposited or base metal by passing a direct current through an electrolyte solution containing some salt of the coating metal.

The article to be electroplated is first immersed in hot alkaline solution to remove oil and grease. It is then treated with diluted hydrochloric acid or sulphuric acid to remove scales and oxides. The main elements involved are:

Electrolyte, an anode (positive terminal), a cathode (negative terminal), a low voltage DC terminal. The work piece to be plated is suspended and immersed in the electrolyte filled in a tank; electrically it is connected to the negative terminal to act as cathode. The material to be deposited to provide coating is connected to positive terminal. When direct current is passed, the coating metal ions move to the cathode and get deposited there.

Commonly used metals for electroplating are copper, nickel, chromium, cadmium, zinc, tin, silver, gold, platinum, brass, bronze, tungsten etc.

It is extensively used for decorative purpose and for preventing corrosion. It is widely used in aircraft and automobile industries.

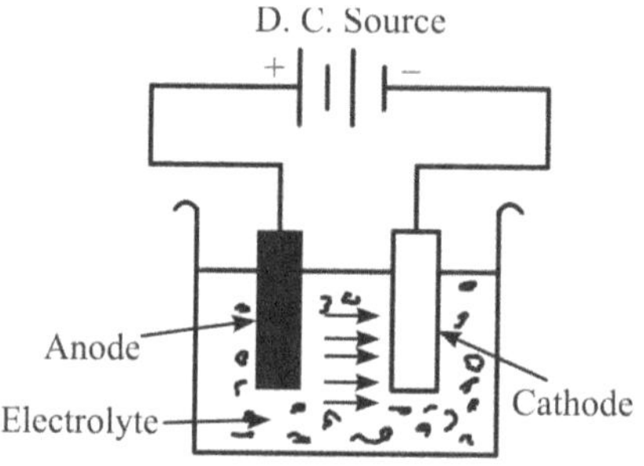

Fig. 10.8

10.4 Galvanizing

Zinc plating is known as galvanizing. It is highly corrosion resistant. It is done for steel sheets and steel parts by dipping them in a bath of molten zinc.

The parts to the galvanized are first cleaned by pickling with diluted sulphuric acid, to remove any scale, rust and impurities. After washing and drying, it is dipped in a bath of molten zinc maintained at about 450 °C. The surface of the bath is kept covered with a molten flux layer usually of zinc ammonium chloride to prevent oxide formation. The duration of immersion in zinc bath depends on the size of the part to be coated. It is then taken out from the bath and passed through a pair of rollers to remove any excess zinc. The part is cooled and dried.

It is used widely for sheets for roofs, paper, containers, fencing materials, nails, wires, light poles etc.

Questions

1. Enumerate the main characteristics of metal powders.

2. What are the different steps of powder metallurgy? Show with the help of a flow diagram.

3. What are the different methods of producing metal powders?

4. Write short notes on

   (a) Morphology

   (b) Compressibility

   (c) Agglormeration

   (d) Electrolysis

5. Describe about the following

   (a) Atomisation

   (b) Chemical production

6. What are the advantages and disadvantages of powder metallurgy?

7. How many types of plastics are there? What are the various types of plastics.

8. Differentiate between thermoplastics and thermosetting plastics.

9. Enumerate different types of thermoplastics along with their uses.

10. Write short notes on

    (a) Phenolic plastics

    (b) Amino plastics

    (c) Cellulose plastics

    (d) Polyethylene

11. What is the difference between compression moulding and transfer moudling?

12. Write short notes on

    (a) Injection moulding

    (b) Extrusion moulding

    (c) Blow moulding

    (d) Slush moulding.

13. What are various plastic processing methods? Write short notes on

    (a) Calendring

    (b) Laminating.

14. What is electroplating? Why it is used in automobile and aeronautical industry.

15. Write short notes on

    (a) Galvanising

    (b) Electroplating.